Energy, Environment, and Sustainability

Series editors

This books series publishes cutting edge monographs and professional books focused on all aspects of energy and environmental sustainability, especially as it relates to energy concerns. The Series is published in partnership with the International Society for Energy, Environment, and Sustainability. The books in these series are editor or authored by top researchers and professional across the globe. The series aims at publishing state-of-the-art research and development in areas including, but not limited to:

- Renewable Energy
- Alternative Fuels
- Engines and Locomotives
- Combustion and Propulsion
- Fossil Fuels
- Carbon Capture
- Control and Automation for Energy
- Environmental Pollution
- Waste Management
- Transportation Sustainability

More information about this series at http://www.springer.com/series/15901

Kaushik Saha · Avinash Kumar Agarwal
Koushik Ghosh · Sibendu Som
Editors

Two-Phase Flow for Automotive and Power Generation Sectors

Springer

Editors
Kaushik Saha
Centre for Energy Studies
Indian Institute of Technology Delhi
New Delhi, India

Koushik Ghosh
Department of Mechanical Engineering
Jadavpur University
Kolkata, India

Avinash Kumar Agarwal
Department of Mechanical Engineering
Indian Institute of Technology Kanpur
Kanpur, Uttar Pradesh, India

Sibendu Som
Argonne National Laboratory
Lemont, IL, USA

ISSN 2522-8366 ISSN 2522-8374 (electronic)
Energy, Environment, and Sustainability
ISBN 978-981-13-3255-5 ISBN 978-981-13-3256-2 (eBook)
https://doi.org/10.1007/978-981-13-3256-2

Library of Congress Control Number: 2018962380

This Springer imprint is published by the registered company Springer Nature Singapore Pte Ltd.
The registered company address is: 152 Beach Road, #21-01/04 Gateway East, Singapore 189721, Singapore

Preface

Energy demand has been rising remarkably due to increasing population and urbanization. Global economy and society are significantly dependent on the energy availability because it touches every facet of human life and its activities. Transportation and power generation are two major examples. Without the transportation by millions of personalized and mass transport vehicles and availability of 24×7 power, human civilization would not have reached contemporary living standards.

The International Society for Energy, Environment and Sustainability (ISEES) was founded at Indian Institute of Technology Kanpur (IIT Kanpur), India, in January 2014 with the aim of spreading knowledge/awareness and catalysing research activities in the fields of energy, environment, sustainability and combustion. The society's goal is to contribute to the development of clean, affordable and secure energy resources and a sustainable environment for the society and to spread knowledge in the above-mentioned areas and create awareness about the environmental challenges, which the world is facing today. The unique way adopted by the society was to break the conventional silos of specializations (engineering, science, environment, agriculture, biotechnology, materials, fuels, etc.) to tackle the problems related to energy, environment and sustainability in a holistic manner. This is quite evident by the participation of experts from all fields to resolve these issues. ISEES is involved in various activities such as conducting workshops, seminars and conferences in the domains of its interest. The society also recognizes the outstanding works done by the young scientists and engineers for their contributions in these fields by conferring them awards under various categories.

The second international conference on "Sustainable Energy and Environmental Challenges" (SEEC-2018) was organized under the auspices of ISEES from 31 December 2017 to 3 January 2018 at J N Tata Auditorium, Indian Institute of Science Bangalore. This conference provided a platform for discussions between eminent scientists and engineers from various countries including India, USA, South Korea, Norway, Finland, Malaysia, Austria, Saudi Arabia and Australia. In this conference, eminent speakers from all over the world presented their views related to different aspects of energy, combustion, emissions and alternative energy resources for sustainable development and a cleaner environment. The conference presented five high-voltage plenary talks from globally renowned experts on topical themes, namely "Is it Really the End of Combustion Engines and Petroleum?" by Prof. Gautam Kalghatgi, Saudi Aramco; "Energy Sustainability in India: Challenges and Opportunities" by Prof. Baldev Raj, NIAS Bangalore; "Methanol Economy: An Option for Sustainable Energy and Environmental Challenges" by Dr. Vijay Kumar Saraswat, Hon. Member (S&T), NITI Aayog, Government of India; "Supercritical Carbon Dioxide Brayton Cycle for Power Generation" by Prof. Pradip Dutta, IISc Bangalore; and "Role of Nuclear Fusion for Environmental Sustainability of Energy in Future" by Prof. J S Rao, Altair Engineering.

The conference included 27 technical sessions on topics related to energy and environmental sustainability including 5 plenary talks, 40 keynote talks and 18 invited talks from prominent scientists, in addition to 142 contributed talks, and 74 poster presentations by students and researchers. The technical sessions in the conference included Advances in IC Engines: SI Engines, Solar Energy: Storage, Fundamentals of Combustion, Environmental Protection and Sustainability, Environmental Biotechnology, Coal and Biomass Combustion/Gasification, Air Pollution and Control, Biomass to Fuels/Chemicals: Clean Fuels, Advances in IC Engines: CI Engines, Solar Energy: Performance, Biomass to Fuels/Chemicals: Production, Advances in IC Engines: Fuels, Energy Sustainability, Environmental Biotechnology, Atomization and Sprays, Combustion/Gas Turbines/Fluid Flow/Sprays, Biomass to Fuels/Chemicals, Advances in IC Engines: New Concepts, Energy Sustainability, Waste to Wealth, Conventional and Alternate Fuels, Solar Energy, Wastewater Remediation and Air Pollution. One of the highlights of the conference was the rapid-fire poster sessions in (i) Energy Engineering, (ii) Environment and Sustainability and (iii) Biotechnology, where more than 75 students participated with great enthusiasm and won many prizes in a fiercely competitive environment. More than 200 participants and speakers attended this four-day conference, which also hosted Dr. Vijay Kumar Saraswat, Hon. Member (S&T), NITI Aayog, Government of India, as the chief guest for the book release ceremony, where 16 ISEES books published by Springer, Singapore, under a special dedicated series "Energy, Environment, and Sustainability" were released. This is the first time that such significant and high-quality outcome has been achieved by any society in India. The conference concluded with a panel discussion on "Challenges, Opportunities & Directions for Future Transportation Systems", where the panellists were Prof. Gautam Kalghatgi, Saudi Aramco; Dr. Ravi Prashanth, Caterpillar Inc.; Dr. Shankar Venugopal, Mahindra and Mahindra; Dr. Bharat Bhargava, DG, ONGC Energy Centre; and Dr. Umamaheshwar, GE Transportation, Bangalore. The panel discussion was moderated by Prof. Ashok Pandey, Chairman, ISEES. This conference laid out the road map for technology development, opportunities and challenges in energy, environment and sustainability domains. All these topics are very relevant for the country and the world in the present context. We acknowledge the support received from various funding agencies and organizations for the successful conduct of the second ISEES conference SEEC-2018, where these books germinated. We would therefore like to acknowledge SERB, Government of India (special thanks to Dr. Rajeev Sharma, Secretary); ONGC Energy Centre (special thanks to Dr. Bharat Bhargava); TAFE (special thanks to Sh. Anadrao Patil); Caterpillar (special thanks to Dr Ravi Prashanth); Progress Rail, TSI, India (special thanks to Dr. Deepak Sharma); Tesscorn, India (special thanks to Sh. Satyanarayana); GAIL, Volvo; and our publishing partner Springer (special thanks to Swati Mehershi).

The editors would like to express their sincere gratitude to a large number of authors from all over the world for submitting their high-quality work in a timely manner and revising it appropriately at short notice. We would like express to our special thanks to Profs. Pedro Marti, Pallab Sinha Mahapatra, Prof. Sudipto Mukhopadhyay, Somnath Roy, Achintya Mukhopadhyay, Swarnendu Sen, Dipankar Sanyal, Dr. Le Zhao, Dr. Souvick Chatterjee, Dr. Kausik Nandi, Dr. Chetankumar Patel and Mr. Uddalok Sen, who reviewed various chapters of this book and provided very valuable suggestions to the authors to improve their manuscript.

The book covers different aspects of the two-phase flow problems pertinent to the automotive and power generation sectors. The contributions cover a range of topics starting from fundamental aspects to application-oriented research work, which are of interest to both the industry and the academia. The variety of topics included liquid–gas two-phase interactions, nucleate and film boiling, role of two-phase problems in performance assessment of internal combustion engines, gas turbines, fuel cells, nuclear reactor and MEMS.

New Delhi, India — Kaushik Saha
Kanpur, India — Avinash Kumar Agarwal
Kolkata, India — Koushik Ghosh
Lemont, USA — Sibendu Som

Contents

Editors and Contributors

About the Editors

Kaushik Saha is Assistant Professor at Centre For Energy Studies, IIT Delhi. He was previously working at Bennett University, India, as Assistant Professor. Prior to that, he worked at the Energy Systems Division of Argonne National Laboratory, USA. He completed his M.S. and Ph.D. in mechanical engineering from the Universities of Connecticut and Waterloo, respectively. His research interests are flash boiling for gasoline injectors (GDI), cavitation in fuel injectors, spray atomization and droplet evaporation, advanced combustion strategies for internal combustion engines, urea SCR for NO_x reduction and material processing in thermal plasmas. He is Member of Institute for Liquid Atomization and Spray Systems, Society of Automobile Engineering, American Society of Mechanical Engineering, Canadian Institute of Combustion Science, Combustion Institute, USA, and has authored 11 journal articles, 19 conference articles and 1 chapter.

Avinash Kumar Agarwal is Professor in the Department of Mechanical Engineering at Indian Institute of Technology Kanpur. His areas of interest are IC engines, combustion, alternative fuels, conventional fuels, optical diagnostics, laser ignition, HCCI, emission and particulate control, and large bore engines. He has published 24 books and more than 230 international journal and conference papers. He is Fellow of SAE (2012), ASME (2013), ISEES (2015) and INAE (2015). He received several awards such as the prestigious Shanti Swarup Bhatnagar Award-2016 in engineering sciences; Rajib Goyal Prize-2015; NASI-Reliance Industries Platinum Jubilee Award-2012; INAE Silver Jubilee Young Engineer Award-2012; SAE International's Ralph R. Teetor Educational Award-2008; INSA Young Scientist Award-2007; UICT Young Scientist Award-2007; INAE Young Engineer Award-2005.

Koushik Ghosh is Professor in the Mechanical Engineering Department at Jadavpur University, India, and has previously worked as Scientific Officer in the Reactor Safety Division at Bhabha Atomic Research Centre, India. His research interests include heat transfer, multiphase flow, boiling and condensation, and transport in porous media. He has published more than 2 chapters and 80 research articles in journals and national and international conferences.

Dr. Sibendu Som has over a decade of experience in enabling technologies for more efficient engine combustion using computational tools. He leads a CFD team at Argonne National Laboratory with a research focus on the development of nozzle flow, spray and combustion models, using HPC tools for piston engines and gas turbine applications. His team is responsible for developing predictive simulation capabilities for OEMs to develop advanced high-efficiency, low-emission engines. He is Co-Founder of Argonne's Virtual Engine Research Institute and Fuels Initiative (VERIFI) programme which is aimed at providing predictive simulations for OEMs. He has authored more than 110 journal and peer-reviewed conference papers with more than 3100 citations.

Contributors

Avinash Kumar Agarwal Engine Research Laboratory, Department of Mechanical Engineering, Indian Institute of Technology Kanpur, Kanpur, Uttar Pradesh, India

Choongsik Bae Engine Lab, Department of Mechanical and Aerospace Engineering, Korea Advanced Institute of Science and Technology (KAIST), Daejeon, South Korea

R. Banerjee Department of Mechanical & Aerospace Engineering, Indian Institute of Technology Hyderabad, Kandi, Sangareddy, Telangana, India

Michele Battistoni Mechanical Engineering, University of Perugia, Perugia, Italy

Aranyak Chakravarty Department of Mechanical Engineering, Jadavpur University, Kolkata, India

Souvick Chatterjee Department of Mechanical and Industrial Engineering, University of Illinois at Chicago, Chicago, IL, USA

Dipak Chandra Das Department of Mechanical Engineering, National Institute of Technology, Agartala, Tripura, India

Priyankan Datta Department of Mechanical Engineering, Jadavpur University, Kolkata, India

Ranjan Ganguly Department of Power Engineering, Jadavpur University, Kolkata, India

Caroline L. Genzale Georgia Institute of Technology, Atlanta, GA, USA

Koushik Ghosh Department of Mechanical Engineering, Jadavpur University, Kolkata, India

G. Giustini Mechanical Engineering Department, Imperial College, London, UK

Joonsik Hwang Engine Lab, Department of Mechanical and Aerospace Engineering, Korea Advanced Institute of Science and Technology (KAIST), Daejeon, South Korea

Rakesh Kale Department of Mechanical & Aerospace Engineering, Indian Institute of Technology Hyderabad, Kandi, Sangareddy, Telangana, India

Seong-Young Lee Mechanical Engineering-Engineering Mechanics, Michigan Technological University, Houghton, MI, USA

Xianguo Li Mechanical and Mechatronics Engineering, University of Waterloo, Waterloo, ON, Canada

Gina M. Magnotti Argonne National Laboratory, Argonne, IL, USA

Pallab Sinha Mahapatra Department of Mechanical Engineering, IIT Madras, Chennai, India

Pedro Martí-Aldaraví Univesitat Politécncia de Valéncia, Valencia, Spain

Achintya Mukhopadhyay Department of Mechanical Engineering, Jadavpur University, Kolkata, India

K. Nandi Reactor Projects Division, Bhabha Atomic Research Centre, Mumbai, India

José Manuel Pastor Univesitat Politécncia de Valéncia, Valencia, Spain

Chetankumar Patel Engine Research Laboratory, Department of Mechanical Engineering, Indian Institute of Technology Kanpur, Kanpur, India

Rolf D. Reitz University of Wisconsin-Madison, Madison, WI, USA

Somnath Roy Department of Mechanical Engineering, Indian Institute of Technology Kharagpur, Kharagpur, West Bengal, India

Kaushik Saha Centre for Energy Studies, Indian Institute of Technology Delhi, New Delhi, India

Dipankar Sanyal Department of Mechanical Engineering, Jadavpur University, Kolkata, India

Swarnendu Sen Department of Mechanical Engineering, Jadavpur University, Kolkata, India

Uddalok Sen Department of Mechanical and Industrial Engineering, University of Illinois at Chicago, Chicago, IL, USA

Sibendu Som Energy Systems Division, Argonne National Laboratory, Lemont, IL, USA

Tibin M. Thomas Department of Mechanical Engineering, IIT Madras, Chennai, India

Raman Vedarajan Centre for Fuel Cell Technology, International Advanced Research Centre for Powder Metallurgy, Taramani, Chennai, India

Zongyu Yue University of Wisconsin-Madison, Madison, WI, USA; Argonne National Laboratory, Lemont, USA

Le Zhao Mechanical Engineering-Engineering Mechanics, Michigan Technological University, Houghton, MI, USA

Chapter 1
Introduction: Two-Phase Flow for Automotive and Power Generation Sectors

Kaushik Saha, Avinash Kumar Agarwal, Koushik Ghosh and Sibendu Som

Abstract Two-phase problems play vital roles in several industrial applications in power generation sectors. The energy crisis is one of the major concerns in today's world. Fuel sprays are still widely used in the automotive industry because of their high energy content on a volumetric basis and favorable storage. Two-phase problems related to the automotive sprays, whether it is inside the fuel injector or as an emerging spray in the combustion chamber, continue to garner tremendous interest in the scientific community. Different industrial applications, such as fuel cells or ferrofluid propulsion systems, encounter two-phase problems, which need in-depth understanding to improve energy efficiency and ensure durability. Apart from experimental investigations, researchers also need to resort to numerical simulations to cater to the new challenges and innovative product design as well as to better understand the underlying science. Running large-scale simulations using high-performance computing (HPC) unravels unforeseen insights. Therefore, gathering knowledge of HPC systems and usage is a prerequisite to carry out industry-relevant numerical simulation studies. Boiling and condensation are commonly encountered in nuclear energy

K. Saha (✉)
Centre for Energy Studies, Indian Institute
of Technology Delhi, New Delhi 110016, India
e-mail: Kaushik.Saha@ces.iitd.ac.in

A. K. Agarwal
Department of Mechanical Engineering, Indian Institute
of Technology Kanpur, Kanpur 208016, Uttar Pradesh, India
e-mail: akag@iitk.ac.in

K. Ghosh
Department of Mechanical Engineering, Jadavpur University, Kolkata 700032, India
e-mail: kghoshjdvu@gmail.com

S. Som
Energy Systems Division, Argonne National Laboratory, Lemont, IL 60439, USA
e-mail: ssom@anl.gov

K. Saha et al. (eds.), *Two-Phase Flow for Automotive and Power Generation Sectors*, Energy, Environment, and Sustainability,
https://doi.org/10.1007/978-981-13-3256-2_1

and other power generation devices. Both experimental and numerical efforts have been reported in the literature for the advancement of the fundamental understanding of these processes. This book will cover all the above-mentioned aspects regarding two-phase problems, which are part and parcel of different power generation sectors.

Keywords Two-phase flows · Spray atomization · Vaporization · Fuel cells Gas diffusion layer · High-performance computing · Ferrofluids · Boiling Condensation

Two-phase problems are widely common in many industrial and day-to-day applications in today's world. Among the various necessities in today's world, energy demand is one of the most concerning issues and is expected to be a major hurdle for years to come in the future. Energy demand comes from transportation sectors, power plants, household appliances, etc. Applications of two-phase problems are seen in many of these sectors, such as liquid fuel spray evaporating in a combustion chamber, multiphase flows in fuel cells and microchannels, boiling and condensation problems in various thermochemical and nuclear engineering applications. Plethora of studies has been carried out at fundamental levels as well as in application-oriented problems in the area of two-phase phenomena. This book has been divided into three parts.

The first part revisits the development in experimental and numerical aspects of spray atomization and droplet dynamics in the field of automotive engineering. Recent advancements and findings in these areas have been also documented by the contributing authors. Internal combustion engine and gas turbine combustors happen to be among the leading sectors meeting the global power demands. In both these sectors, liquid fuels (primarily fossil fuels) have been a vital source of power because of their ability to produce a substantial amount of energy on a volumetric basis. The fossil fuels in liquid form when injected either into the combustor or upstream of the combustion engine intake port need to get vaporized to form a suitable fuel (vapor)–air mixture to form the basis for exothermal reactions yielding tremendous power. Accurate estimation of fuel–air distribution is a prerequisite for the assessment of the quality of the combustion phenomenon. Spray–wall interaction is also a vital area where the accurate estimation of fuel droplets interacting with the solid wall is needed to assess the generation of unburnt hydrocarbons. Consequently, a detailed understanding of fuel spray behavior and characterization has been warranted over the last century. The modern inventions in the engine technologies, a multitude of options for using the liquid fuels (early injections, split injections dual fuel, alternative fuels, etc.), generate endless scopes for fundamental and applied research in the transportation sector. Two-phase problems not only arise when the fuel is injected in the form of a spray. Prior to the fuel leaving the fuel injector, a liquid fuel may encounter cavitation (mainly in diesel injectors) and flash boiling (mainly gasoline direct injectors). Both the cases are examples of two-phase problems where the fuel is transforming from liquid to vapor phase due to sudden depressurization. Inside the fuel injectors, the liquid fuel moves at a velocity of few hundred meters per second, through flow passages with a hydraulic diameter of couple hundred microns. Consequently, the number of experimental studies addressing in-nozzle flow characteristics is limited. Computational fluid dynamics studies have been prevalent in this area to provide the details, which are otherwise extremely difficult to obtain through experiments. Understanding spray atomization (disintegration of a high-speed liquid jet into ligaments and blobs, followed by further breakups into smaller droplets) has been really vital. Several experimental and modeling studies have been undertaken to unravel the physics of spray formation. Intricate phenomena, such as breakup, collision, coalescence, collectively form one of the most complex research problems. Analysis of spray atomization is useful even in the exhaust sections of heavy-duty vehicles, where the urea–water solution is sprayed upstream of SCR section to form the mixture of ammonia and isocyanic acid to eventually reduce harmful NO_x emissions to nitrogen gas, in the catalyst layers. Many experimental techniques, such as shadowgraphy, Mie scattering, spray PIV, and others, have been used to analyze spray characteristics with different fuel injection systems. Fuel sprays through single hole and multi holes have been examined to provide insights into a varied range of spray characteristics. Several modeling approaches have been proposed, implemented, and tested to represent the spray atomization processes. In some cases, sprays are represented by the assembly of Lagrangian particles/droplets, where right at the exit of the injector holes the spray is initiated with either a big drop (size similar to the hole diameter) or a group of droplets with an assigned size distribution (e.g., Rosin-Rammler). Such approaches have been widely implemented and utilized, despite the limitation of unphysical representation of actual fuel flow out of the injector holes. It is not physically correct to assume that spray is emerging as a drop right after the exit of the injector holes. However, with the speed of the flow out of the holes, the aerodynamic forces, typical in combustors, acting upon the liquid jet, it is safe to assume within a very short span of time the liquid jets break down to droplets. Several breakup models are used in vogue in engine spray combustion community, such as Kelvin–Helmholtz, Rayleigh–Taylor, and others. There have been also some studies on diesel sprays where it is assumed that the spray is not right away emerging as a drop or blob at the hole exit and the entire spray is simulated in the Eulerian framework. Such approaches require extremely fine grid resolution in the computational domain, resulting in substantial computational cost. The behavior of spray plumes changes considerably depending on the application. For example, in diesel injectors, the individual plumes do not typically interact, since the angle of injection of the spray plumes with respect to the injector axis is very wide. For gasoline direct injection, the angle of injection of spray plumes is noticeably smaller compared to that in diesel injection, leading to higher chances of plume-to-plume interactions. Additionally, flash boiling occurs in gasoline direct injection engines at part load, when the in-cylinder pressure is subatmospheric and the injected fuel is hot enough to be superheated. Such a phenomenon leads to an early collapse of spray plumes rendering the overall spray to be much narrower and sudden burst of enhanced momentum toward the downstream directions. In these days of fast-depleting fossil fuel reserves, alternative fuels and biofuels are growing areas of research. Significant differences in physical and chemical properties of the alternate fuel compared to the conventional fuels lead to fresh scopes of research in the automotive community, whether it is a spray characterization in a non-reacting environment or in engine combustion scenarios. As mentioned before, the first part of this book sums up the latest status in the spray atomization and droplet dynamics of the automotive sectors.

The second part of the book will focus on research in two-phase flows carried out in different industrial and power generation sectors. The concerns surrounding pollutant emissions and emerging global warming issues are paving the way of alternate power generation systems for, e.g., fuel cells. Polymer electrolyte membrane fuel cells (PEMFCs) are gaining popularity due to high power density, but water management is a major concern. To mitigate this problem, research is going on to understand the two-phase flow heat transfer across the gas diffusion layers (GDLs). Excessive water accumulation leads to lowering of overall efficiency, and hence, innovative strategies of water removal have become essential and recent advancements have made tremendous progress in this regard. Ferrofluids, which are colloidal suspensions of single domain magnetic nanoparticles in a non-magnetic carrier fluid, are currently being projected as potential candidates for micropropulsion sectors. Utilization of ferrofluids requires analysis of the electrospray emissions and their inherent instabilities. Advancements in this new emerging technology are also documented in the second section of this book.

The second part also covers the aspect of the recent rise in large-scale computational research undertaken across the globe. Two-phase problems are of concerns in numerous other industrial applications, and in many of these applications, the flow is turbulent in nature. Due to characteristic length- and timescales of such problems, performing experimental investigations in real-scale devices is extremely challenging. High-fidelity simulations resolving the necessary length- and timescales are a viable approach with the advent of high-performance computing facilities across the world. For accurate simulations of the fluid flow, large eddy simulation (LES) and direct numerical simulation (DNS) are becoming more common among the scientific community. Moreover, in case of problems dealing with two-phase flow aspects, additional precautions are needed in terms of grid refinement to capture the interactions between the two phases in bubbly or droplet-laden flows. To track such complex problems on HPC machines, several parallelization techniques are used, such as MPI, GPU-accelerated computing with CUDA. The two-phase problems have reached a juncture where interdisciplinary research has become extremely warranted.

The third part deals with the application of boiling and condensation related to power and automotive sectors. This includes the boiling problem in debris coolability in nuclear power reactor in the accidental scenario. The nuclear debris might be formed following a MFCI accident. The types of formation of prototype beds have been discussed in great details. The beds contain the heat-generating porous debris along with water and vapor. The coolability and the liquid–vapor convection and dry-out heat flux, which is a measure of ultimate coolability, are the main parameters of a debris bed. First, in the known geometry, the problem is solved for single-phase heat-generating porous bed for natural and mixed convection. Then, boiling simulation results are given with an estimation of dry-out heat flux for porous debris. The direct contact condensation of steam and water is a very complex and interesting problem in the power sector, particularly relevant in the conventional and nuclear power plant. The detailed review of direct contact condensation of steam in water and water in steam has been carried out by the group of authors in one of the chapters. The different regime mapping (condensation of steam in water) and its different modes and cycles have been explained in details. This includes oscillatory bubble regime, conical jet regime, oscillatory bubble regime, and chugging regime. The underlying mechanism of the chugging regime has been explained in greater details. In addition, the condensation-induced water hammer (CIWH) has been addressed in a systematic manner, where the peak pressure rise as a mechanism of stratified to slug flow transition has been discussed. The occurrence of CIWH from different sources has been reported along with the proper physical explanation.

The film boiling studies have its relevance to numerous researchers through decades in power and automotive sectors owing to its complexity and involved analysis. Although the mechanism of heat and mass transfer is clear through experiments and theory by researchers, there are still uncertainties in the prediction of heat transfer in film boiling. The mixed convection subcooled film boiling is taken up by the authors in a subsequent chapter, and a combined integral regression analysis is performed to predict heat transfer within 10% of experimental data. To perform this, the first two existing instability models have been incorporated into 'two-phase boundary layer integral model.' The model shows an uncertainty in comparing the existing experimental data. The regression model is proposed with a new instability length as a function of relevant non-dimensional parameters. The proposed model successfully captures the existing experimental data of numerous researchers within 10% variation.

In the last chapter, the numerical modeling of boiling is reviewed by the authors. The mechanism of boiling, the pool boiling curve, the flow boiling curve, and the single bubble growth pattern are discussed first. The bulk convection, superheated layer, and the role of microlayer evaporation are explained by the authors. Then in the framework of numerical analysis, the two different approaches, namely interpenetrating continuum approach and single field formalism, are elaborated by the authors. The interpenetrating continuum approach needs the proper heat flux partitioning model along with the reliable and implementable closure relations. The role of interfacial source terms and their treatments have been discussed. The single fluid formalism is based on more fundamental principle of a marker property/indicator function, and volume of fluid (VOF) and level set methods fall under this category. A continuum surface force model is being used to incorporate the surface tension. The simulation results for each of the methods are provided by the authors.

Part I
Spray Atomization and Droplet Dynamics

Chapter 2
Review of an Eulerian Σ-Y Spray Atomization Model for Nozzle Flow and Near-Field Diesel Spray Modeling

Pedro Martí-Aldaraví and José Manuel Pastor

Abstract In order to contribute to a more environment-friendly community, a lot of research is still needed in the field of fossil fuels and internal combustion engines. In those applications, fuel injection systems are one of the key subsystems. However, due to their small characteristic sizes and timescale experiments are difficult to carry out. Thus, computational fluid dynamics (CFD) has been a very successful tool to improve engine efficiency during the last years. Several models have been successfully developed to accomplish that goal. One of the latest is the Eulerian or Eulerian–Lagrangian spray atomization model, which has proved to be able to deal with multi-phase flow physics taking place during fuel injection. The key feature of this model is that it is able to seamlessly simulate both the nozzle internal flow and the subsequent spray development into the ambient gas. In this chapter, a review of this model with examples of its applications is performed. Nozzle flow parameters such as fuel mass flow rate and momentum flux are accurately predicted. The flow pattern (pressure, velocity, and temperature) is then analyzed to give ideas about how to improve the nozzle design. At the same time, fuel atomization and mixing with the surrounding gas can also be studied. Spray macroscopic parameters penetration length (both liquid and vapor) and spray angle are again precisely calculated when compared with experimental measurements. Additionally, this model could be also used to analyze microscopic parameters such as droplet size and distribution. This is done by the calculation of the interphase surface density with the addition of a new transport equation. Even though this model has shown great potential in the field of multi-phase flows for engine applications, there is still room for improvement for its sub-models and programming.

Keywords Spray · CFD · Multiphase · Atomization · Eulerian approach
Internal Combustion Engines

P. Martí-Aldaraví (✉) · J. M. Pastor
Univesitat Politécncia de Valéncia, Camino de Vera, S/N, 46022 Valencia, Spain
e-mail: pedmar15@mot.upv.es

J. M. Pastor
e-mail: jopasen@mot.upv.es

K. Saha et al. (eds.), *Two-Phase Flow for Automotive and Power Generation Sectors*, Energy, Environment, and Sustainability,
https://doi.org/10.1007/978-981-13-3256-2_2

2.1 Introduction

In the latest decades, both environmental regulations and fuel economy requirements have become more restrictive all around the world. As an example, fuel costs used in transportation means have been continuously increasing and are expected to do so in the near future. As a result, investigations on modern engines must focus on improving combustion efficiency and decreasing pollutant emissions. To accomplish these two broad goals, the understanding of every phenomenon taking place in an engine is basic. And that is why the research in internal combustion engines is in turn divided into multiple areas: air management, thermal management, fuel injection, combustion. Among all of them, understanding fuel injection processes and subsequent fuel–air mixture formation is crucial due to its impact in combustion development and also pollutant formation.

Some of the phenomena taking place during the injection process, such as primary atomization and nozzle cavitation, are not fully understood (Desantes et al. 2016a). Numerical simulations combined with experimental tools make advancements in this challenging field of science possible. Furthermore, simulations can provide additional information about the underlying problem, which may be difficult or even impossible to obtain with experiments.

For fuel injection studies (both experimental and computational), the problem is usually decomposed in two steps, depending on the area of interest and composition of the fluid: internal nozzle flow and spray external flow. Internal flow studies investigate the impact of the nozzle geometry on the flow pattern, the cavitation phenomenon, the needle lift, its eccentricity, and other manufacturing issues, while external flow studies deal with fuel atomization, breakup, collision, evaporation, fuel–air mixing, and finally combustion processes. This division is made because of the different flow nature: In the internal part, the flow is continuous, single-phase liquid (or multi-phase if cavitation is considered), and in the external part, the fluid evolves from a dense multi-phase flow in the near-nozzle region to a dispersed flow far downstream.

Focusing on the liquid atomization process in current injection systems (high injection pressure sprays, high Reynolds and Weber numbers, typical of direct injection engines), it occurs at extremely small length scales and high speeds, which complicates both the experimental investigation and modeling of spray flow, especially in the dense near-nozzle region where the flow contains large density ratios, phase change, and even near-supercritical conditions. In fact, those scales are beyond the spatiotemporal capabilities of current imaging systems. The lack of optical accessibility, except by means of special diagnostic techniques (such as X-ray radiography), hinders the flow characterization and the development of predictive primary atomization models. It gets even more complicated when dual-fuel or multiple injections strategies are used to improve combustion efficiency.

In order to improve understanding of primary atomization process for diesel-like sprays, several modeling frameworks, from Lagrangian–Eulerian to Eulerian–Eulerian and from Reynolds-averaged Navier–Stokes (RANS) to direct numerical simulations (DNS) have been employed over the years to capture and explain experimentally observed spray details (Battistoni et al. 2018).

Due to the multi-phase, multi-physics, and multi-scale nature of the spray development in direct injection engines, DNS are reported to be extremely expensive, pushing both industry and laboratories toward the development of primary atomization models. Numerous of these models have been received widespread attention, some of them with limited predictive capabilities, requiring inputs such as initial drop size and mass and momentum flux. For more than 30 years, the most commonly employed spray modeling approach utilizes a Lagrangian framework for the liquid phase while using an Eulerian framework for the gaseous one, i.e., the classical Lagrangian-discrete droplet method (DDM) approach. However, this DDM method presents some well-known drawbacks for dense two-phase flow modeling, which more recent Eulerian modeling approach overcome (Battistoni et al. 2018; Desantes et al. 2016a). Furthermore, they are not necessarily appropriate from a physical point of view for describing the near-nozzle dense region of the spray. Some basic hypothesis, such as low liquid volume fraction or homogeneously distributed parcels in the computational cells, are not valid. In addition, the majority of existing drag, collision, breakup, and vaporization sub-models are based on assumptions of near-spherical droplets in a sparse spray. The validity of these isolated drop-based models is hardly justified in this region which is characterized by strong interaction between phases. Moreover, in order to assure numerical stability, it is often necessary to use grid sizes larger than the orifice diameter resulting in inadequately resolved flow structures. This issues usually require a "best-practice" approach when using this method (Garcia-Oliver et al. 2013). Thus, DDM are well-suited for only low liquid volume fraction flow, even though they are able to properly reproduce the macroscopic spray behavior.

Because the interface between the fuel and surrounding gas is so complex, it is an over-simplification to think of the dense spray core as being comprised of isolated droplets. Multi-phase Eulerian approach fits better to this region. With extreme levels of computational effort, the interface details can potentially be resolved through interface tracking (volume-of-fluid and/or level-set methods). However, in a typical engineering calculation, the mesh resolution is considerably coarser than in those high-fidelity computations. Assuming that these interfacial details are far smaller than the mesh size, smoothing its features over at least one cell, the end result is a diffuse-interface treatment in an Eulerian framework. This framework is naturally extensible to near critical or supercritical regimes. Under these conditions, the gas–liquid interface "disappears" and spray simulation becomes entirely an exercise in modeling turbulent mixing (Pandal et al. 2016). This explains why homogeneous single-fluid Eulerian models (Vallet et al. 2001) have emerged for spray simulations as a promising alternative to classical Lagrangian models. They have both numerical (i.e., reduced number of source terms and grid convergence) and physical (i.e., the intact core length can be simulated) advantages. Following this approach, instead of tracking droplets or the interphase surface, a "more general notion" of interfacial area density is used. This metric, with dimensions of reciprocal length, represents how much interfacial area is present per unit volume. In addition, simulating the liquid phase using also an Eulerian reference frame allows for the solution to include both the external spray and the internal nozzle flow, thus capturing the effects of the nozzle geometry and then increasing model predictive capabilities.

In the present chapter, some of the research carried out by the authors[1] on diesel-like sprays using this single-fluid homogeneous Eulerian model is summarized, from nozzle flow calculations to combustion analysis, and from RANS to large eddy simulations (LES). Most of the work is done in the context of the engine combustion network (ECN), an international collaboration nexus between research institutes and universities around the world that addresses the study of the engine performance, focusing on the injection–combustion processes, with reliable high standards. Experimental data to compare with simulation results and validate the models is obtained thanks to this network.

The chapter is divided into six different sections. After this first introductory part, the common computational methodology is explained in Sect. 2.2. Particularities of the different models or implementations are also mentioned. Sect. 2.3 focuses on the results inside the injector nozzle. Compressibility and thermal effects are studied. Following, Sect. 2.4 deals with the near-field spray simulations. Spray characteristics such as penetration or liquid distribution and concentration are examined. This forms, so to speak, the main area of interest since these types of models are developed to analyze that region. Logically, next Sect. 2.5 inspects the uses of these model on evaporative sprays and the combustion process. And finally, general conclusions are given in Sect. 2.6 together with some ideas of future work.

2.2 Review of the Computational Methodology

As commented in Sect. 2.1, separation of the large-scale flow features, such as mass transport, from the atomization process occurring at smaller scales can be assumed for high Reynolds and Weber numbers (Desantes et al. 2017; Vallet et al. 2001). That is the case of current sprays in the automotive industry. Then, large-scale liquid dispersion can be modeled as the turbulent mixing of a variable density homogeneous flow. For small-scale atomization, the surface density concept is introduced in order to evaluate the mean size of liquid fragments and droplets, considering that interfacial details are smaller than the mesh size, and that there is a small difference in velocity between phases.

These diffuse-interface Eulerian spray models share two common elements: a model for the transport of liquid, denoted by Y, and a model for the evolution of the interfacial surface area, denoted by Σ. This chapter deals with the strictly Eulerian model, named as Σ-Y model, in contrast to the Eulerian–Lagrangian spray

[1]As commented also in the Bibliography section, the reader should be aware that there are other interesting and significant works dealing with Eulerian–Lagrangian spray atomization models whose reading is recommended. For the present chapter, it was decided to focus on the work of the same research institution.

atomization (ELSA) which includes transition to Lagrangian particle tracking (in other words, ELSA includes both, Eulerian single-fluid model for the dense part and a classic DDM for the disperse spray).

From the point of view of the model, the transport of the liquid employs mass (or volume)-averaged convection along with turbulent mixing. Thus, the accuracy of the liquid fraction transport is largely dependent on the accuracy of the two-phase turbulent modeling. Despite the challenges of such modeling, there is extensive theoretical basis to properly establish the closure terms. However, the model for the interface evolution is somewhat more especulative, with several unclosed terms (Pandal et al. 2017c). There are available in the literature of several modeling approaches that have been applied to sprays.

2.2.1 *Favre-Averaged Transport Equations*

In order to ensure mass conservation, the continuity equation (Eq. 2.1) for the mixture is solved.

$$\frac{\partial \bar{\rho}}{\partial t} + \frac{\partial (\bar{\rho}\tilde{u}_i)}{\partial x_i} = 0 \tag{2.1}$$

Being $\tilde{Y}$ the mass-averaged liquid mass fraction, continuity equation of this phase yields in Eq. 2.2, where u' denotes the turbulent fluctuations in velocity, Y' the turbulent fluctuations in liquid mass fraction, and $\tilde{\dot{Y}}_{evap}$ the evaporation source term (see Sect. 2.2.4). Due to the hypothesis of the model, the turbulent dispersion of the injected liquid into the gaseous ambient $\widetilde{u_i'Y'}$ can be modeled with a turbulent diffusion flux given by Fick's law[2] of Eq. 2.3, where Sc is the Schmidt number and μ_t the turbulent viscosity (thus, a turbulent viscosity model is required). By the Kolmogorov hypothesis for small-scale features of the flow, $Sc = 1$ can be assumed, although the value $Sc = 0.9$ is also generally used with good results (Garcia-Oliver et al. 2013).

$$\frac{\partial\left(\bar{\rho}\tilde{Y}\right)}{\partial t} + \frac{\partial\left(\bar{\rho}\tilde{u}_i\tilde{Y}\right)}{\partial x_i} + \frac{\partial\left(\bar{\rho}\widetilde{u_i'Y'}\right)}{\partial x_i} = -\bar{\rho}\tilde{\dot{Y}}_{evap} \tag{2.2}$$

$$\bar{\rho}\widetilde{u_i'Y'} = -\frac{\mu_t}{Sc}\frac{\partial \tilde{Y}}{\partial x_i} \tag{2.3}$$

[2]More sophisticated closures recently developed remain to be tested, specially under low ambient densities and downstream positions where the model has shown poorer accuracy, probably due to the relevance of interfacial dynamics.

For those cases where phase change and/or combustion are considered, additional mass fraction transport equations are required. Similarly to Eq. 2.2, Eq. 2.4 can be written for each considered phase i. In this case, two different source terms are included, one to take into account liquid evaporation and the other related to the chemical reactions during the combustion process.

$$\frac{\partial\left(\tilde{\rho}\tilde{Y}_i\right)}{\partial t} + \frac{\partial\left(\tilde{\rho}\tilde{u}_i\tilde{Y}_i\right)}{\partial x_i} = \tilde{\rho}\tilde{\dot{Y}}_{i,evap} + \tilde{\rho}\tilde{\dot{Y}}_{i,chem} \tag{2.4}$$

Under the assumption that the three or more phases (i.e., liquid, vapor, and gas) form an immiscible mixture, the mass-averaged density of the mixture is related to the mass fraction of all phases by Eq. 2.5. For those simulations without phase change, the same equation can be applied taking $Y_i = 0$.

$$\frac{1}{\tilde{\rho}} = \frac{\tilde{Y}}{\rho_l} + \frac{\tilde{Y}_g}{\rho_g} + \sum \frac{\tilde{Y}_i}{\rho_i} \tag{2.5}$$

An equation of state is then assigned to each phase. This is discussed later in Sect. 2.2.3.

While the approach of this Σ-Y model assumes that the resolved momentum of the liquid–gas mixture can be characterized by a single bulk velocity, the term $\widetilde{u_i'Y'}$ captures the effect of the relative velocity between the two phases; thus, the slip velocity (though small) can be derived, resulting in Eq. 2.6.

$$\tilde{u}_{i,l} - \tilde{u}_{i,g} = \frac{1}{\tilde{Y}\left(1-\tilde{Y}\right)}\widetilde{u_i'Y'} \tag{2.6}$$

The mass-averaged velocity of the flow (the homogeneous mixture) is then obtained by the momentum equation in the form of Eq. 2.7, where τ_{ij} represents the Reynolds shear-stress tensor and its particular value depends on the turbulent approach (i.e., RANS Desantes et al. 2014 or LES Desantes et al. 2017) and model (i.e., k-ε Pandal et al. 2017a or SST k-ω Payri et al. 2017a). Note that surface tension effects and volume forces are generally neglected (not without controversy) in this equation for fuel injection applications because their impact on the solution is small in comparison with turbulence terms.

$$\frac{\partial(\tilde{\rho}\tilde{u}_i)}{\partial t} + \frac{\partial\left(\tilde{\rho}\tilde{u}_i\tilde{u}_j\right)}{\partial x_j} = -\frac{\partial\tilde{p}}{\partial x_i} + \frac{\partial\tau_{ij}}{\partial x_j} \tag{2.7}$$

Regarding the energy equation, either the mass-averaged total enthalpy $\tilde{h}_0$ (Eq. 2.8) (Pandal et al. 2017a) or the mass-averaged static enthalpy $\tilde{h}$ (Eq. 2.9) (Payri et al. 2017a) can be considered, where in both cases α_{eff} is the effective turbulent thermal diffusivity, $\tau_{ij}\,\partial u_j/\partial x_i$ is the viscous dissipation, and $K = 0.5\tilde{u}_i^2$ is the mean kinetic energy.

$$\frac{\partial\left(\tilde{\rho}\tilde{h}_0\right)}{\partial t}+\frac{\partial\left(\tilde{\rho}\tilde{u}_i\tilde{h}_0\right)}{\partial x_i}-\frac{\partial}{\partial x_i}\left(\alpha_{eff}\frac{\partial\tilde{h}_0}{\partial x_i}\right)=\frac{\partial\tilde{p}}{\partial t}+\tilde{u}_i\frac{\partial\tilde{p}}{\partial x_i}+\tau_{ij}\frac{\partial\tilde{u}_j}{\partial x_i} \tag{2.8}$$

$$\frac{\partial\left(\tilde{\rho}\tilde{h}\right)}{\partial t}+\frac{\partial\left(\tilde{\rho}\tilde{u}_i\tilde{h}\right)}{\partial x_i}-\frac{\partial}{\partial x_i}\left(\alpha_{eff}\frac{\partial\tilde{h}}{\partial x_i}\right)=\frac{\partial\tilde{p}}{\partial t}-\frac{\partial K}{\partial t}-\frac{\partial(\tilde{\rho}\tilde{u}_iK)}{\partial x_i}+\tau_{ij}\frac{\partial\tilde{u}_j}{\partial x_i} \tag{2.9}$$

The mass-averaged temperature evolution is calculated from the transported enthalpy applying a bulk mixture enthalpy Eq. 2.10, under the assumption of local thermodynamic equilibrium, where h_l, h_g, and h_i denote the enthalpy of the liquid fuel, the ambient gas, and each of the species in gas phase, respectively.

$$\tilde{h}\,(p,T)=\tilde{Y}h_l\,(p,T)+\tilde{Y}_gh_g\,(p,T)+\sum\tilde{Y}_ih_i\,(p,T) \tag{2.10}$$

A thermodynamic model is then assigned to each phase. This is discussed later in Sect. 2.2.3.

The solution of the above equations fully characterizes the large-scale bulk motion of the flow. Nevertheless, several other options exist in the literature for obtaining closure terms in the above system of equations. Conversely, the small-scale atomization is modeled by solving a transport equation for the evolution of the density of interfacial surface area Σ (Vallet et al. 2001) depicted in Eq. 2.11, where D_Σ is a suitable diffusion coefficient usually taken as the turbulent viscosity (ν_t) over the Schmidt number (now Sc_Σ). A detailed explanation of the different terms (and possible alternatives for the source terms) in this equation and how relevant quantities for spray characterization, such as Sauter mean diameter (SMD), are obtained is given in Sect. 2.2.5.

$$\frac{\partial\tilde{\Sigma}}{\partial t}+\frac{\partial\left(\tilde{u}_i\tilde{\Sigma}\right)}{\partial x_i}-\frac{\partial}{\partial x_i}\left(D_\Sigma\frac{\partial\tilde{\Sigma}}{\partial x_i}\right)=(A+a)\,\tilde{\Sigma}-V_s\tilde{\Sigma}^2-\tilde{\dot{\Sigma}}_{evap} \tag{2.11}$$

The model and all of its variants are implemented using the open-source OpenFOAM library. The advantages of this library are that polyhedral mesh and parallelism are intrinsically supported. The implementation of the governing equations has to be carefully analyzed because:

- The proper sequence of equations and updates improves the stability and the computational cost of the solution (Payri et al. 2015).
- Consistency and conservation must be ensured, if not corrections need to be added (Garcia-Oliver et al. 2013).

- Selection of the right linear solvers and numerical algorithms allows to improve the accuracy, stability, and parallel efficiency of the solver (Pandal et al. 2016).

2.2.2 *Pressure–Velocity Coupling*

Up to date, three different approaches (in addition to the original one Vallet et al. 2001) have been successfully applied to carry out the pressure–velocity coupling and then compute the pressure field.

2.2.2.1 Compressible Pressure Equation

A standard pressure projection step may be implemented through Eq. 2.12 (Salvador et al. 2014), where ϕ represents the flux at cell faced, a_p the diagonal coefficient from the momentum equation, and the symbol ϕ^* the flux without the pressure gradient. The compressiblity Ψ, calculated by Eq. 2.13 where a_l is the speed of sound of the liquid phase, accounts for the effects of the turbulent mixing, Mach (compressiblity) and thermal expansion.

$$\frac{\partial(\Psi \tilde{p})}{\partial t} + \frac{\partial \phi^*}{\partial x_i} - \frac{\partial}{\partial x_i}\left(\frac{1}{a_p}\frac{\partial \tilde{p}}{\partial x_i}\right) = 0 \tag{2.12}$$

$$\Psi = \frac{\partial \rho}{\partial p} = \tilde{Y}_l \frac{\tilde{\rho}}{\rho_l}\frac{1}{a_l^2} + \sum Y_i \frac{\tilde{\rho}}{\rho_i}\frac{1}{R_i T} \tag{2.13}$$

2.2.2.2 Total Derivative of Density

For the case of multi-phase, compressible flow the velocity divergence is nonzero and could be split between the effects of the turbulent mixing, Mach (compressibility) and thermal expansion by applying the chain rule to the continuity equation. In order to guarantee the consistency between the definitions of density and mass fraction, a small penalty function in the pressure projection step is included, which relaxes the density calculated from the continuity equation toward the value stipulated by Eq. 2.5 (Garcia-Oliver et al. 2013), denoted as ρ_{EOS}. The final form of the formula is given in Eq. 2.15, where the constant multiplier K_r represents the approximate number of time steps for the relaxation to the correct density, and Δt is the simulation time step.

$$\frac{\partial \phi^*}{\partial x_i} - \frac{\partial}{\partial x_i}\left(\frac{1}{a_p}\frac{\partial \tilde{p}}{\partial x_i}\right) = -\left(\frac{\tilde{Y}\psi_l}{\rho_l} + \frac{1-\tilde{Y}}{\tilde{p}}\right)\left(\frac{\partial \tilde{p}}{\partial t} + u_i\frac{\partial \tilde{p}}{\partial x_i}\right) +$$
$$+\frac{1-\tilde{Y}}{T}\left(\frac{\partial T}{\partial t} + u_i\frac{\partial T}{\partial x_i}\right) - \tag{2.14}$$
$$-\left(\frac{1}{\rho_l} - \frac{1}{\rho_g}\right)\frac{\partial\left(\widetilde{\bar{\rho}u_i'Y'}\right)}{\partial x_i} - \frac{\rho_{EOS} - \tilde{\rho}}{\Delta t\, K_r \tilde{\rho}} + \tilde{\rho}\tilde{\dot{Y}}_{evap}$$

2.2.2.3 Coupled, Density-Based Algorithm

In this type of solvers, density and liquid mass fraction are strongly correlated due to the high density ratio between phases. And this generates large pressure fluctuations. Therefore, a density-based approach seems well-suited for this Σ-Y models (Pandal et al. 2016). Under this approach, the pressure equation is derived from an equation of state. If the gas phase (in this case there is only one, the ambient gas) obeys an ideal gas law while the liquid phase is assume to have linear compressiblity (Eq. 2.18), the final explicit pressure equation is Eq. 2.15.

$$\tilde{p} = \frac{\tilde{\rho} - \tilde{Y}\left(\rho_{l,0} - \Psi_l p_0\right)}{\frac{1}{R_g\, T}\left(1 - \tilde{Y}\right) + \tilde{Y}\Psi_l} \tag{2.15}$$

2.2.3 *Equations of State and Thermodynamic Sub-Models*

The gaseous phases (e.g., the fuel vapor or the ambient gas) are assumed to obey an ideal gas law (Eq. 2.16) in all published works.

$$\rho_i = \frac{p}{R_i\, T} \tag{2.16}$$

Under this hypothesis, the relation between enthalpy and temperature is straightforward given by Eq. 2.17, where $C_{p,i}$ is the phase specific heat capacity at constant pressure. If these gaseous species are also considered perfect gases (Payri et al. 2017a), $C_{p,i}$ is assumed to be constant. Otherwise (for combustion studies), specific heat capacity as function of temperature can be evaluated from 7-coefficient NASA polynomials (Pandal et al. 2017a) or from a set of coefficients taken from JANAF tables of thermodynamics (Desantes et al. 2016b).

$$\mathrm{d}h_i = C_{p,i}\, \mathrm{d}T \tag{2.17}$$

On the other hand, there is still some controversy about the equation of state and thermodynamic model of the liquid phase. It can be assumed to be incompressible with constant heat capacity (Vallet et al. 2001), but also to have a linear compressibility (Garcia-Oliver et al. 2013) denoted by ψ_l such as in Eq. 2.18, where $\rho_{l,0}$ and p_0 denote the reference density and the reference pressure, respectively, about which the equation of state is linearized.

$$\rho_l = \rho_{l,0} + \psi_l \left(p - p_0\right) \tag{2.18}$$

More recent works (Desantes et al. 2016b; Pandal et al. 2017a, b) employ the Hankinson–Brobst–Thomson (HBT) correlation to estimate the liquid density as function of temperature and pressure; and a Rowlinson–Bondi equation to relate the liquid specific heat with the temperature. There is also the possibility of using polynomial fittings (Payri et al. 2017a; Salvador et al. 2014) to liquid phase properties experimentally measured as function also of temperature and pressure. These last two approaches reproduce better the thermodynamic behavior of the liquid phase and allow to capture expansion and thermal effects inside the nozzle (Desantes et al. 2016b; Payri et al. 2017a).

2.2.4 Liquid Evaporation Sub-Model

The source term in Eq. 2.2 for liquid–vapor transport equations is calculated in terms of a rate needed to achieve local adiabatic saturation conditions. This can be written as Eq. 2.19, where $\tilde{Y}_v$ is the local vapor mass fraction, $Y_{v,sat}$ is the value of vapor mass fraction under adiabatic saturation conditions, and τ_{evap} is a relaxation time. So far, the latter parameter has been set equal to the computational time step.

$$\tilde{\dot{Y}}_{evap} = \frac{Y_{v,sat} - \tilde{Y}_v}{\tau_{evap}} \tag{2.19}$$

The term $Y_{v,sat}$ in Eq. 2.19 is calculated by means of locally homogeneous flow approach (Garcia-Oliver et al. 2013). According to that, state relationships are applied to describe spray thermodynamic conditions under the assumption of local thermodynamic equilibrium.

In those cases where combustion is also simulated (Pandal et al. 2017a), there could be several gaseous species additional to the ambient gas. Notwithstanding, the liquid is assumed to be mono-component, and thus, the evaporation source term is different from zero only for the fuel species (i.e., n-dodecane $C_{12}H_{26}$).

2.2.5 Primary Atomization Sub-Model

The terms A and a of Eq. 2.11 are inverse timescales that define the rate at which surface area is produced. Specifically, the A term models the creation of surface area via the stretching of the interface by mean of velocity gradients. The original model (Vallet et al. 2001) takes this term to be proportional to the same timescale as that used in the production of kinetic energy in the standard k-ε turbulence model, as shown in Eq. 2.20.

$$A = \alpha_0 \frac{\widetilde{u'_i u'_j}}{k} \frac{\partial \tilde{u}_i}{\partial x_j} \tag{2.20}$$

The a term accounts for small-scale interface area production. Here there are several possibilities, but if it is assumed that the dominant mechanism is related to the collision and breakup of droplets, the inverse of droplet collision timescale of Eq. 2.21 may be used, where l_t is the turbulent length scale.

$$a = a_{coll} = \frac{\alpha_1}{(36\pi)^{\frac{2}{9}}} (l_t \Sigma)^{\frac{2}{3}} \left(\frac{\rho_l}{\bar{\rho}\tilde{Y}}\right)^{\frac{4}{9}} \frac{\varepsilon}{k} \tag{2.21}$$

The V_s term captures the effects of interface destruction by coalescence. It is determined in Eq. 2.22 by solving the value of V_s that provides an equilibrium value of Σ set by a predicted equilibrium droplet radius given by Eq. 2.23.

$$V_s = \frac{a_{coll} \rho_l r_{eq}}{3\bar{\rho}\tilde{Y}} \tag{2.22}$$

$$r_{eq} = \alpha_2 \frac{\sigma^{\frac{3}{5}} l_t^{\frac{2}{5}}}{k^{\frac{3}{5}}} \frac{\left(\bar{\rho}\tilde{Y}\right)^{\frac{2}{15}}}{\rho_l^{\frac{11}{15}}} \tag{2.23}$$

Finally, the $\dot{\tilde{\Sigma}}_{evap}$ term models the effects on interfacial surface produced by evaporation. Obviously, it is related to the source term for the fuel liquid or vapor transport equations, $\dot{\tilde{Y}}_{evap}$ through Eq. 2.24.

$$\dot{\tilde{\Sigma}}_{evap} = \frac{2}{3} \Sigma \frac{\dot{\tilde{Y}}_{evap}}{\tilde{Y}} \tag{2.24}$$

Together with the mass-averaged liquid volume fraction X, the interface surface area density can be used to calculate the local SMD of the spray, as shown in Eq. 2.25, and the drop number density, in Eq. 2.26.

$$\text{SMD} = \frac{6\bar{\rho}\tilde{Y}}{\rho_l \tilde{\Sigma}} \tag{2.25}$$

$$n = \frac{\rho_l^2 \tilde{\Sigma}^3}{36\pi\bar{\rho}^2\tilde{Y}^2} \tag{2.26}$$

The coefficients α_0, α_1, and α_2 are modeling constants which, so far, have to be calibrated (Pandal et al. 2017b) for each case (see Sect. 2.4.2 as an example, but also to check the sensibility of the model to these coefficients).

2.2.5.1 Alternative Modeling Option

Previously presented approach includes source terms to account for generation due to the growth of fluid instabilities (i.e., Kelvin–Helmholtz) and the destruction of surface due to droplet coalescence (in the case of dispersed flow). However, nowadays the most common form for the combination of these two source terms is the restoration to an equilibrium value (Σ_{eq}) or critical surface density to which the local surface density is driven (Desantes et al. 2017). Equation 2.11 can be rewritten in the form of Eq. 2.27.

$$\frac{\partial \tilde{\Sigma}}{\partial t} + \frac{\partial\left(\tilde{u}_i \tilde{\Sigma}\right)}{\partial x_i} - \frac{\partial}{\partial x_i}\left(D_\Sigma \frac{\partial \tilde{\Sigma}}{\partial x_i}\right) = C_\Sigma \tilde{\Sigma}\left(1 - \frac{\tilde{\Sigma}}{\Sigma_{eq}}\right) + S_{\Sigma_{init}} - \tilde{\dot{\Sigma}}_{evap} \quad (2.27)$$

Term Σ_{eq} could be set by a suitable equilibrium droplet radius (Eq. 2.23) (Pandal et al. 2017b) in the form of Eq. 2.28.

$$\Sigma_{eq} = \frac{3\bar{\rho}\tilde{Y}}{\rho_l r_{eq}} \quad (2.28)$$

This term can also be computed from an equilibrium Weber number, instead of using an equilibrium droplet radius, in order to avoid any kind of assumption of spherical droplets (Desantes et al. 2017), resulting in Eq. 2.29.

$$\Sigma_{eq} = \alpha_2 \frac{\left(\rho_l + \rho_g\right)\tilde{Y}\left(1 - \tilde{Y}\right)k}{\sigma} \quad (2.29)$$

Then the coefficient C_σ is modeled as the inverse of the turbulent timescale as given by Eq. 2.30.

$$C_\Sigma = \alpha_1 \frac{\varepsilon}{k} \quad (2.30)$$

Note the presence of two modeling constants α_1 and α_2, by default equal to 1.0. However, as mentioned above, they should be calibrated.

Until now all the source terms involved in this equation are proportional to the interface surface density. As a result, there is not production/destruction if there is no interface. Therefore, a proper initialization should be made by means of the term $S_{\Sigma_{init}}$. This initialization corresponds to a production due to the liquid-gas mixing process, whose modeling is represented in Eq. 2.31 (Desantes et al. 2017).

$$S_{\Sigma_{init}} = 2\tilde{\rho}\frac{\nu_t}{Sc}\frac{6\tilde{\rho}}{\rho_l\rho_g l_t}\frac{\partial\tilde{Y}}{\partial x_i}\frac{\partial\tilde{Y}}{\partial x_i} \quad \text{if } \tilde{Y}\left(1-\tilde{Y}\right) \leq 0.001$$

$$S_{\Sigma_{init}} = 2\frac{\nu_t}{Sc}\frac{\tilde{\Sigma}}{\tilde{Y}\left(1-\tilde{Y}\right)}\frac{\partial\tilde{Y}}{\partial x_i}\frac{\partial\tilde{Y}}{\partial x_i} \quad \text{if } \tilde{Y}\left(1-\tilde{Y}\right) > 0.001 \tag{2.31}$$

There is also an alternative for this initialization source term (Pandal et al. 2017b), which consists of considering a minimum value of $\tilde{\Sigma}$ in any computational cell which is not filled with pure liquid or gas. This minimum value is estimated as $V^{\frac{1}{3}}$ where V is the volume of the cell mesh. The source term only takes a positive value if the interface field is lower than this Σ_{min}, resulting in Eq. 2.32.

$$S_{\Sigma_{init}} = \frac{\Sigma_{min} - \tilde{\Sigma}}{\Delta t} \quad \text{if } \Sigma_{min} - \tilde{\Sigma} > 0$$

$$S_{\Sigma_{init}} = 0 \quad \text{if } \tilde{Y}\left(1-\tilde{Y}\right) \leq 0 \tag{2.32}$$

2.2.6 Combustion Sub-Model

The combustion modeling approach proposed for diesel-like reacting sprays can be categorized as an unsteady flamelet/progress variable (UFPV) model, using the approximated diffusion flamelet (ADF) model with the aim of decreasing the computational cost of the generation of the laminar flamelet manifolds (Pandal et al. 2017a).

As a basis for the model, a transport equation for the mass-averaged mixture fraction $\tilde{Z}$, Eq. 2.33, and the mixture fraction variance $\tilde{Z}''^2$, Eq. 2.34, are needed.

$$\frac{\partial\left(\tilde{\rho}\tilde{Z}\right)}{\partial t} + \frac{\partial\left(\tilde{\rho}\tilde{u}_i\tilde{Z}\right)}{\partial x_i} - \frac{\partial}{\partial x_i}\left(\frac{\mu_t}{Sc}\frac{\partial\tilde{Z}}{\partial x_i}\right) = \tilde{\rho}\tilde{\dot{Y}}_{evap} \tag{2.33}$$

$$\frac{\partial\left(\tilde{\rho}\tilde{Z}''^2\right)}{\partial t} + \frac{\partial\left(\tilde{\rho}\tilde{u}_i\tilde{Z}''^2\right)}{\partial x_i} - \frac{\partial}{\partial x_i}\left(\frac{\mu_t}{Sc}\frac{\partial\tilde{Z}''^2}{\partial x_i}\right) = 2\frac{\mu_t}{Sc}\left(\frac{\partial\tilde{Y}_{vap}}{\partial x_i}\right)^2 - \tilde{\rho}\tilde{\chi} \tag{2.34}$$

In Eq. 2.34, the mean scalar dissipation rate[3] is modeled as shown in Eq. 2.35, where the turbulent dissipation rate ε and the turbulent kinetic energy k are directly obtained from the turbulence model. The constant C_χ is calibrated in terms of inert spray measurements.

[3]The presented combustion model is design and has been applied for two-equation RANS turbulence models.

$$\tilde{\chi} = C_{\chi}\frac{\varepsilon}{k}\tilde{Z}''^2 \tag{2.35}$$

The turbulence–chemistry subgrid interaction is accounted for by means of a presumed probability density function (PDF) approach. A tabulation technique is adopted to store pre-calculated turbulent flamelet solutions in order to allow the use of detailed chemical mechanisms at reasonable computational cost. The mechanism proposed to describe the n-dodecane chemistry consists of 255 species and 2289 reactions. Out of the full set of chemical mechanism species, only CO, CO_2, H, H_2, H_2O, OH, C_2H_2, and CH_2O are transported in the solver by means of an equation of the type of Eq. 2.4.

The main interaction between the fluid dynamics solver and the combustion model is reviewed next.[4] Mixture fraction average and variance, scalar dissipation rate, and CO and CO_2 are retrieved by the combustion model from the corresponding transport equations. Progress variable is then reconstructed as $Y_C = Y_{CO} + Y_{CO_2}$, which together with a derived stoichiometric scalar dissipation rate enables the calculation of the values of the pre-integrated tabulated species $\tilde{Y}_i^{tab}$ at the subsequent time step $t + \Delta t$. Finally, these species are combined with those retrieved from the solver at time t, so that the source term from the transport equation is given by Eq. 2.36.

$$\tilde{\dot{Y}}_{i,chem} = \frac{\tilde{Y}_i^{tab}(t + \Delta t) - \tilde{Y}_i(t)}{\Delta t} \tag{2.36}$$

Up to date and to the best of the authors' knowledge, the presented combustion sub-model (Pandal et al. 2017a) is the only one employed under an Eulerian description of the spray. Therefore, the effect of the combustion model cannot be analyzed. It is proposed as future work.

2.3 Nozzle Flow Results

2.3.1 Validation

Effects of nozzle geometry on spray development and combustion efficiency have been widely reported in the specialized literature. To obtain geometrical details of nozzles, silicone mold technique has widely been used over the years. Nevertheless, new powerful techniques such as X-ray tomography have been also successfully employed. In fact, this last technique allows not only obtaining the sizing of the nozzle with a resolution of microns, but also the time resolved position of the needle and measurements in the dense region of the spray.

[4] For a full description of the interactions, and also the layout of the complete combustion sub-model, please consult the reference (Pandal et al. 2017a).

Table 2.1 Nozzle geometry and injection conditions of the validation simulations

Orifice inlet diameter	140	μm
Orifice outlet diameter	112	μm
Orifice fillet radius	42	μm
Orifice length	0.988	mm
Fuel	Standard	Diesel elite$^+$
Injection pressure	30	MPa
Fuel temperature	306	K
Ambient gas pressure	3.6	MPa
Ambient gas temperature	306	K

The information obtained with these techniques, in addition to mass flow rate ($\dot{m}$) and momentum flux measurements ($\dot{M}$), is used to validate the model presented in Sect. 2.2 for nozzle flow simulations (Desantes et al. 2014; Salvador et al. 2014). Validation also includes the proper selection of turbulence model and boundary conditions.

Results of the single-hole nozzle described in Table 2.1 simulations are presented next[5] as an example of validation procedure.

As comparison metrics, steady-state discharge coefficient (C_d), velocity coefficient (C_v) and area coefficient (C_a) at the orifice outlet are used, as defined in Eqs. 2.37–2.39, respectively.

$$C_d = \frac{\dot{m}}{\rho_l U_{theory} A_o} = \frac{\dot{m}}{A_o \sqrt{2\rho_l \Delta p}} \tag{2.37}$$

$$C_v = \frac{U_{eff}}{U_{theory}} = \frac{\dot{M}}{\dot{m}\sqrt{\frac{2\Delta p}{\rho_l}}} \tag{2.38}$$

$$C_a = \frac{A_{eff}}{A_o} = \frac{\dot{m}^2}{A_o \rho_l \dot{M}} \tag{2.39}$$

The parametric study of boundary conditions is presented in first place. Inlet boundary condition, place further upstream the needle position at its maximum lift, is changed from constant pressure (nominal value) to time-varying velocity (obtained from experiments). Outlet boundary condition, which is non-reflective pressure, is located 80 mm far downstream the orifice outlet. This type of boundary condition can also be used at inlet. Although this boundary condition is programmed to avoid pressure wave reflections, there is always certain grade of bouncing which can be increased (better stability) or decreased (higher accuracy). The grade of reflection is

[5]Details of the meshes and numerical setup of all cases are omitted in this chapter for the sake of brevity. Nevertheless, that information is well described in the referenced documentation.

Table 2.2 Steady-state and non-dimensional comparison metrics of the inlet boundary conditions study in the validation case

Boundary	$\dot{m}$ [kg/s]	$\dot{M}$ [N]	C_d [−]	C_v [−]	C_a [−]
Experimental	1.654	0.372	0.888	0.953	0.846
Constant pressure	1.638	0.347	0.843	0.941	0.793
Time-varying velocity	1.687	0.370	0.872	0.938	0.817
Low reflection pressure	1.687	0.370	0.872	0.941	0.799
Medium reflection pressure	1.712	0.370	0.869	0.941	0.819
High reflection pressure	2.098	0.567	1.074	0.946	1.016

changed from low (standard value) to medium and high (inlet boundary condition is constant pressure). Table 2.2 shows the results with the standard compressible k-ε turbulence model. All conditions predict with high accuracy (error lower than 10%) the corresponding experimental value except for the non-reflective boundary condition with high grade of reflection. This conditions gives a continuous growing mass flow rate, coming from a continuous growing of the inlet pressure value. Large values of grade of reflection are not correct and must be dismissed. Since there are not significant differences among the rest of options, the time-varying velocity conditions is then suggested simply because allows more accurate representation of the transient behavior of the nozzle.

When running RANS, a turbulence model needs to be selected. Five of them[6] are tested with non-reflective inlet boundary condition (with low grade of reflection) and not with the time-varying velocity as suggested above because simulations were performed at the same time than previous study. Results are summarized in Table 2.3. SST k-ω and realizable k-ε models clearly overestimate all parameters. The other two k-ε models predict will with a maximum error of 3% all parameters.

Figure 2.1 shows the velocity profile at the exit section of the orifice. Standard and high density ratio k-ε models have the same profile: small area at the center of the orifice with constant velocity and large area with a parabolic velocity profile, what is characteristic of laminar/low Reynolds number flows. With the RNG k-ε the parabolic region is reduced, and even further reduced with the SST k-ω or realizable k-ε models. These last two show a typical profile of turbulent/high Reynolds number flow, and that is why they should be used although they clearly overestimate the mass

[6]Once again, these models are not described in this chapter for the sake of brevity. Reader can consult the specialized literature to gain insights into each model. All tested models are the standard ones of the OpenFOAM libraries except the high density ratio k-ε (Garcia-Oliver et al. 2013)

Table 2.3 Steady-state and non-dimensional comparison metrics of the RANS turbulence model study in the validation case

Boundary	$\dot{m}$ [kg/s]	$\dot{M}$ [N]	C_d [−]	C_v [−]	C_a [−]
Experimental	1.654	0.372	0.888	0.953	0.846
Standard k-ε	1.700	0.370	0.869	0.942	0.819
High density ratio k-ε	1.701	0.375	0.874	0.943	0.824
RNG k-ε	1.749	0.398	0.903	0.938	0.847
Realizable k-ε	1.992	0.500	0.998	0.967	0.965
SST k-ω	1.980	0.491	0.984	0.974	0.959

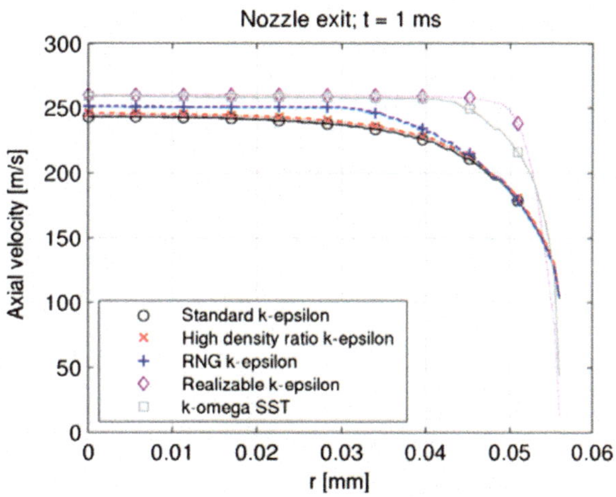

Fig. 2.1 Velocity profile at the orifice exit of the RANS turbulence model study in the validation case

flow rate (proper selection of inlet boundary condition and values can overcome this issue). Comparing both, the realizable k-ε predicts high turbulence levels at the core flow of the orifice, while the SST k-ω presents higher values of eddy viscosity next to walls. Because of those reasons, that one is the final recommendation.

With the goal of proving the validity of these decisions, the same nozzle is simulated with the SST k-ω turbulence model and a time-varying velocity at inlet boundary conditions. Spray results of the simulation are depicted in Fig. 2.2. For each time step, the upper part corresponds to the liquid mass fraction contour and the lower part to the experimental contour obtained via backlight illumination technique. Very good agreement is found, the only aspect of the spray that is not well captured is the angle, wider in the simulations, especially near the nozzle.

This validation is just an example to show the methodology. The model presented in Sect. 2.2 captures the trends when nozzle geometry and injection conditions are varied (Desantes et al. 2014). Once the model is validated, it can be applied to study different aspects of the nozzle flow.

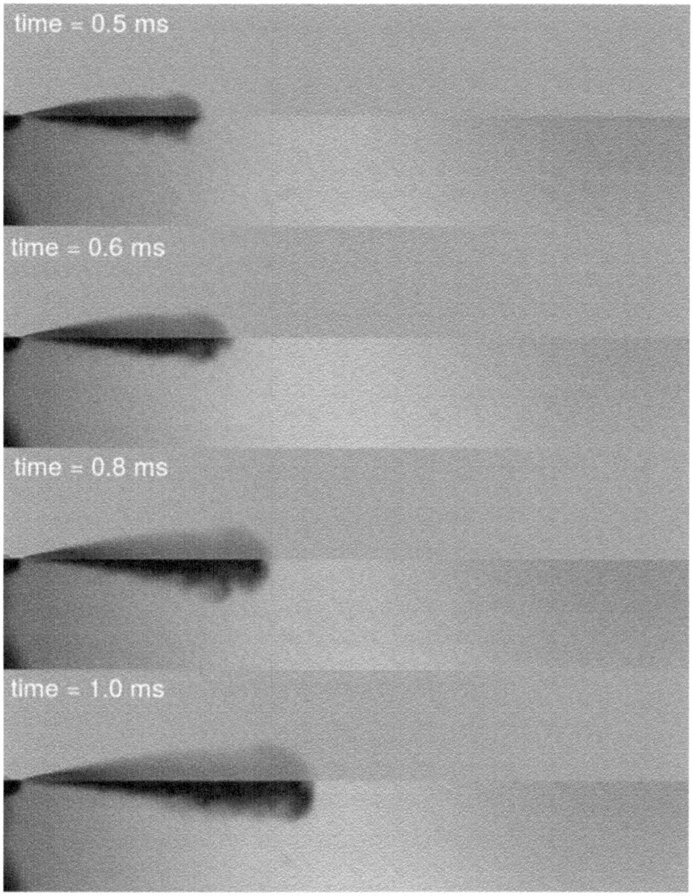

Fig. 2.2 Spray evolution in the validation case. Comparison with experimental results

2.3.2 Effects of Orifice Eccentricity

For this analysis, the ECN Spray A conditions are simulated and summarized in Table 2.4. However, these conditions lead to an evaporative spray and combustion, which is not the focus of the simulation at this time. Therefore, for the results shown in this section the ambient temperature is decreased to 500 K to avoid evaporation of droplets. Ambient pressure is also decreased to 3.4 MPa in order to keep the same ambient density value, and 0% of oxygen concentration is selected at the discharge chamber (N_2 as ambient gas).

The nozzle internal geometry was measured by means of X-ray tomography, as commented in Sect. 2.3.1. Two particularities were found: Orifice is not on the same axis than the needle, and in other words, there is certain eccentricity that ranges from 20 to 53 μm (in the case of study is 37 μm), and the orifice section is not exactly

Table 2.4 Specifications for the Spray A operating conditions of the Engine Combustion Network (ECN)

Ambient gas temperature	900 K
Ambient gas pressure	6.0 MPa
Ambient gas density	22.8 kg/m^3
Ambient gas oxygen (by volume)	15% O_2 (reacting); 0% O_2 (non-reacting)
Ambient gas velocity	Near-quiescent, less than 1 m/s
Common rail fuel injector	Bosch solenoid-activated, generation 2.4
Nominal orifice outlet diameter	90 μm
Nozzle K-factor	$K = 1.5$ μm
Nozzle shaping	Hydro-eroded
Mini-sac volume	0.2 mm^3
Discharge coefficient	0.86
Number of holes	1 (single-hole)
Orifice orientation	Axial (0° full included angle)
Fuel injection pressure	150 MPa, prior to start of injectio
Fuel	n-dodecane
Fuel temperature at nozzle	363 K
Injection duration	1.5 ms
Injection mass	3.5–3.7 mg

circular. Desantes et al. (2014) show how the dynamic viscosity distribution near the wall is slightly affected by the eccentricity. The boundary layer grows wider in the direction of the eccentricity due to slightly larger flow acceleration in that area. This leads to not completely axisymmetric mass distributions on the near-field spray region, as it has been experimentally observed. Nevertheless, these effects are small and only local.

2.3.3 *Effects of Fuel Temperature*

Pressure gradient inside injector nozzles accelerates the fuel converting its flow work into kinetic energy. This expansion comes with certain fluid cooling. In the process, there are losses. One can assume that those losses are converted into heat (turbulent kinetic energy is neglected) which increases fuel temperature. Therefore, it is possible to estimate the temperature change as function of the losses, measured for example with the velocity coefficient C_v, as shown in Fig. 2.3.

Injection temperature may have an impact on this process since its modifies the fuel thermodynamic properties (density, viscosity, and compressibility). To computationally study this effect, the fuel and nozzle wall temperatures are varied from 263 K (cold start conditions) to 363 K (standard engine operation conditions).

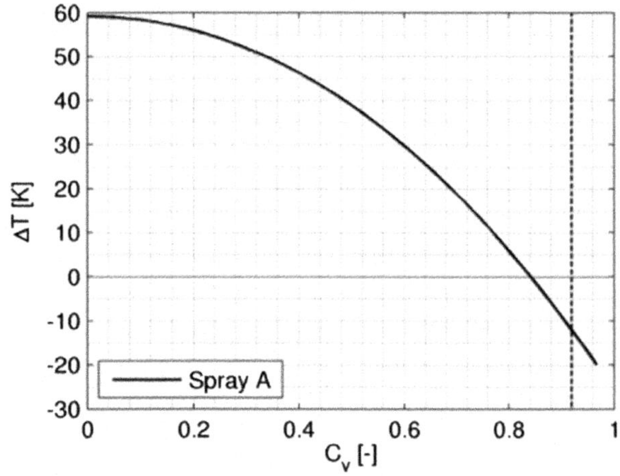

Fig. 2.3 n-dodecane temperature change as function of the velocity coefficient (losses) for ECN Spray A injection conditions. Results of a 0D model

Table 2.5 Steady-state and non-dimensional comparison metrics of ECN Spray A injection conditions for different wall and initial fuel temperatures. Experimental data and simulation with adiabatic walls are also included

	C_d [−]	C_v [−]	C_a [−]
Experimental	0.90	0.92	0.98
Adiabatic (T_f = 363 K)	0.90	0.92	0.98
T_f = 263 K	0.90	0.92	0.98
T_f = 273 K	0.91	0.92	0.98
T_f = 303 K	0.89	0.91	0.98
T_f = 323 K	0.89	0.91	0.98
T_f = 343 K	0.90	0.92	0.98
T_f = 363 K	0.90	0.91	0.98

Additionally, the effect of using adiabatic walls instead of constant temperature boundary conditions can also be analyzed (Payri et al. 2017a). Table 2.5 shows the discharge coefficient (C_d), the velocity coefficient (C_v) and the area coefficient (C_a) as defined in previous Sect. 2.3.1. When comparing simulations with adiabatic walls and constant temperature walls, no difference is found. A reduction in the discharge coefficient as the fuel temperature decreases is reported in the literature. This effect is associated with a wider boundary layer inside the orifice that reduces the effective area and velocity. However, simulations predict negligible variation in the nozzle flow coefficients. And this is because at high injection pressures (such as in Spray A conditions), the Reynolds number is higher than the critical value where the boundary layer transitions to turbulent and the discharge coefficient turns constant, regardless the fuel temperature.

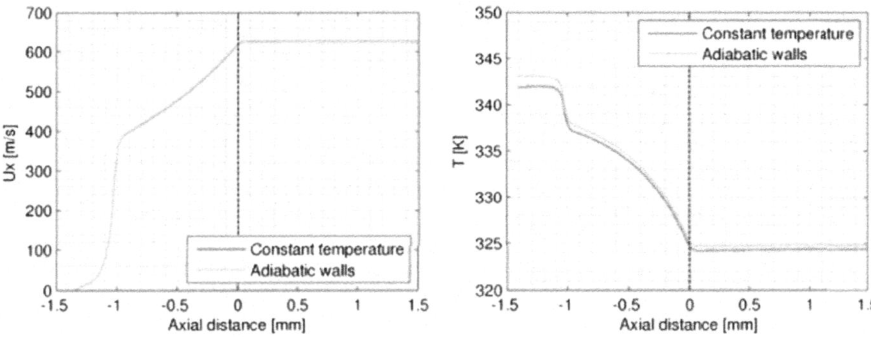

Fig. 2.4 Profiles of axial velocity and temperature along the nozzle axis at time 0.5 ms after start of injection for two different wall boundary conditions at ECN Spray A conditions. The dashed vertical line represents the orifice outlet position

In order to further investigate nozzle flow differences between the adiabatic and constant temperature walls, Fig. 2.4 shows the core flow properties inside the nozzle. As commented before, differences are small. Inside the sac, there is almost stagnated flow (low velocity) at constant temperature established by wall and initial temperature value. Large and fast acceleration occurs at the orifice inlet section, which is coupled to a sudden decrease in temperature due to expansion. Inside the orifice, the core flow accelerates and cools down as the section decreases down to the outlet diameter. After exiting the nozzle, velocity and temperature are kept constant a few millimeters inside the intact core length.

Fluid properties, such as density, do change in a significant amount when the temperature is varied (following the equation of state presented in Desantes et al. 2016b and Payri et al. 2017a), but this has no consequences in spray momentum as shown in Table 2.5, therefore, neither in the spray development. Furthermore, all the tested conditions fall in the complete atomization regime, supporting the result of negligible impact on spray characteristics. There is still open questions, whether or not this affects the evaporation and combustion processes and how much, because different fuel temperatures at the orifice exit would lead to different droplet size and evaporation rates during the injection event.

2.3.3.1 Effects of Needle Lift

It is known that needle movement affects spray behavior. Needle lift strategies can influence on air utilization and mixing process and, thus, in combustion efficiency. Needle position variation in time can be measured by means of X-ray tomography. Nonetheless, some of the frequencies of the signal may be lost during the acquisition, resulting in a peaky shape needle velocity time evolution. This may lead to unstable and unrealistic internal nozzle flow pattern, so the signal could be filtered using a moving average filter. Simulations of the original and the filtered signals of needle velocity were carried out with a morphing mesh strategy for the ECN Spray A injection conditions (Payri et al. 2017b). The filtering process seems to have negligible effects, so it is not recommended.

Differences between simulations and experiments are specially large at start of injection. This is related to the initial conditions. In this case (Payri et al. 2017b), the nozzle is initially filled with stagnated liquid at injection pressure. That initialization leads to higher mass flow rate values in the early stages of the injection. This issue can be solved by initializing the orifice filled with gas at ambient conditions and the rest of the nozzle filled with liquid at injection conditions.

A significant effect of the needle transient is found at the temperature field inside the nozzle. In the case of fixed needle performed by Payri et al. (2017b), there is a temperature drop on the core of the flow of about 20 from 343 K, which is the fuel temperature set at boundaries. A small increase of temperature is observed next to walls. However, in simulations with needle movement, the temperature inside the sac and orifice goes up to almost 400 K at least until 0.5 ms after start of injection. That increase in temperature is due to the viscous heating of the flow trespassing the needle seat when the lift is small, and for longer times after start of injection diminishes, reaching temperature levels similar to the ones observed in the simulation with fixed needle position.

That increase in temperature modifies the fuel properties (reduces density) at the orifice outlet, reducing the spray momentum an therefore its penetration. Although the effect is small, it improves the computational prediction in comparison with experiments (Payri et al. 2017b).

2.4 Near-Field Spray Results

As commented in Sect. 2.1, one of the main advantages of the Eulerian models is that they allow simulating the nozzle flow and the spray seamlessly. This type of simulation (from now on, coupled simulation) is compared to two different spray simulations (decoupled, because only the spray is considered) (Desantes et al. 2016b). The first simulation was conducted using as an inlet boundary condition, and the fields obtained at the nozzle exit in the coupled simulation. The second was made using as an inlet boundary condition a top-hat (TH) radial profile of axial velocity obtained from mass flow rate and momentum flux measurements. The projected mass density along the transverse direction comparing the simulations and X-ray radiography data at 0.1, 2.0, and 6.0 mm downstream of the nozzle exit is shown in Fig. 2.5. The projected mass density is interesting because it is a time-resolved measurement of liquid concentration. Being a line-of-sight parameter allows easy comparison between experiments and simulations, and also observe some of the ligaments and flow structures near the orifice outlet. Radial dispersion is quite similar regardless the simulation type, and only at the first axial location, the coupled simulation is slightly narrower. More differences appear when comparing peak projected mass density. At the three locations, the profiles predicted with the flat inlet time-varying velocity

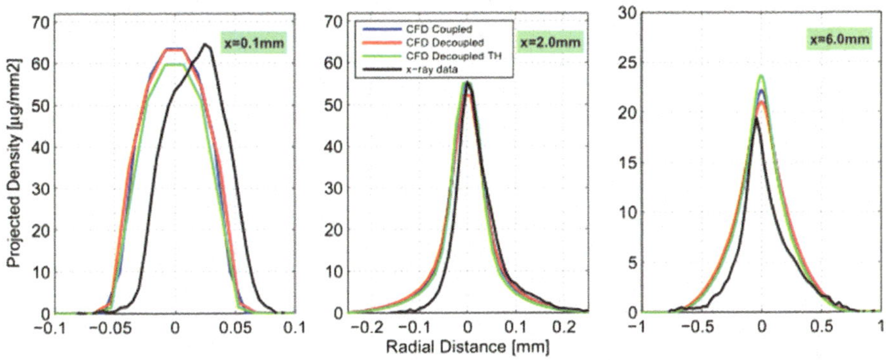

Fig. 2.5 Computed and experimentally measured profiles of projected mass density $\mu g/mm^2$ at 0.5 ms after start of injection at different axial locations downstream of the nozzle orifice exit for ECN Spray A conditions

Fig. 2.6 Computed and experimentally measured mass-averaged spray velocity along the axis at 0.5 ms after start of injection for different types of coupled/decoupled simulations for ECN Spray A injection conditions

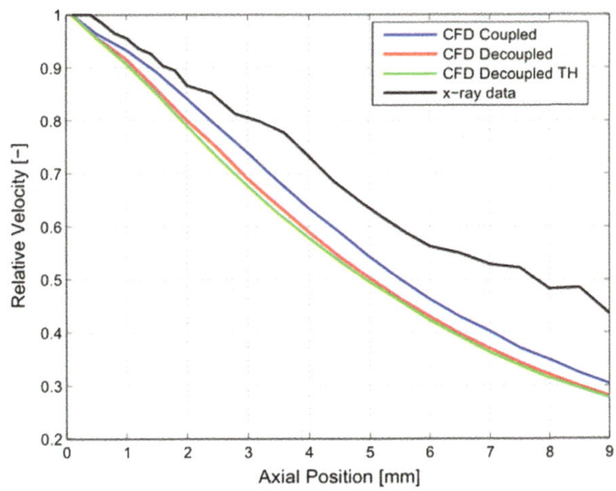

achieve a slightly worse value, being at 6 mm downstream the one which diverges more from the experimental measurements.

The relative velocity axial profile as derived from the transverse integrated mass is shown in Fig. 2.6. Once again, the three profiles are quite similar and the trends are well captured. The coupled simulation achieved the best match with the experimental measurements and predicted a less diffusive spray as indicated by slower relative velocity decay.

Thus, a lower accuracy in the near region (within 10 mm) is achieved by the simulations without the nozzle geometry, although agreement is still quite remarkable with the experimental measurements in the case of the projected mass density. Nevertheless, the effects of the internal nozzle flow characteristics are shown in the profiles of axial velocity in this near region of the spray. The different velocity profile and the subsequent induced turbulent viscosity modify spray dispersion and then liquid volume fraction profiles in the near-nozzle region. Further downstream, effects vanish probably due to the fact that mass and momentum flux of both simulations are similar. The main drawback of the coupled simulation is its 10 times higher computational cost mainly due to the reduction of the simulation time step (for the same CFL number) caused by the high-speed nozzle flow (where small cells are needed).

Direct comparison of the Σ-Y with classical DDM is explained in Sect. 2.4.1. Additionally, following the spirit of the Engine Combustion Network (ECN), the performance of this model is evaluated against other possibilities which are currently being used for research (Battistoni et al. 2018):

- Lagrangian–Eulerian formulation in CONVERGE (by Georgia Institute of Technology).
- LES Volume-of-fluid (VOF) methodology in CONVERGE (by Università degli Studi di Perugia).
- High-fidelity simulations with a CLS-VOF framework (by Sandia National Laboratories).

For this study, non-reacting cold ECN Spray D injection conditions are employed. They differ from Spray A (Table 2.4) only in the nozzle geometry: The nominal orifice outlet diameter is 180 μm. Battistoni et al. (2018) show how all models predict similar axial velocities of the liquid phase, particularly within the first 4 mm of the spray.

Additionally, all models predict less surface area than suggested by the experimental data at the spray centerline. Along the spray periphery, the CLS-VOF approach captures the experimentally observed projected surface area profile. The Eulerian Σ-Y model seems to predict a much more aggressive atomization process (because of a faster reduction in SMD) than is predicted by the other models. Careful calibration of the Eulerian Σ-Y model would allow for improved prediction of droplets formed from the atomization process (see Sect. 2.4.2).

Nonetheless, the higher-fidelity modeling approaches (Σ-Y, VOF-LES, and CLS-VOF) are in general able to capture both the peak projected density at the spray centerline, as well as width of the spray within the near-nozzle region.

These results indicate that although different degrees of atomization are predicted, similar spray structures are suggested by the models.

LES has also been carried out with the Σ-Y model (Desantes et al. 2017). A 1/7th power law mean velocity profile (top-hat) with a turbulent intensity between 3 and 5% is used as boundary condition at orifice outlet. Simulations results have been averaged from 0.5 ms after start of injection, during the quasi-steady period, as in X-ray experiments. The projected mass density from time-averaged LES data is presented in Fig. 2.7. This figure also includes results using the same mean velocity profile and turbulent intensity (5%) for RANS. It is shown that LES captures the projected mass density at the nozzle exit, but under-predicts it at 2 mm downstream. Better results are obtained at 6 mm, contrary to RANS. This indicates that spray dispersion is over-predicted in the near nozzle (to about 2 mm) in LES (which, as shown before, can be improved if the nozzle is included in the computational domain-coupled simulation). Notice that the same mass and momentum flux are injected for

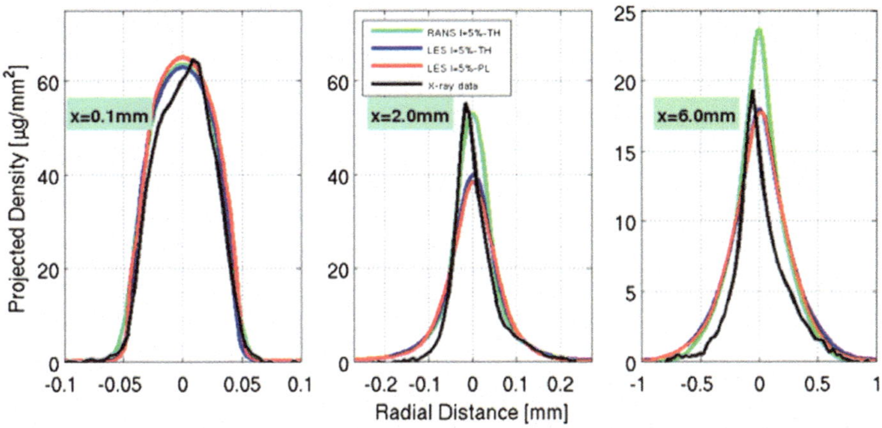

Fig. 2.7 Computed and experimentally measured projected mass density at different positions downstream the orifice exit for RANS and LES simulations based on ECN Spray A injection conditions

all cases. According to this result, the impact of fluctuations intensity is assessed in LES. Nevertheless, noticeably sensitivity to inflow turbulence intensity levels is observed, which requires further insight.

Another of the parameters used by researches to study the dense region of the spray is the intact core length (IL). For the simulations, it is defined as the maximum axial distance where the liquid mass fraction is greater than 99.9%. Its value for ECN Spray A is about 1.30 mm (Desantes et al. 2014). When this value is compared to 0D models and correlations, the difference is lower than 5%. Nevertheless, when experimental mass distributions are observed, as it is done in this section, the length with high density or fuel concentration values seems larger than the IL predicted in the simulations (although this is just a qualitative result which still requires real quantification).

2.4.1 DDM Versus Σ-Y Eulerian Atomization Model

Desantes et al. (2016a) carried out a comparison between DDM and Σ-Y Eulerian atomization model. At 0.1 mm from the orifice outlet, both models predict almost the same peak projected mass density, but noticeable differences in width could be detected with a narrower DDM profile. At 0.6 mm downstream of the nozzle exit, although DDM profile width is quite similar to experiments, the peak value is under-predicted. However, predictions by the Σ-Y model are really close to the experimental values, showing a better spatial behavior reproducing the near-nozzle flow structure. The largest differences can be observed at 2.0 mm. Fuel mass projected mass density predictions by DDM model are largely under-predicted in terms of peak value, while Σ-Y model ones reproduce very well the experimental trend both in peak value and radial dispersion. Finally, at 4.0 mm downstream location, both models predict reasonable values of peak projected mass density, but it has to be noted that Σ-Y model reproduces the decrease from 2.0 to 4.0 mm, while the DDM model keeps the value almost constant and predicts more radial dispersion. Overall, the Σ-Y model provides the best match with the experimental data and can capture the trend of the internal structure of a diesel spray in the near field fairly well. Nevertheless, DDM approach is generally an order of magnitude faster due to lower resolution grids (although the computational cost of the DDM greatly depends on the number of injected parcels).

2.4.2 Optimization of the Interfacial Surface Density Sub-Model

One of the most challenging topics of these Eulerian models of sprays is precisely the calibration of the surface density equation. As commented before, this metric, with dimensions of reciprocal length, represents how much interfacial area is present per unit volume. This calibration procedure can be made in terms of numerical comparison with DNS results or from a more practical point of view, by comparison with experimental measurement of SMD, projected mass density or projected surface area.

Statistical design of experiments (DoE) is applied because of its high reliability and accuracy in the results. The other great advantage of this statistical analysis is the huge reduction in the number of simulations needed to achieve the optimum set of values for the input parameters, which exactly define the number of iterations. Finally, an important outcome from these studies, apart from the optimum configuration, is the gained knowledge about mutual effects between the variables and their individual effect on the final results.

Firstly, the Σ-Y model results are compared to DNS (Martí-Aldaraví 2014). Four factors are selected for the statistical analysis (modeling constants α_0, α_1, and α_2 of Sect. 2.2.5 and the Schmidt number Sc_Σ). Each of them with two levels, high and low, but Sc_Σ, which has three levels. By doing so, linear dependency is assumed between the variable of interest and each factor. Nonetheless, a parabolic dependency between the error and the diffusive coefficient Sc_Σ is considered. This type of design leads to 24 simulations. The mathematical model is established with 95% confidence. Axial profiles and radial profiles of air–fuel surface density $\tilde{\Sigma}$ at three different axial positions are used as comparison metrics. The final optimum values are shown in Table 2.6. Contrary to the optimization explained later in this section, only one constant, Sc_Σ, is statistically significant. However, tunning the present model does not lead to small errors when comparing with DNS results as shown in the axial $\tilde{\Sigma}$ profiles in Fig. 2.8, so maybe it is recommended to change the model (as done in

Table 2.6 Optimum values of the modeling constants obtained with the multilevel factorial statistical analysis for the comparison with DNS

Constant	α_0 [−]	α_1 [−]	α_2 [−]	Sc_Σ [−]
Original	1.00	1.00	1.00	1.00
Optimization	1.25	0.75	3.125	0.75

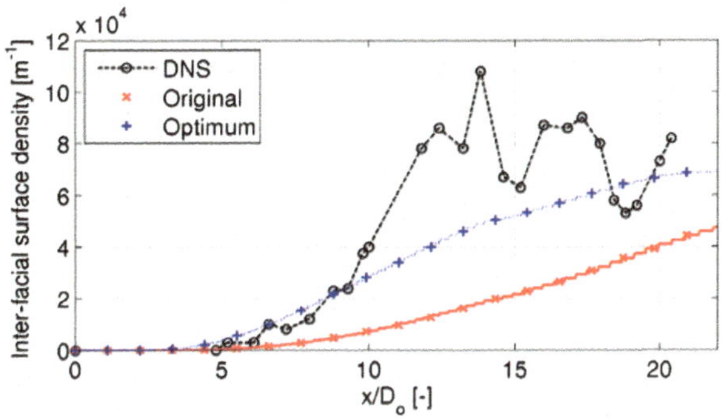

Fig. 2.8 Axial profiles of the interfacial surface density distribution on the axis of the spray for the statistical study comparing the original case with the optimum one and DNS results

Sect. 2.2.5.1, model used for the second statistical study). Still, an over-prediction in spray width is expected due to the nature of the model.

Secondly, the technique known as response surface method (RSM) is used in order to obtain an optimum set of modeling constants. The optimization process uses three inputs (modeling constants α_1, α_2 of Sect. 2.2.5.1 and the Schmidt number Sc_Σ) and a central composite design (CCD), which results in a test plan of 15 simulations. The output parameter of RSM is the mean error between the measurements along the axis and the calculated SMD, which is computed at a time late enough to ensure quasi-steady-state predictions. Because slightly different trends in SMD evolution where experimentally found depending on injection pressure, the optimization is carried out for two different injection conditions: the standard non-evaporative reference ECN Spray A (see Table 2.4), called here HP; and for lower injection pressure of 100 MPa, called MP (Pandal et al. 2017c). The mathematical model is again established with 95% confidence, keeping only the significant terms in both cases. At the end, the minimum error output is found with the constant values shown in Table 2.7. The same optimum value for the Sc_Σ parameter. Additionally, this parameter has negligible effect on the error as long as its value is below 1, slightly increasing the error on the contrary. Also interesting is the result of α_2 for high injection pressure optimization. Increasing its value above 2, the error is hugely increased as a consequence of too much coalescence. This is not the case for medium injection pressure.

Third and finally, another methodology is applied to minimize the overestimation of interfacial surface density obtained with the default values of the model constants. A parametric study is carried out in order to investigate the effects on the results of the constant values individually (Pandal et al. 2017b). The atomization sub-model is again the one described in Sect. 2.2.5.1, and following previously presented results,

Table 2.7 Optimum values of the modeling constants obtained with the RSM for ECN Spray A conditions

Constant	α_1 [−]	α_2 [−]	Sc_Σ [−]
HP optimization	0.96	1.459	0.9468
MP optimization	0.77	2.482	0.9468

only two of its constants are analyzed (concretely α_1 and α_2). The results indicate that a higher α_1 value makes the transition toward the equilibrium faster and vice versa. Regarding α_2, it seems that a value lower than 3.5 overestimates the interphase density, in comparison with the experiments, while a higher value (e.g., $\alpha_2 = 4.0$) under-predicts the surface formation between the liquid and the gas. After analyzing the trends of the parametric study, up to four different combinations of values are proposed. In light of the predictions, the simulation with $\alpha_1 = 0.8$ and $\alpha_2 = 3.5$ is able to match the measurements with great deal of accuracy, and as a result, it is chosen as the optimum for the following calculations. Using this configuration, experimental trends varying injection conditions (injection and ambient pressures) are well reproduced. In fact, predictions of this set of optimum parameters are remarkably close to the experimental data, providing a clear improvement with respect to the default values.

2.5 Evaporative Sprays and Combustion Results

Experimental measurements and simple 1D models show that characteristic vaporization lengths in direct injection diesel sprays can be predicted by the means of mixing-controlled assumptions (Desantes et al. 2016a). An implication of these finding is that under current diesel injection conditions, turbulent mixing and gas entrainment may be the dominant phenomena with respect to fuel vaporization.

Once again, ECN Spray A specifications (see Table 2.4) were selected as the base case to evaluate the model predictions by García-Oliver et al. (2013). Their results show good agreement for both liquid and vapor penetration. In both cases, predictions are within the experimental error interval of measured values. The accuracy in maximum liquid length predictions confirms that the evaporation process under Spray A conditions is mainly mixing controlled. In order to obtain those results, the round-jet correction of the RANS k-ε model had to be employed, which also improved the accuracy of non-evaporating sprays.

Regarding radial distribution of mixture fraction, the shape of the normalized profiles is also adequately predicted by the model. There is a slight bias toward narrower radial profiles in the calculations, which should indicate less radial dispersion and hence slightly lower entrainment. This is also coherent with the fact that the on-axis mixture fraction is always in the upper part of the experimental error interval, and the same applies to spray tip penetration trends. In any case, the overall agreement is remarkable.

Parametric studies with injection conditions variations, as well as with different ambient density and temperature, were also performed (Garcia-Oliver et al. 2013). Trends of decreasing vapor penetration with decreasing injection pressure and increasing penetration when decreasing ambient density are captured by the model. In general, good agreement between calculations and experiments is obtained, with most of the predicted results within experimental uncertainties. Effects of ambient density and temperature on quasi-steady values of liquid length are also well predicted. Departures tend to be noticeable at the lowest ambient density conditions. These results also confirms that evaporation model hypotheses presented in Sect. 2.2 are valid over a wide range of operating conditions of current compression ignition engines.

As done in Sect. 2.4.1 for the near field, the performance of this homogeneous single-fluid Eulerian model can be compared to the traditional DDM approach for evaporative sprays (Desantes et al. 2016a). After a proper validation of both methodologies, there is a good agreement for both models in terms of liquid and vapor penetration. In both cases, predictions seem to fall within the experimental uncertainty of measured values. Nevertheless, only the Σ-Y model is capable of predicting the initial evolution of the penetration. This can be justified because of the fact that the Eulerian model uses a much finer grid to improve spatial resolution, which is not possible with the DDM model.

Although in terms of liquid length both models achieve reasonable predictions, the DDM solver has a different trend with respect to the mixture fraction. While up to 22 mm the predicted values on the axis are higher than both measured as well as Σ-Y ones, then DDM results become equal to both experimental and Σ-Y ones to finally end up with lower values, almost out of the confidence interval. Regarding its radial dispersion, the shape of the profiles is adequately predicted by both models. At 25 mm, both models essentially produce the same results, in agreement with the centerline evolution, but it is not the same at 45 mm. Although simulation profiles are wider compared to the experimental ones at 45 mm, the Σ-Y predictions are slightly narrower, that is, less radial dispersion, which may be the explanation of the higher values shown on the centerline profile. Considering all the results, the overall agreement of both models is quite remarkable.

Apart from the nominal ECN Spray A condition, additional operating conditions were selected to evaluate the performance of both models (Desantes et al. 2016a). For higher ambient temperature ($T_a = 1100\,\mathrm{K}$), vapor penetration predictions are within the experimental uncertainty of measured values for both models. However, the predicted liquid penetration in the case of the Σ-Y model is slightly underestimated, while with the DDM model it is over-predicted. Once again, in the early evolution of the vapor penetration, the Σ-Y model achieves the best agreement with the experimental measurements. When the injection pressure is changed (to values of 50 MPa and 100 MPa), both codes are also capable of making an accurate prediction of vapor penetration. Nonetheless, predictions of liquid length of the DDM solver are sensitive to injection pressure variations, instead of keeping it almost invariable as the Σ-Y model, which depicts a good agreement with experimental measurements.

The global combustion indicators that usually characterize transient reacting diesel sprays are ignition delay (ID) and lift-off length (LoL). From the computational point of view, ID is defined as the time spent from start of injection until the maximum gradient in temperature takes place. On the other hand, LoL is defined as the minimum axial distance to the nozzle where 14% of the maximum value of Favre-average OH mass fraction in the domain is reached. Figure 2.9 shows both computational predictions and experimental measurements of these metrics. Experimental trends followed by both parameters are well captured by the model. LoL values' deviations from experiments are relatively small for all three conditions, with a maximum difference of around 2 mm for standard Spray A conditions. On the other hand, ID is clearly over-predicted, with deviations being very large for both low temperature conditions, similarly to the literature. This disagreement is mainly due to the strong role of chemical mechanism on the exact ignition timing, and means that for the present chemical mechanism, the transition from the low to the high temperature stages is slower than in experiments.

Finally, the flame structure provided by the computational model is evaluated by comparison with PLIF measurements at quasi-steady state. It is depicted in Fig. 2.10 for standard ECN Spray Conditions. In the figure, experimental measurements are shown at the top and computational results at the bottom. Additionally, red indicates zones where OH is detected by the PLIF technique, while green corresponds to regions where PLIF provides signal due either to formaldehyde (CH_2O) or to

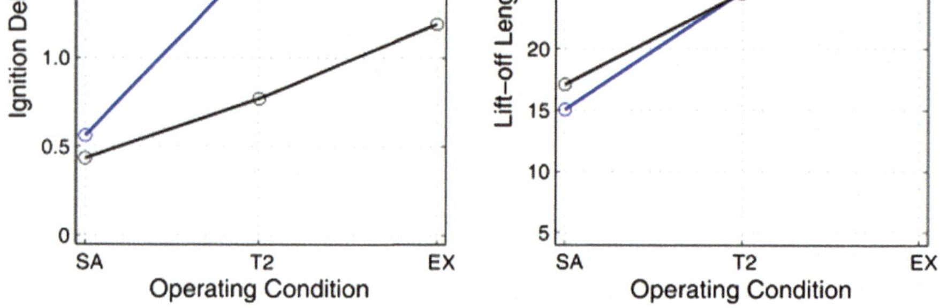

Fig. 2.9 Computed and experimentally measured ignition delay (on the left) and lift-off length position (on the right) for different operating conditions based on ECN Spray A. SA refers to standard conditions; T2 means a decrease in ambient temperature to $T_b = 780$ K; and EX adds a reduction of oxygen concentration to 14%

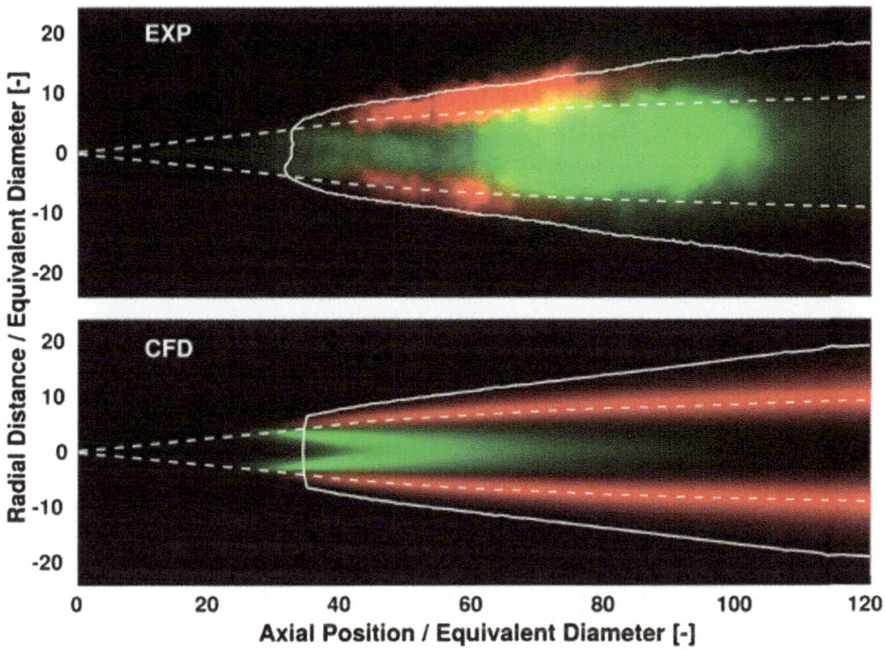

Fig. 2.10 Comparison of predicted CH_2O (green) and OH (red) with PLIF imaging at a quasi-steady state for ECN standard Spray A condition. Color areas normalized by the maximum of each species, respectively

polycyclic aromatic hydrocarbons (PAHs), and the white solid line is the contour of the OH* image. The white dashed line is shown both on experimental and modeling results corresponding to the stoichiometric isolines from calculations in order to have a spatial reference enabling easy comparison between both maps. Formaldehyde location is reasonably captured by the model around 25 equivalent diameters downstream the nozzle exit. This specie is considered as an indicator of low-intermediate temperature chemical reactions (cool-flame), and because of that, it appears slightly upstream of the first OH location. In the case of experiments, some signal can be observed upstream simulations, which is mainly due to light reflections on the liquid length. Modeled formaldehyde disappears at 60 equivalent diameters downstream due to the transition to the high-temperature chemistry within the flame. However, experiments show a strong measured signal, which is most likely due to the presence of PAHs. Regarding modeled OH distribution, location is consistently predicted close to the stoichiometric location, but radial spreading is narrower in comparison with the experiment. Discrepancies in the axial extent downstream of 100 equivalent diameters are probably due to the laser sheet dimensions limit in the measurement.

2.6 Conclusions

This chapter describes a single-fluid diffuse-interface spray modeling approach, the so-called Σ-Y model. This approach assumes high Reynolds and Weber numbers, found at current engine fuel injection systems, and is formulated in an Eulerian framework, which is better suited for seamlessly simulating nozzle flow and the near-nozzle dense spray region. In fact, this new solver provides a quite fair performance, being able to predict and explain the main changes in the flow pattern experimented even under reacting conditions.

Model governing equations, both for RANS and LES turbulence modeling frameworks, have been described as well as some of the numerical aspects for its computational implementation. An overview of model assessment indicates that it is able to properly simulate nozzle and near-nozzle flow under diesel spray injection conditions. This is why commercial CFD codes such as StarCD (some years ago) and CONVERGE (this present year) incorporated it to their standard libraries.

The combination of proper selection of turbulence model and inlet boundary conditions leads to a solution with an error in simulations lower than 5% when compared to experimental values for all internal flow parameters and spray tip penetration. However, the obtained value of the spray angle is slightly wider than experimental values.

According to model hypothesis, spray dispersion depends on turbulence modeling accuracy and closure terms. RANS-based simulations provide good results when using turbulence model constants adapted to round-jet flow requirements, while recent LES simulations do not need any specific calibration. Simple closures have been used for liquid mass fraction transport equation with fair results, indicating low relative velocities for the conditions analyzed.

The validation using X-ray measurements was particularly valuable, because neither the experiments nor the models relied on the assumption of distinct droplets. Only further downstream, when primary atomization is complete, is such an assumption appropriate.

Concerning atomization modeling, it has been shown that after calibration, surface density equation is able to capture near-nozzle interfacial surface or downstream droplet diameter. Nevertheless, the observed deviations, mainly in SMD but also in projected mass density, with respect to the experimental measurements indicate that further efforts are required for the surface density area model in order to improve predictive capabilities for different injection systems and operating conditions.

In addition to a better interfacial modeling which would require less calibration (work currently being done by some researchers in the field), a complete simulation from nozzle flow with needle movement to combustion is still missing. Based on the information presented on this chapter, other suggestions for future work could be the inclusion of cavitation and flash-boiling sub-models, simulation of multi-hole nozzle and spray interaction and the analysis of pre- and post-injections inside and outside the nozzle.

This chapter has been written based on published papers in which authors have collaborated in one way or another, although some other significant works have also been mentioned. Therefore, the list of references presented here is not a complete collection of the work related with the topic of discussion. The reader should be aware that there are other interesting and significant works dealing with Eulerian-Lagrangian Spray Atomization models whose reading is recommended.

Acknowledgements Authors of this chapter would like to acknowledge the strong, continuous, and productive collaboration of the Engine Combustion Network (ECN) members. Without them, this work could not have been possible.

References

Battistoni M, Magnotti GM, Genzale CL, Arienti M, Matusik KE, Duke DJ, Giraldo J, Ilavsky J, Kastengren AL, Powell CF, Marti-Aldaravi P (2018) Experimental and computational investigation of subcritical near-nozzle spray structure and primary atomization in the engine combustion network spray D. SAE technical paper (2018-01-0277), pp 1–15. https://doi.org/10.4271/2018-01-0277

Desantes J, García-Oliver J, Pastor J, Pandal A, Naud B, Matusik K, Duke D, Kastengren A, Powell C, Schmidt D (2017) Modelling and validation of near-field Diesel spray CFD simulations based on the Σ -Y model. In: Proceedings ILASS–Europe 2017. 28th conference on liquid atomization and spray systems, 6–8 Sept 2017. https://doi.org/10.4995/ILASS2017.2017.4715

Desantes J, Payri R, Gimeno J, Marti-Aldaravi P (2014) Simulation of the first millimeters of the diesel spray by an Eulerian spray atomization model applied on ECN Spray A injector. SAE technical papers, vol 1. https://doi.org/10.4271/2014-01-1418

Desantes JM, Garcia-Oliver JM, Pastor JM, Pandal A (2016a) A comparison of diesel sprays CFD modeling approaches: DDM versus Σ-Y Eulerian atomization model. Atomization Sprays 26(7):713–737

Desantes JM, Garcia-Oliver JM, Pastor JM, Pandal A, Baldwin E, Schmidt DP (2016b) Coupled/decoupled spray simulation comparison of the ECN spray a condition with the Σ-Y Eulerian atomization model. Int J Multiph Flow 80:89–99. https://doi.org/10.1016/j.ijmultiphaseflow.2015.12.002

Garcia-Oliver JM, Pastor JM, Pandal A, Trask N, Baldwin E, Schmidt DP (2013) Diesel spray CFD simulations based on the Σ-Υ Eulerian atomization model. Atomization Sprays 23(1):71–95. https://doi.org/10.1615/AtomizSpr.2013007198

Martí-Aldaraví P (2014) Development of a computational model for a simultaneous simulation of internal flow and spray break-up of the Diesel injection process. Ph.D. thesis, Universtitat Politècnica de València, Valencia. https://doi.org/10.4995/Thesis/10251/43719

Pandal A, Pastor JM, Garcia-Oliver JM, Baldwin E, Schmidt DP (2016) A consistent, scalable model for Eulerian spray modeling. Int J Multiph Flow 83:162–171. https://doi.org/10.1016/j.ijmultiphaseflow.2016.04.003

Pandal A, García-Oliver JM, Novella R, Pastor JM (2017a) A computational analysis of local flow for reacting diesel sprays by means of an Eulerian CFD model. Int J Multiph Flow 99:257–272. https://doi.org/10.1016/j.ijmultiphaseflow.2017.10.010

Pandal A, Pastor JM, Payri R, Kastengren A, Duke D, Matusik K, Giraldo JS, Powell C, Schmidt D (2017b) Computational and experimental investigation of interfacial area in near-field diesel spray simulation. SAE Int J Fuels Lubricants **10**(2), 2017–01–0859. https://doi.org/10.4271/2017-01-0859

Pandal A, Payri R, García-Oliver JM, Pastor JM (2017c) Optimization of spray break-up CFD simulations by combining Σ-Y Eulerian atomization model with a response surface methodology under diesel engine-like conditions (ECN Spray A). Comput Fluids 156:9–20. https://doi.org/10.1016/j.compfluid.2017.06.022

Payri R, Gimeno J, Martí-Aldaraví P, Alarcón M (2017a) A new approach to compute temperature in a liquid-gas mixture. Application to study the effect of wall nozzle temperature on a diesel injector. Int J Heat Fluid Flow 68:79–86. https://doi.org/10.1016/j.ijheatfluidflow.2016.12.008

Payri R, Gimeno J, Martí-Aldaraví P, Alarcón M (2017b) Numerical simulation of needle movement nozzle flow coupled with spray for a diesel injector using an Eulerian spray atomization model. J Braz Soc Mech Sci Eng **39**(7). https://doi.org/10.1007/s40430-017-0801-1

Payri R, Ruiz S, Gimeno J, Martí-Aldaraví P (2015) Verification of a new CFD compressible segregated and multi-phase solver with different flux updates-equations sequences. Appl Math Model 39(2):851–861. https://doi.org/10.1016/j.apm.2014.07.011

Salvador FJ, Gimeno J, Pastor JM, Martí-Aldaraví P (2014) Effect of turbulence model and inlet boundary condition on the diesel spray behavior simulated by an Eulerian spray atomization (ESA) model. Int J Multiph Flow 65:108–116. https://doi.org/10.1016/j.ijmultiphaseflow.2014.06.003

Vallet A, Burluka AA, Borghi R (2001) Development of a Eulerian model for the "atomization" of a liquid jet. Atomization Sprays 11(6):619–642. https://doi.org/10.1002/fld.1650080906

Chapter 3
Spray Collapse in a Multi-hole GDI Injector and Its Effect on In-Cylinder Combustion

Rakesh Kale and R. Banerjee

Abstract Gasoline direct injection (GDI) system has been shown to have several advantages over the more conventional Port Fuel Injection system in a SI engine. However, it has also been reported that GDI engines have higher particulate emissions. One of the possible reasons for this higher particulate emission is collapse of the spray emanating from the injector and wetting on the piston head, particularly when the engine is operating under high-load conditions. In this study, a detailed analysis has been performed to understand spray collapse and its effect on in-cylinder combustion for three different fuels: isooctane, *n*-butanol, and isobutanol. Initially, spray studies in a constant volume chamber were performed. Parameters like liquid and vapor penetration lengths, droplet size, and velocity distribution were estimated from image analysis of high-speed videography and phase Doppler particle analyzer (PDPA). To mimic in-cylinder conditions, the injector body temperature was raised such that injected fuel was also at elevated temperature. Spray collapse was observed at higher fuel temperature conditions, and this resulted in higher axial liquid penetration and finer droplet size distribution. Similar experiments were then performed in an optically accessible engine, and it was observed that spray collapse at higher fuel temperature leads to wetting of the piston wall. It also leads to formation of pool fire over the piston head which may lead to particulate emission. Effect of engine operating conditions like start of injection on wall wetting and formation of pool fire has been quantified using image analysis of high-speed videography.

Keywords Flash-boiling · Spray collapse · Alcohols · Fuel impingement · Pool fire · GDI engine

R. Kale (✉) · R. Banerjee
Department of Mechanical & Aerospace Engineering, Indian Institute of Technology Hyderabad, Kandi, Sangareddy 502285, Telangana, India
e-mail: me13p1007@iith.ac.in

K. Saha et al. (eds.), *Two-Phase Flow for Automotive and Power Generation Sectors*, Energy, Environment, and Sustainability,
https://doi.org/10.1007/978-981-13-3256-2_3

3.1 Introduction

Fuel injection and air–fuel mixture preparation are of prime importance for the direct injection spark ignition (DISI) engines. Higher injection pressure results in finer liquid fuel atomization and therefore reduces the time of droplet evaporation. However, system component requirement for such a higher injection pressure system substantially increases the overall vehicle cost. Additionally, such fuel injection systems are mainly tuned to atomize gasoline-like fuels and selecting different fuels may not give desired effect. Flash boiling is an effective way to provide fine spray even at lower injection pressure (Xu et al. 2013; Kale and Banerjee 2018). Under realistic engine running conditions, heat transfer from the combustion process heats up the cylinder wall and fuel injector. This eventually increases the fuel temperature before it is injected inside the combustion chamber. These situations give rise to the flash boiling phenomenon. When hot fuel is injected into pressure conditions lower than the saturation pressure of the fuel, vapor bubbles are generated inside the liquid. Subsequent expansion of these vapor bubbles during injection process causes catastrophic disintegration of the liquid jet that leads to fuel atomization with finer droplets (Wu et al. 2017).

Fuel temperature and ambient pressure are the main defining parameters in case of flash boiling sprays. Degree of superheat and the non-dimensional parameter of ambient to fuel saturation pressure (P_a/P_s) ratio are generally used to determine the strength of flash boiling phenomenon. When P_a/P_s is greater than 1, the condition is considered as non-flash boiling condition. As the pressure ratio is reduced, vapor bubbles start appearing within the liquid phase and bubble growth subsequently increases with decreasing P_a/P_s value. The regime, when the ratio is $1 > P_a/P_s > 0.3$, is called 'transitional flash boiling' phase. In this regime, spray atomization results in smaller droplets compared to non-flashing sprays. When P_a/P_s is less than 0.3, the condition is known as 'flare flashing' stage (Zeng et al. 2012). Under such condition, highly atomized sprays are observed. At the flare flash boiling stage, the droplets in the whole spray are uniform and small. This indicates that flash boiling is an effective approach for producing fine droplets with uniform size distribution (Shen et al. 2016).

Flash boiling can significantly alter the spray plume structure due to spray collapse. For a multi-hole GDI injector, spray structure starts collapsing when P_a/P_s was less than 0.3 (Zhang et al. 2014). Researchers (Wu et al. 2016) have proposed that low pressure region may exist near the centerline of the injector, which directs the fuel vapor and small droplets toward the centerline. 2-D flow field experiments using PIV showed that, at sufficiently high degree of superheat, vortex structures are generated at the front of the injector. Inward velocities cause the spray structure to collapse toward the injector centerline. Small droplets and fuel vapor are observed to have accelerated toward the inner side (Zhang et al. 2013). When $P_a/P_s < 1$, individual plume thickness grows due to vapor bubble growth and their explosion. At flare flash boiling condition, plume thickness of individual spray is sufficiently high to create spray interaction with the neighboring plumes. Under such conditions, spray structure loses its original signature and form a close connected hollow cone structure. This affects air entrainment within the spray plumes and no ambient gas could enter the central region. This led to the formation of a low pressure zone near the injector centerline. Due to this, smaller droplets were driven inside this low pressure zone, creating a collapsed spray structure (Wu et al. 2016).

Spray collapse ultimately alters the spray morphology under flash boiling conditions. Overall spray cone angle reduces significantly with increasing fuel temperature. Additionally, spray penetration length has observed to be higher under flare flashing condition compared to transitional and non-flash boiling condition (Guo et al. 2017; Huang et al. 2014). Although flash boiling favors atomization process, however, spray collapse has its own disadvantages. Due to spray collapse, all the plumes accumulate into one single spray structure. Moreover, increase in penetration length increases the propensity of piston wall wetting, leading to the formation of a localized liquid film. As a result of this, fuel film will take longer time to evaporate and may end up with higher soot emissions (Schulz and Beyrau 2017). To tackle this issue, some researchers like Wang et al. (2017) have proposed to use split injection strategy to overcome fuel impingement. Considerable reduction in penetration length was reported when split injection strategy was implemented as compared to single injection strategy. However, the end effects are still unclear.

It is evident that understanding of flash boiling and its effect on engine combustion are very important for GDI engine. Therefore, this chapter deals with systematic investigation of flash boiling and its effect on spray characteristics and engine combustion. Discussion will also include the effect of different biofuels. Isooctane along with *n*-butanol and isobutanol is compared for different fuel temperature conditions. Detailed analysis of macroscopic as well as microscopic properties such as liquid and vapor penetration length; spray cone angle and droplet size distribution will be presented. Finally, link will be established between flash boiling sprays and their effect on piston wetting and pool fire combustion using optically accessed engine.

3.2 Equipment and Methods

Parameters like liquid penetration length, vapor penetration length, spray cone angle, and droplet size distribution are very important to characterize a spray. Vapor penetration indicates the amount of air entrained due to fuel injection process and the resultant formation of air and fuel vapor mixture. Since all the combustion reactions take place in the gas phase, visualization of the vapor phase in a spray is very important. Schlieren is commonly used technique to visualize the evolution of the vapor plume in spray and atomization process. Liquid penetration and spray cone angle give an estimation of the dispersion quality of a spray. Longer liquid length and wider dispersion angle are useful for higher air entrainment which ultimately helps in better air–fuel mixing. However, longer liquid penetration can also lead to in-cylinder wall wetting. Backlight shadowgraphy, Mie scattering, and Schlieren shadowgraphy are commonly used for liquid phase visualization. Figure 3.1 shows the schematic of

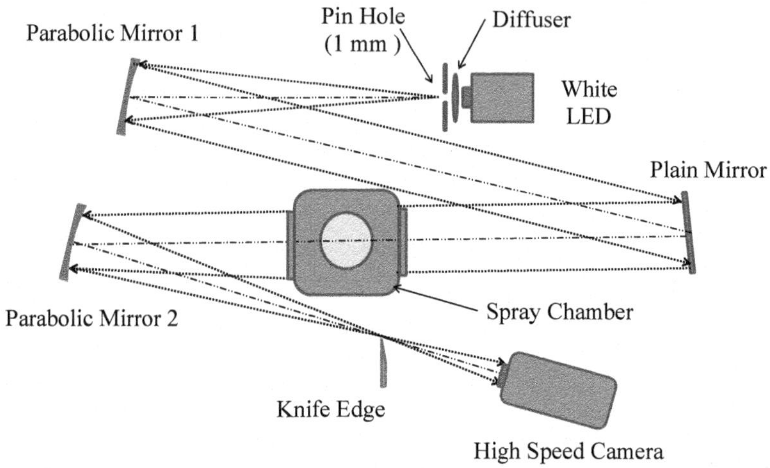

Fig. 3.1 Experimental setup for Schlieren and Shadowgraph technique

the Schlieren–Shadowgraph technique used for the spray visualization. Advantage of this technique is same experimental setup can be used for vapor as well as liquid phase detection. When employed with knife edge, the technique gives the vapor phase information. On the other hand, liquid phase of the spray is captured without the knife edge. Details of these techniques can be found in Settles (2001).

Phase Doppler particle analyzer (PDPA) is used for simultaneous measurement of droplet size and its velocity. This technique can typically measure droplet size ranging from 0.5 μm up to few hundred microns (~300). The measurement method relies on light scattering interferometry. It is a non-intrusive, single point measurement technique where probe volume is defined by the intersection of two laser beams. As a droplet passes through this probe volume, it scatters light from the laser beam. These scatter signals are then captured by the light-detecting receiver which then finds the phase shift between the Doppler burst signal. The phase shift of the signal is proportional to the size of the spherical droplet. On the other hand, frequency of the Doppler burst signal multiplied by the fringe spacing gives the droplet velocity. Schematic of the PDPA system is shown in Fig. 3.2.

Constant volume spray studies provide in-depth insights to spray development process without any disturbance due to external factors. Although optical access engine is an important tool for in-cylinder studies, however, due to its limited optical access, it is difficult to perform comprehensive experiments which are necessary for spray characterization. To perform series of experiments at reproducible boundary conditions, spray studies in a constant volume chambers have been extensively used by researchers. Depending upon the experimental conditions, various designs of the high pressure and temperature chambers have been used by different groups. Fuel conditioning is also very important, particularly when studying flash boiling sprays. In many studies, researches have used injector heating method to create elevated fuel

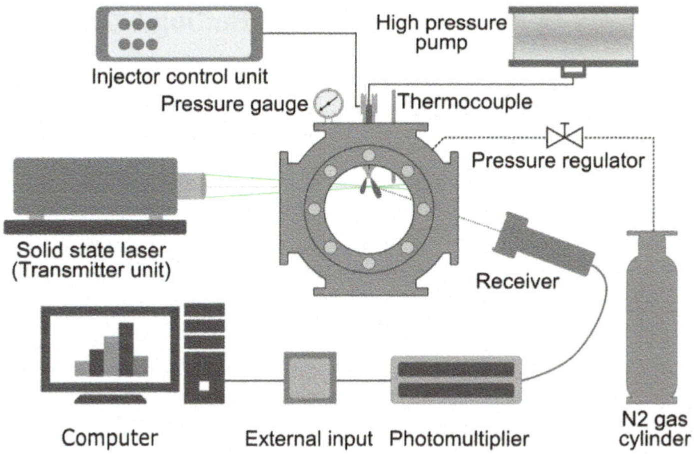

Fig. 3.2 Experimental setup for droplet size measurement using PDPA

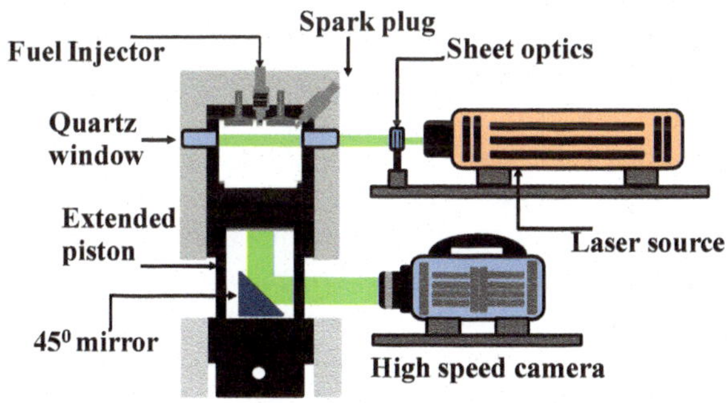

Fig. 3.3 Optical engine setup

temperature conditions. Since very small quantity of fuel is required per injection, generally it is assumed that fuel inside the injector is in thermal equilibrium with the injector body temperature.

Engine with optical access to the combustion chamber is a very sophisticated experimental setup to visualize and study different in-cylinder processes like internal air motion, fuel injection, and combustion, etc. Typically, access to the combustion chamber is provided using hollow extended piston arrangement or transparent cylinder liners. Schematic of the optical engine facility used in this study is shown in Fig. 3.3. Axial and orthogonal optical access was available through an extended hollow piston window and cylinder liner window, respectively. Hollow piston was provided with sapphire window of 40 mm diameter. Side windows were also made of sapphire. This arrangement is useful for side illumination using LED or laser-based system.

3.3 Fuel Spray Behavior Under Flash Boiling Conditions

3.3.1 Effect of Flash Boiling on Spray Morphology

Figure 3.4 shows changes in spray structure of isooctane, *n*-butanol, and isobutanol at different fuel temperature conditions. Schlieren and Shadowgraph images for vapor and liquid phase of fuel at $t = 1$ ms are shown here. It is evident that that spray morphology significantly changes with increase in fuel temperature. At 298 K fuel temperature, three distinct spray plumes can be observed for all the test fuels. As the temperature is raised, it is found that the plumes start coming closer to each other and finally collapses into a single jet. Slight variation in spray structure can be noticed among three fuels; however, it can be clearly seen that spray cone angle decreases with increase in fuel temperature. Additionally, vapor plume seems to be significantly longer at high fuel temperatures. From the liquid data, it can be observed that liquid core is gradually diminishing with increase in fuel temperature. This is because improved evaporation at higher fuel temperature decreases the droplet lifetime hence reducing the liquid core. Some evidence of liquid is seen for *n*-butanol and isobutanol at 523 K. However, no liquid is observed coming out of the injector in case of isooctane.

Form the spray cone angle results in Fig. 3.5, it can be seen that for all the three fuels, spray cone angle decreases with increase in fuel temperature. The reason for sprays cone angle reduction is most probably due to flash boiling. For the selected fuels, the temperature range is less than the critical temperature of the respective fuel (see Table 3.1). Therefore, it is reasonable to expect that the fuels are in their liquid phase inside the fuel injector. However, when the fuel is injected in a chamber that is at near ambient conditions, the liquid experiences a rapid decompression. This results in a rapid generation of vapor bubbles within the liquid volume which then expand due to further decompression. This expansion tends to give individual jets a radial velocity component which in turn leads to interaction with adjacent jets. As fuel temperature increases, the liquid phase thermodynamic conditions move closer to the critical point and the decompression process becomes more rapid. This leads to a catastrophic breakup of the liquid jet. Individual jets tend to lose their momentum leading to the formation of a single jet along the injector axial direction as observed in Figs. 3.4.

From the spray cone angle data, it can be observed that under all the conditions, isooctane gives the lowest spray cone angle compared to the butanol isomers. Reduced temperature ($T_r = T/T_c$) of a fuel is given by the fuel temperature (T) normalized by its critical temperature (T_c). At 523 K, reduced temperature for isooctane, *n*-butanol, and isobutanol is 0.96, 0.93, and 0.96, respectively. Though there is no significant difference in the reduced temperature values of the three fuels, in particular between isobutanol and isooctane, however, the two alcohols have higher latent heat of vaporization leading to marginal higher latent cooling during phase change. This maybe the reason why a small liquid core is seen for the two butanol isomers at 523 K and no liquid was observed for isooctane in Fig. 3.4. It can also be observed

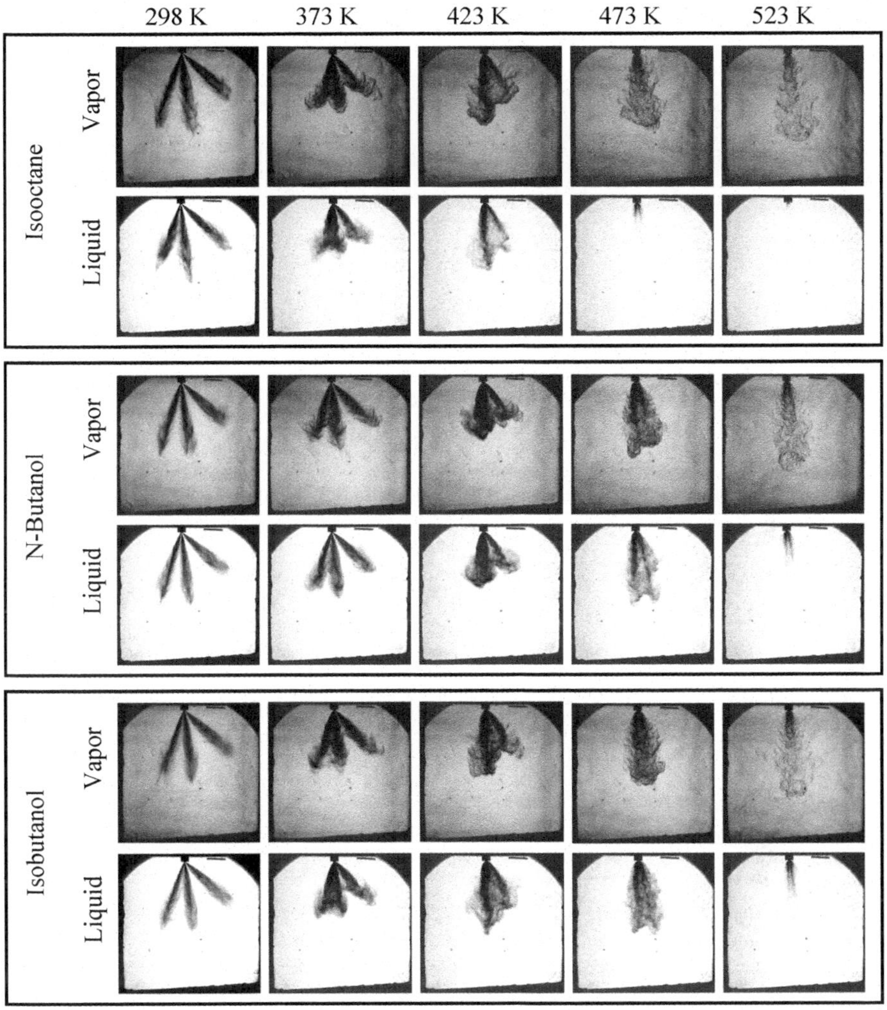

Fig. 3.4 Spray morphology of isooctane and butanol fuels for different fuel temperature at $t = 1$ ms after start of injection

that for isooctane and *n*-butanol, spray cone angle decreases monotonically. However, in case of isobutanol, the cone angle increases when the temperature is raised from 423 to 473 K. At 523 K, spray cone angle for isooctane could not be plotted, because no liquid was experienced at the nozzle exit using shadow imaging.

Spray morphology significantly changes with increasing fuel temperature, therefore affecting vapor and liquid penetration lengths. Figure 3.6 shows vapor penetration of isooctane and butanol isomers at different fuel temperature conditions. It is evident that vapor penetration increases with increase in fuel temperature. However, this increment is not a straight function of fuel temperature. Vapor penetration length

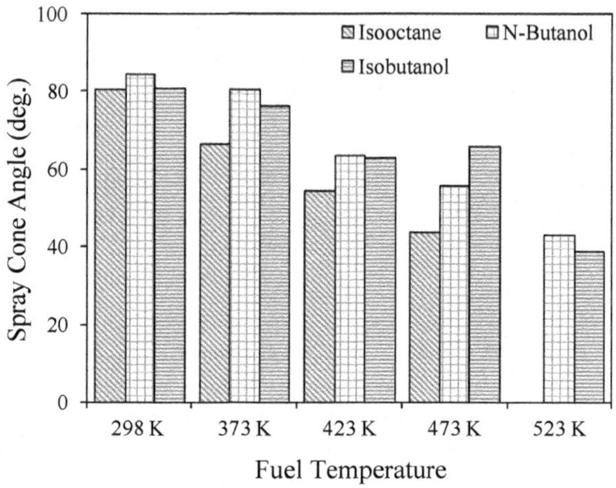

Fig. 3.5 Spray cone angle for different fuels at various fuel temperature conditions

Table 3.1 Fuel thermo-physical properties

Fuel	Isooctane	*n*-Butanol	Isobutanol
Molecular formula	C_8H_{18}	C_4H_9OH	C_4H_9OH
Molecular weight (g/mol)	114.22	74.12	74.12
Density (g/cm^3)	0.70	0.81	0.80
Refractive index	1.391	1.397	1.396
Boiling point (K)	372	391	381
Surface tension (mN/m @ 293 K)	14.7	25.4	23
Viscosity (mPa s @ 293 K)	1.04	3	4
Vapor pressure (kPa @ 293 K)	5.12	0.64	1.2
Research octane number	100	96	94
Critical temperature (K)	543.9	563	547
Critical pressure (bar)	25	44.23	43
Lower heating value (MJ/kg)	44.3	33.07	33
Heat of vaporization (kJ/kg @ 293 K)	305.4	637.7	578.4

is highest when the fuel temperature is 523 K. In addition, it can also be observed that 373 K fuel temperature case has the lowest vapor penetration for all the test fuels. This is because, at 298 and 373 K, spray maintains it course along the individual orifice axis. However, fuel evaporation rate is faster at 373 K. Due to this, penetration length first decreases at 373 K. As fuel temperature is further increased, complete collapse of the spray causes change in the directionality of the spray plumes. Individual spray plumes shift toward the injector axis. Direction vectors for the individual spray plume at non-collapsed condition contribute for both the axial and radial direction. However, with increase in flashing strength, they contribute more to the axial

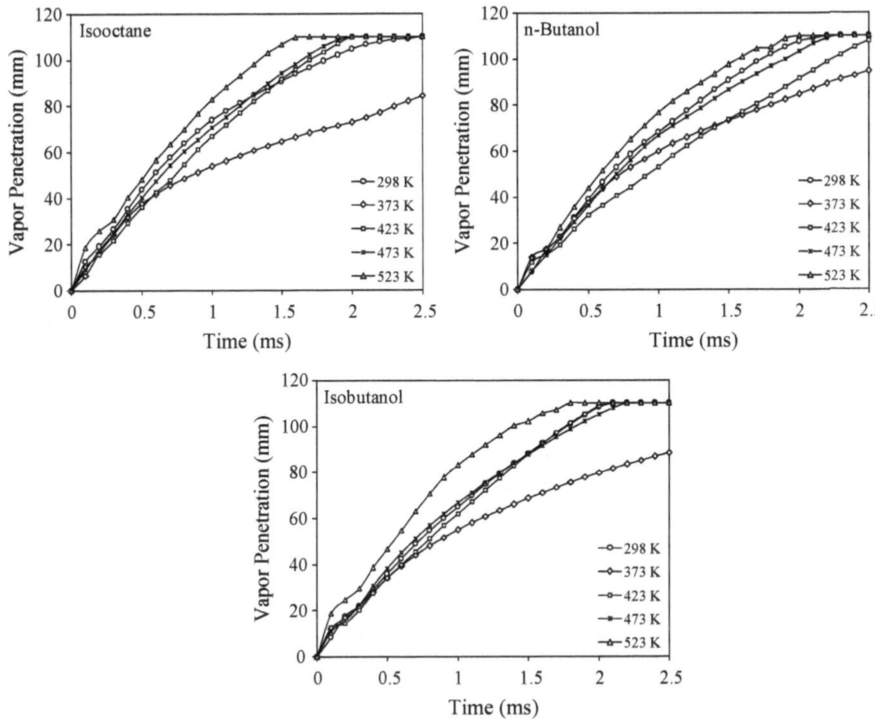

Fig. 3.6 Vapor penetration length for selected fuels at different fuel temperature conditions

direction of the injector. This makes the fuel vapor to move faster in axial direction of the injector and results in higher vapor penetration.

From the liquid penetration data in Fig. 3.7, it can be noticed that liquid penetration length decreases with increase in fuel temperature. *N*-butanol and isobutanol show the lowest penetration for 523 K fuel temperature, whereas lowest penetration for isooctane is recorded at 473 K fuel temperature. This is because, as noticed in shadowgraph images in Fig. 3.4, no liquid was seen at 523 K, whereas vapor plume can be clearly seen at this temperature for isooctane. Since at this temperature, all the fuels are close to their respective critical temperature, there is no distinct phase change process during supercritical injection; such sprays are very similar to dense gas phase injection. Thermo-physical properties of the spray change significantly and therefore conventional atomization and evaporation do not exist under supercritical injection conditions (Zhang et al. 2016). Although fuel temperature is slightly lower than the critical temperature of isooctane, no liquid isooctane is observed at the nozzle exit. On the other hand, liquid plumes are clearly observed for butanol isomers. It is a common observation for evaporating sprays that the liquid penetration decreases with increasing ambient or fuel temperature. From Fig. 3.7, it can be observed that 373 K fuel temperature case is giving the lower penetration compared to 423 and

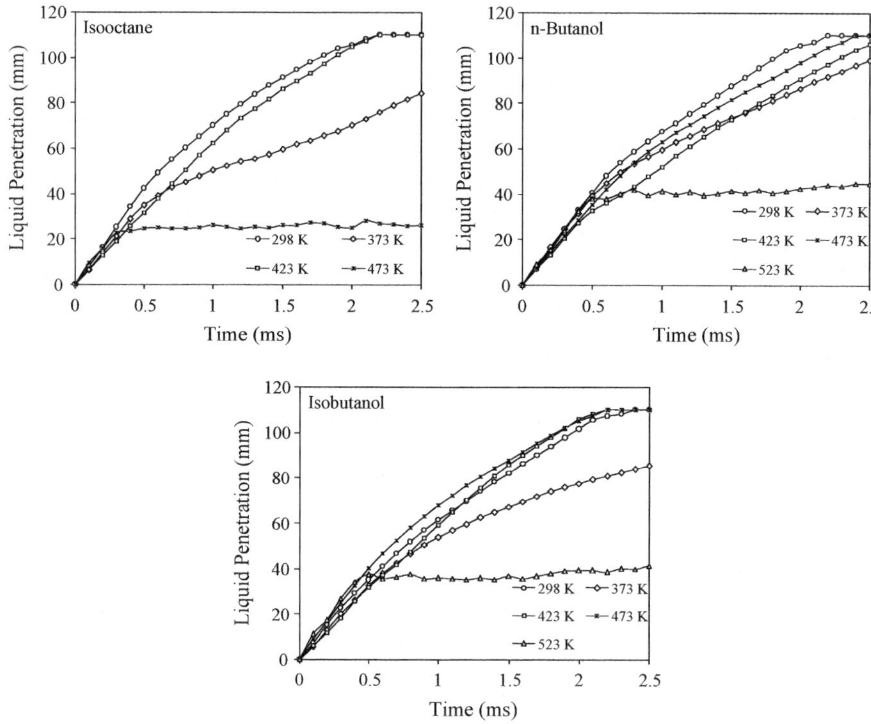

Fig. 3.7 Liquid penetration length for selected fuels at different fuel temperature conditions

473 K fuel temperature case (For *n*-butanol and isobutanol). As explained above, this happens due to change in directionality of the spray.

3.3.2 Effect of Flash Boiling on Atomization Quality of the Spray

SMD, AMD, and Droplet range for different fuels at different fuel temperature is summarized in Fig. 3.8 to understand the quality of the atomization process of a spray under flash boiling conditions. It can be clearly seen that for all the fuels, SMD and AMD show monotonic reduction in the droplet diameter with increase in fuel temperature. The droplet SMD at 423 K, reduced by approximately 64.87% for isooctane, 58.45% for *n*-butanol and 54.51% for isobutanol compared to 298 K fuel temperature case. Additionally, droplet AMD reduced by 74.24% for isooctane, 65.58% for *n*-butanol and 57.12% for isobutanol. For all the fuel temperature conditions, *n*-butanol shows the highest SMD and AMD, whereas isooctane has the lowest values. This might be because of the higher viscosity and surface tension of

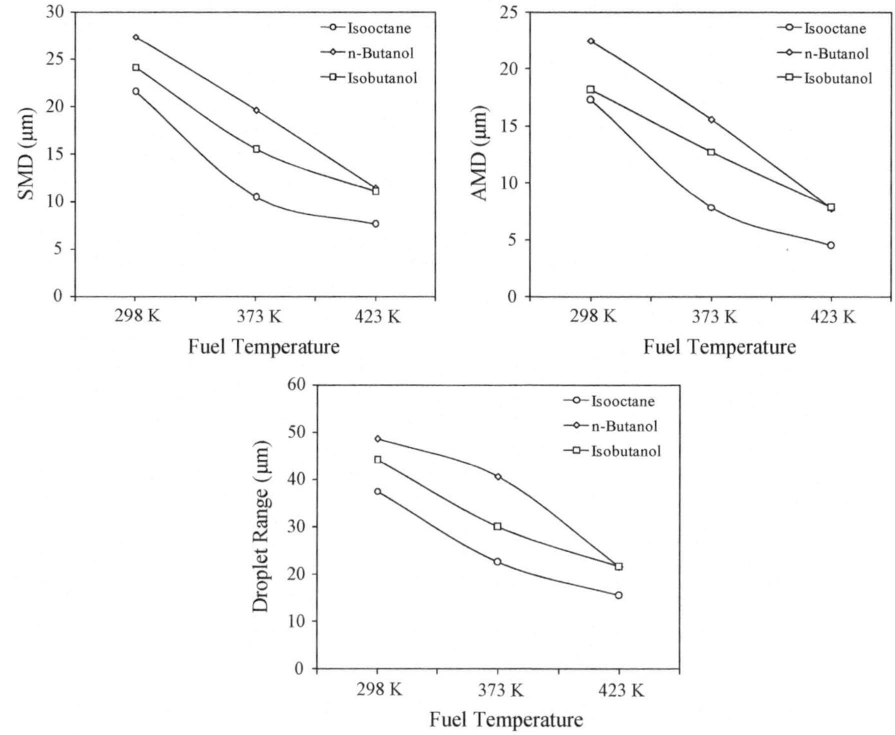

Fig. 3.8 Effect of fuel temperature on SMD, AMD and droplet range for selected fuels

n-butanol. Additionally, butanol isomers have higher latent heat of evaporation compared to isooctane. Droplet range or the maximum diameter in the spray is one of the most important factors for automotive sprays. It tells about the occurrence of bigger droplet in the atomization process. These larger droplets are expected to contribute to soot formation and un-burnt hydrocarbon emission. Due to their bigger size, these droplets fail to evaporate and mix completely with the surrounding air within the required engine cycle duration (Zhao et al. 1999; Schroeter and Meinhart 2017). Due to flash boiling effect, droplet range also gets reduced significantly. When compared to 298 K fuel temperature condition, droplet range at 423 K, reduced by 58.66% for isooctane, 55.67% for *n*-butanol and 51.13% for isobutanol. Generally, it is recommended that droplet SMD for the GDI engines should be below 25 μm to ensure complete evaporation and the mixing within the required engine cycle duration. The above droplet statistics suggest that flash boiling can effectively help in enhancing the liquid atomization to ensure improved in-cylinder combustion characteristics.

3.3.3 Effect of Fuel Temperature on In-Cylinder Spray Structure

Constant volume spray experiments are primarily performed to understand the basic nature of spray and its atomization as there is precise control over boundary conditions. However, in-cylinder air motion due to piston movement, flow past inlet and exhaust valves and the resultant thermodynamic conditions can significantly deviate the spray behavior under engine running conditions. Optical engine with transparent cylinder and piston arrangement as shown in Fig. 3.3 is very helpful for spray studies in such dynamic conditions. Planer Mie scattering using laser sheet illumination gives the qualitative understanding of the liquid mass distribution in a certain cross-sectional plane. Figure 3.9 shows the crank resolved spray pattern for different fuel temperature in an illuminated plane that is 25 mm below the engine head. The fuel injection was initiated at 60° after suction TDC, and images are color mapped for better representation. It should be noted that injector used here was the 6 holes GDI injector and therefore six distinct patterns were expected. However, only four spray plume patterns are being observed at the starting phase. This is because other two plumes have wider spray orientation angle, and they did not intersect with the laser sheet.

From Fig. 3.9, it is evident that spray pattern changes significantly with increase in fuel temperature. At lower fuel temperature conditions, i.e. 298 K, four distinct plumes can be observed and as the fuel injection progressed, plumes show considerable plume-to-plume interaction. This is probably due to the in-cylinder charge motion during the intake stroke. However, spray pattern changes significantly when fuel temperature is further increased. At 423 K, stronger plume interaction can be observed at the later stage of isooctane spray. Compared to isooctane, butanol isomers in Fig. 3.10 showed stronger plume-to-plume interaction. At 423 K, all the plumes merged together to form a single spray print. As discussed previously, in case of constant volume spray experiments, the spray collapse is due to flash boiling which causes all plumes to merge together. For butanol isomers, gradual increase in spray width (at 66 CA) is very clear as the fuel temperature is increased from 298 to 423 K. This confirms the fact that vapor bubble explosion causes increase in individual spray plume width and finally the interaction with the neighboring plumes.

3.3.4 Effect of Fuel Temperature on In-Cylinder Combustion

Figure 3.11 shows the crank resolved natural flame luminosity images of isooctane for different fuel temperature and at two different injection timings. The images were captured using the high-speed color camera. The natural light emission or the chemiluminescence radiated by the hot carbon atoms can be a good measure to quantify the quality of combustion (Gaydon 2012). Generally, premixed flames are blue in color, and they are less luminous compared to diffusion flames which are

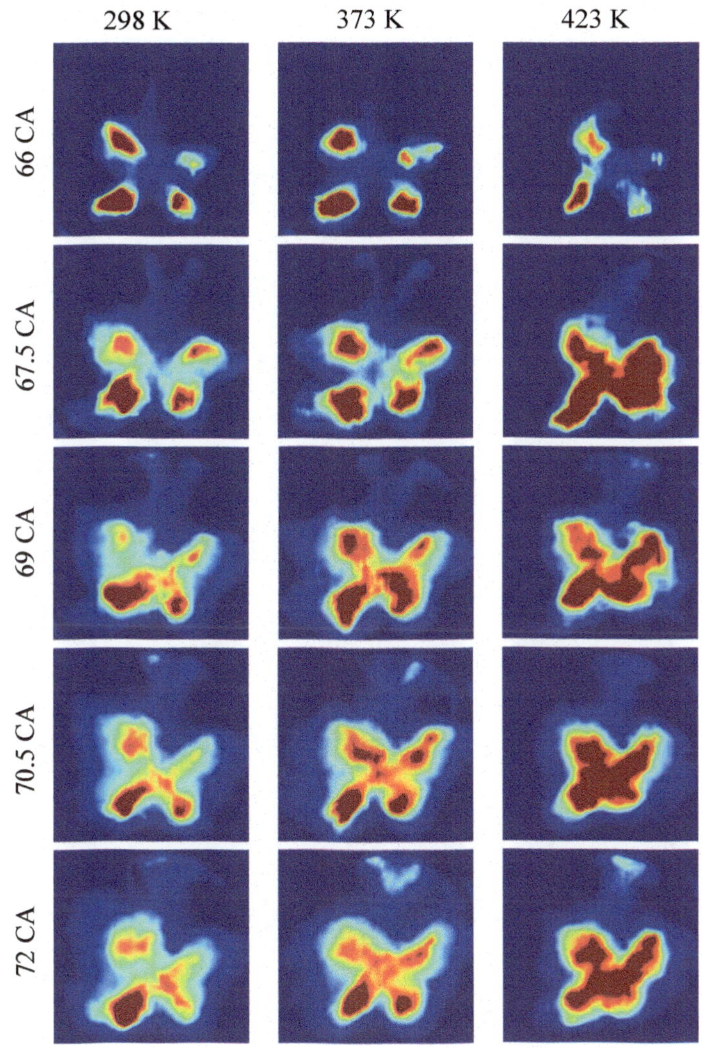

Fig. 3.9 Crank resolved spray patternation for Isooctane for different fuel temperatures

yellowish in nature. From the figure, it is evident that injection timing has a major role in in-cylinder combustion. For an SI engine, it is expected that air–fuel mixture is homogeneous and therefore the mixture burns as a premixed flame. However, in case of GDI engines, it is difficult to satisfy this assumption. Since the fuel is injected directly inside the cylinder, it takes a finite amount of time for droplet evaporation and complete mixing. Additionally, early injection timings increase the chances of fuel impingement on the piston crown. This creates a fuel film on the piston crown and it is a source of tail-pipe emission. Therefore, during the combustion stroke, this

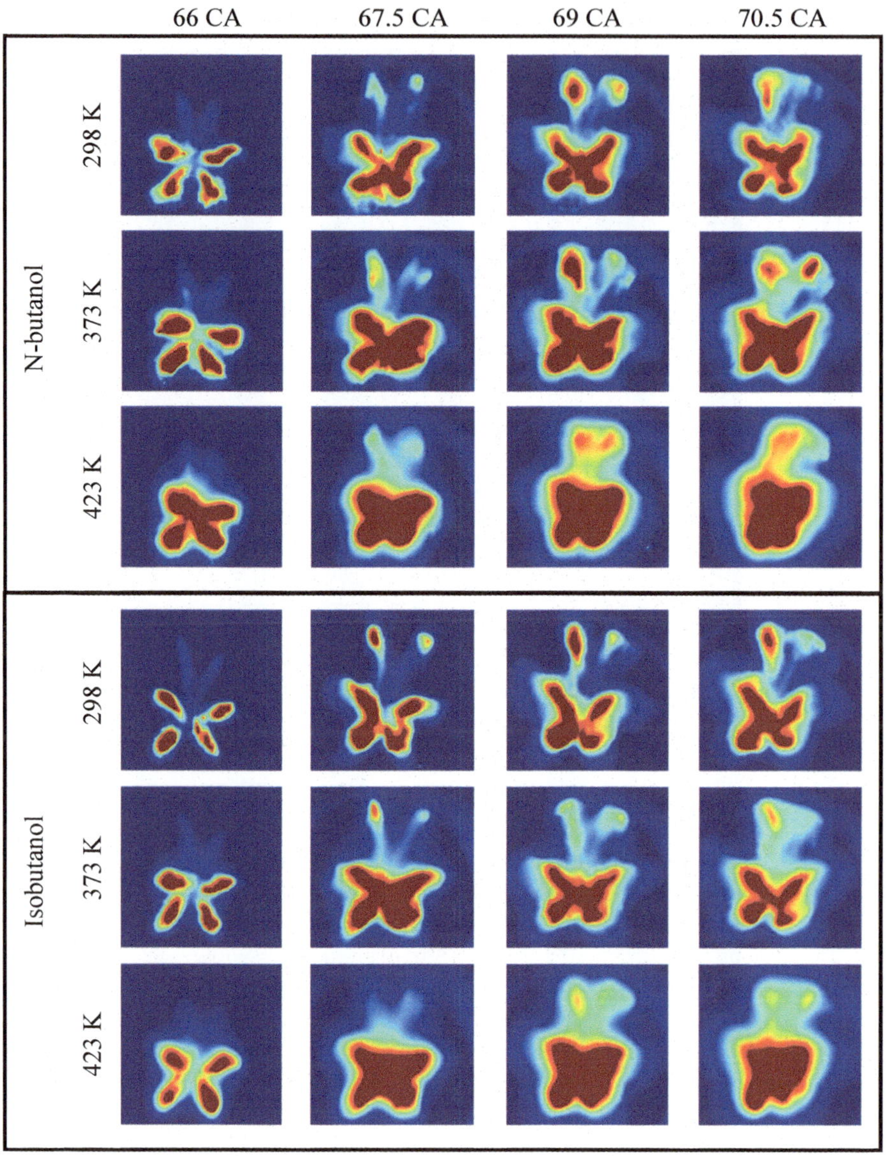

Fig. 3.10 Crank resolved spray pattern for *n*-butanol and isobutanol for different fuel temperatures

fuel layer burns as a diffusion flame and contributes to higher hydrocarbon and soot emission. Due to this, at 298 K fuel temperature and injection timing of 30° after suction top dead center (ASTDC), shows higher light intensity compared to injection timing of 60° ASTDC. This natural flame luminosity can be used as an estimate for soot emission.

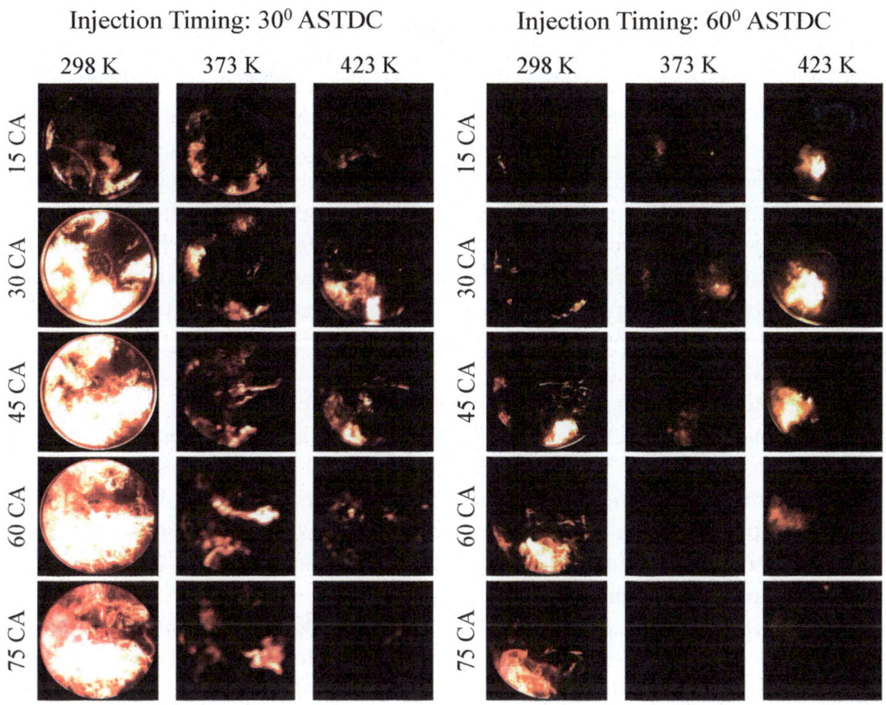

Fig. 3.11 Effect of fuel temperature (Isooctane) on in-cylinder combustion, (left) injection timing 30° ASTDC (right) 60° ASTDC

From the crank resolved images in Fig. 3.11, it is clear that fuel temperature has a significant effect on natural flame luminosity. At 30° ASTDC injection timing, gradual reduction in flame luminosity is evident as fuel temperature is raised. This is because, although fuel impingement is certain in all the cases, however, due to higher fuel temperature, liquid film evaporates quickly and therefore there is a reduction in the overall combustion luminosity. When fuel injection timing was switched to 60° ASTDC, significant reduction in flame luminosity can be observed compared to 30° ASTDC injection timing. This is because, as the fuel injection timing is retarded, the propensity of piston head wetting reduces due to relatively longer distance between the fuel injector and the piston head. However, flame luminosity at 423 K is higher compared to 373 K fuel temperature case. This might be because of the spray collapse at higher fuel temperature condition which causes localized wall wetting effect. Therefore, a single bright spot can be observed at 60° ASTDC injection timing and 423 K fuel temperature. It should be noted that camera settings are crucial for such kind of experiments. Too much light from the combustion creates CMOS sensor saturation; therefore, need lower exposure time or lower lens aperture. On the other hand, larger aperture is required when light intensity is less.

Figure 3.12 shows the effect of fuel temperature on in-cylinder combustion for *n*-butanol, and isobutanol. Results are shown here for only two fuel temperature conditions, i.e., 373 and 423 K. No combustion occurred at 298 K due to low volatility of butanol isomers and therefore not reported here. From the crank resolved images, it is evident that combustion flame luminosity is decreasing with increase in fuel temperature. However, distinct differences can be observed between the flame structure of isooctane and butanol isomers at 423 K fuel temperature. More localized combustion can be observed for isooctane as observed in the images given in Fig. 3.11. However, the flame is well spread across the optical window in case of butanol isomers. This might be because spray collapse for butanol isomers is less prone compared to isooctane. From the spray cone angle data in Fig. 3.5, it can be observed that at 423 K, spray cone angle of butanol fuels is approximately 10° higher compared to isooctane. Additionally, spray structure of butanol fuel at 423 K is approximately similar to the spray structure of isooctane at 372 K as can be observed from the Schlieren images in Fig. 3.4. These factors imply that spray collapse for isooctane is higher compared to spray collapse of butanol isomers. Therefore, butanol isomers have lower propensity to produce localized fuel film on the piston crown. Due to this, it can be clearly seen that flame emission intensity decreases with increase in the fuel temperature for butanol fuels.

To quantify the sooty pool fire due to wall wetting, spatially integrated natural flame luminosity (SINL) was compared for different fuels and fuel temperatures in Fig. 3.13. It should be noted that camera aperture (F. stop no.) varies with respect to experimental conditions; however, results can still be useful for qualitative understanding. As discussed previously, flame luminosity significantly decreases when injection timing is retarded. For the same crank angle, SINL shows significant reduction with increase in fuel temperature. These observations are clearly distinguishable for injection timing of 30° ASTDC. Similar trend is observed for butanol isomers although injection timing was changed to 60° ASTDC. However, some differences can be seen for isooctane. SINL at 423 K was recorded to be higher compared to 373 K fuel temperature. This might be due to total collapse of isooctane at 423 K which results in increase in localized wall wetting, which in turn increases the pool fire on the piston crown. It should be noted that SINL signal of isooctane at 298 K is not presented here because direct comparison between these two is inappropriate due to difference in camera aperture.

3.4 Summary and Concluding Remark

This chapter discusses spray characteristics of isooctane and butanol isomers under flash boiling conditions. Systematic experimental study showed that spray behavior significantly changes under flash boiling conditions and it has a remarkable effect on engine combustion. The major outcomes of the study are as follows:

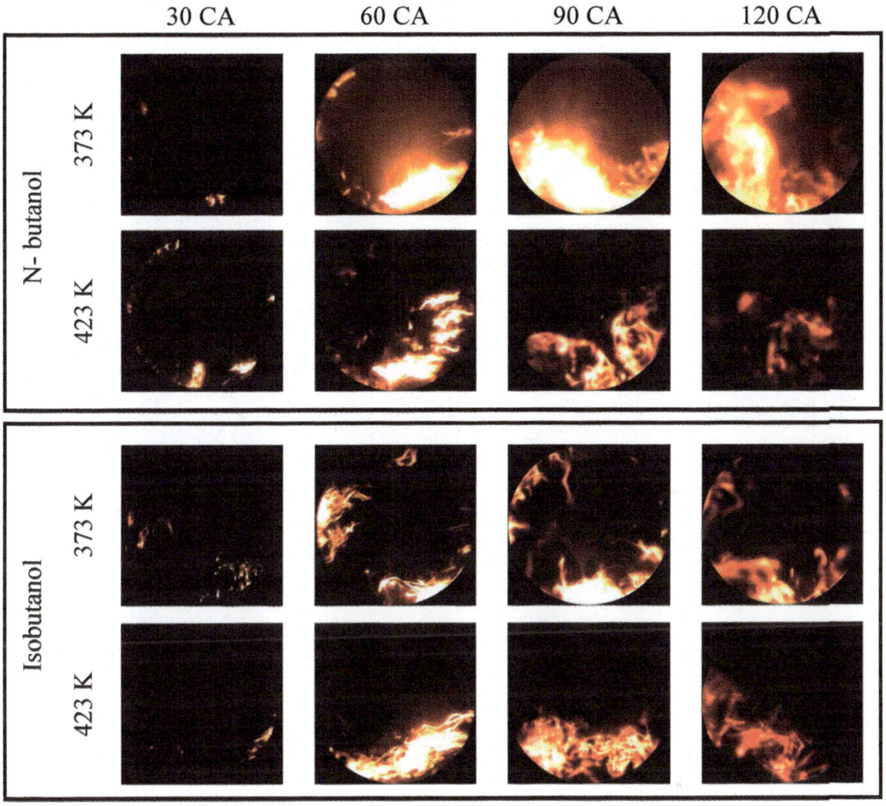

Fig. 3.12 Effect of fuel temperature on butanol combustion at injection timing of 60° ASTDC

- Spray structure changes significantly with increase in fuel temperature. At sufficiently high temperature, multiple spray plumes collapsed into one single spray structure.
- Major changes due to spray collapse are: reduced overall spray cone angle; increased liquid penetration up to certain fuel temperature; and increased vapor penetration.
- Flash boiling sprays show significant reduction in droplet AMD and SMD. Highly viscous liquids such as butanol isomers also show significant improvement in their atomization quality.
- Spray collapse was also observed at elevated temperatures when injected inside an optically accessible engine. Spray patternation show remarkable plume-to-plume interactions due to increased individual plume width.
- Piston wall wetting is strongly related to fuel injection timing. Early injection can significantly increase pool fire on the piston top.

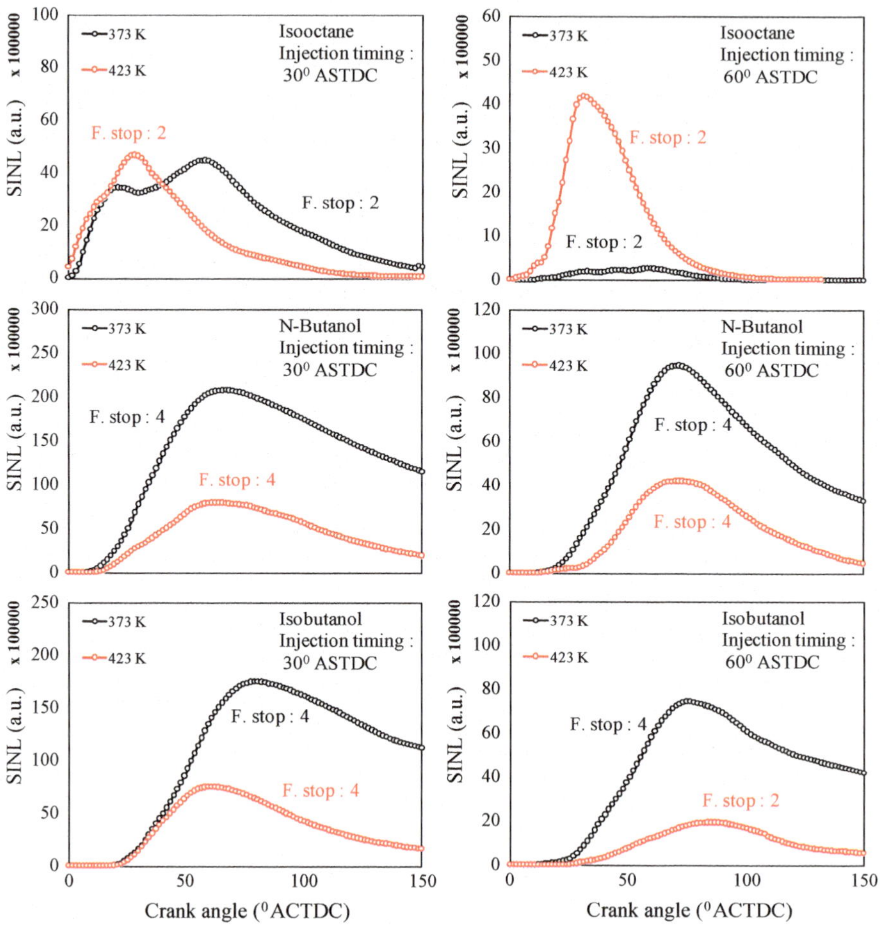

Fig. 3.13 SINL comparison for different fuel temperature and injection timings

- Higher fuel temperature can be helpful in reducing sooty combustion due to wall wetting. However, spray collapse can increase the susceptibility of pool fire due to localize wall wetting.
- Piston wetting changes significantly while selecting different fuel. This is because reduction in spray cone angle is different for different fuels due to differences in their thermo-physical properties.
- Flash boiling is very effective for better atomization and air–fuel mixing. However, spray collapse due to flash boiling is not desirable.

References

Gaydon A (2012) *The Spectroscopy of Flames*. Springer science & business media.
Guo H, Ma X, Li Y, Liang S, Wang Z, Xu H, Wang J (2017) Effect of flash boiling on microscopic and macroscopic spray characteristics in optical GDI engine. Fuel 190:79–89. https://doi.org/10.1016/j.fuel.2016.11.043
Huang Y, Huang S, Deng P, Huang R, Hong G (2014) The effect of fuel temperature on the ethanol direct injection spray characteristics of a multi-hole injector. SAE Int J Fuels Lubricant 7:792–802. https://doi.org/10.4271/2014-01-2734
Kale R, Banerjee R (2018) Investigation of macroscopic as well as microscopic spray behavior of multi-hole GDI injector under engine like hot injector body conditions. In: SAE technical paper. SAE Technical Paper
Schroeter RA, Meinhart M (2017) Optical methodology for characterization of a gasoline direct injection closing event droplet distribution. In: SAE technical paper. SAE International. https://doi.org/10.4271/2017-01-0858
Schulz F, Beyrau F (2017) The influence of flash-boiling on spray-targeting and fuel film formation. Fuel 208:587–594. https://doi.org/10.1016/j.fuel.2017.07.047
Settles GS (2001) Schlieren and Shadowgraph techniques: visualizing phenomena in transparent media. Springer, Berlin
Shen S, Jia M, Wang T, Lü Q, Sun K (2016) Measurement of the droplets sizes of a flash boiling spray using an improved extended glare point velocimetry and sizing. Exp Fluids 57(4):56. https://doi.org/10.1007/s00348-016-2147-3
Wang Z, Jiang C, Xu H, Badawy T, Wang B, Jiang Y (2017) The influence of flash boiling conditions on spray characteristics with closely coupled split injection strategy. Applied Energy 187(Supplement C):523–533. https://doi.org/10.1016/j.apenergy.2016.11.089
Wu S, Xu M, Hung DLS, Li T, Pan H (2016) Near-nozzle spray and spray collapse characteristics of spark-ignition direct-injection fuel injectors under sub-cooled and superheated conditions. Fuel 183(Supplement C):322–334. https://doi.org/10.1016/j.fuel.2016.06.080
Wu S, Xu M, Hung DLS, Pan H (2017) In-nozzle flow investigation of flash boiling fuel sprays. Appl Thermal Eng 117:644–651. https://doi.org/10.1016/j.applthermaleng.2016.12.105
Xu M, Zhang Y, Zeng W, Zhang G, Zhang M (2013) Flash boiling: easy and better way to generate ideal sprays than the high injection pressure. SAE Int J Fuels Lubricant 6:137–148. https://doi.org/10.4271/2013-01-1614
Zeng W, Xu M, Zhang G, Zhang Y, Cleary DJ (2012) Atomization and vaporization for flash-boiling multi-hole sprays with alcohol fuels. Fuel 95:287–297. https://doi.org/10.1016/j.fuel.2011.08.048
Zhang G, Hung DLS, Min X (2014) Experimental study of flash boiling spray vaporization through quantitative vapor concentration and liquid temperature measurements. Exp Fluids 55(8):1804. https://doi.org/10.1007/s00348-014-1804-7
Zhang G, Xu M, Zhang Y, Hung DLS (2013) Characteristics of flash boiling fuel sprays from three types of injector for spark ignition direct injection (SIDI) engines. In: Proceedings of the FISITA 2012 world automotive congress. Springer, Berlin Heidelberg, pp 443–454
Zhang M, Pei Y, Liu Y, Zhang Y (2016) Study on subcritical/supercritical spray characteristics of a multi-hole gasoline direct injector. In: SAE technical paper. SAE International. https://doi.org/10.4271/2016-01-0849
Zhao F, Lai V, Harrington DL (1999) Automotive spark-ignited direct-injection gasoline engines. Prog Energy Combust Sci 25(5):437–562. https://doi.org/10.1016/S0360-1285(99)00004-0

Chapter 4
Recent Progress in Primary Atomization Model Development for Diesel Engine Simulations

Gina M. Magnotti and Caroline L. Genzale

Abstract Predictive engine simulations are key for rapidly exploring and optimizing the design of cleaner burning and more fuel-efficient engines. Injection strategies in advanced engine concepts are resulting in the injection and atomization of fuel under a wide range of operating conditions in order to meet stringent emission regulations. However, the physics governing the breakup of an injected liquid fuel jet into droplets under these conditions have not been well studied or experimentally characterized to date. It is uncertain whether existing atomization and spray breakup models, historically developed to study conventional diesel operation, can be directly applied within engine CFD simulations to study new advanced engine concepts. This chapter summarizes recent progress made in developing an improved physics-based primary atomization model for use in diesel engine simulations across a broad range of in-cylinder conditions. Physical mechanisms that are likely to contribute to the atomization of diesel sprays are first reviewed, with a particular focus on aerodynamic wave growth on the fuel jet surface and turbulence generated in the injector. Then, recent advances in spray diagnostics that have informed characteristic length scales of primary atomization are highlighted. The chapter concludes with the presentation and validation of a newly developed hybrid spray breakup model, the "KH-Faeth" model, capable of representing both aerodynamic and turbulent breakup mechanisms in the atomization of non-cavitating diesel sprays.

G. M. Magnotti (✉)
Argonne National Laboratory, 9700 S. Cass Ave., Argonne, IL 60439, USA
e-mail: gmagnotti@anl.gov

C. L. Genzale
Georgia Institute of Technology, 801 Ferst Drive, Atlanta, GA 30332, USA
e-mail: Caroline.Genzale@me.gatech.edu

K. Saha et al. (eds.), *Two-Phase Flow for Automotive and Power Generation Sectors*, Energy, Environment, and Sustainability,
https://doi.org/10.1007/978-981-13-3256-2_4

Nomenclature

Abbreviations

APS	Advanced photon source
$aTDC$	After top dead center
$bTDC$	Before top dead center
CAD	Crank angle degree
CFD	Computational fluid dynamics
KH	Kelvin–Helmholtz
KH-ACT	Kelvin–Helmholtz aerodynamic cavitation turbulence
LTC	Low-temperature combustion
$PDPA$	Phase Doppler particle analysis
PMD	Projected density
SMD	Sauter mean diameter
SMD_i	Initial Sauter mean diameter formed from turbulent breakup process
SOI	Start of injection
TDC	Top dead center
$USAXS$	Ultra-small-angle X-ray scattering
XRR	X-ray radiography

Greek Symbols

ϵ_0	Dissipation rate at the nozzle exit
Λ_{KH}	Wavelength at maximum growth rate
μ_f	Dynamic liquid viscosity
Ω	Maximum wave growth rate
ω	Growth rate
ρ_f	Liquid density
ρ_g	Ambient gas density
σ	Surface tension
τ_t	Turbulent time scale
τ_{KH}	KH characteristic breakup time

Roman Symbols

$\bar{U}_0$	Average jet exit velocity
a	Radius of the jet
B_0	KH breakup droplet size constant
B_1	KH time constant
C_μ	Model constant from standard k-ϵ turbulence model
C_τ	Faeth turbulent breakup time constant
C_{sa}	Empirical constant for aerodynamic-enhanced breakup
C_{si}	Empirical constant for droplet size at onset of breakup

d_j	Injector nozzle diameter
I_{scat}	Scattered X-ray light intensity
k	Wave number
K_0	Turbulent kinetic energy at the nozzle exit
K_c	Nozzle form factor
l_i	Characteristic length scale of droplet-forming eddy
L_t	Turbulent length scale
N	Number of droplets
Oh	Ohnesorge number
q	X-ray scattering vector
r_c	Radius of "child" droplet
Re_f	Liquid Reynolds number
T	Taylor number
T_g	Ambient gas temperature
U_{rel}	Relative velocity between the liquid and gas phases
v_{li}	Radial velocity of an eddy of size l_i
We_g	Gas Weber number

4.1 Motivation

The diesel engine has remained the preferred power source for ground-based transportation due to its high performance in terms of thermal efficiency and power output. However, due to the nature of the non-premixed combustion process where high peak temperatures and locally rich mixtures are formed, diesel engines suffer from high levels of NO_x and particulate matter production. With increasingly stringent emission standards for NO_x and particulate matter, currently regulated up to 12 and 2% of their 1990 levels (United States Environmental Protection Agency 2003), respectively, both in-cylinder combustion control strategies and aftertreatment management systems must be employed.

In order to control emissions directly within the engine combustion chamber or within the exhaust stream, fuel injection timing with respect to top dead center (TDC) has been utilized as an important tool to control fuel–air mixing and auto-ignition processes, and thereby pollutant formation. Diesel particulate aftertreatment systems often rely on post-injections late in the cycle during the expansion stroke, between 60 and 130 crank angle degrees (CAD) after TDC (aTDC), in order to control the thermodynamic state and chemical composition of the exhaust stream as needed to regenerate the system (Genzale et al. 2010). However, due to the added expense and complexity of these systems, in-cylinder methods have been explored to directly minimize the production of emissions. Low temperature combustion (LTC) concepts are a large class of advanced combustion strategies that leverage direct in-cylinder control of emissions. As opposed to conventional diesel operation with fuel injections near TDC, LTC concepts utilize fuel injections earlier in the engine cycle, either during the intake or compression stroke between 20 and 300 CAD before

TDC (bTDC), as shown in Fig. 4.1. Controlled timing of the start of injection (SOI) with respect to TDC allows for premixing, while in-cylinder temperatures are still low (Pickett and Kook 2009; Kodavasal et al. 2014; Dempsey et al. 2016). As a result, for future engines, fuel injections can be expected to occur over a wide range of conditions throughout the cycle to meet emission regulations.

However, our understanding of the physics controlling fuel injection and spray development, and their effect on combustion and ultimately pollutant formation, has been predominantly focused under conventional diesel operating conditions near TDC, which are characterized by high temperature, T_g, and density, ρ_g, in-cylinder environments. In order to characterize the vaporization process for diesel sprays, Siebers developed a scaling law for the maximum penetration distance of liquid-phase fuel, more commonly referred to as the liquid length (Siebers 1999). By applying gas jet theory as a simplified model of the fuel spray, a scaling law was developed for the liquid length that accounted for the influence of injector, fuel, and ambient conditions on vaporization. Comparison between liquid length measurements for a range of fuels, injection and in-cylinder conditions (Siebers 1998) and scaling law predictions revealed good agreement under conventional diesel conditions, as shown in the gray region of Fig. 4.2. Because Siebers' gas jet model predictions of mixing-controlled vaporization showed good agreement with the experimental measurements for fuel injection near TDC, vaporization was convincingly hypothesized to be controlled by turbulent mixing, or entrainment, of hot ambient gases with the liquid fuel spray, as opposed to atomization or heat and mass transfer at droplet interfaces (Siebers 1999). However, as the SOI is advanced or retarded with respect to TDC to conditions with relatively lower ρ_g, as shown in the yellow region of Fig. 4.2, larger discrepancies are seen between Siebers' scaling law and the experimental data. One proposed hypothesis for these discrepancies is that the details of droplet breakup

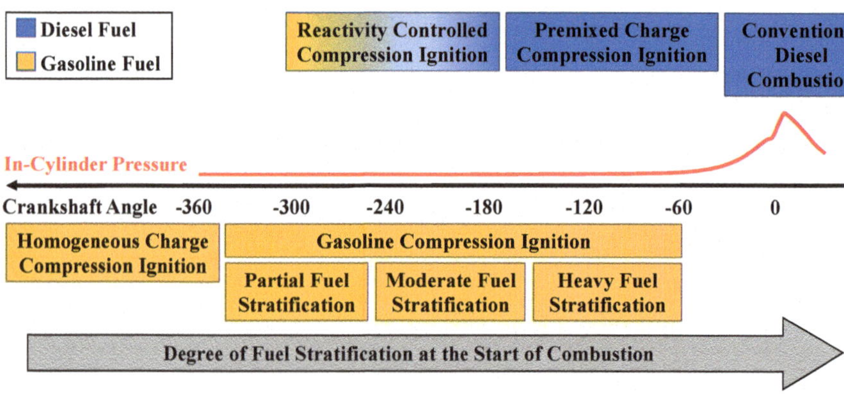

Fig. 4.1 Range of advanced compression ignition combustion strategies using gasoline and/or diesel fuel to achieve low-temperature combustion. Combustion strategies are ordered according to their respective fuel injection timing with respect to top dead center (0 CAD). Modified from Dempsey et al. (2016)

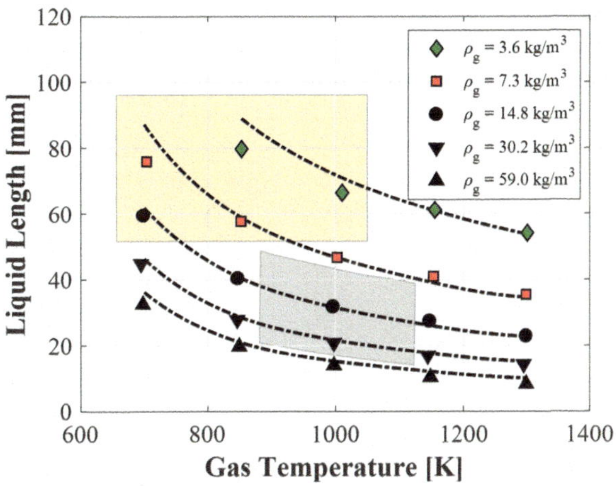

Fig. 4.2 Comparison of Siebers' dense gas jet model predictions with diesel spray liquid length measurements. Modified from Siebers (1999)

affect vaporization rates at low ρ_g conditions (ρ_g less than ~7 kg/m^3) (Siebers 1999; Iyer et al. 2000). Therefore, atomization processes may control vaporization within the range of in-cylinder conditions relevant to LTC strategies.

If computational design tools are to be used to guide the use of direct injection strategies for cleaner and more fuel-efficient engines, the physics underpinning atomization must be better understood to ensure the development of accurate models and predictive simulations of fuel–air mixing and vaporization within the engine. However, the fuel injection and spray breakup processes for engine computational fluid dynamics (CFD) simulations are a challenging computational problem due to the multi-phase, multi-physics, and multi-scale nature of the flow. Several modeling approaches have been used to represent the liquid and gas phases and the exchange of mass, momentum, and energy, but the most commonly employed method for engine simulations is the Lagrangian–Eulerian framework, as schematically shown in Fig. 4.3. In this method, the gas phase is resolved on the Eulerian grid, while the liquid phase is modeled by tracking discrete parcels and their evolution using a Lagrangian formulation. Using the "blob" injection method developed by Reitz and Diwakar (1987), the injection event is represented by a train of discrete injected parcels which start with a droplet size on the order of the nozzle diameter. Each computational parcel statistically represents a number of droplets, N, that share identical droplet properties (size, temperature, etc.) (Dukowicz 1980). Because the liquid phase is not directly resolved on the grid, there is a need to employ submodels to represent the unresolved physics, such as primary and secondary breakup, coalescence, evaporation.

It is uncertain whether existing atomization and spray breakup models, historically developed to study conventional diesel operation, can be directly applied within engine CFD simulations to evaluate new advanced engine concepts. The most widely employed spray breakup model used within nearly all engine CFD codes, such as KIVA (Amsden et al. 1989; Fluent 2005), CONVERGE (Richards et al. 2013) and

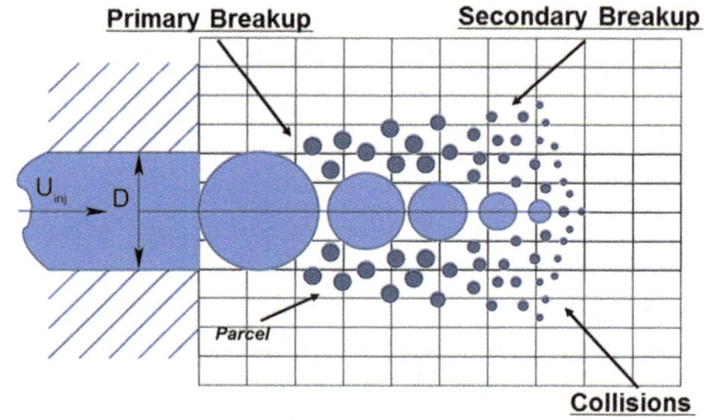

Fig. 4.3 A Lagrangian–Eulerian modeling framework is used to describe the spray formation process, where the gas phase is resolved on the grid and the liquid phase is represented with Lagrangian computational parcels (Magnotti 2017)

OpenFOAM (Weller et al. 1998), assumes that the growth of hydrodynamic instabilities, formed due to the velocity difference at the liquid–gas interface of a fuel spray, is the sole mechanism driving the primary breakup process (Reitz and Bracco 1982; Reitz 1987; Beale and Reitz 1999). It stands to reason that aerodynamic inertial and drag forces should strongly influence the spray breakup process when ρ_g is relatively large, as is the case for fuel injection near TDC. However, as ρ_g decreases for injections earlier in the cycle, such as those employed in advanced combustion engines, aerodynamic inertial forces are expected to decrease, bringing into question if aerodynamic-induced spray breakup should still remain the dominant mechanism. Indeed, if the fuel injection timing is sufficiently advanced such that in-cylinder ambient densities approach atmospheric conditions, recent measurements have shown that spray breakup characteristics scale with turbulence properties at the injector nozzle exit (Wu and Faeth 1993), suggesting that turbulence formed within the injector may govern the spray breakup process under such conditions. Therefore, when computationally investigating injection strategies ranging from early to late cycle fuel injection timings, it seems unlikely that a spray model assuming a single breakup mechanism would be capable of yielding reliable predictions to guide design evaluation and optimization. A key premise of the research presented in this chapter is that a hybrid spray breakup modeling approach, which considers the contributions of several influential breakup mechanisms for the conditions of interest, is needed for use in design evaluation and optimization. This work aims to reassess the appropriateness of the physics underlying existing spray breakup models for the range of conditions relevant for current and future engine design, and determine pathways toward improving these models.

The remainder of this chapter details the physical processes that govern fuel injection and spray formation under engine-relevant conditions, and reviews previous experimental and computational investigations characterizing different primary atomization mechanisms within the spray literature. Recent experimental characterization of the fuel spray structure is then summarized. Finally, comparison with theoretical scalings is used to develop an improved hybrid spray breakup model, and the recent implementation and validation of this model is reviewed.

4.2 Background and Literature Review

The development of a spray under engine-relevant conditions, as illustrated in Fig. 4.4, can be divided into four different processes: injection, spray formation and atomization, ambient entrainment, and vaporization. The order of events leading to combustion of the air–fuel mixture includes development of a turbulent, possibly cavitating, flow within the injector, primary breakup of a liquid jet into droplets, secondary breakup of droplets into smaller droplets, and simultaneous entrainment of air and vaporization of fuel until critical air–fuel ratio and temperature conditions are achieved for ignition and combustion. The mechanisms of spray breakup are important to understand because they determine critical parameters, such as spray geometry, initial droplet size, and number distribution, and serve as initial conditions for vaporization and subsequent downstream processes. However, the spray formation process is difficult to analyze both computationally and experimentally due to the multi-scale, multi-dimensional, and multi-physics nature of the problem.

In order to appreciate the challenges associated with studying the spray breakup process under engine-relevant conditions, this section outlines the historical development of theoretical, computational, and experimental efforts focused on the study of spray atomization. First, the theoretical basis of different physical processes believed to govern the primary breakup of a fuel spray will be presented. Next, the experimental techniques and spray measurements that have informed the fundamental basis of atomization and development of spray breakup models will be discussed. Then, commonly employed spray breakup models in current engine CFD codes will be reviewed.

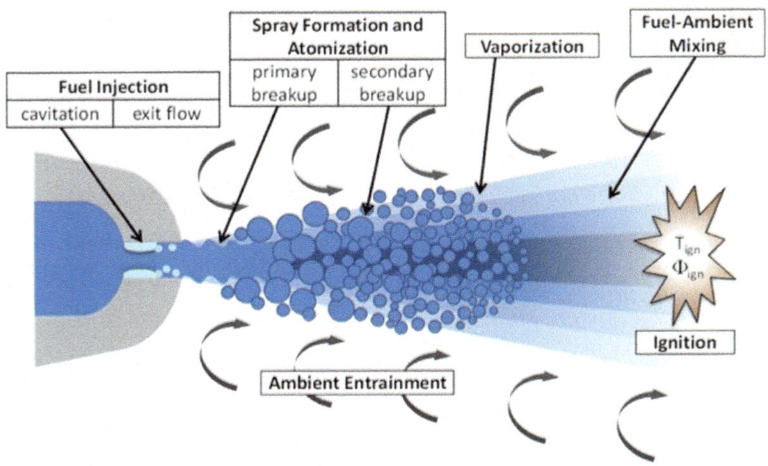

Fig. 4.4 Physical processes influencing fuel spray development in a direct injection engine-relevant environment (SPhERe Lab Homepage 2012)

4.2.1 Spray Breakup Theory

Current understanding of diesel spray breakup was developed from the body of research centered on the breakup of low-velocity round liquid jets (Levich 1962; Sterling and Sleicher 1975; Rayleigh 1878; Weber 1931; Grant and Middleman 1966; Haenlein 1932; Borodin and Dityakin 1951; Castleman 1932). Depending on the relative velocity of the liquid jet with respect to the ambient gas, the breakup of a liquid jet is governed by different physical mechanisms (Reitz and Bracco 1982; Castleman 1931; Reitz 1978). As shown in the jet breakup regime diagram developed by Reitz (1978) in Fig. 4.5, four main regimes of spray breakup are observed: Rayleigh, first wind-induced, second wind-induced, and atomization. In an effort to gain insight into the spray breakup process within the atomization regime, Reitz and Bracco sought a unifying theory to explain the spray breakup characteristics of the Rayleigh, first and second wind-induced regimes (Reitz and Bracco 1982). They hypothesized that if the aerodynamic effects are the dominant factor for the stability of a jet with a Reynolds number beyond that of the second wind-induced regime, as suggested by the experimental work by Castleman (1931), then an extension of such a framework could help provide insight into the dominant forces governing breakup within the atomization regime.

Assuming that the dominant mechanism driving the spray breakup process within the Rayleigh, first and second wind-induced regimes was the growth of disturbances due to hydrodynamic instabilities (Drazin and Reid 2004), Reitz and Bracco evaluated the linear stability of a round liquid jet issuing into a quiescent gaseous environment(Reitz 1987; Reitz and Bracco 1986). Figure 4.6 provides a schematic of the modeled primary breakup process proposed by Reitz. The stability analysis yields a dispersion relation,

$$\omega^2 + 2\nu_f k^2 \omega F_1(ka) = \frac{\sigma k a}{\rho_f a^3} F_2(ka) + \frac{\rho_g}{\rho_f}(U - c)^2 k^2 F_3(ka) \tag{4.1}$$

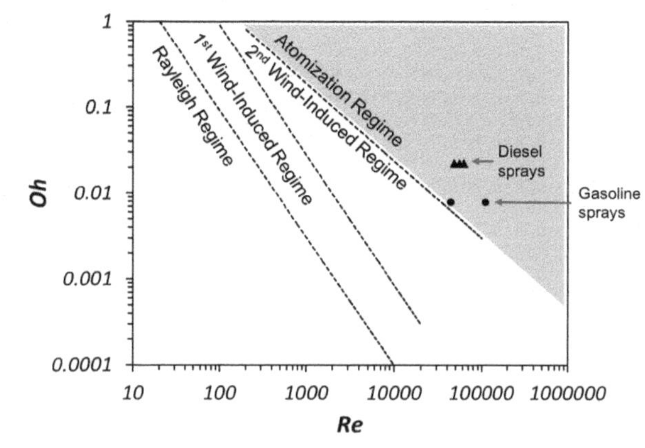

Fig. 4.5 Four main regimes of round jet breakup (adapted from Reitz and Bracco 1982), namely the Rayleigh regime, the first wind-induced regime, the second wind-induced regime, and the atomization regime. Fuel sprays are characterized by high Re_f and therefore typically reside within the atomization regime

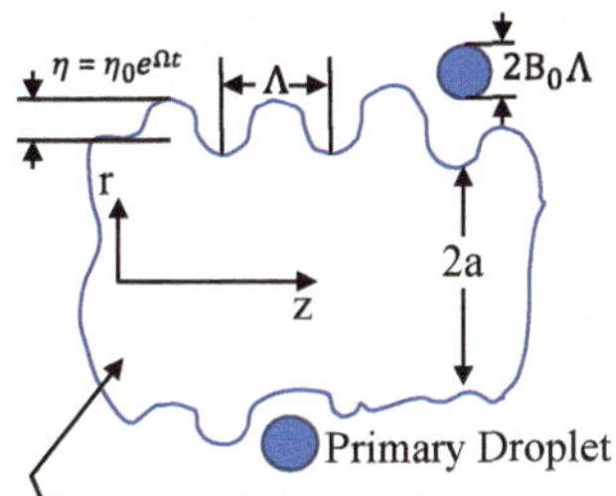

Fig. 4.6 Schematic depicting primary breakup of a liquid fuel jet due to the aerodynamic growth of waves. Modified from Reitz and Bracco (1982)

which relates the growth rate, ω, of an initial linear perturbation of wave number $k = 2\pi/\lambda$, traveling with phase velocity c, to jet and ambient properties. The dispersion relation defines the stability of the jet in terms of non-dimensional ratios, F_i of modified Bessel functions of the first and second kind, which depend on k and the radius of the jet, a. The jet and ambient properties can be formulated in terms of non-dimensional numbers, such as the gas Weber number, We_g, liquid Reynolds number, Re_f, and Ohnesorge number, Oh,

$$We_g = \frac{\rho_g U_{rel}^2 a}{\sigma} \tag{4.2}$$

$$Oh = \frac{\sqrt{We_f}}{Re_f} = \frac{\mu_f}{\sqrt{\rho_f \sigma a}} \tag{4.3}$$

where ρ_g and ρ_f are the gas and liquid densities, respectively, U_{rel} is the relative velocity between the liquid and gas phases, σ is the surface tension, and μ_f is the dynamic liquid viscosity. We_g is an indicator of the relative importance of gas inertia to surface tension, whereas Oh is an indicator of the relative importance of viscous forces to both liquid inertial and surface tension forces.

Once the stability criteria of the jet are determined from evaluation of the dispersion relation (Eq. 4.1), salient characteristics of the spray can be identified for the three breakup regimes. For low-velocity jets within the Rayleigh regime, a low Reynolds number jet will undergo a capillary instability, where the destabilizing nature of the capillary pinching overcomes the stabilizing surface tension forces, and result in the formation of droplets that are larger than the diameter of the jet (Reitz and Bracco 1982). When the Reynolds number of the jet is increased, breakup occurs within the first wind-induced regime. Under these conditions, relative velocities between the liquid and gas phases increase to the point where aerodynamic inertial and drag forces become important. Growing disturbances distort the jet to form ligaments, upon which aerodynamic forces can act to form droplets on the order of the jet diameter or smaller. Further increases in the jet Reynolds number result in breakup within the second wind-induced regime, where hydrodynamic instabilities grow on the liquid–gas interface and ultimately lead to the formation of droplets even smaller than those produced in the first wind-induced regime. For high Reynolds number jets, such as gasoline and diesel sprays, breakup occurs within the atomization regime, and results in the formation of droplet much smaller than the jet diameter.

Within the atomization regime, conditions which are relevant for fuel sprays in direct injection engines, the dominant mechanisms driving the spray breakup process are unknown and have remained a major open question within the spray research community. Several sources of jet breakup have been proposed, including liquid supply oscillations (Giffen and Muraszew 1970), cavitation (Bergwerk 1959; Sadek 1959), velocity profile rearrangement due to the changing boundary conditions at the nozzle exit (Rupe 1962; Hooper and Eisenklam 1958; McCarthy and Molloy 1974), turbulence generated in the nozzle (DeJuhasz 1931; Schweitzer 1937; Sitkei 1959), and the growth of aerodynamic-induced disturbances (Castleman 1931, 1932; Ranz 1956). To test the ability of these proposed mechanisms to characterize jet breakup in the atomization regime, Reitz performed a set of experiments to image the spray and study its response to changes in fuel viscosity, nozzle geometries, injection, and ambient conditions (Reitz 1978). Fourteen different single-hole nozzles were used, each with a nozzle diameter of 340 μm but varying internal geometries, as characterized by the length-to-diameter ratio (L/d_j) and inlet radius of curvature of the nozzle, in order to evaluate a range of nozzle exit flow conditions. Due to the coarse resolution of the camera, the breakup process could not be directly imaged and evaluated. As a result, the behavior of the spray was defined using the diameter of the jet at the nozzle exit and the divergence angle of the spray.

Out of the five evaluated mechanisms, none of them were able to explain all of the experimentally observed trends. Breakup induced from liquid supply oscillations was discounted as a potential breakup mechanism because breakup of the jet was found to occur even when the liquid injection pressure was held constant. Velocity profile rearrangement was reasoned not to be a contributing atomization mechanism because laminar nozzle exit flow conditions were found to be the most stable. However, aerodynamic-induced breakup was able to explain the majority of the spray behavior, with the exception of nozzle geometry effects. The theoretical basis of aerodynamic-induced breakup was evaluated through comparison with the measured divergence of the spray, θ. The theoretical divergence of the spray was related to the initial flight path of a formed droplet,

$$\tan\theta = \frac{v}{u} \tag{4.4}$$

with axial and transverse components of velocity, u and v, respectively. Using surface wave growth theory to define the droplet velocity components in terms of the growth rate and wavelength of the fast-growing wave, good agreement was achieved between the measured and predicted trends, although calibration of the model was required for every nozzle considered.

Based on the observed trends of the spray with respect to changes in nozzle L/d_j and inlet radius of curvature, Reitz reasoned that nozzle-generated cavitation and turbulence likely augment the aerodynamic breakup process (Reitz 1978). Although no single mechanism could explain all of the experimentally observed trends, a hybrid spray breakup mechanism including the influence of aerodynamics, cavitation, and turbulence was thought to well describe the breakup within the atomization regime.

Although the seminal work from Reitz demonstrated the likelihood of aerodynamics, turbulence, and cavitation influencing the breakup process in diesel sprays, none of these mechanisms have been directly observed in diesel sprays under engine-relevant conditions. As a result, existing scalings in the literature describing aerodynamic-, turbulence-, and cavitation-induced breakup have only been indirectly validated through their ability within a spray simulation to predict experimentally observed trends in spray penetration (Beale and Reitz 1999; Som and Aggarwal 2010; Huh et al. 1998), spreading angle (Huh and Gosman 1991; Som 2009), and far-field droplet size distributions (Reitz and Bracco 1986; Huh et al. 1998; Som 2009). In order to improve fundamental understanding of the physics underpinning spray breakup and their appropriate scalings under diesel-relevant conditions, recent work from Genzale and co-workers (Magnotti and Genzale 2017; Magnotti et al. 2017; Kim et al. 2018) systematically suppressed the dynamic and geometric factors contributing to cavitation inception so that aerodynamic- and turbulence-induced breakup could be studied in isolation. The collective experimental and computational work, summarized in Sects. 4.3.2–4.4.2, is used to construct a hybrid spray breakup model, capable of representing non-cavitating diesel spray formation under a broad range of conditions expected in future engines. The remainder of this section details the current understanding of the theory underpinning aerodynamic and turbulent breakup mechanisms.

4.2.1.1 Aerodynamic Breakup

Using aerodynamic breakup theory, Reitz was able to explain many of the experimentally observed responses of the spreading of the spray to changes in fuel viscosity, injection, and ambient conditions (Reitz 1978). These conclusions were only possible by developing a relationship between the surface wave growth theory and the droplet formation process. Building off the work of Ranz (1956), Reitz postulated that the maximum wave growth rate and the corresponding wavelength characterize the fastest growing waves on the liquid surface, and that these waves are ultimately responsible for primary breakup of the liquid jet. The size of the droplet formed during primary breakup was assumed to be proportional to the size of the fastest growing wave. To characterize the fastest growing waves, the dispersion relation defined in Eq. 4.1 was solved numerically. For a set of We_g and Oh condition, the solutions predicted a non-dimensional wave growth rate, $\omega\sqrt{\rho_f a^3/\sigma}$ in terms of a non-dimensional wavelength, $\rho_g U_{rel}^2 \lambda/\sigma$. The results indicate that there is a maximum wave growth rate, $\omega = \Omega$, which occurs at a wavelength of $\lambda = \Lambda_{KH}$. Curve fits of the numerical solutions to Eq. 4.1 for the maximum growth rate, Ω, and corresponding wavelength, Λ_{KH}, were generated as functions of We for both phases, Oh and Taylor number, T, defined below:

$$\frac{\Lambda_{KH}}{a} = 9.02\frac{\left(1 + 0.45\sqrt{Oh}\right)\left(1 + 0.4T^{0.7}\right)}{\left(1 + 0.87We_g^{1.67}\right)^{0.6}} \tag{4.5}$$

$$\Omega\left[\frac{\rho_f a^3}{\sigma}\right]^{0.5} = \frac{\left(0.34 + 0.38We_g^{1.5}\right)}{(1 + Oh)\left(1 + 1.4T^{0.6}\right)} \tag{4.6}$$

$$T = Oh\sqrt{We_g} = \frac{\mu_f U_{rel}}{\sigma}\sqrt{\frac{\rho_g}{\rho_f}} \tag{4.7}$$

where T represents the contributions of viscosity, surface tension, and the relative inertia of the ambient gas and the liquid jet. For a given condition, if the relative velocity between the liquid and gas phases is known, the expected aerodynamic droplet size and breakup timescale can be determined.

The robustness of the aerodynamic wave growth theory is ultimately limited by its inability to capture geometric nozzle effects and their influence on the initial state of the jet as it exits the injector nozzle. In the linear stability analysis used to derive the dispersion relation in Eq. 4.1, it is assumed that the jet exit conditions are single phase and laminar and that the gas–jet interface is deformed by an infinitesimally small disturbance (Reitz and Bracco 1982; Drazin and Reid 2004). Experimental work, such as the image of a turbulent jet shown in Fig. 4.7 (Taylor and Hoyt 1983), provides evidence that mechanisms other than the growth of aerodynamic-induced instabilities are responsible for atomization. More specifically, for the conditions shown in Fig. 4.7 where water is injected into atmospheric conditions, no significant gas inertial forces are expected to act on the jet. In spite of reduced aerodynamic effects, disturbances on the surface of the jet are observed to grow that result in the formation of droplets. These results provide additional evidence that mechanisms other than aerodynamic-induced breakup can contribute to primary atomization. Out of the possible mechanisms considered by Reitz (1978), turbulence generated in the nozzle is the most likely mechanism to augment the primary atomization process for non-cavitating diesel sprays. As a result, there is a need to consider how turbulence can drive the spray breakup process under the wide range of operating conditions characterizing advanced engine technologies. The predominant theories describing turbulence-induced breakup are detailed in the next section.

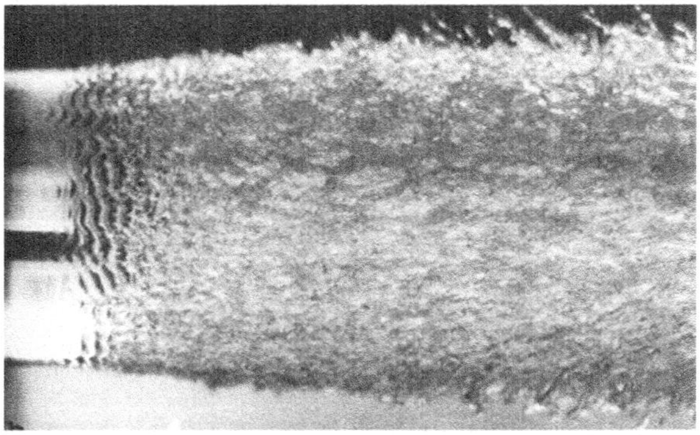

Fig. 4.7 Photograph from Taylor and Hoyt (1983) of a turbulent water jet injected into quiescent atmospheric environment. Reproduced with permission of Springer Nature, from Taylor and Hoyt (1983)

4.2.1.2 Turbulence-Induced Breakup

Several theories have been put forth to explain the role of nozzle-generated turbulence on the primary breakup process (DeJuhasz 1931; Schweitzer 1937; Sitkei 1959; Faeth et al. 1995; Wu and Faeth 1993). Schweitzer proposed that turbulence generated in the nozzle serves to augment the aerodynamic breakup process (Schweitzer 1937). This theory was evaluated by examining images of fuel jet breakup conducted by Lee and Spencer (1933), as shown in Fig. 4.8, and Schweitzer (Schweitzer 1937). Fuel was injected into evacuated and pressurized chambers across a range of Reynolds numbers (Re_f ~1500–9000). Schweitzer found that complete atomization of the jet could be suppressed if the spray was injected into rarefied gas or if the nozzle exit conditions of the jet were laminar. These results led to the hypothesis that the radial component in turbulent pipe flow could cause disturbances on the surface of the jet beyond the nozzle exit, which then grow according to aerodynamic wave growth. However, without sufficient spatial and temporal resolution of their imaging setup to characterize the length and timescales of the primary atomization process, the proposed theory could not be directly validated. In spite of this, this conceptual framework forms the foundation for the majority of existing turbulence-induced breakup models used to study diesel spray formation, as will be discussed in further detail in Sects. 4.2.3.2 and 4.2.3.3.

Due to advancements in imaging technology since the work of Schweitzer, Faeth and co-workers were able to propose and validate a phenomenological model for turbulence-driven spray atomization using pulsed shadowgraphy and high-magnification holographic imaging (Wu et al. 1992, 1995; Wu and Faeth 1993; Faeth et al. 1995). Wu and co-workers postulated that droplets formed from turbulent breakup were due to turbulent kinetic energy overcoming the surface tension energy at the liquid–gas interface. This hypothesis was tested by systematically isolating the influence of turbulence-induced breakup from other known breakup mechanisms, such as cavitation through careful design of the injection system, and aerodynamics by injecting into relatively low ambient density environments. By injecting into conditions where the liquid-to-gas density ratio (ρ_f/ρ_g) was large, the magnitude of

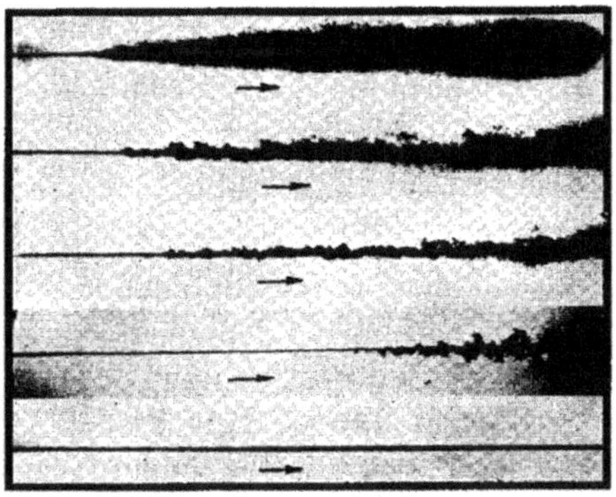

Fig. 4.8 Photographs detailing the effect of ambient pressure on fuel jet breakup. Modified from Lee and Spencer (1933)

the inertial force acting on the jet was expected to be minimized. For ρ_f/ρ_g greater than 500, aerodynamic effects were observed to be diminished and have little effect on the droplet formation process (Wu and Faeth 1993). It was hypothesized that because droplet-forming eddies only needed enough turbulent kinetic energy to surpass the surface energy present at the liquid–gas interface, initially formed droplet sizes, SMD_i, should only scale with jet properties at the nozzle exit. Indeed, analysis of the experimental images confirmed that SMD_i scaled with We_f alone (Wu et al. 1992),

$$\frac{SMD_i}{d_j} = 77 We_{fd_j}^{-0.74} \tag{4.8}$$

where d_j is the injector nozzle diameter. In comparison with length scales characterizing the turbulence spectrum, SMD_i was found to be larger than estimated Kolmogorov length scales, but smaller than integral length scales. As a result, Faeth and co-workers hypothesized that because turbulent eddies formed in the injector convect downstream and dissipate energy while doing so, the reduced size of droplet-forming eddies, l_i, likely exist within the inertial subrange of the turbulence spectrum. This hypothesis is supported by the scaling of SMD_i in Eq. 4.8, which is dependent on both the dimension and velocity of the flow.

However, the sizes of ligaments and droplets were observed to be influenced by aerodynamic effects when the spray was injected into ρ_f/ρ_g conditions less than 500. Wu and co-workers proposed that aerodynamic effects can enhance the spray breakup process by reducing the energy required to form a droplet. It was hypothesized that acceleration of gas over a ligament can reduce the local pressure, akin to flow over a sphere (Munson et al. 2013). Faeth and co-workers modeled the enhanced aerodynamic effects as a mechanical energy, $C_{sa}\rho_g\bar{U}_0^2 l_i^3$, which together with the kinetic energy from the turbulent velocity fluctuations, $\rho_f v_{li}^2 l_i^3$, balances the surface energy, $C_{si}\sigma l_i^2$, at the instant of droplet formation, as mathematically defined below,

$$\left(\rho_f v_{li}^2 + C_{sa}\rho_g \bar{U_0^2}\right) l_i^3 = C_{si}\sigma l_i^2 \tag{4.9}$$

where v_{li} is the radial velocity of an eddy of size l_i, $\bar{U}_0$ is the average jet exit velocity, and C_{sa} and C_{si} are coefficients that incorporate the effects due to ellipticity, non-uniform pressure variation over the ligament surface and non-uniform velocities within the eddy. Even under conditions where aerodynamics augmented the spray breakup process, the size of droplet-forming eddies, l_i, were still found to scale with eddies within the inertial subrange of the turbulence spectrum.

Through the evaluation of images characterizing the formation of ligaments and the resultant droplets for fully developed turbulent jets across a wide range of liquid-to-gas density ratio ($\rho_f/\rho_g \sim 104 - 6230$), Reynolds number ($Re_f \sim 1.5 \times 10^5$ to 5.3×10^5), and Weber number ($We_f \sim 7 \times 10^4$ to 4.1×10^5) conditions, three different primary breakup regimes were identified, as depicted in Fig. 4.9. Non-aerodynamic primary breakup is found to occur for high ρ_f/ρ_g conditions where aerodynamic effects are suppressed, and turbulence is the only mechanism driving the formation of droplets. Transition between non-aerodynamic (turbulent) and aerodynamically enhanced primary breakup was determined to be a function of ρ_f/ρ_g alone. The critical ρ_f/ρ_g condition defining this transition was proposed to be 500, although Wu and Faeth acknowledged that more experimental work was needed to better define the breakup regime boundaries (Wu and Faeth 1993). Within the aerodynamically enhanced breakup regime, turbulence is the primary mechanism governing the spray formation process, although aerodynamics serve to reduce the energy required to form droplets. As a result, smaller primary droplets are observed within this regime relative to those formed in the non-aerodynamic regime.

For ρ_f/ρ_g conditions less than 500, Wu and Faeth hypothesized that for sufficiently large injection velocities, the secondary droplet breakup process would become fast enough such that the primary and secondary breakup processes become effectively merged and indistinguishable from one another (Wu and Faeth 1993). Under such conditions, the measured droplet sizes were thought to be highly influenced by aerodynamic secondary breakup processes. The transition between aerodynamically enhanced and merged aerodynamic primary and secondary breakup regimes was defined using the relative timescales of ligament formation to secondary breakup timescale ratios (τ_R/τ_b), where the critical timescale ratio was selected to be 4. In contrast to the non-aerodynamic primary breakup regime, measured droplet sizes in the aerodynamic primary and secondary breakup regimes are smaller and have a strong dependence on the secondary breakup mechanism.

Although the phenomenological framework developed by Faeth and co-workers to explain the role of turbulence in the primary atomization process is strongly supported through comparison with direct observation and measurements, questions remain about the applicability of these findings to sprays formed from practical diesel injectors. The set of experimental data supporting their theory considers jets issued from large idealized nozzles, with nozzle diameters ranging from 3.6 to 9.5 mm and with long enough nozzle L/d_j to ensure fully developed turbulence conditions at the nozzle exit. In general, diesel injectors utilize nozzles with small diameters and short length-to-diameter ratios, typically with $d_j < 1$ mm and $L/d_j < 12$ (Kastengren et al. 2012). However, results from Wu and co-workers suggest that for conditions where aerodynamic forces have a minimal influence on the spray ($\rho_f/\rho_g > 100$),

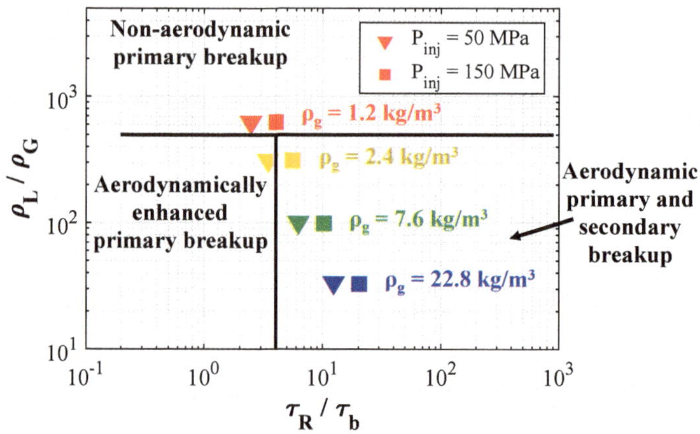

Fig. 4.9 Injection and ambient conditions of interest, overlaid on the turbulent primary breakup regime diagram, as originally proposed by Wu and Faeth (1993)

the condition at the onset of turbulent breakup and the size of primary and secondary droplets were relatively independent of L/d_j (Wu et al. 1995). Therefore, discrepancies between the L/d_j of diesel injectors and those considered by Faeth and co-workers may not affect the applicability of their results to diesel sprays.

In order to assess the applicability of aerodynamic- and turbulence-induced breakup theories to the broad range of conditions characterizing diesel sprays in current and future engines, high-fidelity quantitative spray measurements, capable of characterizing spray formation, are needed. In the next section, available spray measurement and droplet sizing techniques will be discussed, along with the inherent limitations of these methods to quantify the spray breakup process under engine-relevant conditions.

4.2.2 Spray Diagnostics

The physical mechanisms governing atomization and spray formation are still largely unknown due to the difficulty in directly observing this multi-scale and multi-physics process. In order to assess the validity of applying aerodynamic and turbulent spray breakup theory to high-pressure fuel sprays under engine-relevant conditions, high-fidelity quantitative spray measurements are needed. In this section, several imaging and spray measurement techniques are presented, along with the inherent limitations which prevent the direct quantification of the primary breakup process in diesel sprays.

A range of imaging techniques have been applied to sprays in order to directly image and observe global spray characteristics, as well as the initial jet breakup and droplet formation processes in sprays. For example, in the work by Reitz (1978); Reitz and Bracco (1982), previously discussed in Sect. 4.2.1.1, the shadowgraph technique was used to characterize the spray and its response to changes in injection and ambient conditions. In its most rudimentary form, the shadowgraph optical configuration only requires a light source and a recording plane to detect the shadow of a given flowfield, which is formed due to the refraction of incident light away from its initial undeflected path. Although this technique does not yield a quantitative description of the spray, it can provide qualitative characterization of the spray geometric features, such as the spreading angle. The images obtained from Reitz and Bracco were limited in their spatial resolution ($\sim \mathcal{O}(100\,\mu\text{m})$) and as a result were not able to resolve features related to the primary atomization process, such as ligaments or droplets. As a result, their study was limited to indirectly relating measurable spray parameters, such as the spreading angle of the spray, to the proposed mechanisms driving the atomization process.

Since the work of Reitz and Bracco, digital camera resolution, as defined by the number of pixels in the image sensor, has increased by more than two orders of magnitude (Borenstein 2012), resulting in significant improvements in spatial resolution capabilities. Additionally, the use of pulsed light sources, either using lasers (Wu and Faeth 1993) or LEDs (Crua et al. 2015; Zaheer 2015), has improved both the spatial and temporal resolution of imaging techniques. For example, Wu and co-workers used a pulsed ruby laser to obtain single-pulse shadowgraphs capable of resolving the primary spray breakup process in the near-nozzle region (Wu and Faeth 1993; Faeth et al. 1995). Feature extraction and analysis of these images enabled the quantification of the characteristic length and timescales governing the turbulent primary atomization process, as previously detailed in Sect. 4.2.1.2. However, the maximum injection velocity evaluated by Wu and co-workers was approximately 67 m/s (Wu et al. 1992), which is slower than typical diesel sprays traveling with convective speeds greater than 300 m/s. In general, imaging techniques are limited to slow to moderate jet speeds ($\bar{U}_0 \sim 50-100\,\text{m/s}$) due to competing needs of spatiotemporal resolution and contrast required to image the ligament and droplet formation process. Utilizing a state-of-the-art high-speed camera, Zaheer found that the minimum spatial resolution of a feature traveling at a velocity of $100\,\text{m/s}$ was roughly $3\,\mu\text{m}$ or larger; for faster features on the order of $500\,\text{m/s}$, the resolution capabilities drop to approximately $18\,\mu\text{m}$ (Zaheer 2015). Further development of current imaging technology is needed to improve the simultaneous temporal and spatial resolution required to resolve primary droplets produced from high-pressure fuel sprays, which are on the order of $1\,\mu\text{m}$ and travel with convective speeds of $300\,\text{m/s}$ or greater.

In the absence of sufficiently resolved images to visualize spray development, droplet sizing measurements are needed to characterize the outcomes of the spray breakup process. For example, phase Doppler particle analysis (PDPA) measurements (Behrendt et al. 2006; Soare 2007; Payri et al. 2011; Pastor et al. 2012) can provide detailed point-wise spray measurements, such as local droplet size and velocity, that are quite valuable for spray model validation. However, such measurements have had limited success in characterizing primary breakup droplet sizes under engine-relevant conditions. High droplet number density conditions in the near-nozzle region complicate the use of PDPA due to the sampling requirements of isolated single droplets within the probed volume (Behrendt et al. 2006; Soare 2007). In general, PDPA measurements have been conducted far downstream of the nozzle exit ($x/d_j \sim 200-400$). Measured droplet sizes from PDPA are typically much larger than those indicated from more recent near-field measurements (Kastengren et al. 2017; Powell et al. 2013), suggesting that coalescence may influence the droplet size distribution at these locations (Payri et al. 2011; Munnannur 2007). These factors highlight the challenges of using such measurements for the validation of primary breakup theories.

In order to quantify details of the spray in the near-nozzle region, alternative diagnostics to conventional imaging and droplet sizing techniques must be employed. X-ray radiography (Wang 2005; Yue et al. 2001) is an absorption-based technique which quantifies the path-integrated liquid fuel mass distribution in a spray, commonly referred to as projected density. Unlike light in the visible spectrum, X-rays are absorbed rather than scattered by the liquid structures and droplets, which allows such techniques to probe dense portions of the spray. Although X-ray radiography cannot directly quantify droplet size or number density, it does provide measurements of liquid mass distributions, particularly in regions of the spray where primary breakup is expected to occur.

Although X-ray radiography can provide valuable information within the near-nozzle region, droplet sizing measurements are still needed to advance fundamental understanding of the primary breakup process. Recent advances of the X-ray beamline have leveraged existing projected density measurements to quantify droplet sizes using the ultra-small-angle X-ray scattering (USAXS) measurement technique (Kastengren et al. 2017; Powell et al. 2013). This measurement affords a unique opportunity to use the Sauter mean diameter (SMD) of droplet size distributions, particularly in the near-nozzle region, to evaluate primary breakup droplet sizes. Sections 4.3.3 and 4.3.4 detail the salient features of the spray structure that can be observed using these measurements, while Sects. 4.4.1 and 4.4.2 summarize how these experimental findings have been used to inform the development of an improved hybrid spray breakup model.

4.2.3 Computational Spray Breakup Models

As previously mentioned in Sect. 4.1, the most commonly employed method for modeling sprays in engine simulations is the Lagrangian–Eulerian framework. Because it is not computationally feasible to directly resolve the liquid phase in the context of an engine simulation, the evolution of the spray due to primary and secondary breakup, coalescence, and other processes is instead represented with physics-based submodels. Details of the spray submodels employed in today's engine CFD codes, as they relate to the spray atomization theory discussed previously, are presented in the next section.

4.2.3.1 Kelvin–Helmholtz (KH)

As previously mentioned in Sect. 4.1, the KH model is the most widely used spray atomization submodel within engine CFD codes. The model is based on the aerodynamic breakup theory from Reitz and Bracco (1982), as previously detailed in Sect. 4.2.1.1, and is used to describe how the Lagrangian parcels change in size due to the primary breakup process (Beale and Reitz 1999). The primary breakup of the injected fuel is represented in the spray model through the decrease in size of "parent" droplets, a, and formation of "child" droplets of size r_c via KH aerodynamic instabilities, as physically modeled with the following equations:

$$\frac{\mathrm{d}a}{\mathrm{d}t} = -\frac{a - r_c}{\tau_{KH}} \tag{4.10}$$

$$\tau_{KH} = \frac{3.726 B_1 a}{\Lambda_{KH}\Omega} \tag{4.11}$$

$$r_c = B_0 \Lambda_{KH} \tag{4.12}$$

where τ_{KH} is the characteristic breakup time, and Ω and Λ_{KH} are the maximum growth rate and corresponding wavelength of the most unstable surface wave, as numerically solved from linearized stability theory previously described in Eqs. (4.5) and (4.6). The KH breakup model employs two primary empirical constants, namely the breakup time constant, B_1, and droplet size constant, B_0, which are typically calibrated to achieve agreement between modeled and measured liquid-phase penetration (Beale and Reitz 1999). Although B_0 is typically set to a value of 0.61, a wide range of B_1 model constants have been employed with primary breakup models, from 1.76 to 40 (Ning 2007; Som 2009; Beale and Reitz 1999; Liu et al. 1993; Kong et al. 1995; Eckhause and Reitz 1995; Patterson et al. 1994) in order to improve agreement between model predictions and measured spray parameters of interest, such as spray penetration, spreading angle, and far-field droplet size distributions. The need for arbitrary calibration of the KH spray breakup model to match experimental spray data highlights the limitation of the model in capturing the pertinent physics in the primary breakup process.

4.2.3.2 Huh–Gosman

In order to link the internal nozzle flow development with the primary atomization process and reduce the need for excessive tuning seen for the KH model, Huh and Gosman developed a hybrid primary breakup model that incorporated the effects of both aerodynamic- and turbulence-induced instabilities (Huh and Gosman 1991). This model is built on the assumption posited by Schweitzer (1937) that turbulence-induced breakup is controlled by the production of large-scale turbulent fluctuations within the injector, which create the initial disturbances on the liquid–gas interface. These disturbances then grow according to KH instabilities and ultimately control the time to form droplets, as depicted in Fig. 4.10.

In contrast to the theory put forth by Faeth et al. (1995), the turbulent fluctuations responsible for droplet formation are assumed to exist within the energy-containing range of the turbulence spectrum. These fluctuations are represented using a turbulent integral scaling, which are characterized by a turbulent length scale, L_t, and timescale, τ_t,

$$L_t = C_\mu \left(\frac{K_0^{1.5}}{\epsilon_0} \right) \tag{4.13}$$

$$\tau_t = C_\mu \left(\frac{K_0}{\epsilon_0} \right) \tag{4.14}$$

where K_0 and ϵ_0 are the turbulent kinetic energy and dissipation rate at the nozzle exit, and C_μ is a model constant from the standard k-ϵ turbulence model (Huh and Gosman 1991; Som 2009; Som and Aggarwal 2009, 2010). Turbulence levels at the nozzle exit can be determined by predictions of the turbulent flow field at the nozzle exit from high-fidelity internal nozzle flow simulations (Som 2009; Xue et al. 2014; Bode et al. 2014). Similar to the KH primary breakup model, the breakup of the jet is then represented in the spray model through the effective decrease in size of the "parent" drop, a, due to primary breakup as modeled below,

$$\frac{\mathrm{d}a}{\mathrm{d}t} = k_1 \frac{L_A}{\tau_A} \tag{4.15}$$

where L_A and τ_A are the turbulent atomization length and timescales, and k_1 is the main model calibration constant. The breakup length scale, L_A, is then modeled as proportional to L_t and occurs over a timescale, τ_A, that is a weighted sum of τ_{KH} and τ_t.

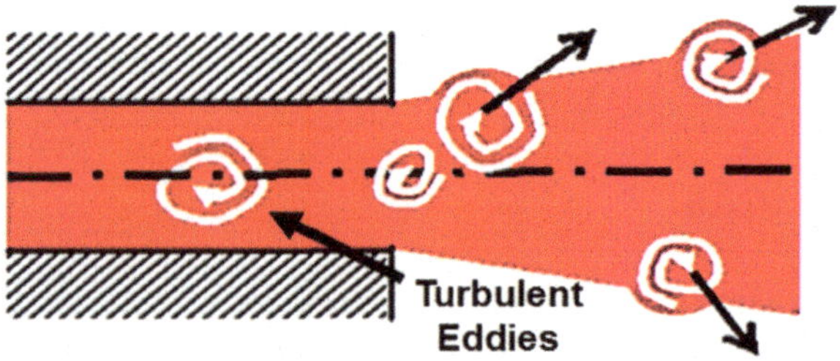

Fig. 4.10 Schematic of turbulence primary breakup model, modified from Som and Aggarwal (2010). Turbulent fluctuations formed within the injector create disturbances at the liquid–gas interface, which grow and result in the formation of primary droplets. Reproduced with permission of Elsevier, from Som and Aggarwal (2010)

In the absence of detailed internal nozzle simulations, K_0 and ϵ_0 can also be estimated using a force balance between the pressure force exerted on the fluid at the nozzle exit and turbulent stress within the nozzle, as detailed in the work of Huh and Gosman (1991); Huh et al. (1998). An order of magnitude analysis was conducted to determine the relevant forces governing the spray atomization process. The possible candidates included surface tension, σ/d_j, gas inertia, $\rho_g U_{inj}^2$, turbulent stress in the jet, $\rho_f u_f^2$, viscous stress in the jet, $\mu_f U_{inj}/L$, viscous stress in the gas, $\mu_g U_{inj}/L$, and gravity, $\rho_f g d_j$, where L is a relevant length scale for each force and u_f and u_g are the turbulent fluctuating velocities in the jet and gas. The analysis led to the conclusion that the dominant forces acting on the jet during the atomization process are the forces due to the gas inertia ($\rho_g U_{inj}^2$) and the turbulent jet internal stress ($\rho_f u_f^2$).

The turbulent jet internal stress was estimated from a force balance

$$\rho_f u_f^2 \pi d_j L = \Delta p_{noz} \frac{\pi d_j^2}{4} \tag{4.16}$$

that equated the resultant wall shear stress to the nozzle pressure drop, Δp_{noz}. Δp_{noz} is obtained by considering the contributions from the total pressure drop, Δp_{tot}, loss pressure drop, Δp_{form}, and acceleration pressure drop, Δp_{acc},

$$\Delta p_{tot} = \Delta p_{noz} + \Delta p_{form} + \Delta p_{acc} \tag{4.17}$$

The pressure loss terms are determined with the following relations,

$$\Delta p_{tot} = \frac{1}{c_d^2} \frac{\rho_f U_{inj}^2}{2} \tag{4.18}$$

$$\Delta p_{form} = K_c \frac{\rho_f U_{inj}^2}{2} \tag{4.19}$$

$$\Delta p_{acc} = (1 - s^2) \frac{\rho_f U_{inj}^2}{2} \tag{4.20}$$

where c_d is the discharge coefficient, K_c is the form factor due to the nozzle inlet radius of curvature for a fixed nozzle diameter, d_j, and s is the area ratio that accounts for the pressure loss due to flow acceleration in the contracting nozzle. Rearrangement of Eq. 4.17 results in the following expression for Δp_{noz}:

$$\Delta p_{noz} = \frac{1}{c_d^2} \frac{\rho_f U_{inj}^2}{2} - (1 - s)^2 \frac{\rho_f U_{inj}^2}{2} - K_c \frac{\rho_f U_{inj}^2}{2} \tag{4.21}$$

Substitution of Eq. 4.21 and the definition of u_f into Eq. 4.16 yields the following expressions for K_0 and ϵ_0:

$$K_0 = \frac{U_{inj}}{8L/d_j}\left[\frac{1}{c_d^2} - K_c - (1 - s^2)\right] \tag{4.22}$$

$$\epsilon_0 = K_\epsilon \frac{U_{inj}^3}{2L}\left[\frac{1}{c_d^2} - K_c - (1 - s^2)\right] \tag{4.23}$$

where K_ϵ is a calibration model constant set to 0.27 (Huh et al. 1998). These relations for K_0 and ϵ_0 can then be substituted into Eqs. 4.13 and 4.14 to characterize the turbulent integral length and timescales.

Although the Huh–Gosman model requires the calibration of three model constants, which control the relative contributions of turbulence and KH to the atomization timescale and the ultimate breakup rate of the spray, the implemented physics were deemed to be validated through replication of the experimentally observed trends for the spreading angle from four different nozzles (Reitz and Bracco 1979; Hiroyasu and Kadota 1974; Yule et al. 1985). Subsequent evaluation of the model was conducted through the comparisons of predicted and measured spray tip penetration and far-field droplet size measurements along the spray centerline and periphery at distances of 40 nozzle diameters or larger from the nozzle exit (Huh et al. 1998).

Although the model was noted to predict the spray observables well, the assumed role of turbulence in the atomization process was never directly validated. It should be noted that the assumed turbulent breakup scaling within the Huh–Gosman model is inconsistent with the body of experimental work from Faeth and co-workers (Wu and Faeth 1993; Faeth et al. 1995). As noted in Sect. 4.2.1.2, the analysis of spray breakup images indicated that primary droplets scale with smaller turbulent length scales, more specifically those within the inertial subrange.

4.2.3.3 Kelvin–Helmholtz Aerodynamic Cavitation Turbulence (KH-ACT)

While the Huh–Gosman model utilizes a hybrid spray breakup approach that includes the influence of both aerodynamics and turbulence on the primary breakup process, the relative contributions of each of the mechanisms on the resultant spray were not evaluated. However, the role of the selected primary atomization model on the predicted spray metrics was extensively studied throughout the body of work conducted by Som 2009; Som and Aggarwal 2009, 2010; Som et al. 2009. In particular, the primary atomization process was characterized by the resultant distribution of liquid mass and droplet dispersion. Such comparisons were only possible through the use of X-ray radiography measurements, which enabled the quantification of liquid mass distributions, particularly in dense regions of the spray (Wang 2005; Yue et al. 2001), as detailed in Sect. 4.2.2. Through the comparison of measured and predicted instantaneous liquid mass distributions at various locations in the spray, the KH model was found to underpredict droplet dispersion, as indicated by the relatively narrower mass distributions in comparison with the experimental data. The underprediction in droplet dispersion was attributed to the insufficient formation of child droplets from the primary atomization process (Som et al. 2009). As a result, the inclusion of additional primary atomization mechanisms, such as turbulence- and cavitation-induced breakup, was motivated by the need to improve model predictions of droplet dispersion. The addition of these physics was further supported through the inability of the KH model to predict the expected trends of injector nozzle geometry on droplet dispersion (Som et al. 2009; Som 2009).

The KH aerodynamic cavitation turbulence (KH-ACT) model improved upon the hybrid spray breakup formulation from the Huh–Gosman model. Because cavitation has been shown experimentally (Payri et al. 2004) and computationally (Huh and Gosman 1991; Som and Aggarwal 2009; Schmidt 1997) to influence the breakup process in diesel sprays, cavitation-induced breakup was included in the model. As previously mentioned, the focus of this review is on the aerodynamic- and turbulence-induced breakup mechanisms, and their relative importance in diesel spray atomization. Therefore, only the KH-ACT model improvements for aerodynamic and liquid turbulence breakup will be highlighted here.

Firstly, in the Huh–Gosman model, each parcel is assumed to have constant turbulence levels throughout the simulation. In the KH-ACT model, a standard k-ϵ turbulence model formulation is used to model the temporal evolution of turbulence levels in each parcel, $K(t)$ and $\epsilon(t)$, as it convects downstream from the nozzle exit prior to the occurrence of primary breakup. Additionally, the Huh–Gosman model assumes that the size of formed droplets is characterized by the turbulent length scale. The KH-ACT model compares and identifies the maximum breakup rate of aerodynamic- and turbulence-induced breakup in order to select the appropriate atomization length and timescales, L_A and τ_A, as defined below:

$$\frac{L_A}{\tau_A} = \max\left\{\frac{a - r_c}{\tau_{KH}}, \frac{L_t(t)}{\tau_t(t)}\right\} \tag{4.24}$$

One final key difference is the philosophy underlying breakup. In the Huh–Gosman model, the breakup timescale is modeled as an averaged process between the two breakup mechanisms. In the KH-ACT model, it is assumed that the breakup process is ultimately determined by a single mechanism, either aerodynamics or turbulence, at each instant in time. If KH primary breakup is dominant, then the parent parcels evolve according to Eq. 4.10. However, if turbulent primary breakup dominates the atomization process, then the parent parcel decreases in size according to the following relation:

$$\frac{da}{dt} = -C_{T,CAV}\frac{L_A}{\tau_A} \tag{4.25}$$

where $C_{T,CAV}$ is the breakup rate calibration constant.

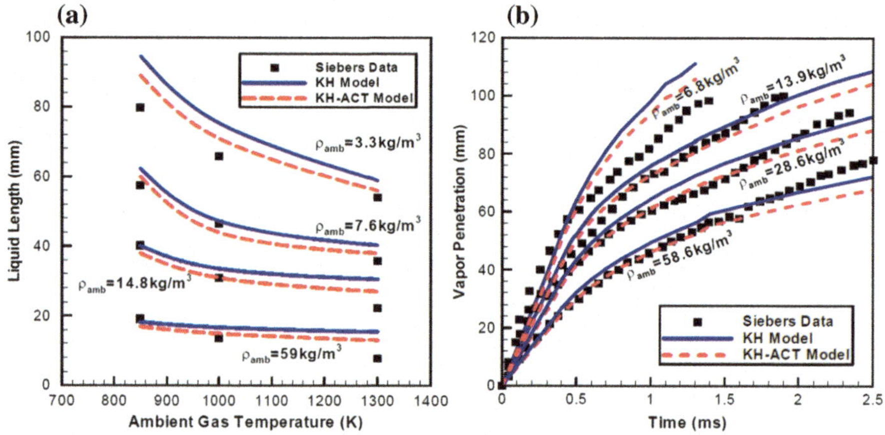

Fig. 4.11 Comparison of KH and KH-ACT model predictions against the Siebers data for **a** liquid length for a range of ambient gas temperature and density and **b** vapor penetration for a range of ambient gas densities (Som 2009). Reproduced with permission of Elsevier, from Som and Aggarwal (2010)

Using the hybrid spray model formulation described above, the KH-ACT model was shown to yield improved prediction of spray characteristics in comparison with the KH model across non-vaporizing, vaporizing, and combusting conditions (Som and Aggarwal 2010; Som 2009). However, the KH-ACT model was only observed to produce marginal improvements over the KH model to predictions of liquid length and vapor penetration for injection into low ambient density environments (ρ_g less than 7 kg/m^3, ρ_f/ρ_g greater than 100), as shown in Fig. 4.11 (Som and Aggarwal 2010). Although the influence of the employed primary atomization model was shown to have diminished influence on the spray formation process under vaporizing conditions (Som 2009; Som and Aggarwal 2010), this predicted discrepancy might suggest that the scalings employed in the turbulence model did not sufficiently enhance droplet formation and droplet dispersion. Indeed, the turbulent atomization process in the KH-ACT model is assumed to scale with the integral scale in the turbulence spectrum (Som 2009). It may be possible that the employment of a different turbulence scaling that results in the formation of smaller droplets could improve the predictive capability of the hybrid spray breakup model under vaporizing conditions.

4.3 Review of X-ray Measurements Applied to Characterize Diesel Spray Structure

In order to assess spray predictions from the existing primary atomization models and determine how the influence of nozzle-generated turbulence should be represented in a spray model, experimental measurements are needed to characterize the spray structure under engine-relevant conditions. This section reviews X-ray spray diagnostics that have been recently employed to guide the assessment and development of an improved spray breakup model. Evaluation of experimental spray observables over a wide range of conditions can provide insight into the governing physical mechanisms of fuel atomization under different engine operating conditions. In particular, newly available spray data at conventional diesel engine conditions ($\rho_f/\rho_g < 100$) are first discussed and ultimately used to assess the predictive capability of the existing primary atomization models premised on aerodynamic wave growth, such as the KH model. Then, experimental spray data under low ambient density conditions ($\rho_f/\rho_g > 300$), where aerodynamic effects are expected to be sufficiently suppressed (Wu and Faeth 1993), are analyzed to assess appropriate scalings for turbulent breakup. The injection and ambient conditions evaluated in the computational and experimental investigations are now presented.

4.3.1 Experimental Test Matrix

In order to better understand the interaction between aerodynamics and turbulence on the primary atomization process in diesel sprays, the dynamic and geometric factors contributing to cavitation inception must be systematically suppressed. As the fuel flows through the injector, the liquid fuel pressure continuously decreases from initially high values in the pressurized fuel supply to the ambient back pressure. Through informed selection of a fuel with a low saturation pressure, such as n-dodecane (NIST 2018), the likelihood of cavitation occurring due to the local reduction in pressure below the fuel vapor pressure can be minimized (Payri et al. 2004; Schmidt 1997). Cavitation can also be suppressed through the careful selection of injectors with converging nozzles having a rounded inlet corner and minimal surface imperfections (Kim et al. 1997; Payri et al. 2004; Schmidt 1997; Duke et al. 2014). Single-hole research-grade diesel injectors, with well-characterized internal nozzle geometries, are available through the Engine Combustion Network (ECN) (Engine Combustion Network 2018). Two classes of injectors provided by ECN that have ideal internal nozzle geometries for suppressing cavitation are the Spray A and Spray D injectors (Duke et al. 2014; Matusik et al. 2016); key geometric features of the injectors are detailed in Table 4.1. As a result, n-dodecane injected from the ECN Spray A and Spray D injectors have been studied so that the influence of cavitation on the resultant spray will be minimized.

Once the influence of cavitation on the spray has been minimized, changes in injection and ambient conditions can be related to the changing turbulent and aerodynamic breakup phenomena. As previously discussed in Sect. 4.2.1.2, a set of experimental conditions can be defined to explore the aerodynamic and turbulent primary breakup regimes proposed by Wu and Faeth (1993). As shown in Fig. 4.9, evaluation of ambient densities, ρ_g, between 1.2 and 22.8 kg/m^3 allows for the proposed non-aerodynamic and merged aerodynamic secondary and primary breakup regimes to be investigated. Modulation of the fuel injection pressure, P_{inj}, from 50 to

Table 4.1 Comparison of engine combustion network Spray A and Spray D injector nozzle geometries (Engine Combustion Network 2018). The total injected mass, injection duration, and nozzle discharge coefficient are given for an injection pressure of 50 MPa and an ambient density of 22.8 kg/m^3 (Payri et al. 2008; Magnotti and Genzale 2017)

Injector parameters	Spray A #210675	Spray D #209133
Nozzle diameter (d_j) [μm]	89.4	186
Nozzle discharge coefficient (C_d)	0.86	0.90
Injection duration [ms]	6.00	4.69
Total injected mass [mg]	15.2	51.6
Nozzle K-factor	1.5	3.7

Table 4.2 Non-vaporizing ambient and injection conditions for the Engine Combustion Network (Engine Combustion Network 2018) Spray A and Spray D nozzles evaluated by Magnotti and Genzale (2017) and Kim et al. (2018) In all cases, n-dodecane at a fuel temperature of 303 K was injected in a pure nitrogen environment at an ambient temperature of 303 K.

Case	ECN injector	Ambient density (ρ_g)[k_g/m^3]	Density ratio (ρ_f/ρ_g)	Injection pressure (P_{inj}) [MPa]
1	A	22.8	32.7	150
2	A	22.8	32.7	50
3	A	7.6	98	150
4	D	22.8	32.7	150
5	D	22.8	32.7	50
6	D	2.4	310.4	150
7	D	2.4	310.4	50
8	D	1.2	620.8	150
9	D	1.2	620.8	50

150 MPa and selection of injectors with different nozzle diameters, d_j, as defined in Table 4.1, allow for the influence of Reynolds on the resultant spray to be assessed. Using these selected ranges for ρ_g, P_{inj}, and d_j, nine different cases were defined, as detailed in Table 4.2. Across the selected range of conditions, the role of aerodynamics and turbulence on the breakup process in diesel sprays can be evaluated using both experimental and computational approaches.

4.3.2 X-ray Diagnostic Techniques

When X-rays interact with steel in fuel injectors or liquid fuel droplets in a spray, the incident light is strongly absorbed and enables the characterization of the medium in the path of the beam. This is in direct contrast to optical diagnostics where visible light is scattered and reduces the detection of transmitted light, thereby limiting the applicability of optical techniques. As a result, X-ray diagnostics provide a unique capability in yielding quantitative information about the injector geometry and liquid mass and surface area distributions, particularly in highly scattering, optically thick regions of the spray. X-ray measurements, conducted by researchers at Argonne National Laboratory at the Advanced Photon Source (APS), enable improved definition for spray model inputs through high-precision quantification of the fuel injector geometry (Matusik et al. 2017) and yield validation data to assess spray breakup model predictions (Kastengren et al. 2017, 2014), particularly in regions of the spray inaccessible to optical techniques. In the subsequent sections, the X-ray experimental techniques and corresponding spray observables are detailed.

4.3.2.1 X-ray Radiography

The X-ray radiography (XRR) technique allows the projected density, PMD, or the path-integrated liquid fuel mass distribution, to be quantified throughout the spray. XRR measurements used to quantify the centerline SMD were performed by Argonne researchers at the 7-BM beamline at the APS (Magnotti and Genzale 2017; Magnotti et al. 2017; Kim et al. 2018). A set of sample transverse distributions of PMD are shown in Fig. 4.12 as scattered markers, and Gaussian fits to the data are overlaid and depicted using the solid lines to highlight differences in the distributions. To obtain two-dimensional maps of PMD, the injector was horizontally mounted in a pressure chamber fitted with a pair of 12 × 30 mm X-ray transparent windows. The chamber was pressurized to the desired back pressure with N_2, which was also used to maintain a continuous purge flow of approximately 4 standard L/min through the chamber to minimize droplet formation on the windows during data acquisition. A diesel common-rail injection system was used to pressurize n-dodecane fuel to

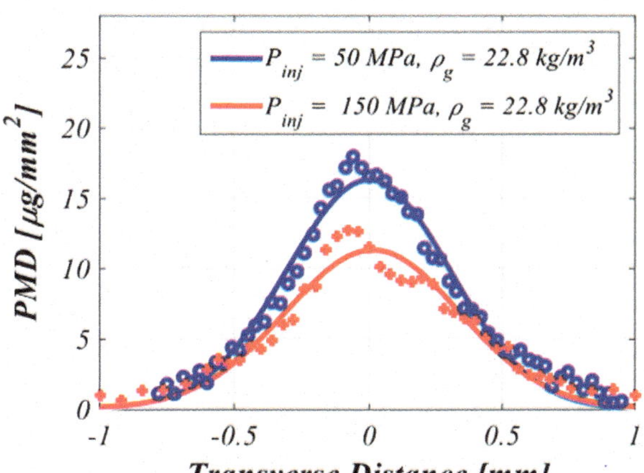

Fig. 4.12 Sample radial distribution of X-ray radiography measurements (Kastengren et al. 2017), at an axial distance of 8 mm from the nozzle exit, for the Spray A injection under the Case 1 and Case 2 conditions

the desired rail pressure. The injector was fired at 3 Hz for a commanded injection duration of 2.0 ms.

Detailed descriptions of the time-resolved radiography measurements may be found in the previous work conducted by Kastengren et al. (2008, 2012, 2014); Kastengren and Powell (2014), but are summarized here. A monochromatic beam at 8 keV energy passed through a set of curved mirrors, which focused the beam to a 5×6 μm point. The incoming beam intensity, I_0, was measured using a diamond X-ray beam monitor placed upstream of the pressure chamber. The outgoing beam intensity, I, downstream of the pressure chamber was measured with a PIN diode. As the X-ray beam passed through the fuel spray, photons were absorbed through the process of photoelectric absorption, attenuating the beam by an amount related to the quantity of fuel in the beam path of length z. When the XRR measurement is normalized by ρ_f, the quantity has been shown to be proportional to the liquid volume fraction, LVF:

$$PMD = \rho_f \frac{V_{liq}}{V} z = \rho_f (LVF) z. \tag{4.26}$$

4.3.2.2 Ultra-small-Angle X-ray Scattering (USAXS)

Researchers at Argonne performed USAXS measurements at the 9-ID beamline of the APS in order to characterize the total surface area per sample volume of the spray (Kastengren et al. 2017). By combining the surface area measured with USAXS and volume of the droplets measured with radiography, the SMD of the droplet size distribution can be determined, where the SMD is defined as

$$SMD = 6\frac{V}{A} \tag{4.27}$$

where V and A are the volume and surface area of the droplets within the measurement volume, respectively.

A schematic of the experimental setup is shown in Fig. 4.13. A beam of X-rays at 21 keV was first shaped into a 50×500 μm H × V spot by a set of high-precision 2D slits. The beam was then collimated using a pair of Si (220) crystals

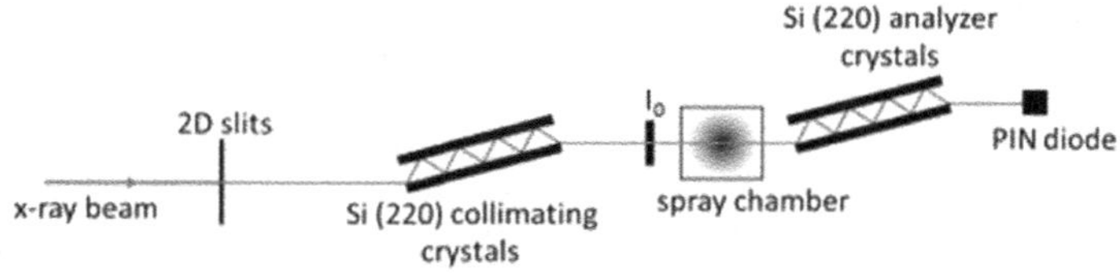

Fig. 4.13 A schematic of the ultra-small-angle X-ray scattering experiment (Kastengren et al. 2017). Reproduced with permission of Elsevier, from Kastengren et al. Kastengren et al. (2017)

before interacting with the spray. As the beam passed through the spray, X-rays were scattered at small angles. The scattered X-rays were filtered downstream with a pair of Si (220) analyzer crystals, and the resulting intensity measured with a detector. The 9-ID beamline is equipped with a Bonse–Hart instrument to measure the scattering intensity, $I_{scat}(q)$, as a function of scattering vector, q (Ilavsky et al. 2009). The pair of analyzer crystals were rotated to vary q between $1 \times 10^{-4}\text{Å}^{-1} < q < 1 \times 10^{-2}\text{Å}^{-1}$ with a step size of $1 \times 10^{-5}\text{Å}^{-1}$ at low q, with increasing step size for larger q. The scattered beam intensity as a function of q was measured at axial distances ranging from 1 to 20 mm downstream of the injection nozzle tip, at the centerline of the spray. Data were recorded in a 1-ms interval during the steady-state portion of the spray event. Background measurements were also recorded over 80 ms before each scan to account for any changes within the measurement domain caused by previous spray events. Once $I_{scat}(q)$ is measured, post-processing is performed using the Irena data analysis package (Ilavsky and Jemian 2009) in order to obtain the surface area per volume of fuel droplets.

In order to find the spray centerline during USAXS measurements, a transverse scan at fixed q was also recorded at each axial location of interest. The spray centerline was taken to be the transverse location at which the beam intensity was a maximum, i.e., the location with the highest droplet density. Radiography measurements were temporally averaged during the steady portion of the spray event for the SMD calculation. The transverse profiles from the USAXS and radiography measurements were each centered about their full width at half maximum in order to index the profiles onto the same coordinate system. Because the transverse location of the USAXS measurement is known at each axial distance, the corresponding radiography data at that location may be found. The USAXS measurement point is assumed to be in the center of the $50 \times 500\,\mu\text{m}$ measurement volume. All measured radiography points that fall within this window are averaged to arrive at one value of the pathlength, with interpolation and appropriate weighted averaging performed to accurately incorporate the edges of the measurement volume. The pathlength of fuel obtained from the radiography measurements provides the line-of-sight integrated volume of droplets in a sample of unit thickness. The USAXS measurements provide the line-of-sight surface area per volume of droplets, likewise in a sample of unit thickness. Thus, the two measurements can be combined as per Eq. 4.27 to arrive at a line-of-sight integrated SMD value at each measured axial location. Based on an uncertainty analysis, the resulting uncertainty in the measured SMD data was estimated to be about 6% (Kastengren et al. 2017). A sample set of USAXS measurements conducted along the spray centerline for the ECN Spray A injector under Case 1–3 conditions is shown in Fig. 4.14.

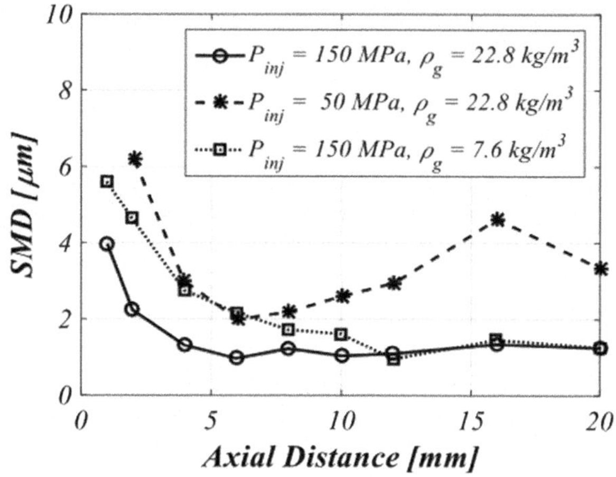

Fig. 4.14 Comparison of axial distributions of USAXS SMD measurements along the spray centerline (Kastengren et al. 2017). Data are shown for a range of ambient density and injection pressure conditions for the ECN Spray A injector. Reproduced with permission of Elsevier, from Magnotti and Genzale (2017)

4.3.3 Measured SMD Profiles Under Conventional Diesel Operating Conditions

Kastengren et al. (2017) employed joint X-ray radiography and USAXS measurements to characterize the spray structure for the ECN Spray A injector under injection and ambient conditions relevant to conventional diesel operation. The USAXS measurements quantifying the SMD distribution along the spray centerline, as shown in Fig. 4.14, were conducted under the Case 1 − 3 conditions defined in Table 4.2. The measurements begin 1 mm from the nozzle ($x/d_j \sim 11$) and extend downstream along the centerline.

Across all three conditions, the SMD measurements show droplet sizes in the near-nozzle region that are much smaller than the nozzle diameter (90 μm). As can be seen in Fig. 4.14 for the Case 1 condition ($P_{inj} = 150$ MPa, $\rho_g = 22.8$ kg/m^3), the measurement indicates a rapid decrease in SMD within the first 4 mm. This suggests that the initial breakup process does not occur immediately at the injector exit, but rather occurs over a finite time and distance. The length of the breakup region has not been well characterized by previous experimental measurements and is a salient spray feature that can be used to significantly improve model validation efforts, as will be explored in Sect. 4.4.1. For example, faithful representation of the spray breakup region over a range of ambient and injection conditions would suggest that the timescale and rate of spray disintegration are well captured by the primary atomization spray model. For the first time, these USAXS measurements provide quantitative details of the atomization process in the near-nozzle region.

Downstream of the breakup region, interesting trends can be observed in the measured SMD profiles. For parametric variations in ambient density (Case 1 vs. Case 3 conditions), the measured SMD reaches a consistent minimum value that remains relatively constant along the spray centerline. However, a decrease in injection pressure (Case 1 vs. Case 2 conditions) is found to provide an increase in SMD

in downstream regions of the spray. These observed trends are due to the complex interplay among primary and secondary breakup, as well as drop–drop interactions via collisions and coalescence. As will be detailed in Sect. 4.4.1, these newly available measurements were used by Magnotti and Genzale (2017) to evaluate the predictive capability of existing spray models.

4.3.4 Measured SMD Profiles Under Low-Temperature Combustion-Relevant Conditions

Experimentalists from Argonne also employed joint X-ray radiography and USAXS measurements to characterize the spray structure for the ECN Spray D injector under injection and ambient conditions relevant to LTC operating conditions ($\rho_f/\rho_g >$ 300) (Magnotti et al. 2017; Kim et al. 2018). USAXS measurements of SMD along the spray centerline for Spray D #209133 are shown in Fig. 4.15 for Case $4-9$ conditions, along with curves fitted to the data to illustrate the general trends in droplet size evolution (Magnotti et al. 2017). In general, the measured SMD is observed to decrease with increasing axial distance from the nozzle exit, indicating continual breakup of the spray and droplets. The SMD along the spray centerline is also seen to increase with decreasing ambient pressure, P_{amb}, and injection pressure, P_{inj}. However, the experimental measurements suggest a transition in droplet formation behavior as ρ_f/ρ_g increases beyond 100, which corresponds to P_{amb} less than or equal to 0.2 MPa. As mentioned in Sect. 4.2.1.2, for ρ_f/ρ_g greater than 500, Faeth et al. have shown that aerodynamic forces do not exert a significant influence on the droplet formation process (Wu and Faeth 1993). The similarity in SMD distributions from the USAXS measurements for P_{inj} of 50 MPa and P_{amb} of 0.1 and 0.2 MPa conditions suggests that changing the aerodynamic inertia by a factor of two does not appreciably change the droplet formation process. However, for P_{inj} of 150 MPa, a larger change in SMD is observed when P_{amb} is increased from 0.1 to 0.2 MPa.

In order to extract more detailed information about the local sensitivity of the SMD to changes in injection and ambient conditions, the axial distribution of SMD was fit to a two-term exponential function for each condition. The curve fit is of the form

$$f(x) = Ae^{Bx} + Ce^{Dx} \tag{4.28}$$

where A, B, C, and D are unique fitting parameters for each condition. An additional point of 186 μm at the nozzle exit ($x = 0$) was added to each data set to capture the rapid decrease in SMD from its initial value of the nozzle outlet diameter. The two-term exponential function captures the data well, with an R^2-value greater than 0.99 for all conditions.

Using these curve fits, the local sensitivity of SMD to changes in injection and ambient conditions was quantified by Magnotti and co-workers by evaluating the percent change in SMD from the selected Case 9 baseline condition ($P_{inj} = 50$ MPa,

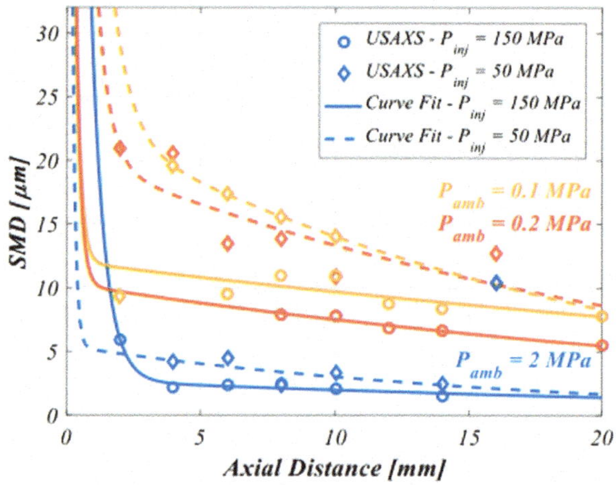

Fig. 4.15 SMD measurements from X-ray measurements conducted at the APS are shown for a range of ambient and injection conditions along the spray centerline (Magnotti et al. 2017). A two-term exponential function is fit to the data (solid and dashed lines)

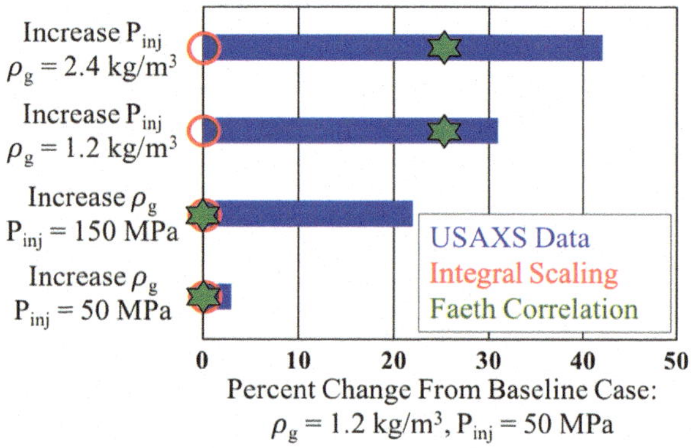

Fig. 4.16 Local sensitivities of SMD to changes in P_{inj} and P_{amb} for ρ_f/ρ_g conditions greater than 300, as indicated by the X-ray measurements and scalings for turbulent length scales (Magnotti 2017)

$\rho_g = 1.2$ kg/m^3) (Magnotti et al. 2017). The local sensitivities of the measured centerline SMD to changes in P_{inj} and P_{amb} were used to better understand the appropriate scaling for the droplet formation process. In particular, an insensitivity to changes in ρ_g would suggest that aerodynamic inertial forces have minimal influence on the resultant SMD and that the droplet formation process is likely controlled by nozzle-generated turbulence. The results of the computed sensitivities for the USAXS measurements are shown in Fig. 4.16. Indeed, these results confirm previous qualitative observations. At a fixed P_{inj} of 50 MPa, the SMD is not strongly influenced by the change in P_{amb} from 0.1 to 0.2 MPa. However, the results suggest a joint sensitivity of SMD to changes in P_{amb} and P_{inj}, as a higher sensitivity to P_{amb} is indicated for $P_{inj} = 150$ MPa. The measured centerline SMD is also observed to respond to changes in P_{inj}. In Sect. 4.4.2, analysis from Magnotti (2017), where the USAXS measurements conducted under $\rho_f/\rho_g > 300$ conditions are directly compared with proposed scalings within the turbulence spectrum, is summarized in order to highlight potential length scales that can characterize the turbulence-induced spray breakup process.

4.4 Development of an Improved Hybrid Spray Breakup Model

Using the experimental data detailed in Sect. 4.3, it is possible to evaluate the predictive capability of existing spray breakup theories and models through their ability to reproduce the observed trends and features in the USAXS measurements. This section summarizes selected computational investigations that have informed the development of an improved hybrid spray breakup formulation.

4.4.1 Assessment of Existing Spray Models Under Conventional Diesel Operating Conditions

The USAXS measurements from Kastengren and co-workers (Kastengren et al. 2017), detailed in Sect. 4.3.3, provide newly available information about the internal structure of the spray under high ambient density conditions relevant to conventional diesel operation. These data were utilized by Magnotti and Genzale (2017) to test the physical appropriateness of the KH primary spray breakup model and the ability of other existing spray submodels to capture the experimentally observed features in the spray structure. The key findings from this study are summarized in this section.

First, the computational investigation revealed the ability of a carefully calibrated KH model to capture the salient features in the SMD profile. As shown in Fig. 4.17, two spray breakup model constants were shown significantly to influence the predicted SMD distributions: B_0, the KH primary breakup droplet size constant, and B_1, the KH breakup time constant, as defined in Eqs. 4.12 and 4.11, respectively. It was noted that increasing B_1 elongates the initial breakup process in the near-nozzle region and spatially delays the location where the minimum SMD is achieved. Better agreement was achieved with the measured drop size decrease within the spray formation region when the B_1 constant was increased to a value of 60. Increasing B_0 from a standard value of 0.61 to a value of 1.0 yielded improved agreement between the predicted and measured minimum SMD along the spray centerline. These result suggested that the internal structure of the spray, as characterized by their centerline SMD profile, can be well represented when droplet sizes formed by KH breakup are assumed to be directly proportional to Λ_{KH}.

Second, the predictive capability of the calibrated KH spray model, acting together with other spray and droplet models, was tested and compared to the USAXS data over a range of injection and ambient conditions. In particular, the trends in predicted SMD along the spray centerline were evaluated when secondary droplet breakup via

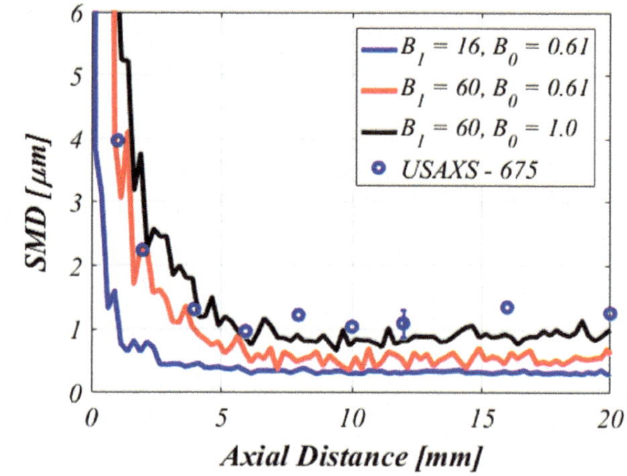

Fig. 4.17 Comparison of calibrated KH-RT spray models with measured axial distributions of SMD at the Case 1 condition ($P_{inj} = 150\,\text{MPa}$ and $\rho_g = 22.8\,\text{kg/m}^3$) along the spray centerline (Magnotti and Genzale 2017). The spray model predictions do not include the effects of collisions. Reproduced with permission of Elsevier, from Magnotti and Genzale (2017)

the RT breakup mechanism (Beale and Reitz 1999) and droplet collisions and coalescence, modeled using the O'Rourke formulation (O'Rourke and Bracco 1980), were included or excluded in the spray model. Comparisons of the measured SMD distributions along the spray centerline with predictions from the calibrated model, as shown in Fig. 4.18 for the (a) Case 1, (b) Case 2, and (c) Case 3 conditions, highlight the ability of a model to capture parametric variations in injection and ambient conditions. Overall, the KH-RT spray model without the influence of collisions (solid line) was found to represent the measured centerline SMD distribution across all conditions. Of particular note, when the injection pressure was decreased to 50 MPa (Fig. 4.18b), the model predicted a slight increase in SMD that is consistent with the measurements. Because no droplet-droplet collisions were modeled, increasing SMD with axial distance was due to the momentum of larger droplets overtaking slower neighboring droplets. When the effects due to droplet collisions were included (dotted line), a substantial increase in droplet size was predicted that was inconsistent with the experimental measurements across the range of evaluated conditions. Although a collision model might be able to replicate experimentally measured SMD at a single condition, it was concluded that collisions could not explain the experimentally observed trend in centerline SMD distribution with respect to changes in injection pressure. Additionally, the KH-RT model without the influence of collisions was found to reproduce the experimentally observed features when the ambient density was parametrically varied, namely the insensitivity of the minimum SMD and increased length of the breakup region with decreasing ambient density.

Although the calibrated KH-RT spray model was found to capture many of the experimentally observed trends, the findings from Magnotti and Genzale do not confirm that the initial spray breakup process is completely described by the Kelvin–Helmholtz mechanism under all conditions. In fact, several details in the model predictions suggest that KH is not likely to be the only mechanism influencing the primary breakup process. Although the initial spray formation region is well captured

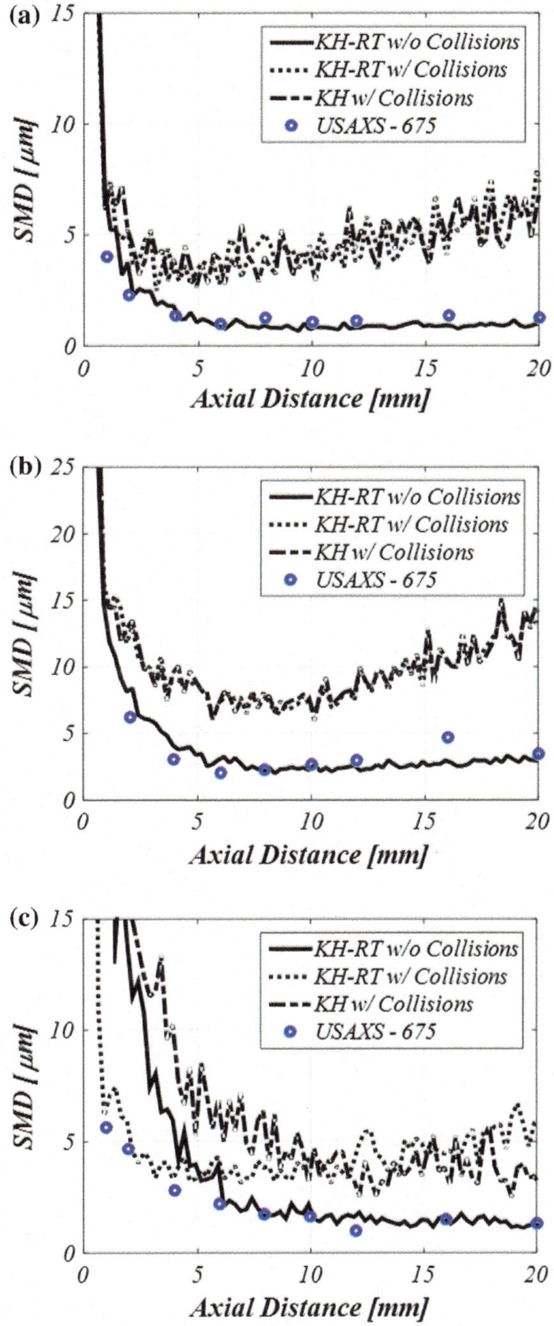

Fig. 4.18 Comparison of different selected spray models with measured axial distributions of SMD at the **a** Case 1, **b** Case 2, and **c** Case 3 conditions (Magnotti and Genzale 2017). The comparison is conducted along the spray centerline for the spray model employing the KH spray model constants $B_1 = 60$ and $B_0 = 1.0$. Reproduced with permission of Elsevier, from Magnotti and Genzale (2017)

by the carefully calibrated model ($B_1 = 60$ and $B_0 = 1.0$) for the Case 1 and Case 2 conditions, as shown in Fig. 4.18a and b, discrepancies can be seen between the predicted and measured spray structure in the near-nozzle region for the Case 3 condition with a lower ambient density, as shown in Fig. 4.18c. While the minimum SMD and length of the spray breakup region are well captured by the model, the rate of droplet size decrease within the spray breakup region is underpredicted and results in the overprediction of SMD in the near-nozzle region relative to the measurements.

These results suggest that a spray breakup model premised on aerodynamic breakup alone, such as the KH model, is not fully predictive in representing the spray structure, particularly at ambient density conditions away from conventional diesel operation. It is likely that the KH mechanism cannot predict the correct trends in isolation and that other primary breakup mechanisms (e.g., turbulence-induced breakup) augment the breakup process in diesel sprays. By analyzing the sensitivity of the USAXS measurements presented in Sect. 4.3.4 under low ambient density conditions, Magnotti was able to identify an improved formulation for describing the size of droplet formed for turbulence-induced breakup (Magnotti 2017). The details of this work are summarized in the next section.

4.4.2 *Evaluation of Turbulent Length Scale Relations Under Low-Temperature Combustion-Relevant Conditions*

As previously mentioned, comparison of measured sensitivities from the USAXS measurements with theoretical turbulent scalings was conducted by Magnotti (Magnotti 2017) to provide clearer insight into the mechanisms driving the turbulent breakup process. For example, the majority of spray models that consider the influence of nozzle-generated turbulence on the spray breakup process assume that the size of primary droplets scales with the largest eddies in the turbulence spectrum, such as the Huh–Gosman and KH-ACT models described in Sect. 4.2.3 (Huh and Gosman 1991; Som and Aggarwal 2010; Huh et al. 1998). In these models, the resultant droplet size scales with the dimension of the flow and is independent of the Reynolds number and ambient environment properties (Pope 2000). This can be demonstrated by evaluating the dependence of the turbulent integral length scale, L_t, on injection and ambient parameters. L_t can be defined using Eq. 4.13. Estimations of the turbulent kinetic energy and dissipation rates at the nozzle exit, K_0 and ϵ_0, respectively, can help determine the functional dependencies of L_t. Equations 4.22 and 4.23 can then be inserted into the integral scaling as defined in Eq. 4.13 to yield a relationship between L_t and the injection and nozzle parameters:

$$L_t \propto \frac{d_j}{(L/d_j)^{0.5}} \left[\frac{1}{c_d^2} - K_c - (1 - s^2) \right]^{0.5} \tag{4.29}$$

Equation 4.29 provides insight into the physical dependencies of the size of eddies within the energy-containing range. If cavitation does not influence the flow conditions at the nozzle exit, c_d will remain nominally constant across all P_{inj} and P_{amb} conditions (Payri et al. 2004; Schmidt 1997). Therefore, L_t is independent of U_{inj} or ρ_g and is only a function of geometric features of the nozzle, such as the L, d_j, and s of the nozzle, as well as the inlet nozzle radius of curvature. Other researchers have noted that for a fixed flow configuration, increases in U_{inj} (Re_{fd}) do not largely affect integral scales, but do increase the available kinetic energy to distort the liquid–gas interface (Wu et al. 1992; Ruff et al. 1987; Tennekes and Lumley 1972).

As a result, the size of droplets formed from the largest eddies within the energy-containing range would be insensitive to changes in P_{inj} and P_{amb}, as shown in Fig. 4.16. This theoretical scaling suggests a larger insensitivity with respect to changes in the injection and ambient conditions than indicated by the USAXS measurements in Fig. 4.16. Therefore, the centerline distributions of SMD are not likely formed by turbulent eddies within the energy-containing range. As a result, any turbulent breakup model that assumes that the resultant droplets are proportional in size to the integral length scale, such as the Huh–Gosman or KH-ACT spray models (Huh and Gosman 1991; Som and Aggarwal 2010; Huh et al. 1998), is not likely able to predict the experimentally observed trends for diesel sprays, particularly for injection into ambient environments deviating from conventional diesel operating conditions ($\rho_f/\rho_g > 100$).

The experimental measurements conducted by Wu and Faeth suggest that droplet-forming eddies do not exist within the energy-containing range of the turbulence spectrum (Wu et al. 1992; Wu and Faeth 1993). For injection into conditions where $\rho_f/\rho_g > 500$, the size of primary droplets was observed to decrease in size as U_{inj} was increased, although the large-scale disturbances on the jet surface were observed to be similar in size for the range of evaluated injection conditions (Wu et al. 1992). Based on their analysis of the measured liquid surface ligament and droplet properties at the onset of turbulent breakup from holographic imaging, Wu and Faeth developed an empirical correlation to relate the size of droplet-forming eddies, L_{Faeth}, to nozzle exit turbulence properties (Wu and Faeth 1993), as defined below:

$$\frac{L_{Faeth}}{\Lambda} = C_{sx}\left(\frac{x}{\Lambda We_{f\Lambda}^{1/2}}\right)^{2/3} \tag{4.30}$$

where Λ is the radial integral length scale, C_{sx} is an empirical constant, and $We_{f\Lambda}$ is the Λ-based liquid Weber number ($\rho_f U_{inj}^2 \Lambda/\sigma$). Based on experimental data across a wide range of Re_f ($9 \cdot 10^4$–$5.3 \cdot 10^5$) and ρ_f/ρ_g (104–6230) conditions, Wu and Faeth determined that the empirical correlation in Eq. 4.30 best fit the entire experimental data set when C_{sx} was set to 0.65. By estimating U_{inj} using the Bernoulli equation, a relationship among L_{Faeth}, P_{inj}, and P_{amb} can be approximated with the following relation:

$$L_{Faeth} \propto We_{f\Lambda}^{-1/3} \propto U_{inj}^{-2/3} \propto (P_{inj} - P_{amb})^{-1/3}. \tag{4.31}$$

Using the relation in Eq. 4.31, the sensitivity to changes in P_{inj} and P_{amb} can be calculated to characterize the response of the Wu and Faeth correlation for non-aerodynamic primary breakup to changes in injection and ambient conditions. For a fixed P_{inj} of 50 MPa or 150 MPa, a slightly larger insensitivity of the centerline SMD to changes in P_{amb} is indicated than is suggested by the measurements in Fig. 4.16. However, a threefold increase in P_{inj} at a fixed P_{amb} of 0.1 or 0.2 MPa results in an approximately 26% percent decrease in SMD. Comparison with predicted sensitivity of the USAXS measurements with respect to changes in P_{inj}, as shown in Fig. 4.16, reveals improved agreement between the measured and predicted sensitivities. It is therefore possible for a properly calibrated L_{Faeth} correlation, as defined in Eq. 4.30, to well characterize the experimentally observed sensitivities of SMD to changes in injection and ambient conditions for $\rho_f/\rho_g > 300$.

Although Wu and Faeth's phenomenological framework for characterizing turbulent primary atomization was developed from a database of sprays issued from idealized nozzles, with long L/d_j to ensure fully developed turbulent flow at the nozzle exit, comparisons between the empirical correlations from Wu and Faeth and the diesel SMD measurements in Fig. 4.15 suggest that the geometric differences with practical diesel injector hardware do not strongly influence the resultant spray atomization process. These results are in agreement with experimental findings from Wu and co-workers, where the condition at the onset of turbulent breakup and the size of primary and secondary droplets were found to be relatively independent of L/d_j for ρ_f/ρ_g conditions greater than 100 (Wu et al. 1995). As a result, the empirical correlations from Wu and Faeth may serve as a foundation for representing turbulent primary atomization in diesel spray breakup models.

4.4.3 Validation of the KH-Faeth Model

Based on the success of the KH model to represent the detailed spray structure under high ambient density conditions (Magnotti and Genzale 2017), as summarized in Sect. 4.4.1, and the consistent sensitivities indicated by the empirical correlations from Wu and Faeth (Wu and Faeth 1993) and USAXS measurements under low ambient density conditions (Magnotti et al. 2017), as discussed in Sect. 4.4.2, a new hybrid spray breakup modeling formulation was proposed by Magnotti (2017). In the "KH-Faeth" hybrid spray breakup model, the competition between the aerodynamic and turbulent breakup mechanisms is represented in a similar fashion to the KH-ACT model (Som and Aggarwal 2010). At each instant in time, the dominant mechanism is selected based on the maximum breakup rate, as defined below:

$$\frac{L_A}{\tau_A} = \max\left\{\frac{a - r_c}{\tau_{KH}}, \frac{L_{Faeth}}{\tau_{Faeth}}\right\} \tag{4.32}$$

where the breakup parameters for the KH model have been previously defined in Eqs. 4.10–4.12, and the characteristic length scale for the turbulence-induced breakup process, L_{Faeth}, was defined in Eq. 4.30. The characteristic timescale for turbulence-induced breakup, τ_{Faeth}, is described using a Rayleigh timescale from Wu and Faeth (Wu and Faeth 1993), which describes the time required for a droplet to form from a ligament of size L_{Faeth}, as defined below:

$$\tau_{Faeth} = C_{\tau}\sqrt{\rho_f \frac{L_{Faeth}^3}{\sigma}} \tag{4.33}$$

where C_{τ} is an empirical time constant.

Recently, the KH-Faeth model was implemented by (Kim et al. 2018) into the OpenFOAM modeling framework using the user-defined library Lib-ICE, which was developed by researchers at Politecnico di Milano (Lucchini et al. 2011; D'Errico et al. 2014). Validation of the KH-Faeth model against X-ray experimental data is now summarized.

In order to test the predictive capability of the newly developed hybrid spray model, the spray structures predicted by the KH-Faeth model were compared with those predicted by the KH model and those measured via USAXS across a broad range of ambient and injection conditions for the ECN Spray D injector (Magnotti et al. 2017). Comparisons among measured and predicted axial SMD profiles along the spray centerline are shown in Fig. 4.19. Overall, the KH-Faeth model was found to well match the experimental trends. While the SMD profiles predicted by the KH model were found to be overly sensitive to changes in injection pressure and ambient density, the predictions from the KH-Faeth model exhibited a similar response as the USAXS measurements. The improved performance of the KH-Faeth model over the KH model highlights the promise of this new hybrid model formulation for predicting primary atomization across the range of evaluated ambient and injection conditions, which may be found in current and future LTC operating conditions.

Although the KH-Faeth model was found to perform better than the KH model, Kim and co-workers noted some deviations among the USAXS measurements and KH-Faeth model predictions at the highest ambient density conditions, as shown in Fig. 4.19e and f. Based on the findings from Magnotti and Genzale (2017), Magnotti et al. (2017) and Kim et al. (2018), three likely candidates can be identified to explain the discrepancies. First, uncertainty in the injection boundary conditions can influence the primary breakup process and initial spray development. Indeed, Kim and co-workers noted an underprediction of the spray penetration under the Case 5 condition ($P_{inj} = 50\,\text{MPa}$, $P_{amb} = 2\,\text{MPa}$), which indicates a disagreement in the measured and predicted momentum exchange between the liquid and gas phases. Future computational studies that can model the internal flow development and nozzle exit conditions for the ECN Spray D #209133 injector could help reduce uncertainty in the prescribed injection velocity and spray spreading angle. Second, a detailed analysis of KH-Faeth model predictions revealed a rapid primary breakup process and short spray formation region. Careful calibration of the turbulent primary breakup time constant, C_{τ}, as defined in Eq. 4.33, could yield improved agreement among the measured and predicted SMD profiles and spray penetration. More

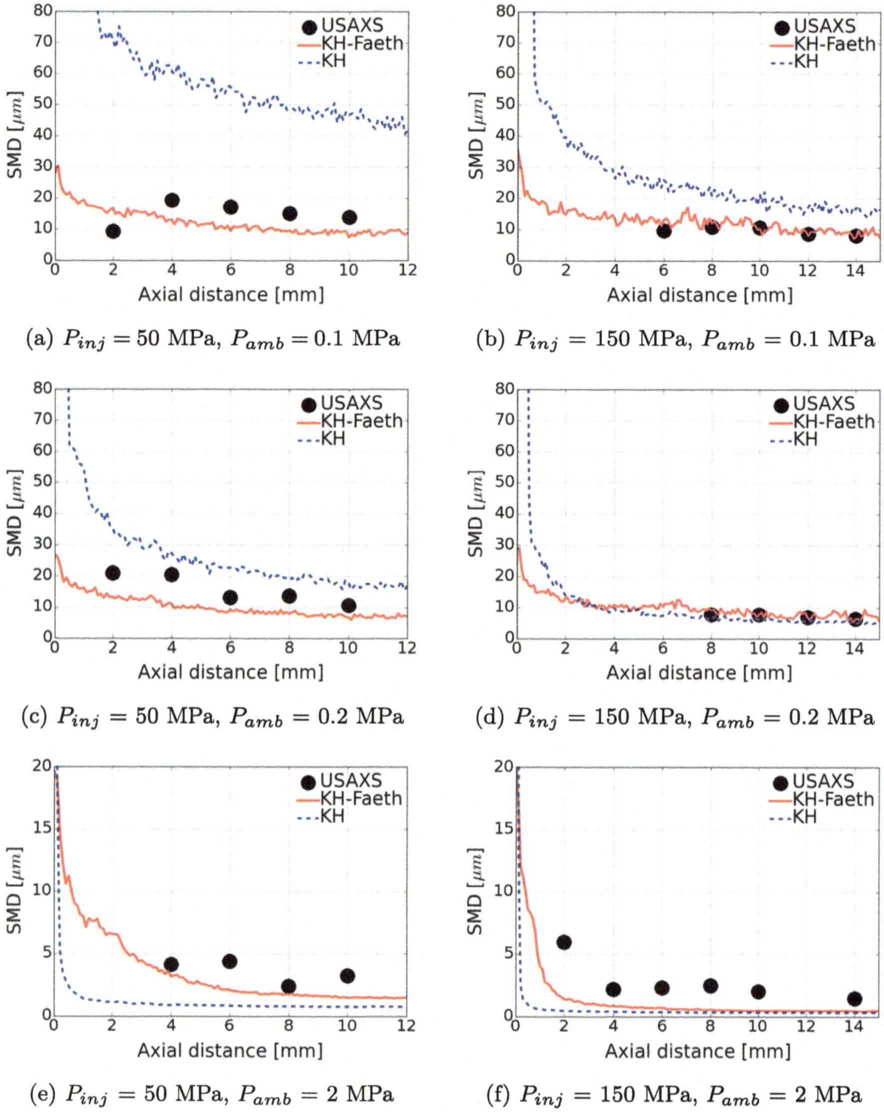

Fig. 4.19 Comparison of axial SMD distributions along the spray centerline measured using the USAXS technique and predicted by the KH and KH-Faeth (Kim et al. 2018). Adapted with permission from Kim et al. (2018)

specifically, increasing C_τ would elongate the primary breakup process and spray formation region, which would serve to increase the SMD values in the near-nozzle region and accelerate spray penetration during the early injection transient. Finally, because it is not possible to directly observe the interaction between competing primary breakup mechanisms, there is great uncertainty in how the hybrid spray breakup process should be represented in a model (Som 2009; Magnotti 2017). Although the KH-Faeth model assumes that only a single breakup mechanism can act upon a computational parcel to decrease its size, other model formulations, such as the Huh–Gosman (Huh and Gosman 1991; Huh et al. 1998) and Bianchi models (Bianchi and Pelloni 1999), represent the outcome of the primary breakup process as a weighted average between the aerodynamic and turbulent breakup mechanisms. Future computational investigations could explore the influence of the representation of the hybrid breakup process on the predicted spray.

References

Amsden AA, O'Rourke P, Butler T (1989) Kiva-ii: A computer program for chemically reacting flows with sprays. Technical report la-11560-ms, Los Alamos National Laboratory

Beale JC, Reitz RD (1999) Modeling spray atomization with the kelvin-helmholtz/rayleigh-taylor hybrid model. Atomization Sprays 9:623–650

Behrendt T, Carl M, Heinze J, Hassa C, Optical measurements of the reacting two-phase flow in a realistic gas turbine combustor at elevated pressures. Atomization and Sprays 16:5

Bergwerk W (1959) Flow pattern in diesel nozzle spray holes. Proc Inst Mech Eng 173:655

Bianchi GM, Pelloni P (1999) Modeling the diesel fuel spray breakup by using a hybrid model. SAE technical paper 1999-01-0226

Bode M, Diewald F, Broll DO, Heyse JF, Chenadec VL (2014) Influence of the injector geometry on primary breakup in diesel injector systems. SAE technical paper, no. 2014-01-1427

Borenstein G (2012) The more pixels law: gigapixel cameras and the 21st century reality effect. http://makematics.com/drafts/more_pixels_law.html

Borodin V, Dityakin Y (1951) Unstable capillary waves on surface of separation of two viscous liquids. N.A.C.A, TM

Castleman RA (1931) The mechanism of the atomization of liquids. US Department of Commerce, Bur Stan

Castleman R (1932) Mechanism of atomization accompanying solid injection. N.A.C.A, Report, p 440

Crua C, Heikal M, Gold M (2015) Microscopic imaging of the initial stage of diesel spray formation. Fuel 157:140–150

DeJuhasz K (1931) Dispersion of sprays in solid injection oil engines. Trans Am Soc Mech Eng 53

Dempsey AB, Curran SJ, Wagner RM (2016) A perspective on the range of gasoline compression ignition combustion strategies for high engine efficiency and low nox and soot emissions: effects of in-cylinder fuel stratification. Int J Engine Res

D'Errico G, Lucchini T, Jangi M, Contino F (2014) Comparison of well-mixed and multiple representative interactive flamelet approaches for diesel spray combustion modelling. Combustion Theory and Modelling 18:65–88

Drazin PG, Reid WH (2004) Hydrodynamic stability. Cambridge university press

Duke D, Swantek A, Tilocco F, Kastengren A et al (2014) X-ray imaging of cavitation in diesel injectors. SAE Int J Eng 7(2):1003–1016

Dukowicz JK (1980) A particle-fluid numerical model for liquid sprays. J Comput Phys 35(2):229–253

Eckhause J, Reitz R (1995) Modeling heat transfer to impinging fuel sprays in direct injection engines. Atomization Sprays

Engine Combustion Network (2018) Engine Combustion Network Experimental Data Archive. http://www.sandia.gov/ECN

Faeth G, Hsiang L-P, Wu P-K (1995) Structure and breakup properties of sprays. Int J Multiph Flow 21:99–127

Fluent (2005) FLUENT 6.2 User's Guide

Genzale C, Pickett L, Kook S (2010) Liquid penetration of diesel and biodiesel sprays at late-cycle post-injection conditions. SAE Int J Engines 3(1):479

Giffen E, Muraszew A (1970) The atomization of liquid fuels. Wiley

Grant R, Middleman S (1966) Newtonian jet stability A.I.Ch.E. 12

Haenlein A (1932) On the disruption of a liquid jet. N.A.C.A, TM, p 659

Hiroyasu H, Kadota T (1974) Fuel droplet size distribution in diesel combustion chamber. SAE technical paper 740715

Hooper P, Eisenklam P (1958) Ministry of supply dggw report. EMR/58/JRL/42

Huh KY, Gosman A (1991) A phenomenological model of diesel spray atomization. in Multiphase Flows. Tsukuba, Japan

Huh K, Lee E, Koo J (1998) Diesel spray atomization model considering nozzle exit turbulence conditions. Atomization Sprays 8(4):453–469

Ilavsky J, Jemian P (2009) Irena: tool suite for modeling and analysis of small-angle scattering. J Appl Crystallogr 42(2):347–353

Ilavsky J, Jemian P, Allen A, Zhang F, Levine L, Long G (2009) Ultra-small-angle x-ray scattering at the advanced photon source. J. Appl Crystallogr

Iyer VA, Post SL, Abraham J (2000) Is the liquid penetration in diesel sprays mixing controlled? Proc Combust Inst 28:1111–1118

Kastengren AL, Tilocco FZ, Powell CF, Manin J, Pickett LM, Payri R, Bazyn T (2012) Engine combustion network (ECN): measurements of nozzle geometry and hydraulic behavior. Atomization Sprays 22(12):1011–1052

Kastengren A, Tilocco F, Duke D, Powell C, Zhang X, Moon S (2014) Time-resolved x-ray radiography of sprays from engine combustion network spray a diesel injectors. Atomization Sprays 24(3):251–272

Kastengren A, Ilavskya J, Vierab JP, Payri R, Duke D, Swantek A, Tilocco FZ, Sovis N, Powell C (2017) Measurements of droplet size in shear-driven atomization using ultra small angle x-ray scattering. Int J Mulitphase Flows

Kastengren A, Powell C (2014) Synchrotron x-ray techniques for fluid dynamics. Exp Fluids 55(3)

Kastengren A, Powell C, Arms D, Dufresne E, Gibson H, Wang J (2012) The 7bm beamline at the aps: a facility for time-resolved fluid dynamics measurements. J Synchrotron Radiat. 19(4)

Kastengren A, Powell C, Riedel T, Cheong S-K, Im K-S, Liu X, Wang Y Wang J (2008) Nozzle geometry and injection duration effects on diesel sprays measured by x-ray radiography. J Fluids Eng 130(4)

Kim S, Magnotti G, Martinez G, Yraguen B, Knox B, Kastengren A, Matusik K, Sforzo B, Powell C, Lucchini T, D'Errico G, Genzale C, Validation of a new turbulence-induced Lagrangian primary breakup model for diesel spray atomization. In: 14th International Conference on Liquid Atomization and Spray Systems (ICLASS) (Chicago, IL), July 2018

Kim J, Nishida K, Hiroyasu H (1997) Characteristics of the internal flow in a diesel injection nozzle. Proceedings of ICLASS 1997

Kodavasal J, Kolodziej C, Ciatti S, Som S (2014) CFD simulation of gasoline compression ignition. In: ASME 2014 Internal Combustion Engine Division Fall Technical Conference. American Society of Mechanical Engineers, pp V002T06A008–V002T06A008

Kong S-C, Han Z, Reitz R (1995) The development and application of a diesel ignition and combustion model for multidimensional engine simulations. SAE Paper 950278

Lee D, Spencer R (1933) Photomicrographic studies of fuel sprays. N.A.C.A, Report No, p 454

Levich V (1962) Physicochemical Hydrodyn. Prentice Hall

Liu A, Mather D, Reitz R (1993) Modeling the effects of drop drag and breakup on fuel sprays. SAE Paper 930072

Lucchini T, D'Errico G, Ettorre D (2011) Numerical investigation of the spray-mesh-turbulence interactions for high-pressure, evaporating sprays at engine conditions. Int J Heat Fluid Flow 32:285–297

Magnotti GM, Genzale CL (2017) Detailed assessment of diesel spray atomization models using visible and x-ray extinction measurements. Int J Multiph Flow

Magnotti GM, Genzale CL (2017) Exploration of turbulent atomization mechanisms for diesel spray simulations. In: Proceedings of the 2017 SAE World Congress (Detroit, MI)

Magnotti GM (2017) Modeling the Influence of Nozzle-Generated Turbulence on Diesel Sprays. Ph.D. thesis, Georgia Institute of Technology, Atlanta, GA

Magnotti GM, Matusik KE, Duke DJ, Knox BW, Martinez GL, Powell CF, Kastengren AL, Genzale CL (2017) Modeling the influence of nozzle-generated turbulence on diesel sprays. in ILASS Americas 29th Annual Conference. Atlanta, GA

Matusik KE, Duke DJ, Kastengren AL, Sovis N, Swantek AB, Powell CF (2017) High-resolution X-ray tomography of Engine Combustion Network diesel injectors. Int J Engine Res

Matusik K, Duke D, Swantek A, Powell C, Kastengren A (2016) High resolution x-ray tomography of injection nozzles. In: 28th ILASS-Americas Conference. Dearborn, MI

McCarthy M, Molloy N (1974) Review of stability of liquid jets and the influence of nozzle design. Chem Eng J 7

Munnannur A (2007) Droplet Collision Modeling in Multi-Dimensional Engine Spray Computations. Ph.D. thesis, University of Wisconsin-Madison,

Munson B, Okiishi T, Huebsch W, Rothmayer A (2013) Fundamentals of fluid mechanics. Wiley, Hoboken, NJ

Ning W (2007) Development of a next-generation spray and atomization model using an Eulerian-Lagrangian methodology. Ph.D. thesis, University of Wisconsin-Madison

NIST (2018) Thermophysical properties of dodecane. http://webbook.nist.gov/cgi/fluid.cgi?ID=C112403&Action=Page

O'Rourke P, Bracco F (1980) Modeling of drop interactions in thick sprays and a comparison with experiments. I Mech E 9:101–116

Pastor J, Payri R, Salavert J, Manin J (2012) Evaluation of natural and tracer fluorescent emission methods for droplet size measurements in a diesel spray. Int J Automot Technol 13(5):713–724

Patterson M, Kong S-C, Hampson G, Reitz R (1994) Modeling the effects of fuel injection characteristics on diesel engine soot and nox emissions. SAE Paper 940523

Payri F, Bermúdez V, Payri R, Salvador F (2004) The influence of cavitation on the internal flow and the spray characteristics in diesel injection nozzles. Fuel 83:419–431

Payri R, Salvador F, Gimeno J, Bracho G (2008) A new methodology for correcting the signal cumulative phenomenon on injection rate measurements. Exp Techn 32(1):46–49

Payri F, Pastor JV, Payri R, Manin J (2011) Determination of the optical depth of a DI diesel spray. J Mech Sci Technol 25:209–219

Pickett L, Kook S (2009) Transient liquid penetration of early-injection diesel sprays. Int J Engines 4970

Pope SB (2000) Turbulent flows. Cambridge University Press

Powell C, Duke D, Kastengren A, Ilavsky J (2013) Measurements of diesel spray droplet size with ultra-small angle x-ray scattering. In 25th ILASS-Americas conference. Pittsburgh, PA

Ranz W (1956) On sprays and spraying. Department of Engineering Research Penn State University Bulletin

Rayleigh W (1878) On the instability of jets. Proc London Mathe Soc 4

Reitz R (1978) Atomization and other breakup regimes of a liquid jet. Ph.D. thesis, Princeton University

Reitz RD, Diwakar R (1987) Structure of high-pressure fuel sprays SAE Technical Paper

Reitz RD (1987) Modeling atomization processes in high pressure vaporizing sprays. Atomization Spray Technol 3(4):309–337
Reitz R, Bracco F (1982) Mechanism of atomization of a liquid jet. Phys Fluids 25(10):1730–1742
Reitz RD, Bracco F (1986) Mechanisms of breakup of round liquid jets. Encycl Fluid Mech 3:233–249
Reitz R, Bracco FB (1979) On the dependence of spray angle and other spray parameters on nozzle design and operating conditions. In: SAE Technical paper series
Richards K, Senecal P, Pomraning E (2013) CONVERGE 2.1.0 Theory Manual, Convergent Science
Ruff G, Sagar A, Faeth G (1987) Structure of the near-injector region of pressure-atomized sprays. AIAA J 27:549–559
Rupe J (1962) On the dynamic characteristics of free-liquid jets and a partial correlation with orifice geometry. J.P.L, Technical report No, p 32
Sadek R (1959) Communication to bergwerk. Proc Inst Mech Eng 173
Schmidt DP (1997) Cavitation in diesel fuel injector nozzles. Ph.D. thesis, University of Wisconsin - Madison
Schweitzer P (1937) Mechanism of disintegration of liquid jets. J Appl Phys 8
Siebers D (1998) Liquid-phase fuel penetration in diesel sprays. SAE Paper 980809
Siebers DL (1999) Scaling liquid-phase fuel penetration in diesel sprays based on mixing-limited vaporization. SAE 01–0528:724
Sitkei G (1959) On the theory of jet atomization. Acta Tech Acad Sci Hungaricae 25
Soare VT (2007) Phase Doppler Measurements in Diesel Dense Sprays: Optimisation of Measurements and Study of the Orifice Geometry Influence over the Spray at Microscopic Level. Ph.D. thesis, Universidad Politecnica de Valencia
Som S (2009) Development and Validation of Spray Models for Investigating Diesel Engine Combustion and Emissions. Ph.D. thesis, University of Illinois at Chicago
Som S, Aggarwal SK (2009) Assessment of atomization models for diesel engine simulations. Atomization Sprays 19(9):885–903
Som S, Aggarwal S (2010) Effects of primary breakup modeling on spray and combustion characteristics of compression ignition engines. Combust Flame 157:1179–1193
Som S, Ramirez AI, Aggarwal SK, Kastengren AL (2009) Development and validation of a primary breakup model for diesel engine applications. SAE technical paper 2009-01-0838
SPhERe Lab Homepage (2012) http://www.spherelab.gatech.edu
Sterling A, Sleicher C (1975) The instability of capillary jets. J Fluid Mech 68:477
Taylor J, Hoyt J (1983) Water jet photography-techniques and methods. Exp Fluids 1(3):113–120
Tennekes H, Lumley J (1972) A first course in turbulence. Press, M.I.T
United States Environmental Protection Agency (2003) EPA420-F-03-022. Fact sheet: diesel exhaust in the United States
Wang J (2005) X-ray vision of fuel sprays. J Synchrotron Radiat 12(2):197–207
Weber C (1931) On the breakdown of a fluid jet. Z.A.M.P. 11:136
Weller H, Tabor G, Jasak H, Fureby C (1998) A tensorial approach to cfd using object oriented techniques. Comput Phys 12(6):620
Wu P-K, Faeth G (1993) Aerodynamic effects of primary breakup of turbulent liquids. Atomization Sprays 3(3):265–289
Wu P-K, Tseng L-K, Faeth G (1992) Primary breakup in gas/liquid mixing layers for turbulent liquids. Atomization Sprays 2:295–317
Wu P-K, Mirand R, Faeth G (1995) Effects of initial flow conditions on primary breakup of non-turbulent and turbulent round liquid jets. Atomization and Sprays 5:175–196
Xue Q, Battistoni M, Som S, Quan S, Senecal P, Pomraning E, Schmidt D (2014) Eulerian cfd modeling of coupled nozzle flow and spray with validation against x-ray radiography data. SAE Int J Engines 7(2014-01-1425):1061–1072
Yue Y, Powell CF, Poola R, Wang J, Schaller JK (2001) Quantitative measurements of diesel fuel spray characteristics in the near-nozzle region using x-ray absorption. Atomization Sprays 11:4

Yule A, Mo S, Tham S, Aval S (1985) Diesel spray structure. Proceedings of the third international conference on liquid atomization and spray systems
Zaheer H (2015) Transient microscopy of primary atomization in gasoline direct injection sprays. Master's thesis, Georgia Institute of Technology

Chapter 5
Modeling of High-Pressure Fuel Injection in Internal Combustion Engines

Zongyu Yue and Rolf D. Reitz

Abstract The internal combustion engine has been a major power plant in transportation and industry, and demands continuously advanced technologies to improve its performance and fuel economy, and to reduce its pollutant emissions. Liquid fuel injection is critical to the combustion process in both compression ignition (CI) diesel engines and gasoline direct injection (GDI) engines. Much effort has been focused on modeling of spray atomization, droplet dynamics, and vaporization using a Lagrangian-drop Eulerian-fluid (LDEF) framework, which has been applied in engine computational fluid dynamic (CFD) simulations with success. However, recent experiments have shown the mixing-controlled characteristics of high-pressure fuel injection under vaporization conditions that are relevant to both gasoline and diesel engines. Under such conditions, instead of being dominated by droplet dynamics, the vaporization process of a liquid spray is limited by the entrainment rate of hot ambient gas and a saturated equilibrium phase is reached within the two-phase region. This suggests that an alternative approach of fuel spray modeling might be applicable. An equilibrium phase (EP) spray model was recently proposed for application to engine combustion simulation, based on this mixing-controlled jet theory and assumption of local phase equilibrium. This model has been applied to simulate both diesel fuel injection and GDI sprays, and has shown excellent predictions for transient vapor/liquid penetrations, spatial distribution of mixture fraction, as well as combustion characteristics in terms of flame lift-off length and soot emission. It has also shown better computational efficiency than the classical LDEF spray modeling approach since the dynamic process of droplet breakup, collision, coalescence, and vaporization is not modeled. The model and results relevant to engine simulation are reviewed in this chapter.

Z. Yue (✉) · R. D. Reitz
University of Wisconsin-Madison, 1500 Engineering Dirve, Madison, WI 53706, USA
e-mail: zyue@anl.gov

R. D. Reitz
e-mail: reitz@wisc.edu

Z. Yue
Argonne National Laboratory, Lemont, USA

K. Saha et al. (eds.), *Two-Phase Flow for Automotive and Power Generation Sectors*, Energy, Environment, and Sustainability,
https://doi.org/10.1007/978-981-13-3256-2_5

5.1 Introduction

In all direct injection engines, such as diesel and gasoline direct injection (GDI) engines, high-pressure fuel spray is an important process that impacts the subsequent steps of fuel/air mixture preparation, auto-ignition, combustion, emission formation, etc. Multiple-injection strategies with shaped rate-of-injection profile have been used as effective means for reduction of combustion noise (Augustin et al. 1991; Schulte et al. 1989), particulate matter, and nitrogen oxides (NO_x) emissions (Han et al. 1996; Nehmer et al. 1994; Pierpont et al. 1995; Su et al. 1996; Tow et al. 1994). In advanced combustion modes, such as reactivity controlled compression ignition (RCCI) (Reitz et al. 2014) and partially premixed compression ignition (PPCI) (Musculus et al. 2013), the stability and controllability of high-efficient combustion are usually achieved by in-cylinder stratification that is tailored by injection strategies. Although the study of liquid fuel spray has been ongoing for decades in engine combustion research community, it is still required to better understand high-pressure fuel injection process in order to improve engine design and meet ever-stringent regulations. Conventional knowledge suggests that the liquid spray is dominated by dynamic processes of liquid breakup, collision, coalescence, and interfacial vaporization. After exiting the nozzle into the combustion chamber, the continuous liquid phase is disintegrated into discrete liquid structure through aerodynamic force, cavitation, and turbulence effects, which is followed by secondary breakup and drop collision. Finite-rate vaporization occurs when the droplet is surrounded by unsaturated vapor.

In computational fluid dynamics (CFD) studies, the techniques for multi-phase flow problem can be generally divided into two categories: the Eulerian–Eulerian and the Lagrangian–Eulerian approaches. In the Eulerian–Eulerian approach, both gaseous and liquid phases are resolved on a computational mesh as continuous Eulerian fluids. For fuel injection simulations, this Eulerian–Eulerian approach usually demands refined mesh resolution of the order of microns in order to fully resolve the internal and/or near-nozzle region or even to resolve the discrete liquid drops. However, the characterization of a sharp interface between liquid and gas still needs to be modeled, such as volume-of-fluid (VOF) methods, level-set methods, or front-tracking methods (Gueyffier et al. 1999). While it provides detailed insight into the processes of internal flow and external spray development (Battistoni et al. 2018; Lacaze et al. 2012; Ménard et al. 2007; Xue et al. 2015; Yue et al. 2018a), the computational cost of this Eulerian–Eulerian approach is prohibitively expensive for application in large-scale problems, e.g., internal combustion (IC) engine simulations.

The Lagrangian–Eulerian approach has been widely used for engine simulations, wherein the liquid phase is represented as Lagrangian parcels and the gaseous phase is resolved as a continuous Eulerian fluid. Specifically, the liquid fuel is introduced into the computational domain as discrete 'blobs' with the prescribed boundary conditions, such as rate of injection, spray cone angle, initial Sauter mean diameter (SMD) (Reitz 1987; Reitz and Diwakar 1987). Following that, the processes of droplet/blob breakup, collisions, coalescence, and vaporization are also modeled (Beale et al. 1999; Munnannur et al. 2009; Ra et al. 2009; Ricart et al. 1997; Som et al. 2010; Su et al. 1996). A hybrid Kelvin–Helmholtz (KH)/Rayleigh–Taylor (RT) instability analysis model is often used to predict drop breakup (Su et al. 1996). In this case, a breakup length determined from Levich's theory (Levich 1962) is often adopted to regulate the competition between KH and RT breakup (Beale et al. 1999; Ricart et al. 1997). Further development of breakup models has led to the inclusion of in-nozzle flow effects, such as cavitation and turbulence (Som et al. 2010). Recent development also enables the coupling of Lagrangian-drop Eulerian-fluid (LDEF) spray modeling with internal flow simulation (Saha et al. 2017, 2018; Wang et al., 2011b), wherein the prescribed fluid condition at the nozzle outlet is not required and internal flow effects can be directly included. The spray modeling usually involves a number of model constants that requires optimization to improve the prediction. Perini et al. (2016) applied a multi-objective genetic algorithm (GA) to study the interactions of various spray model constants for improved prediction accuracy in simulations of a diesel spray. However, accurately predicting the fuel injection and spray process under a wide range of engine-relevant conditions and for different types of fuel remains a very challenging task.

While much effort has been focused on modeling liquid breakup and droplet vaporization, the experimental study (Siebers 1998, 1999) has revealed the mixing-controlled characteristics of high-pressure fuel injection under diesel engine-relevant conditions. In this scenario, the liquid atomization is strong enough that the liquid jet can be considered analogous to a turbulent gas jet and the vaporization process is limited by the rate of hot ambient gas entrainment, rather than by liquid breakup and mass transfer rates at droplet surfaces. It is assumed that local phase equilibrium is achieved and a saturated vapor condition is fully reached at the tip of the liquid length.

Iyer et al. (2000) performed a diesel spray CFD simulation with a two-fluid model where the Eulerian transport equations are solved for both the liquid and the gaseous phases. In their simulation approach, the vaporization is controlled by turbulent mixing by applying the locally homogeneous flow (LHF) approximation with an empirical phase equilibrium model. The results showed reasonably good agreement in the liquid length prediction, again suggesting the mixing-controlled behavior of diesel sprays. Matheis et al. (2017) adopted a detailed thermodynamic phase equilibrium model (Qiu et al. 2014) that consists of phase stability analysis and flash calculation in large eddy simulation (LES) of the Engine Combustion Network (ECN) Spray A condition and concluded that the mixing-controlled two-phase mixture model can be applied for dense or moderately dense spray regimes.

An efficient equilibrium phase (EP) spray model (Yue and Reitz 2017, 2018) has been recently proposed for engine CFD simulations, which is based on mixing-controlled jet theory and the assumption of local phase equilibrium. This model has been applied to simulate diesel-type fuel injection as well as GDI sprays with good predictions in terms of spray and combustion characteristics under a wide range of operating conditions without the need for model constant tuning. Moreover,

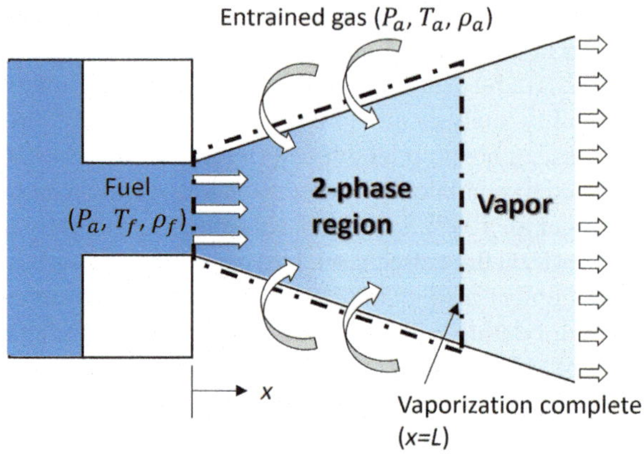

Fig. 5.1 Schematic of the 'idealized' spray model used to develop the liquid length scaling law (Siebers 1999)

the model shows better computational efficiency than the classical spray modeling approach since the consideration of the droplet dynamic processes is bypassed.

In the following section, an idealized model for mixing-controlled sprays and the derivation of a liquid length scaling law are reviewed. Then, the formulation of the EP spray model for CFD simulations and its application to diesel and GDI spray simulations are presented, followed by a brief summary.

5.2 Mixing-Controlled Vaporization of High-Pressure Sprays

Siebers (1998, 1999) studied liquid fuel penetration in a constant volume chamber operated under diesel-like conditions for a number of fuels, injector orifice sizes, and ambient conditions. This study discovered that: (1) The liquid length, which is defined as the maximum penetration distance of liquid fuel, linearly correlates with injector orifice size; (2) the liquid length is insignificantly dependent on injection pressure. These experimental findings can be well explained by mixing-controlled vaporization process, which is examined by applying jet theory.

Siebers' idealized model for high-pressure spray is illustrated in Fig. 5.1, which assumes a quasi-steady flow with a uniform growth rate and perfect adiabatic mixing within the spray boundaries. The dashed lines outline the control volume used for mass, momentum, and energy balances. The downstream side of the control surface ($x = L$, and L is the liquid length) is defined as the axial location where the fuel has just completely vaporized (viz., the liquid length).

Applying integral analysis for the controlled volume with the assumption of local thermodynamic equilibrium, while neglecting the recovery of the kinetic energy in the fuel vaporization region, the following equations are derived from mass and momentum conservation:

$$\dot{m}_f = \rho_f \cdot A_{\text{noz}} \cdot U_{\text{inj}} = \dot{m}_f(x) \propto \rho_f \cdot d^2 \cdot U_{\text{inj}} \tag{5.1}$$

$$\dot{m}_a(x) = \rho_a \cdot A(x) \cdot U(x) \propto \sqrt{\rho_a \cdot \rho_f} \cdot d \cdot x \cdot U_{\text{inj}} \cdot \tan(\theta/2) \tag{5.2}$$

$$\dot{m}_f \cdot U_{\text{inj}} = \dot{m}_f(x) \cdot U(x) + \dot{m}_a(x) \cdot U(x) \tag{5.3}$$

where $\dot{m}_f$ is the injected fuel mass flow rate, ρ_f is the injected fuel density, A_{noz} is the effective nozzle area, U_{inj} is the constant injection velocity, $\dot{m}_f(x)$ is the fuel mass flow rate at any axial location x, $\dot{m}_a(x)$ is the mass flow rate of entrained ambient gas, ρ_a is the ambient gas density, $A(x)$ is the spray cross-sectional area, $U(x)$ is the radially uniform spray velocity, d is the nozzle diameter, and θ is the spray spreading angle.

From jet theory, $\dot{m}_a(x)$ is proportional to d, while $\dot{m}_f(x)$ is proportional to d^2, which results in a linear dependency of liquid length on the nozzle diameter if the vaporization process is mixing controlled. On the other hand, both $\dot{m}_a(x)$ and $\dot{m}_f(x)$ are proportional to U_{inj}, indicating an independency of liquid length on injection velocity or injection pressure.

Equations (5.1)–(5.3) can be rearranged to give a parabolic equation for the fuel/ambient gas ratio, $\frac{\dot{m}_f(x)}{\dot{m}_a(x)}$, and its positive root gives the axial variation of this ratio,

$$\frac{\dot{m}_f(x)}{\dot{m}_a(x)} = \frac{2}{\sqrt{1 + 16 \cdot \tilde{x}^2} - 1} \tag{5.4}$$

Here, $\tilde{x} = x/x^+$ is the axial distance in the spray normalized by $x^+ = \sqrt{\frac{\rho_f}{\rho_a}} \cdot \frac{\sqrt{C_a} \cdot d}{\tan(\theta/2)}$. C_a is the area-contraction coefficient of nozzle.

At the tip of the liquid length ($x = L$), the liquid fuel has just vaporized completely such that

$$\frac{\dot{m}_f(L)}{\dot{m}_a(L)} = B\left(T_a, P_a, T_f\right) \tag{5.5}$$

The term $B\left(T_a, P_a, T_f\right)$ is the mass ratio of liquid fuel and ambient gas in a completely vaporized and saturated condition, which is a thermodynamic equilibrium problem with given ambient temperature (T_a), pressure (P_a), and injected fuel temperature $\left(T_f\right)$.

A liquid length scaling law (Siebers 1999) is then derived by combining Eqs. (5.4) and (5.5) and substituting the definition of x^+,

$$L = C_L \cdot \sqrt{\frac{\rho_f}{\rho_a}} \cdot \frac{\sqrt{C_a} \cdot d}{\tan(\theta/2)} \sqrt{\left(\frac{2}{B\left(T_a, P_a, T_f\right)} + 1\right)^2 - 1} \tag{5.6}$$

Here, C_L is a model constant which has a theoretical value of 0.38. Siebers adjusted this value to 0.62 to match n-hexadecane and heptamethylnonane liquid length data. The adjusted value of 0.62 is also used in current work.

Siebers (1999) compared the scaling law with measured liquid length data for several types of fuels under a very wide range of engine-relevant operating conditions and showed that the scaling law reproduced all major trends in the experimental results. The close agreement between the liquid length scaling law and the measured data suggests that vaporization in a high-pressure diesel injection approaches a limit governed by spray mixing processes, and atomization and local interphase transport processes at droplet surfaces are not limiting factors for fuel vaporization. Parrish (2014) examined the scaling law in GDI measurement by lumping the term $C_{\mathrm{L}} \cdot \frac{\sqrt{C_a} \cdot d}{\tan(\theta/2)}$ into a single constant, which was adjusted through trial and error to match the experimental data. It was found that the lumped constant reached a fixed value at high ambient temperature, indicating that under such conditions the fuel vaporization also becomes mixing controlled in GDI sprays.

5.3 Equilibrium Phase (EP) Spray Model

5.3.1 Governing Equations

In the LDEF approach for two-phase turbulent flow simulations, such as in the KIVA code (Amsden et al. 1989) implementation, the liquid flow and gaseous flow are resolved on two different but connected fields. The gaseous mixture of ambient gas and vaporized fuel is resolved on an Eulerian field where the transport equations are solved for species, momentum, and internal energy. Using the Reynolds-averaged Navier–Stokes (RANS) method, the instantaneous equations are averaged and expressed with mean quantities, as shown below (Amsden et al. 1989).

The continuity equation for species m is

$$\frac{\partial \rho_m}{\partial t} + \nabla \cdot (\rho_m \mathbf{u}) = \nabla \cdot \left(\rho D \nabla \left(\frac{\rho_m}{\rho} \right) \right) + \dot{\rho}_m^c + \dot{\rho}_m^s \delta_{m1} \tag{5.7}$$

where $\mathbf{u}$ is the velocity vector for the mean flow, ρ_m is the mass density of species m, ρ is the total mass density, D is the diffusion coefficient, $\dot{\rho}_m^c$ is a source term due to chemical reactions, $\dot{\rho}_m^s$ is a source term due to the spray, and δ_{m1} is the Dirac delta function so that there is no spray source term for non-fuel species.

The momentum equation for the fluid mixture is

$$\frac{\partial (\rho \mathbf{u})}{\partial t} + \nabla \cdot (\rho \mathbf{u}\mathbf{u}) + \nabla P = \mu \nabla \cdot \left(\nabla \mathbf{u} + (\nabla \mathbf{u})^{\mathrm{T}} - \frac{2}{3} \nabla \cdot \mathbf{u} \mathbf{I} \right) - \nabla \left(\frac{2}{3} \rho \mathrm{k} \right) + F^s \tag{5.8}$$

where P is the fluid pressure, μ is the dynamic viscosity, superscript T denotes transpose, $\mathbf{I}$ is the identity tensor, k is the turbulent kinetic energy, and F^s is the momentum exchange between the liquid spray and the ambient flow.

The internal energy equation is

$$\frac{\partial(\rho I)}{\partial t} + \nabla \cdot (\rho \mathbf{u} I) = -P\nabla \cdot \mathbf{u} + \lambda \nabla^2 T + \rho D \sum_m \left(h_m \nabla^2 \left(\frac{\rho_m}{\rho} \right) \right) + \rho\varepsilon + \dot{Q}^c + \dot{Q}^s \tag{5.9}$$

where I is the specific internal energy, λ is the thermal conductivity, h_m is the specific enthalpy of species m, ε is the dissipation rate of turbulent kinetic energy, $\dot{Q}^c$ is a source term due to chemical heat release, and $\dot{Q}^s$ is a source term due to spray interaction.

For a turbulent flow, the Reynolds stresses are modeled by assuming the same form as the viscous tensor for Newtonian fluids but with a modified viscosity, μ, which is given as

$$\mu = \mu_0 + \mu_t \tag{5.10}$$

where μ_0 is the laminar dynamic viscosity, $\mu_t = C_\mu k^2/\varepsilon$ is the turbulent viscosity, and C_μ is a model constant. Turbulent fluxes of species and enthalpy are modeled by assuming Fick's law diffusion with a single diffusion coefficient, which is given as

$$D = \frac{\mu}{\rho \cdot Sc} \tag{5.11}$$

where Sc is the Schmidt number and a constant value is assumed. Two more transport equations for k and ε are needed in the turbulence model (Han et al. 1995; Launder et al. 1972; Wang et al. 2011a; Yakhot et al. 1992), which will be discussed later in this section.

On the other hand, the liquid fuel is initialized at the location of nozzle exit as Lagrangian parcels with the prescribed rate of injection and initial size equal to an effective nozzle diameter. Each parcel represents an assembly of a number of identical drops with the same temperature, size, momentum, etc. The Lagrangian and Eulerian fluids interact with each other in terms of mass, momentum, and energy transfer, through the source terms $\dot{\rho}_m^s$, F^s, $\dot{Q}^s$ introduced above. In the traditional LDEF approach, the two-phase flow process is primarily governed by the dynamics of the Lagrangian drops, i.e., breakup, collision, and vaporization.

In order to model the fuel spray process with emphasis on mixing-controlled vaporization, it is desired to simulate the liquid as a continuous fluid with an Eulerian treatment. However, this treatment requires the computational mesh to be refined to a micron level such that the near-nozzle flow can be resolved, and its application in engine simulations is limited due to the resulting significant computational cost. In the current EP model, the framework of the LDEF approach is retained such that a practical mesh resolution can be used for engine simulations. However, the non-equilibrium processes of liquid breakup and collision are not considered. Instead, after being injected into the domain, the Lagrangian parcels gradually transition to the Eulerian liquid phase, which is assumed to be homogenously mixed with the

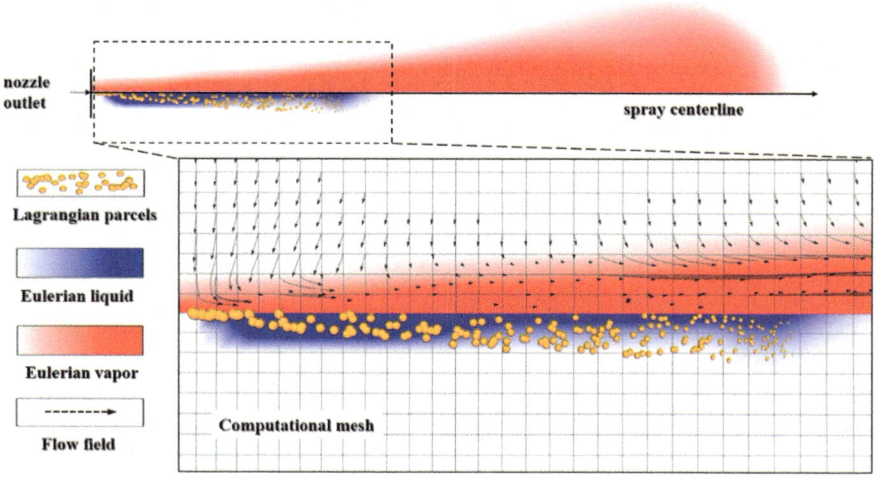

Fig. 5.2 Schematic of EP spray model, generated from CFD results (Yue et al. 2017)

local gaseous phase. The governing equation for the Eulerian liquid can be written as

$$\frac{\partial \rho_l}{\partial t} + \nabla \cdot (\rho_l \mathbf{u}) = \nabla \cdot \left[\rho D \nabla \left(\frac{\rho_l}{\rho} \right) \right] + \dot{S}_{\mathrm{EP}} + \dot{S}_{\mathrm{release}} \tag{5.12}$$

where the subscript l denotes the Eulerian liquid phase, $\dot{S}_E$ describes the mass exchange between the Eulerian liquid and vapor phase assuming phase equilibrium, and $\dot{S}_{\mathrm{release}}$ is the source term that describes the transition rate from Lagrangian drops to the Eulerian liquid. Other governing equations used in the LDEF approach are unchanged, and the Eulerian liquid can be considered as an intermediate step between Lagrangian drops and Eulerian vapor.

Figure 5.2 illustrates a schematic diagram for the current spray modeling approach. The injected Lagrangian parcels are denoted by the yellow spheres with size proportional to the drop mass. The blue contour represents the Eulerian liquid mass fraction, while the red contour represents the vapor mass fraction. The dashed arrows indicate velocity vectors of the Eulerian flow field. As can be seen, the droplets release mass smoothly as they move away from the nozzle. The phase change for the Eulerian fluid $\left(\dot{S}_{\mathrm{EP}}\right)$ is solved for by employing a phase equilibrium solver, and it is a function of the local mixture composition and thermodynamic state. Thus, the process of fuel vaporization becomes controlled by the air entrainment and mixing, as illustrated by the velocity field.

A liquid-jet model was derived based on Siebers' scaling analysis to solve for the term $\dot{S}_{\mathrm{release}}$ in Eq. (5.12). Combining Eqs. (5.1) and (5.4) gives

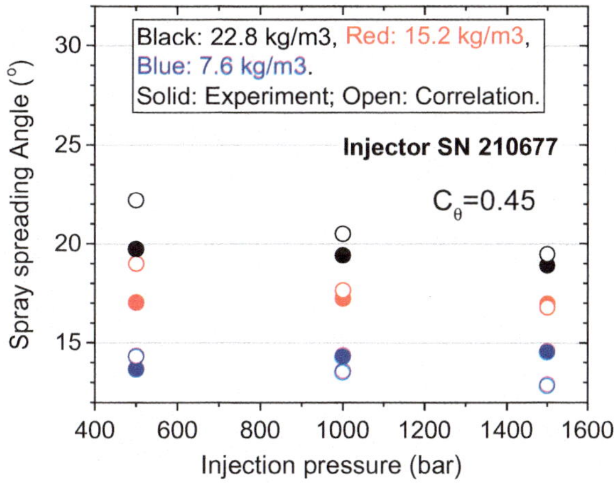

Fig. 5.3 Spray spreading angle as a function of injection pressure at three ambient conditions (Yue et al. 2017)

$$\gamma(x) = \frac{\dot{m}_a(x)}{\dot{m}_a(L)} = \frac{\sqrt{1 + 16\tilde{x}^2} - 1}{\sqrt{1 + 16(L/x^+)^2} - 1} \tag{5.13}$$

The function $\gamma(x)$ is the ratio between the entrainment flow rate at distance x and the one at the tip of liquid length. In a mixing-controlled spray, $\gamma(x)$ also represents the possible upper bound of the proportion of fuel being vaporized at any axial location. Therefore, the conversion rate from the Lagrangian parcels to the Eulerian liquid is modeled by adopting the function $\gamma(x)$ as

$$\dot{S}_{\text{release}} = \left(m_{\text{parcel}} - m_{\text{parcel,initial}} \cdot (1 - \gamma(x))\right)/\text{d}t \tag{5.14}$$

Here, m_{parcel} is the mass of the parcel, $m_{\text{parcel,initial}}$ is the initial mass of the same parcel upon injection, and dt is the computational time step. One should note that $\dot{S}_{\text{release}}$ controls only the process of liquid release from Lagrangian-to-Eulerian phase and not the vaporization process. The vaporization process is controlled by the local phase equilibrium $\left(\dot{S}_{\text{EP}}\right)$.

The spray spreading angle θ is modeled as a function of gas/liquid density ratio, $\frac{\rho_f}{\rho_a}$, Reynolds number, Re_f, and Weber number, We_f, according to the aerodynamic surface wave theory for liquid-jet breakup (Ranz 1958; Reitz et al. 1979):

$$\tan\theta/2 = C_\theta 4\pi\sqrt{\frac{\rho_a}{\rho_f}}\, f\left(\frac{\rho_f}{\rho_a}\left(\frac{Re_f}{We_f}\right)^2\right) \tag{5.15}$$

C_θ is a model constant and is dependent on injector's internal geometry and flow condition. The function, f, is given in the references. The value of C_θ can be determined from the best fit of the experimental data. Figure 5.3 shows a comparison of experiments and model predictions of spreading angles as a function of injection pressure for three ambient densities, for the ECN Spray A injector, with a C_θ of 0.45.

5.3.2 Gas-Jet Model

The Lagrangian and Eulerian fluids are fully coupled in conservation of mass, momentum, and energy. An aerodynamic drag force, F^s, is applied to the Lagrangian parcels when a relative velocity exists between the liquid droplets and the gaseous phase, and leads to momentum transfer as seen in the momentum equation. F^s is modeled as

$$F^s = \frac{3}{8}\frac{\rho}{\rho_f}\frac{\left|\mathbf{u}+\mathbf{v}'-\mathbf{v}\right|}{r_d}\left(\mathbf{u}+\mathbf{v}'-\mathbf{v}\right)C_D C_{\text{distortion}} \tag{5.16}$$

where $\mathbf{v}$ is the velocity vector of droplet, $\mathbf{v}'$ is a turbulent drop dispersion velocity, and r_d is the droplet radius. The effect of droplet distortion on drag is considered by an enhanced coefficient, $C_{\text{distortion}}$ (Liu et al. 1993). The drag coefficient for undistorted droplet, C_D, is modeled as

$$C_D = \begin{cases} \frac{24}{Re_d}\left(1+1/6Re_d^{2/3}\right) & Re_d < 1000 \\ 0.424 & Re_d \geq 1000 \end{cases}, \quad Re_d = \frac{2\rho\left|\mathbf{u}+\mathbf{v}'-\mathbf{v}\right|r_d}{\mu_t} \tag{5.17}$$

However, a practical computational mesh for engine simulation is usually much larger than the injector orifice size and cannot resolve the velocity profile in the near-nozzle region. The estimation of drag force and momentum exchange becomes dependent on mesh size when a mean value of CFD-resolved velocity is used to calculate the relative velocity, and a coarse mesh usually results in overprediction of momentum transfer and slow spray penetration. In order to improve the grid independency of momentum coupling, the gas-jet model (Abani et al. 2008a, b) is employed, wherein a sub-grid scale velocity $\mathbf{u}_{sgs}$ estimated from the turbulent gas-jet theory is used instead of the CFD-resolved velocity.

In the gas-jet model, the solution for a gas jet with same injection velocity, mass flow rate, and injection momentum is used to represent the velocity field of the liquid fuel spray. Accordingly, an equivalent orifice size for the gas jet is determined to be $d_{\text{eq}} = d\sqrt{\rho_f/\rho}$. In a gas jet with transient injection profile, the effective injection velocity $\mathbf{u}_{\text{eff}}(x,t)$ can be written as a function of axial distance, x, and time, t, which represents the convolution of n successive changes in the initial injection velocity, $\mathbf{U}_{\text{inj}}(t_0)$ (Abani et al. 2007),

$$\mathbf{u}_{\text{eff}}(x,t) = \mathbf{U}_{\text{inj}}(t_0) + \sum_{k=1}^{n}\left(\left(1-\exp\left(-\frac{t-t_k}{\tau(x,t_k)}\right)\right) \times \left(\mathbf{U}_{\text{inj}}(t_k) - \mathbf{U}_{\text{inj}}(t_{k-1})\right)\right) \tag{5.18}$$

$$\tau(x,t) = St\frac{x}{\left|\mathbf{U}_{\text{inj}}(t)\right|} \tag{5.19}$$

Here, the jet response time, $\tau(x, t)$, is calculated using the spray jet analogy with a constant value of 3.0 for the particle Stokes number, St (Abani et al. 2007).

Once the effective injection velocity is determined, the transient sub-grid gas-jet velocity can be calculated with given axial distance, radial distance, and time, as shown in Eqs. (5.20)–(5.23) (Perini et al. 2016).

$$\mathbf{u}_{\text{sgs}}(x, r, t) = \frac{f(\chi)\mathbf{u}_{\text{eff}}(x, t)}{\left(1 + \frac{12r^2}{K_{\text{entr}}^2 x^2}\right)^2} \tag{5.20}$$

$$f(\chi) = \begin{cases} 1/\chi, & \chi \geq \varphi \\ \gamma_{\text{max}} - \frac{32\gamma_{\text{max}}^3(\chi-1)^2}{(3+\Delta)^2(3+\Delta-4\gamma_{\text{max}})}, & 1 \leq \chi < \varphi ; \quad \chi = \frac{xK_{\text{entr}}}{3d_{\text{eq}}} \\ \gamma_{\text{min}} + \chi(2-\chi)(\gamma_{\text{max}} - \gamma_{\text{min}}), & \chi < 1 \end{cases} \tag{5.21}$$

$$\Delta = \sqrt{9 - 8\gamma_{\text{max}}} \tag{5.22}$$

$$\varphi = \frac{3 + \Delta}{4\gamma_{\text{max}}} \tag{5.23}$$

The denominator in the right-hand side of Eq. (5.20) describes the velocity decay along radial direction, and K_{entr} is a coefficient of turbulent entrainment rate. The piecewise function, $f(\chi)$, describes the velocity damping profile along axial direction that is defined by γ_{max} and γ_{min}. The values of these model constants are adopted from a multi-objective GA optimization in simulation of diesel spray (Perini et al. 2016).

5.3.3 Phase Equilibrium in High-Pressure Sprays

Liquid fuel injection involves several non-equilibrium phenomena. However, as discussed previously, liquid vaporization is limited by the rate of ambient hot gas entrainment, while local interphase transport rates are so fast due to the small drop sizes that the assumption of local phase equilibrium can be applied to simplify the problem. Therefore, other than the liquid fuel and ambient gas mixing rate, the determination of phase equilibrium is also critical for the determination of spray vaporization.

For a multi-component mixture under ideal conditions, the vapor/liquid equilibrium can be determined using Raoult's law,

$$P_{i,v} = X_{i,l} \cdot P_{\text{sat},i} \tag{5.24}$$

$P_{i,v}$ is the partial pressure of species i in the vapor phase. $X_{i,l}$ is the mole fraction in the liquid phase. $P_{\text{sat},i}$ is the vapor pressure of the pure species for a given temperature, which can be calculated by a number of vapor pressure equations (Reid et al. 1987). The subscripts v and l denote the vapor phase and liquid phase, respectively. It is seen that equilibrium in this case only depends on temperature. However, under high-pressure conditions, Raoult's law is no longer valid, and a real gas equation of state (EoS) is needed to solve the problem.

The Gibbs free energy is a thermodynamic potential that is used to measure the obtainable energy from a closed thermodynamic system, and it is minimized when the equilibrium is reached with a specified temperature, pressure, and mixture composition. Therefore, the equilibrium calculation is a search for the global minimum of the Gibbs free energy in a multi-dimensional space, which is usually achieved by requiring fugacity equality, e.g., in case of a two-phase equilibrium,

$$f_{i,v} = f_{i,l} \tag{5.25}$$

f is the fugacity, which is equal to the pressure of an ideal gas that has the same chemical potential as the real gas at some pressure. However, the fugacity equality is a necessary but insufficient condition for phase equilibrium, because it stands for a local stationary point that does not guarantee a global minimum. While the local extreme given by the fugacity equality could happen to be the global extreme for a simple fluid, this approach sometimes presents a 'false' equilibrium state for a multi-component mixture.

Qiu and Reitz (2014a, b, 2015), Qiu et al. (2014) developed a phase equilibrium solver based on classical thermodynamic equilibrium with a real gas EoS, which is reported to be thermodynamically consistent and is computationally robust and efficient. This solver has been applied to a number of multi-phase flow problems such as condensation and supercritical fluids. In this approach, the phase equilibrium problem is solved in two steps, viz., a phase stability test and a phase-splitting calculation. The phase stability test is performed first to examine whether a phase-splitting calculation is required and to provide initial guess values for the phase-splitting calculation. For a multi-component mixture, the tangent plane distance (TPD) method (Baker et al. 1982; Michelsen 1982) is used to test the phase stability. The TPD function is defined as:

$$\mathrm{TPD}(\boldsymbol{X}) = \sum_{i=1}^{N_c} X_i(\mu_i(\boldsymbol{X}) - \mu_i(\boldsymbol{Z})) \tag{5.26}$$

Here, $\boldsymbol{X}$ is the mole fraction array, $\boldsymbol{Z}$ is the feeding composition array, μ is the chemical potential, and N_c is the number of components. Geometrically, $\mathrm{TPD}(\boldsymbol{X})$ represents the vertical distance from the tangent hyperplane to the molar Gibbs free energy surface of $\boldsymbol{Z}$. It requires $\mathrm{TPD}(\boldsymbol{X})$ to be nonnegative at all stationary points to achieve phase stability. Therefore, the TPD method reduces the search space from the whole domain to the local extremes and hence an exhaustive search of all possible $\boldsymbol{X}$ is avoided. Whenever the phase stability test yields a negative value of $\mathrm{TPD}(\boldsymbol{X})$, the flash calculation is performed based on the Rachford–Rice algorithm (Rachford et al. 1952):

$$\sum_{i=1}^{N_c} \frac{Z_i(k_i - 1)}{1 + \lambda(k_i - 1)} = 0 \tag{5.27}$$

k_i is the phase equilibrium ratio of species i, and λ is the mole fraction of one phase. The outcome of the stability test is used as an initial guess for solving the objective function.

5.3.4 Equation of State

In CFD simulations, the EoS calculation is called numerous times for each computational cell at each time step. Therefore, it is important to maintain a good balance between accuracy and efficiency for EoS calculation in engineering applications. The Peng–Robinson (PR) EoS is a two-parameter efficient cubic EoS, which has the form

$$P = R_u T/(\bar{v} - b_m) - a_m/\left(\bar{v}^2 + 2b_m\bar{v} - b_m^2\right) \tag{5.28}$$

where P is the pressure, T is the temperature,$\bar{v}$ is the molar volume, and R_u is the universal gas constant. a_m and b_m are attraction and repulsion parameters, respectively. For species i, the parameters are expressed as functions of critical properties and acentric factor,

$$\mathrm{a}_i = 0.45724 R_u^2 T_c^2 / P_c^2 \left[1 + f(\omega)\left(1 - \sqrt{T_c/T}\right)\right]^2 \tag{5.29}$$

$$\mathrm{b}_i = 0.07780 R_u T_c / P_c \tag{5.30}$$

where P_c is the critical pressure, T_c is the critical temperature, ω is the acentric factor, and $f(\omega) = 0.37464 + 1.54226\omega - 0.26992\omega^2$. The parameters for a multi-component mixture can be calculated following the van der Waals mixing rule

$$a_m = \sum_i \sum_j x_i x_j (a_i a_j)^{\frac{1}{2}} (1 - k_{ij}) \tag{5.31}$$

$$b_m = \sum_i x_i b_i \tag{5.32}$$

The subscript m denotes the mixture, and i and j denote individual species. x is the species mole fraction. k_{ij} is the binary interaction parameter for species i and j, and its value can be found in reference (Knapp 1982). Table 5.1 lists the properties required for EoS calculation for the major species that is considered in this work. The listed species in total account for more than 99% mole fraction of the mixtures in either a non-reacting or a reacting case in current study and, therefore, are considered sufficient to represent the complete mixture in the EoS calculations. These parameters are also used by the phase equilibrium solver.

Table 5.1 Species and properties considered for the EoS calculation

Species	T_c (K)	P_c (bar)	ω (–)
$nC_{12}H_{26}$	658.0	18.2	0.5764
iC_8H_{18}	543.8	25.7	0.3034
N_2	126.2	33.9	0.0377
O_2	154.6	50.43	0.0222
CO	132.92	34.99	0.0663
CO_2	304.19	73.82	0.2276
H_2O	647.13	220.55	0.3449

Substituting the definition of the compressibility factor $z \underset{=}{\text{def}} P\bar{\upsilon}/R_u T$ into Eq. (5.28), an alternative form of PR EoS is derived,

$$z^3 - (1 - B)z^2 + \left(A - 3B^2 - 2B\right)z - \left(AB - B^2 - B^3\right) = 0 \tag{5.33}$$

where $A = \frac{a_m P}{R_u^2 T^2}$ and $B = \frac{b_m P}{R_u T}$. This cubic equation is numerically solved with Newton's method to yield one or three roots depending on the number of phases.

The internal energy is only a function of temperature when the ideal gas law is applied. When a real fluid is considered at high temperatures and pressures, it requires two independent thermodynamic variables to calculate the non-ideal departure in internal energy. For example, internal energy as a function of temperature and molar volume is expressed as

$$\mathrm{I}(\mathrm{T}, \bar{\upsilon}) = I(T, \infty) + \left(\frac{a_m - T\frac{da_m}{dT}}{w_{\mathrm{avg}} 2\sqrt{2} b_m}\right)\left[\ln\left|\frac{b_m\left(\sqrt{2} - 1\right) - \bar{\upsilon}}{b_m\left(\sqrt{2} + 1\right) + \bar{\upsilon}}\right|\right] \tag{5.34}$$

where w_{avg} is the mixture-averaged molecular weight. The first term, $I(T, \infty)$, on the right-hand side represents the internal energy at the ideal condition where the molar volume approaches infinity. The second term on the right-hand side represents the departure from ideality. This form of internal energy equation can be solved analytically and is therefore implemented in this study.

Figure 5.4 shows the compressibility factors for a simple fluid ($\omega = 0$) calculated by the PR EoS, in comparison with results by the Lee–Kesler-modified Benedict–Webb–Rubin (BWR) EoS (Reid et al. 1987). The BWR EoS is considered a more accurate EoS that applies to a wide range of pressures and temperatures, but is more computationally expensive compared to the cubic PR EoS. The compressibility factor is plotted as a function of reduced temperature T_r and reduced pressure P_r, which are defined as $T_r = T/T_c$ and $P_r = P/P_c$, respectively. A close agreement of predictions by the PR and the BWR EoS can be found at low-reduced pressures with a relative difference less than 5%. The discrepancy increases at higher-reduced pressures, especially at lower-reduced temperatures. The blue symbols in Fig. 5.4 mark the status of major species of engine charge at 900 K, 75 bar, which is relevant to

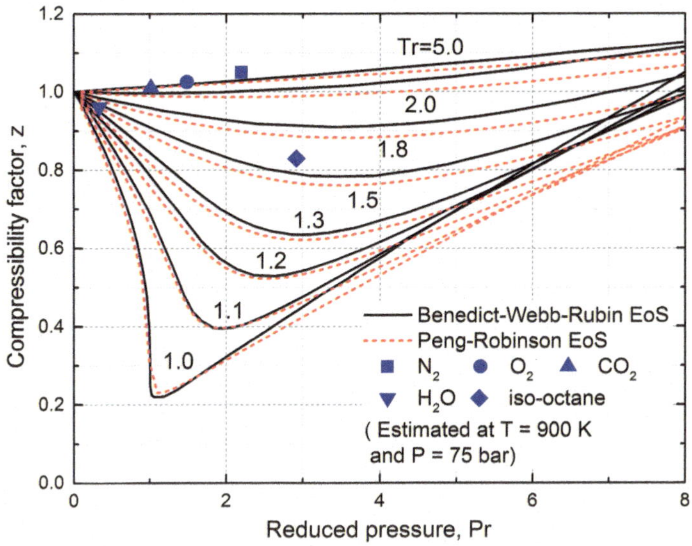

Fig. 5.4 Compressibility factor as a function of reduced pressure and temperature for a simple fluid comparing PR EoS and BWR EoS (Yue et al. 2018b). Black solid lines are BWR EoS, and red dashed lines are PR EoS. The blue symbols are the compressibility factors for selected species at typical engine operating condition (900 K, 75 bar), estimated using this z-chart

engine operating conditions. It is seen that these species locate in a low-reduced pressure and high-reduced temperature region, where the PR EoS provides similar accuracy as the BWR EoS.

The Peng–Robinson EoS was implemented into the KIVA-3vr2 code (Amsden 1999) with the RANS framework, wherein the turbulent viscosity and diffusivity are modeled to provide closures for the Reynolds stresses and turbulent fluxes, respectively. Non-ideal effects in the transport properties are not taken into account here since turbulent transport properties in RANS approach are generally two orders of magnitude larger than laminar transport properties for the turbulent flows in engines.

In the KIVA code, the Navier–Stokes equations are solved by the Semi-Implicit Method for Pressure-Linked Equations (SIMPLE) algorithm. Briefly, a corrected pressure P^c is yielded to eliminate the difference between V^p, the predicted volume based on EoS calculation, and V^c, the corrected volume related to the cell-face velocities (Amsden et al. 1989). This process iterates until the predicted and corrected values converge. Use of a real gas EoS generally requires more iterations since the pressure dependency of thermodynamic derivatives leads to slower convergence compared to the standard KIVA. Therefore, the computational time for an engine CFD simulation is increased by 20–50% when the PR EoS is used (Yue et al. 2018b).

5.3.5 Generalized RNG Model

For turbulence modeling in engine simulations, the RANS approach is currently the most practical and dominant method because of its computational efficiency. In the RANS approach, two more transport equations for the turbulent kinetic energy, k, and its dissipation rate, ε, are solved in addition to the mass, momentum, and energy equations. In this work, a generalized renormalization group (gRNG) k-ε model (Wang et al. 2011a) is used. Compared to the standard two equation models (Launder et al. 1972), the original RNG model (Yakhot et al. 1992) and the modified RNG model (Han et al. 1995), the gRNG model accounts for flow compressibility and has model coefficients that are functions of the flow strain rate. This allows improved predictions in the three types of pure compression of interest in engine flows: a 1D unidirectional axial compression, a 2D cylindrical-radial compression (squish flow), and a 3D spherical compression. In the gRNG model, the transport equations for k and ε are solved as follows

$$\rho\frac{Dk}{Dt} = P_k - \rho\varepsilon + \dot{W}^s + \nabla\cdot[\alpha_k\rho\nu\nabla k] \tag{5.35}$$

$$\rho\frac{D\varepsilon}{Dt} = \frac{\varepsilon}{k}C_1P_k + C_1^{'}\rho\nu\frac{\varepsilon}{k}(\nabla\cdot\mathbf{u})^2 - C_{2n}\rho\frac{\varepsilon^2}{k} - \rho R + C_3\rho\varepsilon(\nabla\cdot\mathbf{u}) + C_s\frac{\varepsilon}{k}\dot{W}^s + \nabla\cdot[\alpha_\varepsilon\rho\nu\nabla\varepsilon] \tag{5.36}$$

where $\dot{W}^s$ is a source term due to interactions with the spray (Amsden 1999). ν is the kinematic viscosity. The production of turbulent kinetic energy P_k and the additional term R from RNG analysis are modeled as

$$P_k = \frac{2}{3}\left[a\rho\nu(\nabla\cdot\mathbf{u})^2 - \rho k(\nabla\cdot\mathbf{u})\right] \tag{5.37}$$

$$R = \sqrt{\frac{2(1+a)}{3}}C_\mu C_\eta\eta\varepsilon|\nabla\cdot\mathbf{u}| \tag{5.38}$$

The model coefficients are summarized in Table 5.2 (Wang et al. 2011a). $\eta = Sk/\varepsilon$ is the ratio of turbulent to mean strain-time scale, and $S = \left(2S_{ij}S_{ij}\right)^{1/2}$ is the magnitude of the mean strain $S_{ij} = \frac{1}{2}\left(\partial u_i/\partial x_j + \partial u_j/\partial x_i\right)$. The terms a and n reflect the 'dimensionality' of the strain rate field, which are defined as

$$a = 3\left(S_{11}^2 + S_{22}^2 + S_{33}^2\right)/(|S_{11}| + |S_{22}| + |S_{33}|)^2 - 1 \tag{5.39}$$

$$n = 3 - \sqrt{2a} \tag{5.40}$$

Table 5.2 Coefficients for gRNG turbulence model

Function	Constant	
$C_1' = a(1 - 2/3C_1)$	C_1	1.42
$C_{2n} = 2.4958 - 0.6857n + 0.1099n^2$	C_μ	0.0845
$C_3 = -\frac{n+1}{n} + \frac{2}{3}C_1 + \sqrt{\frac{2(1+a)}{3}} C_\mu C_\eta \eta (-1)^\delta$, $\delta = \begin{cases} 1, & \text{if } \nabla \cdot \mathbf{u} < 0 \\ 0, & \text{if } \nabla \cdot \mathbf{u} > 0 \end{cases}$	α_k	1.39
	α_ε	1.39
$C_\eta = \frac{\eta(1-\eta/\eta_0)}{1+\beta\eta^3}$	η_0	4.38
	β	0.012

5.4 Diesel Spray Simulation

5.4.1 ECN Spray A

The Sandia combustion vessel (Pickett et al. 2010, 2011; Skeen et al. 2016) simulates a wide range of ambient environments (temperature, density, oxygen concentration, etc.) for high-pressure fuel injection and provides full optical access to help in the understanding of the details of the spray combustion process. The present spray model is validated with the ECN 'Spray A' operating conditions in this section. The Spray A features a single-hole injector. The injected fuel is *n*-dodecane, which is often used as a surrogate for diesel fuel in research. The injection pressure is 150 MPa, and the fuel temperature is 373 K. The injection profile has a rapid start and end with a steady period in between, forming a top hat profile. For the simulations, the rates of injection were determined by a virtual injection rate generator (http://www.cmt.upv.es/ECN03.aspx), as recommended by the ECN. The specifications of the operating conditions are given in Table 5.3.

As can be seen, cases 1–6 represent non-reacting conditions with 0% oxygen, where the transient liquid/vapor penetration and quasi-steady-state mixture fraction were measured. Mie-scatter and Schlieren imaging were used to track the liquid spray and vapor plume, respectively. Rayleigh-scatter imaging was used for the quantitative measurement of vapor mixture fraction (Pickett et al. 2011). Cases 7–11 represent combusting spray conditions with varying oxygen content and ambient temperature. Laser-induced incandescence (LII) imaging combined with line-of-sight extinction technique was applied for the quantitative measurement of soot volume fraction distribution (Idicheria et al. 2005; Manin et al. 2013; Skeen et al. 2016), while OH chemiluminescence was used to indicate the jet lift-off length (Higgins et al. 2001). In the CFD simulation results, the liquid and vapor phase penetrations are determined

Table 5.3 Operating conditions and injector specification for ECN Spray A (Pickett et al. 2010, 2011; Skeen et al. 2016)

	Case #	Nom. temp. (K)	Nom. density (kg/m^3)	O_2 (% vol.)	Nozzle diameter (μm)	Discharge coeff. (–)	Inj. dur. (ms)
Non-reacting	1	900	22.8	0	84	0.89	6
	2	1100	15.2	0	84	0.89	6
	3	700	22.8	0	89.4	0.9	1.54
	4	1200	22.8	0	89.4	0.9	1.54
	5	900	7.6	0	89	0.9	1.54
	6	1400	7.6	0	84	0.89	6
Reacting	7	850	22.8	15	91	0.89	6
	8	900	22.8	13	91	0.89	6
	9	900	22.8	15	91	0.89	6
	10	900	22.8	21	91	0.89	6
	11	1000	22.8	15	91	0.89	6

Table 5.4 Spray model constants

Variable	Name	Value	References
Liquid length scaling law constant	C_L	0.62	Siebers (1999)
Spray spreading angle constant	C_θ	0.45	–
Gas-jet-assumed Stokes number	St	3.0	Abani et al. (2007)
Gas-jet entrainment constant	K_{entr}	0.85	Perini et al. (2016)
Max gas-jet weight near nozzle	γ_{max}	0.7	
Min gas-jet weight near nozzle	γ_{min}	0.6	

as the maximum axial distance between the injector exit and the farthest location with 0.1% liquid volume fraction and with 0.1% vapor mass fraction, respectively. The nearest axial location where the local OH mass fraction reaches 2% of its maximum value is used as the lift-off length. The model constants were kept unchanged for all simulations in this section, as summarized in Table 5.4. For the reacting conditions, the combustion is modeled as a well-stirred reactor. A reduced *n*-dodecane mechanism (Wang et al. 2014) with 100 species and 432 reactions is used. A reduced mechanism for polycyclic aromatic hydrocarbons (PAHs) is included, which enables the use of a semi-detailed soot model (Vishwanathan et al. 2010) that uses pyrene as soot precursor, and also considers acetylene- and PAHs-assisted surface growth, particle coagulation, and soot oxidation by oxygen and OH.

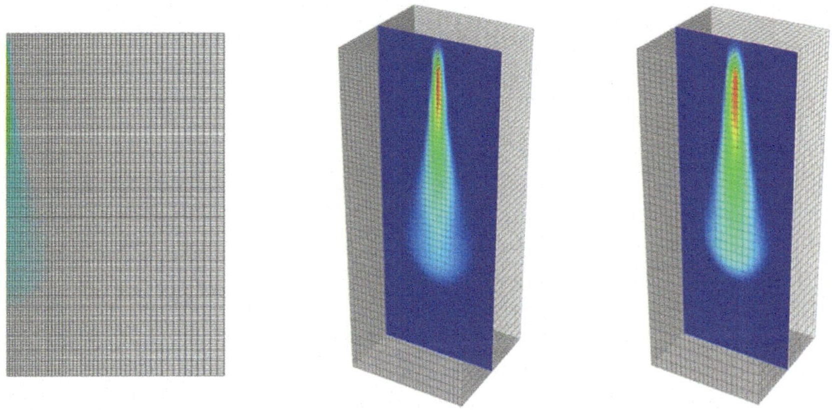

(a) 0.7 mm,2D cylindrical (b) 1 mm, 3D cubic (c) 1.5 mm, 3D cubic

Fig. 5.5 Computational meshes for spray A

5.4.2 Grid Sensitivity

Three computational meshes with different resolutions were used so that the grid sensitivity of the spray model could be tested. The three meshes are: Fig. 5.5a—a 2D cylindrical mesh with a refined resolution of 0.7 mm * 0.7 mm * 0.5° at the nozzle; Fig. 5.5b—a 3D cubic mesh with a refined resolution of 1 mm * 1 mm * 1 mm at the nozzle; and Fig. 5.5c—a 3D cubic mesh with a refined resolution of 1.5 mm * 1.5 mm * 1.5 mm at the nozzle. The 2D cylindrical mesh considers a cylindrical computational domain with a radius of 6.332 cm and a height of 10 cm, which is similar to the geometry of the combustion vessel in the experiments. The 3D cubic meshes consider a domain with a width of 1.8 cm and a height of 10 cm, which is large enough to simulate the spray process.

Figures 5.6 and 5.7 show transient liquid and vapor penetrations and quasi-steady-state mixture fraction distributions for 900 K, 22.8 kg/m^3 and 1100 K, 15.2 kg/m^3 operating conditions. The black lines represent the ensemble average of the experimental measurement, and the shadow area indicates the standard deviation. The red, green, and blue lines are CFD predictions with resolutions of 0.7, 1, and 1.5 mm, respectively. Due to the improved grid independency by using liquid-jet and gas-jet models, the liquid penetration is seen to be almost unaffected by the different mesh sizes. However, deviations are found in the vapor penetrations in the region downstream of the liquid length at 10–20 mm axial distance from the nozzle exit, wherein the penetration rate is lower with the coarser mesh. Although the amount of momentum transfer is accurately predicted, the profiles of gaseous phase velocity and mixture fraction distribution have large radial gradient in the near-nozzle region and cannot be sufficiently resolved by the coarse mesh, which leads to flat spray profile (Figs. 5.6a, b and 5.7a, b) and slow penetration rate. However, as the spray

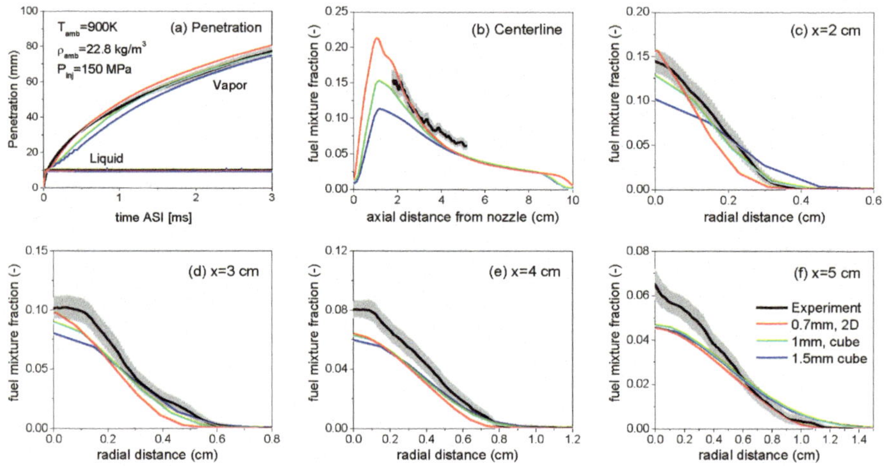

Fig. 5.6 Grid sensitivity for 900 K, 22.8 kg/m^3 operating condition (Yue et al. 2017). Black lines are the experimental results; gray area indicates uncertainty; red lines are results of the 0.5° sector mesh (2D) with a resolution of 0.7 mm; green and blue lines are results of cubic meshes with a resolution of 1 and 1.5 mm, respectively. **a** Vapor and liquid penetrations; mixture fractions **b** along spray centerline, and **c**–**f** along radial directions at four axial distances from the nozzle exit at quasi-steady state

keeps penetrating into farther downstream region, the growth rates of penetration for a different mesh resolution become similar, and the transient penetration profiles are all in good agreement with the experimental data for 1.5–3 ms after the start of injection (ASI). Similarly, in the region of 3–5 cm axial distance, grid convergence is achieved for the prediction of mass fraction distribution, indicating a good grid independency of the current spray model. The 2D mesh with 0.5 cm resolution gives the best match with the experimental measurement under both non-reacting conditions and is used for the rest of this section exclusively.

5.4.3 *Effects of Turbulence Model*

As mentioned in Sect. 5.3.5, a generalized RNG model is used for the turbulence modeling in this work. However, for comparison, simulations were also performed using the modified RNG model (Han et al. 1995) and the standard k-ε model (Launder et al. 1972), and the results are plotted as green lines and blue lines, respectively, in Figs. 5.8 and 5.9. It is seen that the standard k-ε model overpredicts the spray dispersion with slower penetration, which is considered to be a result of an unchanged model constant used for both compressing and expanding flows ($C_3 = -1.0$). For 22.8 kg/m^3, the mRNG model gives a similar result as the gRNG model and only slightly overpredicts the dispersion at downstream locations. This is largely because the compressibility of

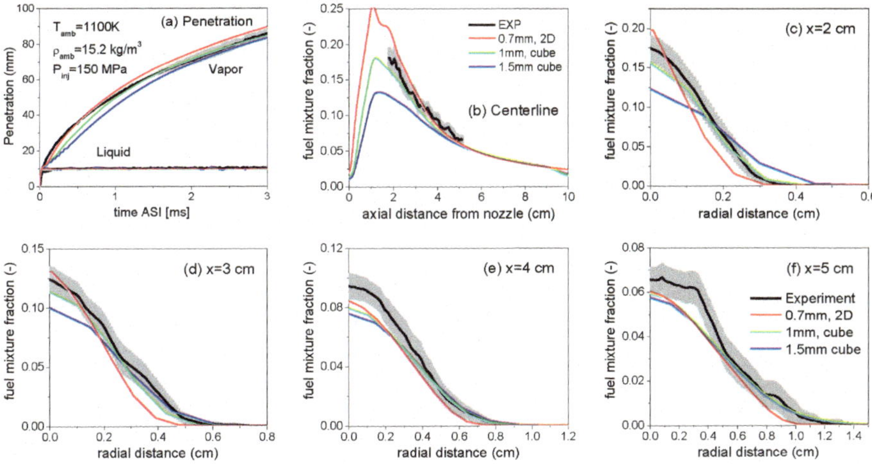

Fig. 5.7 Grid sensitivity for 1100 K, 15.2 kg/m^3 operating condition. See Fig. 5.6 for more explanation

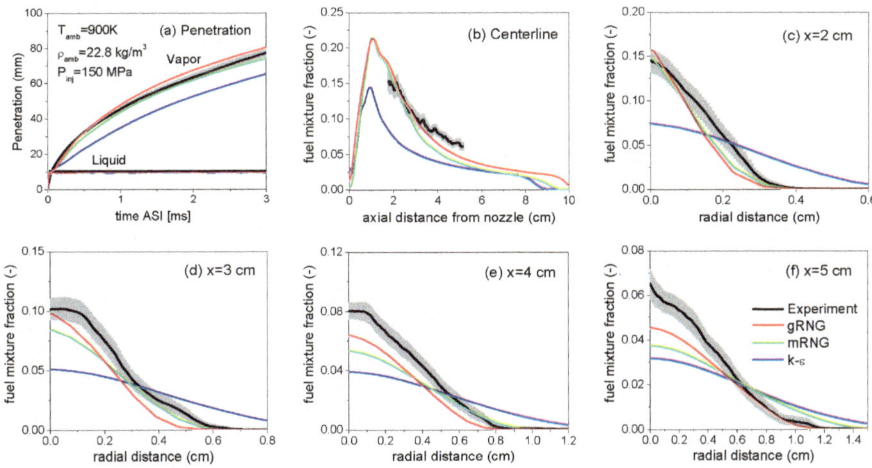

Fig. 5.8 Influence of turbulence models for 900 K, 22.8 kg/m^3 operating condition. Black lines are the experimental results; gray area indicates uncertainty; red, green, and blue lines are CFD results of gRNG, mRNG, and standard k-ε models

flow is also considered in mRNG model $\left(C_3 = \begin{cases} 1.726, \text{ if} \nabla \cdot \mathbf{u} < 0 \\ -0.90, \text{ if} \nabla \cdot \mathbf{u} > 0 \end{cases}\right)$. However, the discrepancy grows for a lower ambient density condition, where mRNG predicts more similarly to the standard k-ε model, as can be seen in Fig. 5.9, indicating that there is better flexibility of the gRNG model in spray modeling for a wide range of operating conditions which is obtained by modeling C_3 as a function of strain rate; see Table 5.2.

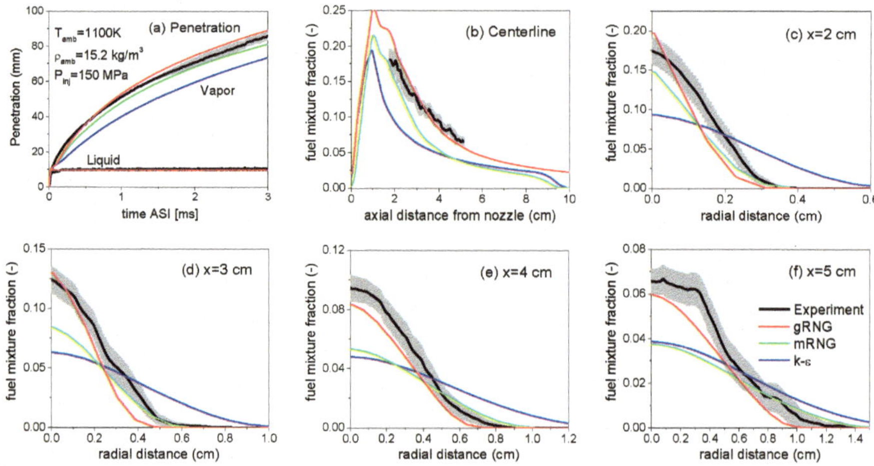

Fig. 5.9 Influence of turbulence models for 1100 K, 15.2 kg/m³ operating condition. See Fig. 5.8 for more explanation

5.4.4 Comparison of the EP and Classical Spray Models

It is also of great interest to compare the EP model with a classical Lagrangian spray modeling approach, which includes the hybrid KH-RT breakup model (Beale et al. 1999), the radius-of-influence (ROI) drop collision model (Munnannur et al. 2009), the gas-jet model (Abani et al. 2007), and the discrete multi-component (DMC) vaporization model (Ra et al. 2009). This integrated approach has been widely employed in simulations of fuel injection in IC engines. In this work, the spray model constants were optimized for condition of 900 K, 22.8 kg/m³ (Wang et al. 2014) and were kept unchanged for all the other cases here.

Figures 5.10 and 5.11 compare the predicted spray penetration and fuel vapor distribution by the EP (red lines) and the classical spray model (green lines), for 900 K, 22.8 kg/m³ and 1100 K, 15.2 kg/m³ operating conditions, respectively. Meanwhile, the KH-RT breakup and ROI collision models were also enabled in couple with the EP model, as shown by the blue lines, to assess the importance of these non-equilibrium processes. The three modeling approaches are seen to give very similar predictions that are also in good agreement with the experimental data. The consideration of drop breakup and collision is shown to have a small influence on the EP model prediction, since these processes neither affect the air entrainment and mixing nor do they have an influence on the Lagrangian-to-Eulerian liquid conversion. The only effect that breakup and collision could have on the EP model prediction is through the calculation of drop drag force, which depends on drop size. However, as seen in Figs. 5.10 and 5.11, this influence is seen to be negligible for such a vaporizing condition with high injection pressure. By contrast, the breakup and collision models are critical in the classical approach since they determine the droplet surface area density and

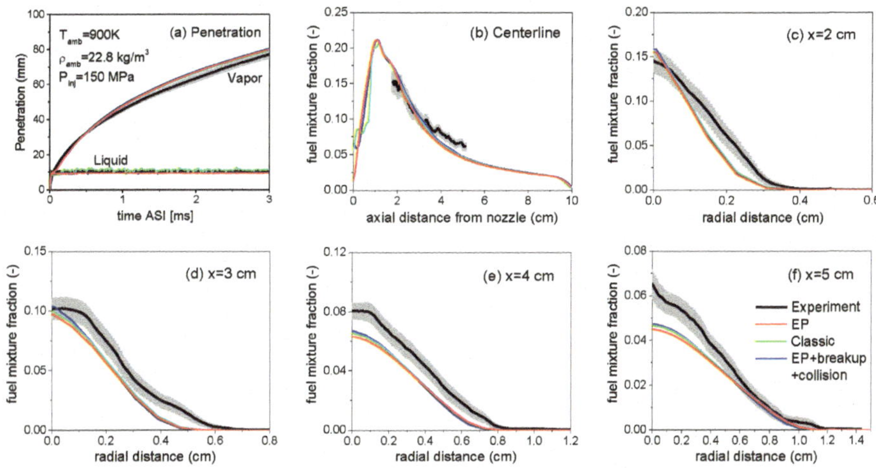

Fig. 5.10 Comparisons of the EP and classical spray models for 900 K, 22.8 kg/m^3 operating condition (Yue et al. 2017). Black lines are the experimental results; gray area indicates uncertainty; red lines are EP model without breakup and collision models; green lines are classical spray model, which features droplet breakup, collision, vaporization models; blue lines are EP model with breakup and collision models

the evaporation rate, which usually have a number of model constants that can be optimized to achieve good prediction for spray simulations. A comparison of liquid lengths predicted by the EP model and the classical model under various ambient temperatures and ambient densities is given in Fig. 5.12. The EP model consistently provides accurate prediction under all conditions, while the classical spray model is shown to be less dependent on the ambient temperature wherein a case-by-case adjustment in the model constants might be necessary to match the experiment under a wide range of conditions. Another advantage of the EP spray model is better computational efficiency since the non-equilibrium processes of breakup, collision, and vaporization are bypassed, as illustrated in Table 5.5. It is also worth to note that the DMC vaporization model used in the classical spray model applies Raoult's law and the Clausius–Clapeyron equation to determine the equilibrium at liquid–gas interfaces, which are less accurate at elevated pressures, e.g., engine operating conditions.

5.4.5 Flame Lift-off Length and Soot Formation

The flame lift-off length and spatial distribution of soot emissions are used to validate reacting sprays of cases 3–7 in Table 5.3. The comparison of CFD predictions and experimental measurement for quasi-steady-state sprays is presented in Fig. 5.13. The spatial distributions of soot volume fraction are shown by the color contour, and

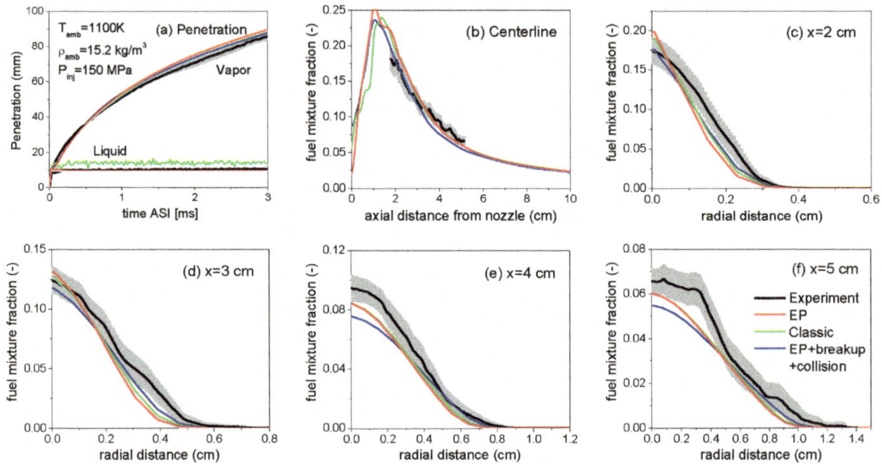

Fig. 5.11 Comparisons of the EP and classical spray models for 1100 K, 15.2 kg/m^3 operating condition. See Fig. 5.10 for more explanation

Fig. 5.12 Quasi-steady-state liquid length for various ambient temperatures and densities (Yue et al. 2017). Red: 22.8 kg/m^3; blue: 15.2 kg/m^3; and black: 7.6 kg/m^3. Solid symbols: experimental results; half symbol: EP spray model; and open symbol: classical spray model

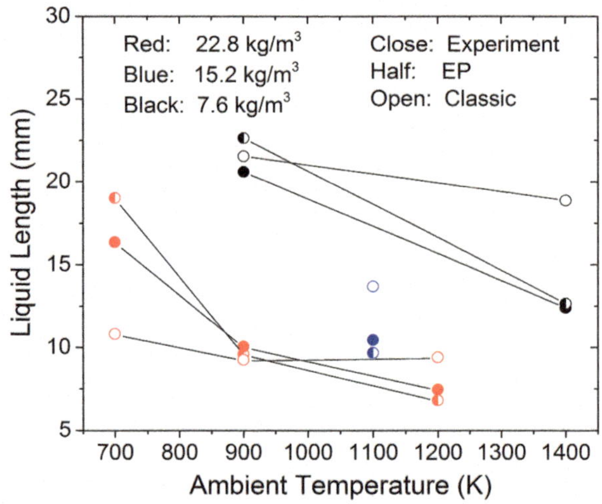

the same scales are applied for both CFD and experimental results. The blue dashed lines mark the locations of flame lift-off lengths. The constants for soot model were kept unchanged for all the operating conditions. At lower ambient temperatures and lower O_2 content, the ignition occurs further downstream with increased lift-off length, which results in better fuel/air mixing and less soot formation. The prediction of soot volume fraction is in good agreement with the experimental data. The location and area of the soot cloud are also well captured, which is attributed to the accurate prediction of mixture fraction in the downstream area.

Table 5.5 Computational time for the EP and classical spray model

Case	CFD time	Time step	Total parcel number	Computational time	
				Classical (min)	EP (min)
900 K, 22.8 kg/m^3	6.5 ms	1 μs	10,000	191.57	139.66
1100 K, 15.2 kg/m^3				147.40	135.05
1400 K, 7.6 kg/m^3				196.14	151.68

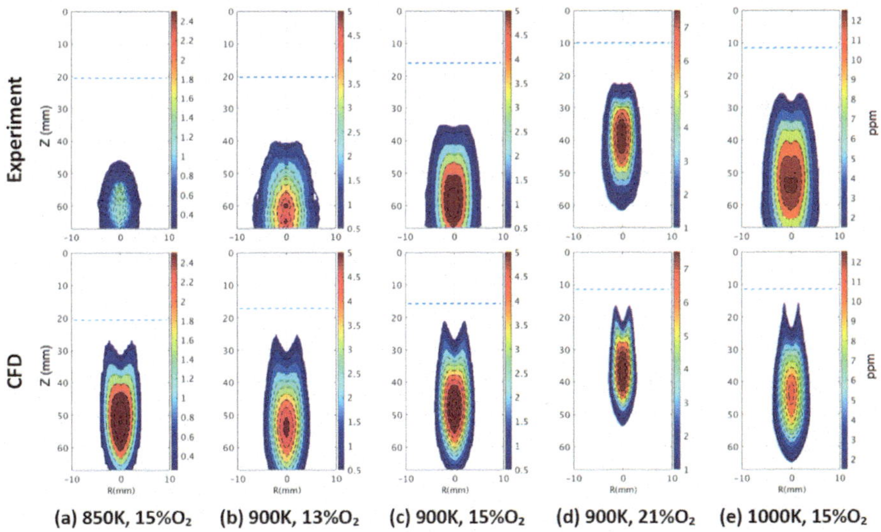

Fig. 5.13 Comparisons of soot from the EP model CFD predictions (Yue et al. 2017) and experimental measurements (Skeen et al. 2016). **a–e** give comparisons of the experiment (left) and CFD (right) for different operating conditions. Color contours indicate the soot volume fraction; blue dashed lines indicate lift-off length

5.5 Gasoline Direct Injection (GDI) Simulation

The previous sections present validation and applications of the EP spray model in simulations of high-pressure, diesel-like fuel sprays. The success of those simulations verifies the mixing-controlled characteristics of high-pressure diesel sprays. On the other hand, GDI engines have drawn increasing attention due to a number of advantages over port fuel injection (PFI) systems, such as accurate fuel delivery, less cycle-to-cycle variation (CCV), better fuel economy with extended knock-limited operating condition, and potential for stratified lean operation (Zhao et al. 1999). Fuel injection is one of the critical processes in a GDI engine, especially for spray-

Table 5.6 Operating conditions and injector specification for GM GDI (Parrish 2014)

Injector	Bosch HDEV 1.2 series
Nozzle holes	8
Orifice diameter	165 μm
L/D ratio	1.8
Plume direction	30° (nominal)/25° (measured)
Fuel	Iso-octane
Injection pressure	20 MPa
Injection duration	766 μs
Ambient temperature	400/500/600/700/800/900 K
Ambient density	3/6/9 kg/m^3

guided combustion systems, which make use of the injection process to form a stably ignitable fuel/air mixture around the spark plug. The multi-hole GDI nozzle has the advantage of providing a stable spray structure, and it is less sensitive to the operating conditions compared to pressure-swirl atomizers (Mitroglou et al. 2007). Despite the similarity to multi-hole diesel injectors, multi-hole gasoline injectors usually feature a narrower drill angle, smaller length/diameter (L/D) ratios, a step-hole design, and a relatively lower fuel pressure (10–40 MPa), with a more volatile, less viscous fuel being injected into a cooler and lower density chamber. In this section, the EP model is applied to simulate sprays of a multi-hole GDI injector and its controlling mechanism is also examined.

The experimental data were taken in a constant volume spray chamber by Parrish (2014), Parrish and Zink (2012). The operating conditions can be found in Table 5.6, showing that three ambient densities were targeted, with temperatures varying from 400 to 900 K. The injector has eight holes with a stepped-hole geometry and a nominal-included angle of 60°. However, the measured value of 50° was used in the simulations. Iso-octane was used as a surrogate for gasoline in the experiments, and the injection pressure was held at 20 MPa for all tested conditions. The duration of the injection was 766 μs, which is shorter and more transient than the ECN Spray A cases. Schlieren and Mie-scatter imagining were used to track the envelopes of the vapor jets and liquid sprays, respectively. Unlike for a single-hole injector, the liquid and vapor penetrations were evaluated for the entire spray rather than for a single plume, and the distance from the nozzle exit to the jet tip was measured along the direction of the injector axis.

For the simulation setup, the computational domain was a cylinder with a diameter of 10 cm and a height of 10 cm. A 45° sector mesh was used, as shown in Fig. 5.14, to simulate one spray plume with the injector located at the top center. A clip plane colored by the mixture fraction is also shown in Fig. 5.14 to illustrate the injector location and spray plume direction. In this case, plume-to-plume interaction is simplified by assuming each plume is identical. The mesh resolution is 0.5 mm * 0.5 mm * 4.5° near the nozzle and transitions to 2.75 mm * 2.75 mm * 4.5° at the sides. As mentioned in Sect. 5.3.1, the model constant C_θ in the spreading

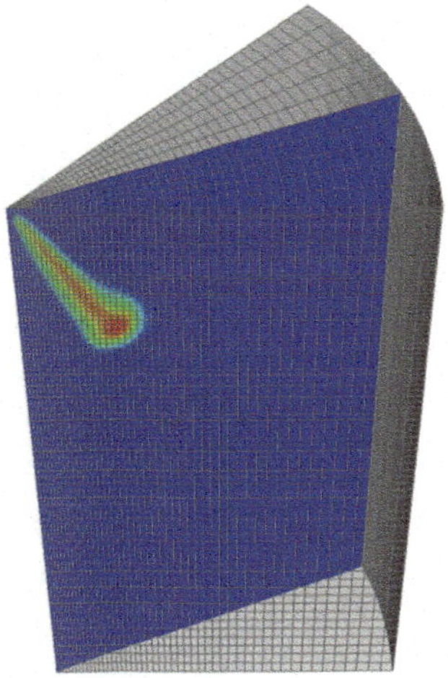

Fig. 5.14 Computational domain for one spray plume simulation

angle correlation depends on the injector configuration and nozzle flow, e.g., cavitation. Therefore, the value of C_θ for a GDI injector is expected to be different than the value used for the Spray A injectors. Due to the lack of the experimental measurement of spreading angles, C_θ was determined to be 0.6, based on a best match of jet penetrations for the 6 kg/m^3 ambient density cases.

The predictions of liquid and vapor envelopes are compared with the experimental measurements for several conditions, as shown in Figs. 5.15, 5.16, and 5.17, for three ambient temperatures, respectively. Three ambient densities are shown for each ambient temperature. For each operating condition, results are compared at 0.5, 1.0, and 1.5 ms after the start of injection in each row. The experimental results are shown on the left in each pair of the experiment/CFD comparison. Red lines and green lines outline the liquid and vapor envelopes, respectively, and the black dashed lines mark the measured spray-included angle. For the CFD predictions, white clouds represent the vapor plume and red clouds represent the liquid jet. The result of a simulated single plume in a sector mesh was replicated and rotated to form an eight-plume spray in order to compare with the experimental results.

The general trend is that both spray penetration and liquid residence time decrease with increasing ambient temperature and increasing ambient density, which is observed in both the experimental results and the CFD predictions. According to the mixing-controlled spray analysis, higher ambient temperature at constant ambient density results in higher internal energy per unit entrained gas and consequently leads to faster vaporization and shorter liquid length. On the other hand, higher ambient density at constant ambient temperature results in a wider spray cone angle that leads to a faster entrainment rate; thus, the vaporization is also accelerated. The included angle and shape of the vapor envelopes seen in the experiments are accurately captured by the simulations at each transient time for each operating condition.

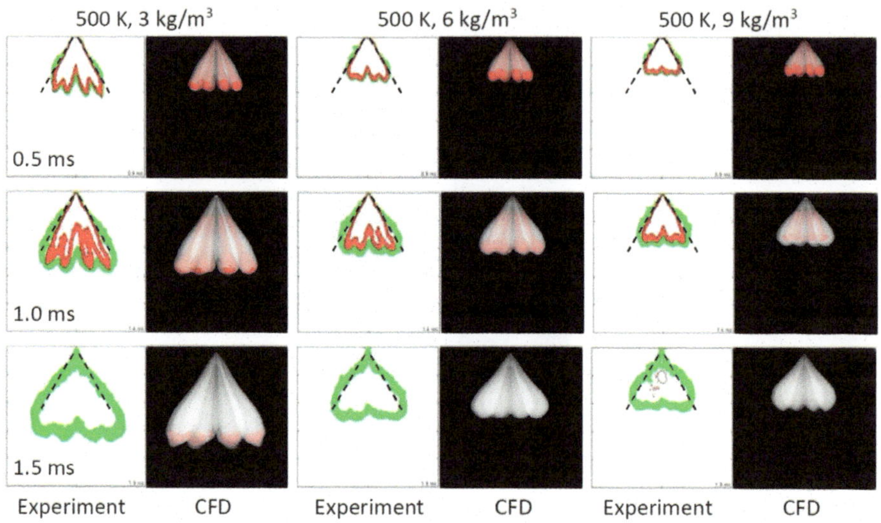

Fig. 5.15 Liquid and vapor envelopes for 500 K conditions. Ambient densities are 3, 6, 9 kg/m^3 from left to right. Each row corresponds to 0.5, 1.0, 1.5 ms after start of injection, respectively. For the experiment, red and green lines outline the envelopes of liquid and vapor phases, respectively. For CFD, red and white clouds represent liquid jet and vapor plume, respectively

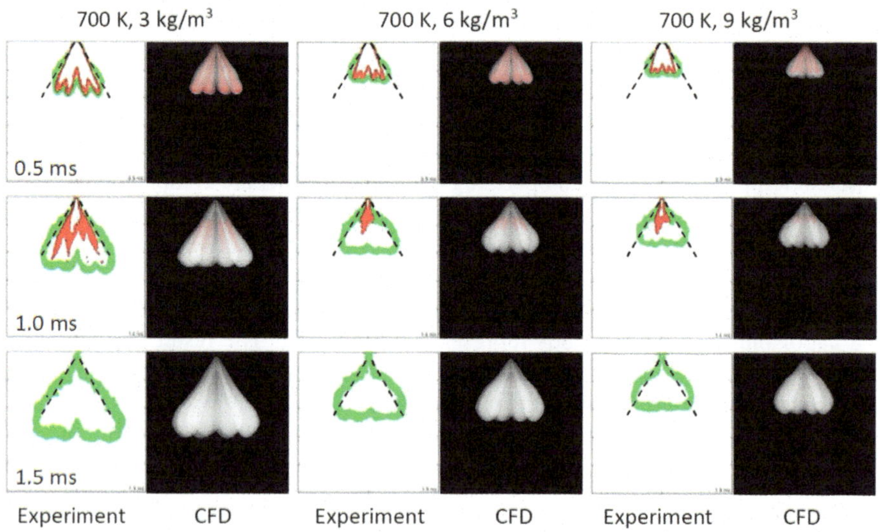

Fig. 5.16 Liquid and vapor envelopes for 700 K conditions. See Fig. 5.15 for more explanations

The fingerlike shape of the liquid jet is also well predicted by the model, with slight overprediction in liquid residence time after the end of injection for low ambient

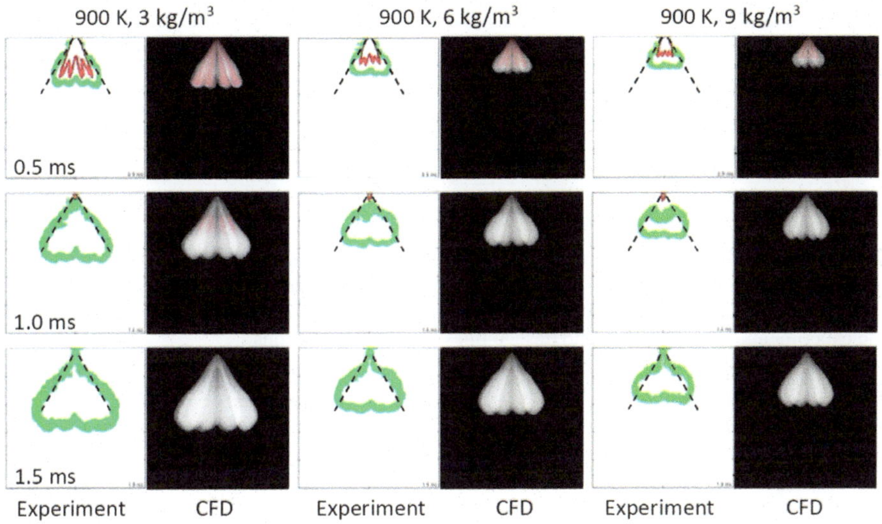

Fig. 5.17 Liquid and vapor envelopes for 900 K conditions. See Fig. 5.15 for more explanations

density conditions, as can be seen at 1.5 ms ASI of the 500 K, 3 kg/m^3 condition and at 1.0 ms ASI of the 900 K, 3 kg/m^3 condition.

The liquid and vapor penetrations are plotted as line graphs in Fig. 5.18. Solid lines are the experimental results, while dashed lines are the CFD predictions and colors indicate different ambient densities. Similar to observations in the 2D spray envelopes, the transient penetration of the vapor plume is accurately predicted by the model predictions with less than 5% error even at 3.5 ms ASI. Quasi-steady state of liquid length is reached after an initial transient period for both the 700 and 900 K cases. However, such a quasi-steady period is not seen for 500 K, indicating the quasi-steady-state liquid length cannot be reached within such a short duration of injection. Even though the current EP spray model is derived based on the experiments of long-duration, quasi-steady-state fuel injection, its application in the present GDI simulations is seen to capture the transient behavior of a short-duration injection very well, as shown in the predictions of liquid penetration for the 6 and 9 kg/m^3 conditions. Furthermore, although the liquid length is considerably overpredicted at low ambient density conditions, the transient vapor penetration is still accurately captured by the model.

Figure 5.19 shows the maximum liquid length as a function of ambient temperature for each ambient density condition. Excellent agreements can be found between the experimental results and the CFD predictions for most cases. However, the error in the prediction of maximum liquid length grows with decreasing ambient density and temperature. In this case, the validity of mixing-controlled assumption is considered not to be the reason of such error, as the phase equilibrium, mixing-controlled vaporization process would be expected to always give a faster vaporization and a shorter liquid length compared to the non-equilibrium vaporization process. Pos-

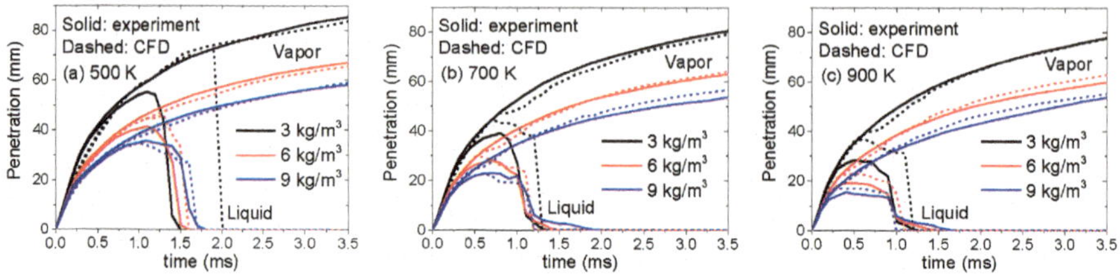

Fig. 5.18 Liquid and vapor penetrations. **a, b,** and **c** are liquid penetrations for 500, 700, and 900 K ambient temperatures, respectively; **d, e,** and **f** are their corresponding vapor penetrations. Black, red, and blue lines correspond to 3, 6, and 9 kg/m^3 ambient densities, respectively. Solid lines are the experimental measurement, and dashed lines are CFD predictions

Fig. 5.19 Liquid length as a function of ambient temperature. Black, red, blue lines correspond to 3, 6, 9 kg/m^3 ambient densities, respectively. Solid triangles are the experimental results; hollow triangles are CFD predictions

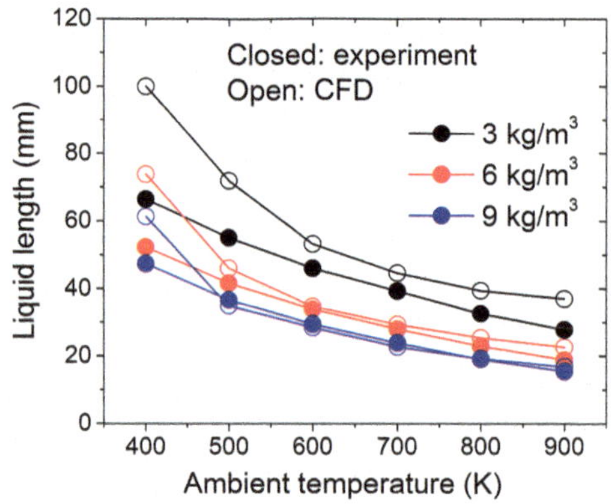

sibly, the constant value of C_L in Eq. (5.6) is questionable. C_L has a theoretical value of 0.38 from the derivation of Siebers' scaling law and was adjusted to 0.62 in order to match the *n*-hexadecane and heptamethylnonane liquid length data (Siebers 1999). The change in C_L is considered to be a compensation for errors introduced by assumptions made when deriving the liquid length scaling law, which should not be expected to be the same when the operating condition changes significantly.

Moreover, the accuracy of the correlation for the spray spreading angle θ can also affect the liquid length prediction. Particularly, liquid cavitation within the nozzle passages is a possible agency in the determination of C_θ in Eq. (5.15) (Reitz et al. 1982), which is not considered in this study. The cavitation number K is usually used to quantify the cavitation transition, which is defined as $K = \left(P_{\text{inj}} - P_{\text{amb}}\right)/(P_{\text{amb}} - P_v)$. P_{inj} is the injection pressure, P_{amb} is the ambient pressure, and P_v is the fuel vapor pressure. For the studied cases in this section, the value of K ranges from 7.5 for the highest ambient pressure condition to 70.4 for the lowest ambient pressure condition, which is estimated using P_v of 0.78 for iso-octane at 363 K. A higher K value indicates higher tendency of cavitation and consequently wider spreading angle. Thus, the consideration of the K factor could mitigate the deviation in liquid length prediction seen at low-pressure conditions.

Nonetheless, the current form of the EP spray model provides satisfying results for most engine-relevant conditions, especially for the vapor plume, which is essential in engine combustion simulations.

5.6 Summary

Experimental studies have shown that the vaporization process of a high injection pressure engine spray is controlled by the hot ambient gas entrainment and the overall fuel/air mixing within the spray, instead of being controlled by transport rates of mass and energy at spray droplet surfaces. This conclusion is valid for fuel injection problem under most of IC engine operating conditions. However, most CFD models that have been applied in IC engine simulations use the LDEF method for the modeling of the two-phase flow process of liquid fuel injection, where the non-equilibrium processes of drop breakup, collision and coalescence, and liquid/vapor interfacial vaporization are modeled by tracking droplets with considerable modeling and computational effort. An equilibrium phase (EP) spray model has been recently developed and implemented into an open-source CFD program, KIVA-3vr2 (Amsden 1999), for application to IC engine simulation, where the role of mixing-controlled vaporization is emphasized by the introduction of an Eulerian liquid phase into the LDEF framework and employing an advanced phase equilibrium solver. The EP spray model was developed based on a jet theory and a phase equilibrium assumption, without the need for modeling drop breakup, collision, and surface vaporization processes.

The integrated model was validated widely in the ECN Spray A, as well as GDI sprays. The validations confirm the good accuracy and grid independency of the EP model in predictions of liquid/vapor penetrations, fuel mass fraction distributions for the non-reacting sprays, and heat release rates, pressure traces, and emission formation for reacting cases, over a wide range of operating conditions (700 K to 1400 K, 7.6–22.8 kg/m^3 for diesel sprays; 500–900 K, 3–9 kg/m^3 for GDI sprays) without the need for tuning of model constants, except for the constant in the spreading angle correlation, which is expected to change for different injector configurations. The accuracy of the liquid length prediction for GDI sprays at low ambient densities could potentially be improved even further by using a comprehensive spray cone angle model that considers in-nozzle flow, cavitation, etc. Moreover, the EP model was also proven to be 37% more computationally efficient than the widely used 'classical' model for the studied cases.

Acknowledgements This work was finished during the Ph.D. project of the corresponding author, Yue, Z., at the University of Wisconsin–Madison. The authors would like to acknowledge the financial support provided by the China Scholarship Council (CSC). The authors are also thankful for Dr. Lu Qiu for sharing the code of phase equilibrium solver and Dr. Randy Hessel for technical discussion.

References

Abani N, Reitz RD (2007) Unsteady turbulent round jets and vortex motion. Phys Fluids 19(12). https://doi.org/10.1063/1.2821910

Abani N, Kokjohn S, Park SW, Bergin M, Munnannur A, Ning W et al (2008a) An improved spray model for reducing numerical parameter dependencies in diesel engine CFD simulations, pp 776–790

Abani N, Munnannur A, Reitz RD (2008b) Reduction of numerical parameter dependencies in diesel spray models. J Eng Gas Turbines Power 130:32809. https://doi.org/10.1115/1.2830867

Amsden AA (1999) KIVA-3V, release 2: improvements to KIVA-3v. Los Alamos, NM

Amsden AA, O'Rourke PJ, Butler TD (1989) KIVA II: A computer program for chemically reactive flows with sprays. Los Alamos, NM

Augustin U, Schwarz V (1991) Low-noise combustion with pilot injection. Truck Technology International

Baker LE, Pierce AC, Luks KD (1982) Gibbs energy analysis of phase equilibria. Soc Petrol Eng J 22(5):731–742. https://doi.org/10.2118/9806-PA

Battistoni M, Magnotti GM, Genzale CL, Arienti M, Matusik KE, Duke DJ, … Marti-Aldaravi P (2018) Experimental and computational investigation of subcritical near-nozzle spray structure and primary atomization in the engine combustion network spray D. SAE technical paper, (2018-01–0277), pp 1–15. https://doi.org/10.4271/2018-01-0277

Beale JC, Reitz RD (1999) Modeling spray atomization with the Kelvin-Helmotz/Raleigh-Taylor hybrid model. Atomization Sprays 9(6):623–650. https://doi.org/10.1615/AtomizSpr.v9.i6.40

Gueyffier D, Li J, Nadim A, Scardovelli R, Zaleski S (1999) Volume-of-fluid interface tracking with smoothed surface stress methods for three-dimensional flows. J Comput Phys 152(2):423–456. https://doi.org/10.1006/jcph.1998.6168

Han Z, Reitz RD (1995) Turbulence modeling of internal combustion engines using RNG κ-ε models. Combust Sci Technol 106(4–6):267–295. https://doi.org/10.1080/00102209508907782

Han Z, Uludogan A, Hampson GJ, Reitz RD (1996) Mechanism of soot and NO_x emission reduction using multiple-injection in a diesel engine. https://doi.org/10.4271/960633

Higgins B, Siebers DL (2001) Measurement of the flame lift-off location on DI diesel sprays using OH chemiluminescence (724). https://doi.org/10.4271/2001-01-0918 http://www.cmt.upv.es/ECN03.aspx (n.d.)

Idicheria CA, Pickett LM (2005) Soot formation in diesel combustion under high-EGR conditions (724). https://doi.org/10.4271/2005-01-3834

Iyer VA, Post SL, Abraham J (2000) Is the liquid penetration in diesel sprays mixing controlled? Proc Combust Inst 28(1):1111–1118. https://doi.org/10.1016/S0082-0784(00)80321-5

Knapp H (1982) Vapor-liquid equilibria for mixtures of low-boiling substances, vol 6. DECHEMA, Frankfurt

Lacaze G, Oefelein JC (2012) A non-premixed combustion model based on flame structure analysis at supercritical pressures. Combust Flame 159(6):2087–2103. https://doi.org/10.1016/j.combustflame.2012.02.003

Launder BE, Spalding DB (1972) Mathematical models of turbulence. Academic Press, New York, NY

Levich VG (1962) Physicochemical hydrodynamics. Prentice-Hall, Englewood Cliffs

Liu AB, Mather D, Reitz RD (1993) Modeling the effects of drop drag and breakup on fuel sprays. SAE Int Congr Exposition 298(412):1–6. https://doi.org/10.4271/93007

Manin J, Pickett LM, Skeen SA (2013) Two-color diffused back-illumination imaging as a diagnostic for time-resolved soot measurements in reacting sprays. SAE Int J Engines 6(4):2013-01–2548. https://doi.org/10.4271/2013-01-2548

Matheis J, Hickel S (2017) Multi-component vapor-liquid equilibrium model for LES of high-pressure fuel injection and application to ECN spray A. Int J Multiph Flow 99:294–311. https://doi.org/10.1016/j.ijmultiphaseflow.2017.11.001

Ménard T, Tanguy S, Berlemont A (2007) Coupling level set/VOF/ghost fluid methods: validation and application to 3D simulation of the primary break-up of a liquid jet. Int J Multiph Flow 33(5):510–524. https://doi.org/10.1016/j.ijmultiphaseflow.2006.11.001

Michelsen ML (1982) The isothermal flash problem. Part I. stability. Fluid Phase Equilib 9(1):1–19. https://doi.org/10.1016/0378-3812(82)85001-2

Mitroglou N, Nouri JM, Yan Y, Gavaises M, Arcoumanis C (2007) spray structure generated by multi-hole injectors for gasoline direct-injection engines (724):776–790. https://doi.org/10.4271/2007-01-1417

Munnannur A, Reitz RD (2009) Comprehensive collision model for multidimensional engine spray computations. Atomization Sprays 19(7):597–619. https://doi.org/10.1615/AtomizSpr.v19.i7.10

Musculus MPB, Miles PC, Pickett LM (2013) Conceptual models for partially premixed low-temperature diesel combustion. Prog Energy Combust Sci 39(2–3):246–283. https://doi.org/10.1016/j.pecs.2012.09.001

Nehmer DA, Reitz RD (1994) Measurement of the effect of injection rate and split injections on diesel engine soot and NO_x emissions (412). https://doi.org/10.4271/940668

Parrish SE (2014) Evaluation of liquid and vapor penetration of sprays from a multi-hole gasoline fuel injector operating under engine-like conditions. SAE Int J Engines 7(2):2014-01–1409. https://doi.org/10.4271/2014-01-1409

Parrish SE, Zink RJ (2012) Development and application of imaging system to evaluate liquid and vapor envelopes of multi-hole gasoline fuel injector sprays under engine-like conditions. Atomization Sprays 22(8):647–661. https://doi.org/10.1615/AtomizSpr.2012006215

Perini F, Reitz RD (2016) Improved atomization, collision and sub-grid scale momentum coupling models for transient vaporizing engine sprays. Int J Multiph Flow 79:107–123. https://doi.org/10.1016/j.ijmultiphaseflow.2015.10.009

Pickett LM, Genzale CL, Bruneaux G, Malbec L-M, Hermant L, Christiansen C, Schramm J (2010) Comparison of diesel spray combustion in different high-temperature, high-pressure facilities. SAE Int J Engines 3(2):2010-01–2106. https://doi.org/10.4271/2010-01-2106

Pickett LM, Manin J, Genzale CL, Siebers DL, Musculus MPB, Idicheria CA (2011) Relationship between diesel fuel spray vapor penetration/dispersion and local fuel mixture fraction. SAE Int J Engines 4(1):2011-01–0686. https://doi.org/10.4271/2011-01-0686

Pierpont DA, Montgomery DT, Reitz RD (1995) Reducing particulate and NO_x using multiple injections and EGR in a D.I. Diesel. SAE technical paper, 950217(950217). https://doi.org/10.4271/940897

Qiu L, Reitz RD (2014a) Investigating fuel condensation processes in low temperature combustion engines. J Eng Gas Turbines Power 137(10):101506. https://doi.org/10.1115/1.4030100

Qiu L, Reitz RD (2014b) Simulation of supercritical fuel injection with condensation. Int J Heat Mass Transf 79:1070–1086. https://doi.org/10.1016/j.ijheatmasstransfer.2014.08.081

Qiu L, Reitz RD (2015) An investigation of thermodynamic states during high-pressure fuel injection using equilibrium thermodynamics. Int J Multiph Flow 72:24–38. https://doi.org/10.1016/j.ijmultiphaseflow.2015.01.011

Qiu L, Wang Y, Jiao Q, Wang H, Reitz RD (2014) Development of a thermodynamically consistent, robust and efficient phase equilibrium solver and its validations. Fuel 115:1–16. https://doi.org/10.1016/j.fuel.2013.06.039

Ra Y, Reitz RD (2009) A vaporization model for discrete multi-component fuel sprays. Int J Multiph Flow 35(2):101–117. https://doi.org/10.1016/j.ijmultiphaseflow.2008.10.006

Rachford HH, Rice JD (1952) Procedure for use of electronic digital computers in calculating flash vaporization hydrocarbon equilibrium. J Petrol Technol 4(10):3–19. https://doi.org/10.2118/952327-G

Ranz WE (1958) Some experiments on orifice sprays. Can J Chem Eng 36(4):175–181. https://doi.org/10.1002/cjce.5450360405

Reid RC, Prausnitz JM, Poling BE (1987) The properties of gases and liquids, 4th edn. McGraw-Hill, New York

Reitz RD (1987) Modeling atomization processes in high-pressure vaporizing sprays. Atomization Spray Technol 3:309–337

Reitz RD, Bracco FB (1979) On the dependence of spray angle and other spray parameters on nozzle design and operating conditions. https://doi.org/10.4271/790494

Reitz RD, Bracco FV (1982) Mechanism of atomization of a liquid jet. Phys Fluids 25(10):1730–1742. https://doi.org/10.1063/1.863650

Reitz RD, Diwakar R (1987) Structure of high-pressure fuel sprays. SAE technical paper, (870598). https://doi.org/10.4271/870598

Reitz RD, Duraisamy G (2014) Review of high efficiency and clean reactivity controlled compression ignition (RCCI) combustion in internal combustion engines. Progr Energy Combust Sci. https://doi.org/10.1016/j.pecs.2014.05.003

Ricart LM, Xin J, Bower GR, Reitz RD (1997) In-cylinder measurement and modeling of liquid fuel spray penetration in a heavy-duty diesel engine. SAE Trans 106(1):1622–1640

Saha K, Quan S, Battistoni M, Som S, Senecal PK, Pomraning E (2017) Coupled Eulerian internal nozzle flow and Lagrangian spray simulations for GDI systems. https://doi.org/10.4271/2017-01-0834

Saha K, Srivastava P, Quan S, Senecal PK, Pomraning E, Som S (2018) Modeling the dynamic coupling of internal nozzle flow and spray formation for gasoline direct injection applications. SAE technical paper, (2018-01–0314), 1–13. https://doi.org/10.4271/2018-01-0314.Abstract

Schulte H, Scheid E, Pischinger F, Reuter U (1989) Preinjection—a measure to influence exhaust quality and noise in diesel engines. J Eng Gas Turbines Power 111(3):445. https://doi.org/10.1115/1.3240274

Siebers DL (1998) Liquid-phase fuel penetration in diesel sprays. SAE technical paper series 107(724):1205–1227. https://doi.org/10.4271/1999-01-0528

Siebers DL (1999) Scaling liquid-phase fuel penetration in diesel sprays based on mixing-limited vaporization. SAE paper 1999-01-0528, (724):1–528. https://doi.org/10.4271/1999-01-0528

Skeen SA, Manin J, Pickett LM, Cenker E, Bruneaux G, Kondo K, … Hawkes E (2016) A progress review on soot experiments and modeling in the engine combustion network (ECN) SAE Int J Engines 9(2). https://doi.org/10.4271/2016-01-0734

Som S, Aggarwal SK (2010) Effects of primary breakup modeling on spray and combustion characteristics of compression ignition engines. Combust Flame 157(6):1179–1193. https://doi.org/10.1016/j.combustflame.2010.02.018

Su TF, Patterson MA, Reitz RD, Farrell PV (1996) Experimental and numerical studies of high pressure multiple injection sprays. SAE technical paper 960861, (412). https://doi.org/10.4271/960861

Tow TC, Pierpont DA, Reitz RD (1994) Reducing particulate and NO_x emissions by using multiple injections in a heavy duty D.I. diesel engine. SAE technical paper, 940897(940897). https://doi.org/10.4271/950217

Vishwanathan G, Reitz RD (2010) Development of a practical soot modeling approach and its application to low-temperature diesel combustion. Combust Sci Technol 182. https://doi.org/10.1080/00102200903548124

Wang B-L, Miles PC, Reitz RD, Han Z, Petersen B (2011a) Assessment of RNG turbulence modeling and the development of a generalized RNG closure model. https://doi.org/10.4271/2011-01-0829

Wang Y, Lee WG, Reitz RD, Diwakar R (2011b) Numerical simulation of diesel sprays using an Eulerian-Lagrangian spray and atomization (ELSA) model coupled with nozzle flow. https://doi.org/10.4271/2011-01-0386

Wang H, Ra Y, Jia M, Reitz RD (2014) Development of a reduced n-dodecane-PAH mechanism and its application for n-dodecane soot predictions. Fuel 136:25–36. https://doi.org/10.1016/j.fuel.2014.07.028

Xue Q, Battistoni M, Powell CF, Longman DE, Quan SP, Pomraning E, … Som S (2015) An Eulerian CFD model and X-ray radiography for coupled nozzle flow and spray in internal combustion engines. Int J Multiph Flow 70:77–88. https://doi.org/10.1016/j.ijmultiphaseflow.2014.11.012

Yakhot V, Orszag SA, Thangam S, Gatski TB, Speziale CG (1992) Development of turbulence models for shear flows by a double expansion technique. Phys Fluids A 4(7):1510–1520. https://doi.org/10.1063/1.858424

Yue Z, Reitz RD (2017) An equilibrium phase (EP) spray model for high pressure fuel injection and engine combustion simulations. Int J Engine Res 1–13. https://doi.org/10.1177/1468087417744144

Yue Z, Reitz RD (2018) Application of an equilibrium phase (EP) spray model to multi-component gasoline direct injection. In: 14th triennial international conference on liquid atomization and spray systems. Chicago

Yue Z, Battistoni M, Som S (2018a) A numerical study on spray characteristics at start of injection for gasoline direct injection. In: 14th triennial international conference on liquid atomization and spray systems. Chicago

Yue Z, Hessel R, Reitz RD (2018b) Investigation of real gas effects on combustion and emissions in internal combustion engines and implications for development of chemical kinetics mechanisms. Int J Engine Res 19(3):269–281. https://doi.org/10.1177/1468087416678111

Zhao F, Lai MC, Harrington DL (1999) Automotive spark-ignited direct-injection gasoline engines. Prog Energy Combust Sci 25(5):437–562. https://doi.org/10.1016/S0360-1285(99)00004-0

Chapter 6
Droplet Impingement and Evaporation on a Solid Surface

Seong-Young Lee and Le Zhao

Abstract An efficient spray injection leads to better vaporization and better air—fuel mixing, resulting in the stable combustion and reduced emissions in the internal combustion (IC) engines. The impingement of liquid fuels on chamber wall or piston surface in IC engines is a common phenomenon, and fuel film formed during the impingement plays a critical role in engine performance and emissions, particularly under cold start conditions. Therefore, the study on the characteristics of spray impingement on the chamber wall or piston surface is necessary. However, first, due to the complexity of the practical fuel injection systems, it is difficult to attain the detailed specific information of the spray impingement from the experiments such as droplet size, mass, number, and velocity distributions in the vicinity of wall region. Second, because of the Lagrangian particle/parcel concept (a particle representing a number of droplets in simulations), the spray–wall interaction model under Eulerian–Lagrangian approach is often developed based on the individual droplet. Therefore, the individual droplet's impingement on wall and the droplet-to-droplet collision have been extensively studied to assist in a profound perception on the spray–wall impingement. In this chapter, the encouraging experimental observations of applying optical diagnostics technology to study droplet–wall impingement are extensively discussed. Single droplet impingement on a solid surface with various conditions was examined to understand the detailed impinging dynamic process. The droplet–wall interaction outcomes, in particular focusing on the splashing criteria, were inspected, and a new correlation of deposition–splashing is developed. Post-impingement characterizations including spreading factor, height ratio, contact line velocity, and dynamic contact angle were then analyzed based on the experimental data at various test conditions. Further, the non-evaporation volume of fluid (VOF) method based on Eulerian approach was used to characterize single droplet impinging on the wall and provide a better understanding of the dynamic impact process. The simulation results of the spreading factor and height ratio matched well with the experimental results during the droplet impingement process. In addition, due

S.-Y. Lee (✉) · L. Zhao
Mechanical Engineering-Engineering Mechanics,
Michigan Technological University, Houghton, MI, USA
e-mail: sylee@mtu.edu

K. Saha et al. (eds.), *Two-Phase Flow for Automotive and Power Generation Sectors*, Energy, Environment, and Sustainability,
https://doi.org/10.1007/978-981-13-3256-2_6

to the evaporation drawing more attention during the engine combustion process, an evaporation VOF (e-VOF) sub-model was developed and applied to multi-droplet impingement on a heated surface to qualitatively and quantitatively analyze the vaporizing process as droplets impacting onto the hot surface. The information obtained from VOF simulations can be applied to improve the spray–wall interaction models in the liquid spray Eulerian–Lagrangian method.

6.1 Introduction

This chapter introduces the characteristics of a single droplet–wall impingement under non-vaporizing and evaporation conditions.

6.1.1 Droplet–Wall Impingement Mechanism Overview

The fluid dynamic phenomena during single droplet (order of millimeters) impingement on a solid surface can be characterized by a few parameters, such as fuel properties (i.e., surface tension, viscosity, wettability, and density) and operating conditions (i.e., surface temperature, surrounding gas temperature, and pressure) (Potham et al. 2017). The impingement occurrence when a liquid droplet interacts with a solid surface (dry, wetted, isothermal, or hot) can be categorized as the phenomena of stick, spread, rebound, splash, and breakup (Bai et al. 2002) as shown in Fig. 6.1. The droplet interaction with a dry or wetted solid surface has been studied by a number of researchers over many decades; all the impingement processes are found to be strongly related to the droplet impact energy (Bai et al. 2002; Habchi et al. 1999). The droplet sticks on the wall when the impact energy is very low, while the liquid spreads and rebounds as the impact energy increases until all the energy is dissipated. The droplet disintegrates within the first instant after impingement as the impact energy further increases; when the droplet interacts with the surface by leaving some liquid on the surface (contributing to the wall-film formation) and splashing back the remaining part, the splashing occurs. This remaining part contains droplets that have different sizes and velocities with respect to the one that originally impinged on the wall (Bai et al. 2002; Habchi et al. 1999; Moreira et al. 2010).

The mechanism of droplet splashing is inherently complex compared with other interaction outcomes, and it largely depends on both the boundary conditions (e.g., liquid droplet velocity and impact angle) and surface topography. Hence, a number of experimental works of the droplet impinging on the dry or wetted surface have been done, and the splash criteria are summarized based on these experimental works. The splashing criterion is usually derived in terms of non-dimensional parameters to characterize the relative magnitude of the forces acting on the droplet. The most general parameters utilized to express the splashing phenomenon are Reynolds number ($Re = \rho D_0 U_0/\mu$), Weber number $\left(We = \rho D_0 U_0^2/\sigma\right)$, Ohnesorge number

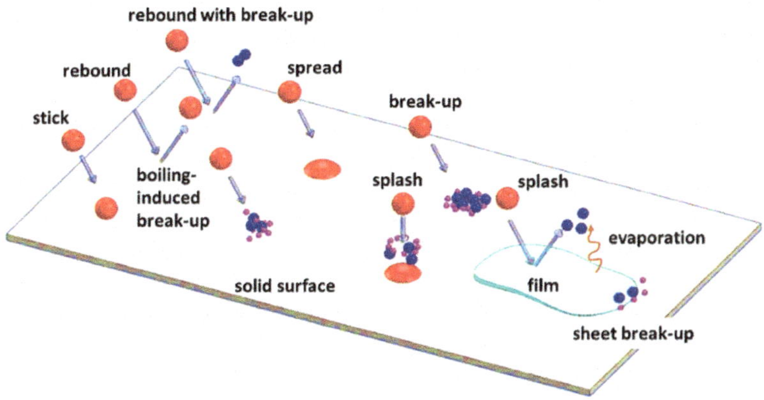

Fig. 6.1 Schematic of the droplet impingement on a solid surface (dry/wetted/isothermal/hot surface) (Zhao et al. 2017)

$\left(Oh = We^{0.5}Re\right)$, and Capillary number ($Ca = We/Re$) ($\rho$, μ, and σ are the density, dynamic viscosity, and surface tension, respectively; D_0 is the initial droplet diameter and U_0 is the impact velocity). It is worthwhile to note that the parameters of *Re, We, Oh*, and *Ca* are calculated from the normal velocity component of the impinging droplet with respect to the surface.

Stow and Hadfield (1981) led one of the earliest experimental studies to understand the water droplets' impingement on a roughened aluminum surface. The dependence of droplet–wall interaction phenomena on the *Re* and *We* of liquid fuel and surface roughness was also performed. They postulated a splashing threshold $K = We^{0.5}Re^{0.25}$, in which the value of K is highly dependent on the surface roughness (Lindagren and Denbratt 2004). Yarin and Weiss (1995) studied the single train of droplets falling on a solid substrate with a thin film at a known impinging frequency (f). They proposed a splashing mechanism and found a splashing threshold as a function of impact parameters of a droplet: *Ca* and non-dimensional viscosity length $\left(\lambda = \left(\frac{\nu}{f}\right)^{0.5}\sigma/\left(\rho\nu^2\right)\right)$, as seen in Eq. (6.1) ($\nu$ is kinematic viscosity), where the dimensionless impact velocity (u) is introduced. They found that splashing threshold does not rely on droplet diameter and is slightly affected by mean surface roughness. They also concluded that the splashing threshold at $u = 17$–18 corresponds to the developed crown instability; a group of secondary droplets will be produced with this splashing threshold.

$$Ca\lambda^{\frac{3}{4}} = Constant = u = U_0\left(\frac{\rho}{\sigma}\right)^{\frac{1}{4}}\nu^{-\frac{1}{8}}f^{-\frac{3}{8}} > 17 \sim 18 \tag{6.1}$$

Nevertheless, as the derived splashing threshold provides an explanation only for the corona splash but not for the prompt splash mechanism, this criterion does not hold true for many cases. Corona splash arises from the instabilities in the rim of the crown (Yarin and Weiss 1995), and prompt splashing arises at the contact line in the beginning of the spreading phase (Rioboo et al. 2001). Additionally, this correlation posed under an assumption of no direct interaction of droplets with the solid dry surface but rather with a thin liquid film. Therefore, it may not be applied for droplet interaction directly with a dry surface.

Another major study to discuss the splashing process of a droplet impingement on a solid surface was done by Mundo et al. (1995). They formulated an empirical model to investigate the deposition and splashing regimes, using the train of monodispersed droplets by varying liquid properties, droplet diameter, and impingement angle. A deposition–splashing criteria as functions of *Oh* and *Re* of the impinging droplet were derived as the splashing threshold $K = OhRe^{1.25} = 57.7$. This splashing threshold was based upon the energy conservation of the impinging droplet, in which the pre-impingement kinetic energy and surface energy of the droplet were conserved into the viscous dissipation and surface energy of droplet spreading. Further, the spreading factor and dynamic contact angle were considered as constant properties for any given liquid and solid in the splashing process.

6.1.2 Detailed Study of Droplet Impingement on a Solid Surface

6.1.2.1 Droplet Impingement on a Cold Surface

In the previous section, most studies of droplet impingement were carried out isothermally through experiments and they have been done for a long time, but the computational study on droplet–wall impingement began long after. Foote (1973) simulated liquid droplet behavior by using a technique on the basis of an extension of the marker-and-cell (MAC) method. This computing method considers the effect of surface tension. He reviewed the theory related to the droplet oscillation problem and predicted the large amplitude oscillation characteristics. His numerical prediction on small amplitude oscillations agreed well with the theory. Trapaga and Szekely (1991) developed a mathematical representation, and simulated droplets impact onto a solid surface under the isothermal condition by using volume of fluid (VOF) method. They found that the spreading times were of the order of microseconds when droplet sizes were in a hundred μm range and droplet velocities were in a hundred m/s range. Fukai et al. (1993) numerically studied the deformation of a spherical liquid droplet impingement on a flat surface by using two liquids (i.e., water and liquid tin). In their work, surface tension during the spreading process was considered. They solved a set of finite element equations built on a theoretical model to accurately simulate the large deformations and characterize the spreading process. The effects of droplet size, impact velocity, liquid and solid properties on the fluid dynamics of the deforming droplet were studied. The results showed that the numerical simulations successfully predicted the droplet recoiling process and mass accumulation around the surface periphery. Bussmann et al. (1999) developed a 3-D model with the VOF method to study a water droplet impinging on an asymmetric surface. In their numerical work, surface tension was modeled as a volume force acting on fluid near the surface, while contact angles were applied as a boundary condition at the contact line. Their simulation results showed a good agreement with the experimental photographs. A 3-D numerical study on a droplet interaction normally with a wall film was presented by Nikolopoulos et al. (2007b). The finite volume solution of the Navier–Stokes equations was coupled with VOF method in their study. Further, an adaptive local grid refinement method was utilized to accurately track the liquid–gas interface. Their results were comparable with the available experimental data for the lamella temporal development. They also found the correlations between *We* and secondary droplets diameter and number after droplet interacting with the surface.

6.1.2.2 Droplet Impingement on a Hot Surface

As results of the droplet evaporation and heat transfer between solid–liquid and solid-surrounding gas, surface temperature level shows additional complexity to the analysis of droplet–surface impingement phenomena. There are four different heat transfer regimes as depicted in Fig. 6.2 (top) identified when a droplet impinges on a heated surface (Kang and Lee 2000; Manzello and Yang 2002; Tamura and Tanasawa 1958).

(I) When the wall temperature is lower than the droplet saturation temperature ($T_w < T_{\mathrm{SAT}}$), the droplet evaporation is predominantly driven by the vapor diffusion and the heat transfer occurs by the conduction and free convection. This regime is known as the natural convection.

(II) When the wall temperature is larger than the droplet saturation temperature but lower than the critical heat flux temperature ($T_{\mathrm{SAT}} < T_w < T_{\mathrm{CHF}}$), the droplet evaporation is primarily driven by the heat transfer from the heated surface to the droplet. This regime is known as the nucleate boiling regime. In this regime, the vapor bubbles form near the hot surface and the buoyancy moves the vapor bubbles toward the liquid-surrounding gas interface. The vaporization removes the heat, the droplet reaches the maximum evaporation rate, and the heat reaches its maximum value at T_{CHF} (Nukiyama 1966).

(III) When the wall temperature is above the critical heat flux temperature but below the Leidenfrost temperature ($T_{\mathrm{CHF}} < T_w < T_{\mathrm{Leidenfrost}}$), the droplet evaporation falls into the transition boiling regime. An insulating vapor layer forms at the solid–liquid interface as the vaporization rate increases. The heat flux reduces to a local minimum value when the Leidenfrost temperature achieves (Leidenfrost 1966).

(IV) When the wall temperature is larger than the Leidenfrost temperature ($T_w > T_{\mathrm{Leidenfrost}}$), the film boiling regime occurs. A thin vapor film forms in this regime and prevents the physical contact between the droplet and the wall. The heat transfer is initially dominated by conduction, but radiation starts to take

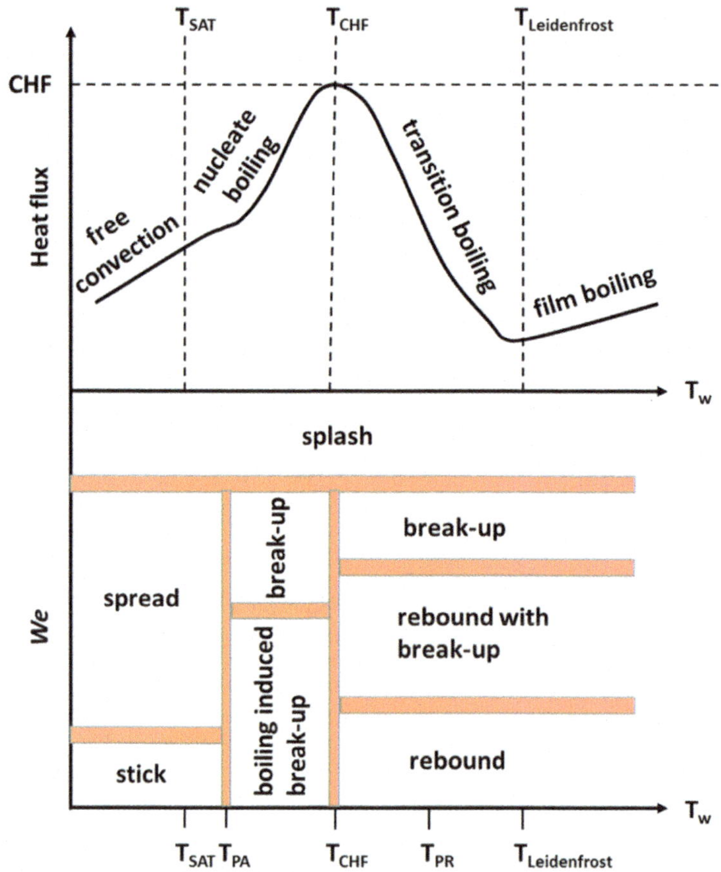

Fig. 6.2 General heat flux and associated boiling regimes (top); the impingement regimes and transition conditions when a droplet impinges on a hot surface (bottom). T_{PA} is the pure adhesion temperature, adhesion happens when T_w is below T_{PA} at low impact energy; T_{PR} is the pure rebound temperature, rebound happens when T_w is above T_{PA} at low impact energy (Zhao 2018)

a major role at higher temperature. Subsequently, the heat flux to the droplet slightly decreases.

As a single droplet impinges on a heated surface, the impingement outcomes become complicated. The various phenomena as observed at the cold impingement must be reconsidered within each boiling regime. Figure 6.1 covers relatively complete impingement outcomes at different surface conditions. Bai et al. (2002) proposed a general representation of the interaction outcomes from the available experimental work as shown in Fig. 6.2 (bottom), which provides a good qualitative explanation of the heat-induced phenomena. In this map, the impingement regimes are described along with the transition regions (shaded regions) in a 2-D space based on *We* and the surface temperature.

In spite of the fact that single droplet impingement is the fundamental aspect of spray impingement and is widely researched, the results of these studies cannot be directly extrapolated to reach an accurate understanding of spray impingement on a solid surface. Multi-droplet impingement comprised of droplet–surface impact and droplet-to-droplet interaction is critical to be studied. The single- and mono-sized droplet train impact onto surfaces with the constant heat flux conditions was studied by Soriano et al. (2010) to obtain the experimental characterization. Laser-induced fluorescence (LIF) technique with a focus on the impingement area was applied in the experiment to characterize the film morphology by measuring film thickness and film wetted area. The surface temperature at the liquid–solid interface was also measured by infrared thermography. The effects of the droplet temperature, droplet frequency, and fluid flow rate on the surface temperature were examined. The results showed that the higher heat flux was affected by multiple droplets with higher fluid flow rate. Lewis et al. (2013) compared the impingement heat transfer of a droplet train and the free surface jets on a wetted hot surface using the VOF method in OpenFOAM. They concluded that droplet train showed the noticeable temporal variations compared with the impinging jets, because the nature of continuous droplet impingement influenced the impingement region, thus, resulting in the increase of an unsteady cooling and heating of the fluid near the wall. On the contrary, the film and the corresponding free surface were stable with the small perturbations for the jet.

6.1.2.3 Post-impingement Characterization

In addition to the study on droplet–wall interaction outcomes and droplet splashing criterion, post-impingement parameters which define liquid–solid interaction such as surface wettability govern the wall-film formation and dynamics. After the droplet impinges on a solid surface, wall surface wettability is a significant factor to determine the whole impact and deformation process. Surface wettability has an effect on the maximum wetting wall-film area and defines whether the impinged droplets in a spray undergo coalescence to form a continuous film on the wall or not. Therefore, it is essential to qualitatively and quantitatively investigate the factors that affect surface wettability.

One of the factors that characterize the surface wettability is the liquid–solid contact angle formed at the solid–liquid-surrounding gas three-phase contact line (Hu et al. 2015). The contact angle formed between the liquid–gas and liquid–solid interface dramatically depends on the flow at three-phase contact line and the corresponding stresses acting on it. The final shape of the deposited droplet is determined by equilibrium contact angle, and the maximum spreading of the droplet is considerably influenced by dynamic contact angle (Vadillo et al. 2009). The dynamic contact angle is known as the contact angle formed at a moving contact line; the dynamic contact angle is usually required as a boundary condition to model the capillary hydrodynamics, for example, the certain stages of the droplet impact problem (Šikalo et al. 2005). Dynamic contact angle is substantially related to the contact line velocity. However, the static equilibrium contact angle as per Young's equation (Schrader

1995) is only a function of surface tension at liquid–gas–solid interfaces. To account for dynamic contact angle variations during droplet impingement, advancing, receding, and equilibrium are differentiated by the motion at the three-phase contact line, therefore, corresponding to the occurrence of dynamic advancing, receding, and the equilibrium contact angles, respectively. On the strength of the experiment, there are various dynamic contact angle models implemented in CFD codes to help predict the underlying physical mechanisms of droplet–wall interaction (Malgarinos et al. 2014).

Further, the flow at three-phase contact line and the dynamic contact angle at the moving contact line influence the spreading rate (Roisman et al. 2008). The dynamic of spreading is characterized into four regions by the impinging droplet *We* and *Oh,* as stated by Schiaffino and Sonin (1997). *We* measures the driving force for droplet spreading, and *Oh* scales the force to resist the spreading. Four regions are described as: inviscid-impact driven (at low *Oh*, high *We*); inviscid-capillarity driven (at low *Oh*, low *We*); highly viscous-capillarity driven (at high *Oh*, low *We*); highly viscous-impact driven (at high *Oh*, high *We*).

6.2 Experimental Methodology

6.2.1 Backlight for Droplet–Wall Impingement

The experimental setup as shown in Fig. 6.3 is for the measurements of a single droplet impingement on a flat solid dry surface. A drop generator, a high-speed camera with appropriate lens systems, and high-intensity light source are used for the test. Single droplet is generated by a precision syringe pump using 0.2 mL/min volume flow rate. The droplet with an initial diameter (D_0) is released and accelerates by gravity before it impinges on the surface and reaches an impact velocity (U_0). The initial droplet diameter varies with different fuels, and the impact velocity changes from 0.72 to 3.0 m/s as the droplet release height varies between 26 and 456 mm in this work. An analog mode LED lamp focused by an iris was passed through a plano-convex lens to generate a collimated cylindrical light. Photron FASTCAM SA 1.1 high-speed camera along with a 200-mm Nikon Nikkor lens and f-stop 4 was placed on the opposite direction of the LED to capture the process of droplet impinging on the plate. The image acquisition frequency was varied between 9000 and 12,000 frames per second (fps), and the exposure time is within a range of 5.6–111 μs varying with the liquid fuel and duration of droplet impingement process. Two different flat plates were used, smooth and roughened, to examine the effect of roughness on the droplet–wall interaction dynamics. The roughened surface has an average roughness of 16 μm and peak-to-peak roughness of 80 μm, which is comparable with a conventional piston surface (Yang and Ghandhi 2007). As well, to study the effect of surface temperature on the dynamic process of the droplet–wall interaction, the smooth flat plate was heated-up using heater controller. In the present work,

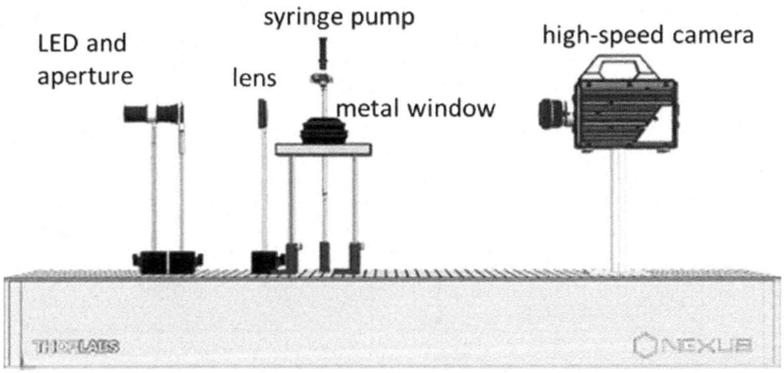

Fig. 6.3 Experimental setup for single droplet–wall impingement (Zhao 2018; Zhao et al. 2018)

the roughened surface is the BK-7 window while the smooth, heated surface is the heat-treated stainless steel.

6.2.2 *Image Processing*

Figure 6.4 (top) shows the schematic of a single droplet placed at a certain location above the impinged plate; Fig. 6.4 (bottom) provides the schematic after the droplet impinging on the surface. The various global parameters, such as the initial droplet diameter (D_0), the impact velocity (U_0), spreading diameter (d), spreading factor (Δ), height ratio (h/D_0), contact line velocity (U_{cl}), and dynamic contact angle (θ), are described to describe the process of droplet impacting on the surface.

Spreading diameter (d) is the distance between the left and right visible three-phase contact points. The three-phase contact points are defined as the points where all three phases (solid, liquid, and surrounding gas) encounter. Spreading factor (Δ) is the ratio between spreading diameter (d) and initial droplet diameter (D_0). Impinged height is defined as the maximum height in the perpendicular direction with respect to the impinged surface; similarly, impinged height ratio (h/D_0) is the ratio of impinged height to initial droplet diameter (D_0). The contact line velocity (U_{cl}) is the rate of change of spreading diameter (d) regarding the time. As described previously, the angle formed between the liquid–gas interface and solid–liquid interface at the three-phase contact point is defined as contact angle. The dynamic contact angle (θ) can be considered as the angle observed at the moving contact line during the droplet impingement process. Three stages based on contact line velocity are observed during a droplet impingement on the surface: advancing, receding, and equilibrium. In current work, if the $U_{cl} > 0$, the dynamic contact angle is known as advancing contact angle; if $U_{cl} < 0$, the dynamic contact angle is known as receding contact angle; and if $U_{cl} = 0$, the droplet becomes stable. In this stage, the equilibrium contact angle is

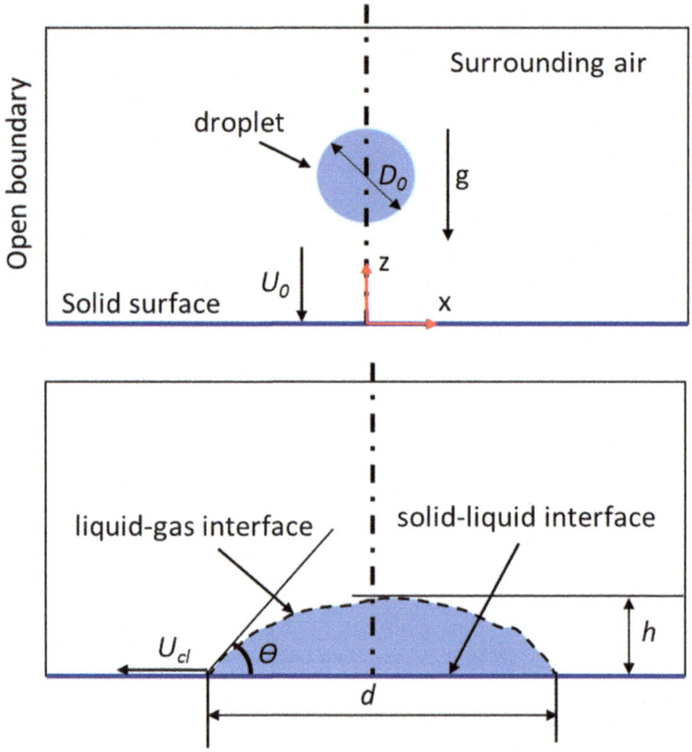

Fig. 6.4 Schematic of droplet impingement on the flat surface (Zhao 2018; Zhao et al. 2018)

known. The averaged contact angle at each phase is calculated by taking the mean of the instantaneous contact angles of respective phases (Zhao 2018; Zhao et al. 2018).

An in-house MATLAB code was used to process the images and analyze the droplet impingement on a flat surface. The procedure of image processing is shown in Fig. 6.5. In Fig. 6.5 (top), first, based on the original image the background was subtracted to remove the unnecessary object other than the droplet. Then, the image was converted into a binary image based on a threshold (Otsu 1979) to aid in accurately predicting the droplet boundary. A sensitive analysis for the threshold value was done on a sample case by increasing and decreasing default threshold by 20%, and the initial droplet diameter shows only ±2% for different threshold values. Additionally, the difference between the horizontal direction and vertical direction in droplet diameter was measured to determine the possible deformation of an impacting droplet caused by drag force. It was found that this difference was within ±1% for all measurements, resulting in that the drag force shows the insignificant impact on the droplet size. Therefore, the image of the droplet is approximated as a circle and the initial droplet diameter is extracted on the basis of the area of this circle.

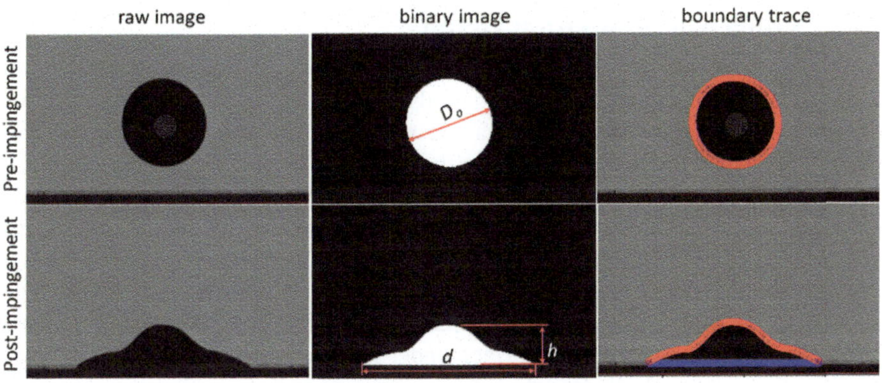

Fig. 6.5 Image processing procedure for single droplet–wall impingement (Zhao 2018; Zhao et al. 2018)

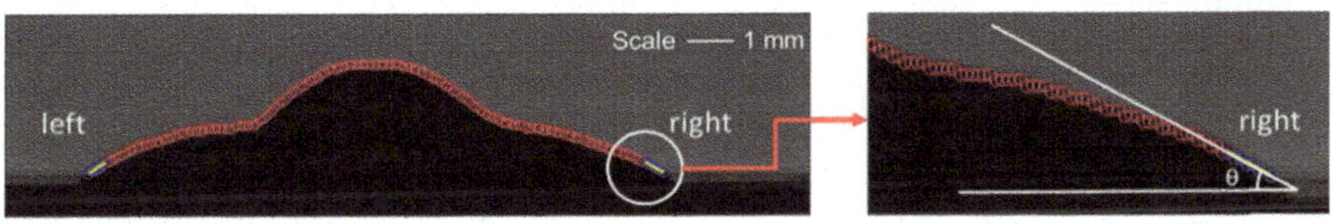

Fig. 6.6 Contact angle measurement technique (Zhao 2018; Zhao et al. 2018)

The processing of post-impingement images is shown in Fig. 6.5 (bottom). The boundary points are separated into two interfaces: solid–liquid interface (blue) and liquid–gas interface (red). The spreading diameter (d) is calculated as the distance between leftmost and rightmost visible three-phase contact points. The spread factor (Δ), ratio of spreading diameter and initial droplet diameter, is then calculated at each time step. Similarly, the height of the impinged droplet is measured as a distance from the topmost point of the droplet to the flat surface, and the impinged height ratio (h/D_0) is found. The contact line velocity (U_{cl}) follows the same way to be obtained.

The dynamic contact angle (θ) is processed as an angle between the tangent to droplet profile at the moving contact line and horizontal solid–liquid interface. The boundary points corresponding to the liquid–gas interface are considered, as shown in Fig. 6.6. Only the pixels, very near to the three-phase contact point on the liquid–vapor interface, are considered to curve-fit a line. This curve-fitted line is used as a tangent to the droplet from the three-phase contact point as shown in Fig. 6.6 (right). The contact angle is finally obtained from the slope of the curve-fitted line. The dynamic contact angle is extracted from each image by averaging the visible left and right contact angles. Moreover, the reference scale in the experiment was obtained by measuring the number of pixels corresponding to a known length; the known length was oriented normal to the camera's line of sight (Zhao 2018; Zhao et al. 2018).

6.3 Numerical Simulation Details

Eulerian–Eulerian approach states two fluids, and transport equations for continuum properties associated with both fluids are resolved. The volume fraction of each phase is weighted based on the terms in the transport equations. Each phase and the interaction between any two phases at any location in the space have to be solved, thus, leading to the large costs of Eulerian–Eulerian models. However, in Eulerian approach, only one set of governing equations need to be resolved for all phases, causing a tremendous cost-saving compared with Eulerian–Eulerian method. VOF method is one of the most widely used Eulerian approaches. As part of this work, droplet impingement on a solid surface under non-evaporation conditions is simulated by the existing VOF model. A VOF modeling technique that can accurately capture evaporation during droplets' impingement on a solid surface is yet to be developed. In particular, modeling evaporation in such complex contact line (encountered in liquid–gas–solid systems) geometries requires an accurate VOF methodology for volume-tracking three-phase systems in 3-D. Therefore, the development, implementation, and validation of a VOF modeling approach including vaporization integrated into CFD codes to provide accurate and predictive simulation of droplet–wall interactions are performed in the current work. This is accomplished by development and inclusion of an evaporation sub-model in existing VOF modeling framework, and it is validated through extensive experimentation of the droplet–wall impingement and droplets collision, spread and vaporization dynamics.

In summary, droplet impingement on a solid unheated surface was simulated by VOF method; evaporation sub-model was implemented based on the existing VOF model. VOF calculations capture important details of droplet impact dynamics onto an unheated or a heated solid surface under non-evaporating and evaporating conditions. For instance, the contact line formed by the impacting droplets in the droplet–wall impingement case is irregular and needs to be captured by a VOF methodology that is capable of robustly reconstructing liquid–gas–solid interfaces. As well, the validated evaporation sub-model in terms of droplet relevant simulations can be further extended to study the spray–wall impingement in the VOF context. Furthermore, with the inclusion of the results of the VOF analysis on droplet/spray–wall impingement, the predictive simulation on the relevant sprays can be eventually achieved with less requirement of extensive parameters tuning.

6.3.1 Eulerian-Based VOF Method

6.3.1.1 Non-evaporation Governing Equations

The conservation laws of mass and momentum are used to describe the fluid motion of isothermal, single-phase flows. Multi-phase flows involving two or more phases require additional equations to define each of the additional phases and the relation between phase properties. These additional equations are transport equations of void fraction variables and are solved to capture the interface. They are solved simultaneously with the mass and momentum conservation equations. The mass conservation is expressed as continuity equation. The momentum equation is attained by balancing the total forces acting on a fluid element with gravity forces, viscous forces, surface tension, and body forces.

The mass conservation and momentum equation for compressible flows are expressed in Eqs. (6.2) and (6.3),

$$\frac{\partial \rho}{\partial t} + \nabla \cdot (\rho \boldsymbol{U}) = 0 \tag{6.2}$$

$$\frac{\partial (\rho \boldsymbol{U})}{\partial t} + \nabla \cdot (\rho \boldsymbol{U} \otimes \boldsymbol{U}) = -\nabla p + \nabla \cdot \left[2\mu S - \frac{2\mu (\nabla \cdot \boldsymbol{U}) I}{3} \right] + f_{\mathrm{st}} + f_{\mathrm{g}} \tag{6.3}$$

where I is identity matrix, p is pressure, μ is dynamic viscosity, f_{st} is surface tension force, and f_{g} is gravity force.

$$S = 0.5\left[\nabla \cdot \boldsymbol{U} + (\nabla \cdot \boldsymbol{U})^T\right] \tag{6.4}$$

The continuity and momentum equations for incompressible flows are obtained by considering the changes in density of an infinitesimally small element as negligible or zero as follows (Potham et al. 2017; Richards et al. 2016):

$$\nabla \cdot (\boldsymbol{U}) = 0 \tag{6.5}$$

$$\rho \left(\frac{\partial U}{\partial t} + \boldsymbol{U} \cdot \nabla \boldsymbol{U} \right) = -\nabla p + \nabla \cdot \left[\mu \left(\nabla \cdot \boldsymbol{U} + (\nabla \cdot \boldsymbol{U})^T \right) \right] + f_{st} + f_{g} \tag{6.6}$$

In VOF, an interface-capturing method, the location of interface is known based on the value of a scalar function called void fraction (α):

$$\alpha = \frac{V_{\mathrm{g}}}{V} \tag{6.7}$$

where V_{g} is the volume of gas phase and V is the total volume of the control volume.

Void fraction α of 1 stands for the liquid phase and α of 0 represents the gas phase, and α between 0 and 1 is known as the interface. Mass of each phase is conserved when the transport equation of its phase fraction is satisfied. Transport equation of void fraction α is given by

$$\frac{\partial \alpha}{\partial t} + \boldsymbol{U} \cdot \nabla \alpha = 0 \tag{6.8}$$

VOF method modeling multi-phase flows neglects the discontinuity and involves obtaining a mixture representation of two or more phases. Transport properties and

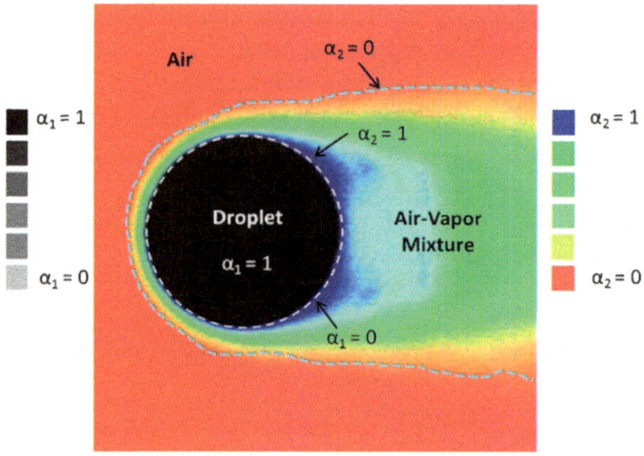

Fig. 6.7 Schematic of liquid and vapor void fractions in the computational domain (Zhao 2018)

velocity of the mixture phase are obtained by volume averaging the properties of individual phases.

$$\rho = \rho_g \alpha + (1 - \alpha)\rho_l \tag{6.9}$$

$$\mu = \mu_g \alpha + (1 - \alpha)\mu_l \tag{6.10}$$

$$\boldsymbol{U} = \boldsymbol{U}_{\boldsymbol{g}} \alpha + (1 - \alpha)\boldsymbol{U}_{\boldsymbol{l}} \tag{6.11}$$

where ρ_g is the gas-phase density, and ρ_l is the liquid density; μ_g is the gas-phase viscosity, and μ_l is the liquid viscosity; $\boldsymbol{U}_{\boldsymbol{g}}$ is the gas-phase velocity, and $\boldsymbol{U}_{\boldsymbol{l}}$ is the liquid velocity.

6.3.1.2 Evaporation Governing Equations

Phase change in VOF is modeled using source terms in continuity, momentum, and phase fraction equations along with the transport equation of temperature. In the current work, multi-phase flows with three phases, including liquid, its vapor phase, and surrounding gas, are considered. Vapor and surrounding gas are modeled as continuum phases without interface separation between them. This continuum phase is referred as gaseous phase. Vapor diffuses in gas; however, both vapor and gas are insoluble in liquid phase. The bulk or advection-based velocities of both gas and vapor phases are identical. Two void fraction variables are used to describe the presence of three phases: liquid void fraction (α_1) and vapor void fraction (α_2). When α_1 is equal to 1, representing only liquid fuel phase, and when α_2 is equal to 1, standing for only vapor fuel phase. The detailed information of liquid and vapor void fractions can be found in Eqs. (6.12) and (6.13) along with Fig. 6.7.

$$\alpha_1 = \begin{cases} 0 & \text{In surrounding air or vapor fuel phase} \\ 0 < \alpha_1 < 1 & \text{At liquid interface} \\ 1 & \text{In liquid fuel phase} \end{cases} \tag{6.12}$$

$$\alpha_2 = \begin{cases} 0 & \text{In liquid or surrounding air phase} \\ 0 < \alpha_2 < 1 & \text{vapor fuel and surrounding air mixed} \\ 1 & \text{In vapor fuel phase} \end{cases} \tag{6.13}$$

In addition, transport properties such as density (ρ), thermal conductivity (k) of individual phases are volume averaged to obtain properties of single mixture phase.

$$\rho = \alpha_1 \rho_l + \alpha_2 \rho_v + (1 - \alpha_1 - \alpha_2)\rho_g \tag{6.14}$$

$$k = \alpha_1 k_l + \alpha_2 k_v + (1 - \alpha_1 - \alpha_2)k_g \tag{6.15}$$

where ρ_l is the liquid fuel density, ρ_v is the vapor fuel density, and ρ_g is the surrounding gas density; k_l is the liquid fuel conductivity, k_v is the vapor fuel conductivity, and k_g is the surrounding air/gas conductivity.

Specific heat (c_p) at constant pressure is obtained by mass averaging the specific heats of individual phases.

$$c_p = \rho_l \alpha_1 c_{p,l} + \rho_v \alpha_2 c_{p,v} + \rho_g (1 - \alpha_1 - \alpha_2) c_{p,g} \tag{6.16}$$

where $c_{p,l}$ is the liquid fuel specific heat, $c_{p,v}$ is the vapor fuel specific heat, and $c_{p,g}$ is the surrounding gas specific heat.

Velocity is modeled as

$$\boldsymbol{U} = \boldsymbol{U}_l \alpha_1 + \boldsymbol{U}_{gp} \alpha_2 + \boldsymbol{U}_{gp}(1 - \alpha_1 - \alpha_2) \tag{6.17}$$

Or simply as

$$\boldsymbol{U} = \boldsymbol{U}_l \alpha_1 + (1 - \alpha_1)\boldsymbol{U}_{gp} \tag{6.18}$$

where $\boldsymbol{U}_{gp}$ is velocity of gaseous phase (including vapor and surrounding gas).

Mass transfer during the phase change in incompressible flows is modeled as addition or removal of liquid or vapor volume, which modifies the continuity equation of Eq. (6.5) as

$$\nabla \cdot (\boldsymbol{U}) = -\dot{m}''' \left(\frac{1}{\rho_l} - \frac{1}{\rho_v} \right) \tag{6.19}$$

where $\dot{m}'''$ is the volumetric rate of mass transfer (Schlottke and Weigand 2008) from liquid phase to vapor phase caused by the temperature and mass fraction gradient. It is calculated as follows:

$$\dot{m}''' = \frac{D_{\mathrm{vg}} * \rho_{\mathrm{gp}}}{1 - Y_{\mathrm{v}}} \nabla Y_{\mathrm{v}} \cdot \nabla \alpha_1 - \frac{k}{h_{\mathrm{v}}} * \nabla T \cdot \nabla \alpha_1 \tag{6.20}$$

where D_{vg} is diffusivity of vapor in gas and ρ_{gp} is density of gaseous phase (including vapor and surrounding gas).

Mass fraction of vapor phase (Y_{v}) is calculated using vapor-phase volume fraction, density of vapor and gaseous phases:

$$Y_{\mathrm{v}} = \frac{\alpha_2}{1 - \alpha_1} \frac{\rho_{\mathrm{v}}}{\rho_{\mathrm{gp}}} \tag{6.21}$$

The momentum equation as shown in Eq. (6.6) is not affected by the evaporation sub-model. Therefore, no source terms are added to momentum equation as their effects are already introduced in continuity equation.

The interface between liquid and gaseous phases is assumed to be at saturation state during phase change. Mass fraction of vapor at interface (Schlottke and Weigand 2008) is given by

$$Y_{\mathrm{v,sat}} = \frac{P_{\mathrm{v,sat}}}{P} * \frac{M_{\mathrm{v}}}{M_{\mathrm{gp,sat}}} \tag{6.22}$$

where $P_{\mathrm{v,sat}}$ is saturated vapor pressure, P is static pressure, M_{v} is molecular weight of vapor, M_{a} is molecular weight of surrounding air, and $M_{\mathrm{gp,sat}}$ is molecular weight of gaseous phase.

$$M_{\mathrm{gp,sat}} = \frac{P_{\mathrm{v,sat}} * M_{\mathrm{v}} + \left(P - P_{\mathrm{v,sat}}\right) * M_{\mathrm{a}}}{P} \tag{6.23}$$

Saturated vapor pressure is calculated using the Wagner equation (Schlottke and Weigand 2008) as follows:

$$P_{\mathrm{v,sat}} = \frac{P_{\mathrm{c}} * T_{\mathrm{c}}}{T} * \left[a * \left(1 - \frac{T}{T_{\mathrm{c}}}\right) + b * \left(1 - \frac{T}{T_{\mathrm{c}}}\right)^{1.5} + c * \left(1 - \frac{T}{T_{\mathrm{c}}}\right)^{3} + d * \left(1 - \frac{T}{T_{\mathrm{c}}}\right)^{6}\right] \tag{6.24}$$

where T_{c} is critical temperature and P_{c} is critical pressure.

Energy equation is introduced to model the heat transfer effect. The source term in the energy equation is the heat transferred due to mass transfer during evaporation.

$$\frac{\partial\left(\rho c_{\mathrm{p}} T\right)}{\partial t} + \nabla \cdot \left(\rho \boldsymbol{U} c_{\mathrm{p}} T\right) = \nabla \cdot (k \nabla T) + h_{\mathrm{v}} \dot{m}''' \tag{6.25}$$

The temperature at the interface is constrained to saturation temperature. As well, surface superheat is not considered.

Transport equations of liquid and vapor volume fractions have source terms to simulate reduction of mass from liquid and addition of mass to vapor phase during evaporation. If $\dot{m}'''$ signifies the volumetric rate of mass transfer from liquid to vapor and an artificial interface compression flux term is introduced, the liquid-phase fraction transport equation can be represented by Eq. (6.26) and the vapor-phase fraction transport equation can be represented by Eq. (6.27),

$$\frac{\partial \alpha_1}{\partial t} + \nabla \cdot (\boldsymbol{U}\alpha_1) + \nabla \cdot ((1-\alpha_1)\boldsymbol{U}_{\mathrm{r}}\alpha_1) = \alpha_1(\nabla \cdot \boldsymbol{U}) - \dot{m}'''\left(\frac{1}{\rho_{\mathrm{l}}} - \alpha_1\left(\frac{1}{\rho_{\mathrm{l}}} - \frac{1}{\rho_{\mathrm{v}}}\right)\right) \tag{6.26}$$

$$\begin{aligned}\frac{\partial \alpha_2}{\partial t} + \nabla \cdot (\boldsymbol{U}\alpha_2) - \nabla \cdot (\alpha_2\boldsymbol{U}_{\mathrm{r}}\alpha_1) = \nabla \cdot \left(D_{\mathrm{vg}}\nabla\alpha_2\right) + \alpha_1(\nabla \cdot \boldsymbol{U}) \\ + \dot{m}'''\left(\frac{1}{\rho_{\mathrm{v}}} + \alpha_1\left(\frac{1}{\rho_{\mathrm{l}}} - \frac{1}{\rho_{\mathrm{v}}}\right)\right)\end{aligned} \tag{6.27}$$

Although this evaporation sub-model is developed for the incompressible flows as described in the above equation, it is found to be also compatible for the compressible flows as the error due to the incompressible assumption is much less than 10% (Tonini and Cossali 2012). Significant improvements in the simulation results will be achieved by considering the evaporation and clarifying the vapor fuel and surrounding gas phases quantitatively.

6.4 Experimental Results

Table 6.1 lists the test conditions of single droplet impingement on a solid surface. Four different fuels were used during droplet–wall impingement test, and Table 6.2 provides their liquid properties; the range of impact *We* and *Re* is also given.

A sequence of droplet shape evolution at various time instants for diesel and water is shown in Fig. 6.8, to describe the dynamic impingement process of a liquid droplet on a smooth solid surface. As the baseline non-splashing condition, the initial droplet-impinged surface height of 52 mm was chosen. Hence, the corresponding impact *We* for diesel is 104 and for water is 53. For splashing condition, the initial droplet-

Table 6.1 Test conditions for single droplet–wall impingement

Parameter	Values
Ambient temperature (°C)	25
Ambient pressure (atm.)	1
Fuel	Diesel, water, *n*-dodecane, *n*-heptane
Height between needle and impinged surface (mm)	26–456
Impact velocity U_0 (m/s)	0.72–3.0
Surface temperature (°C)	25; 150, 185 (heated surface)
Average surface roughness R_a (μm)	1.6 (smooth); 16 (roughened)

Table 6.2 Liquid properties

Parameter	Diesel	Water	*n*-dodecane	*n*-heptane
ρ (kg/m^3)	848	1000	750	684
σ (N/m)	0.024	0.070	0.023	0.019
υ (cSt)	2.6	1.0	1.97	0.38
D_0 (mm)	2.87	3.6	2.86	2.6
We	52–925	26–458	43–833	45–836
Re	789–3300	2562–10,718	1037–4339	4941–20,669

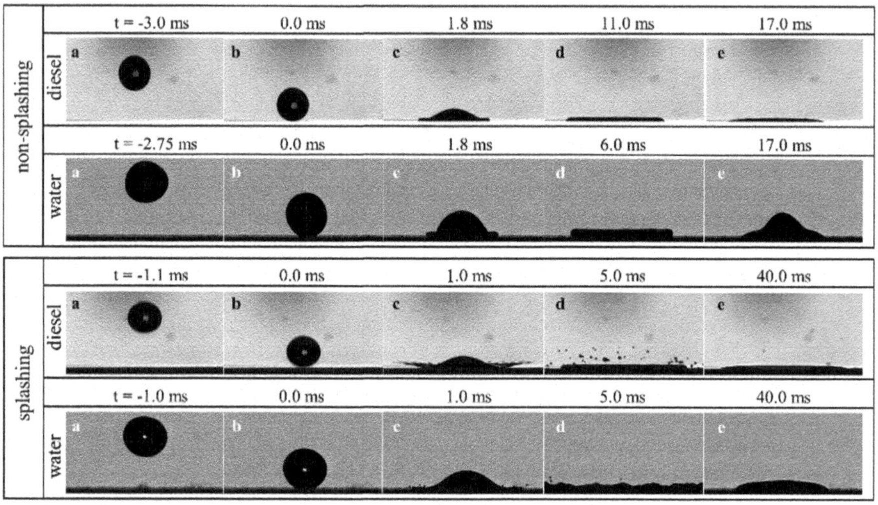

Fig. 6.8 A sequential visualization of droplet–wall impingement experiment for diesel and water: non-splashing (top); splashing (bottom) (Zhao 2018; Zhao et al. 2018)

impinged surface distance of 286 mm was selected. The corresponding impact *We* for diesel is 569 and for water is 289. Since the initial droplet–surface height is much larger compared to the droplet size, the initial location of droplets is not shown in Fig. 6.8; instead, the droplet center to the solid surface is set to be the same distance of 4 mm for all conditions to show the pre-impingement phenomena. In addition, the time stamps are selected with respect to the time when droplet just impacts on the plate (i.e., $t = 0$ ms when droplet interacting with the plate). The time stamps along with each image illustrate slightly variances in water and diesel fuels because of the particular events occurring at the different time, especially after droplet impinging on the surface.

A series of *non-splashing* events for droplet impingement on a smooth solid surface is observed in Fig. 6.8 (top). From left to right, there are (a) pre-impingement, (b) impingement, (c) post-impingement, (d) maximum spreading, and (e) receding. In (a), the initial water droplet size ($D_0 = 3.6$ mm) is observed to be larger than diesel droplet (D_0 =2.87 mm); in (b), the droplet size shows no substantial change before impinging on the surface due to the insignificant effect of the drag force on it, as stated in image processing section; after impingement, it can be clearly seen in (c) that droplet starts spreading radially in the current view, and the diesel droplet spreads more rapidly compared to water droplet at 1.8 ms caused by the larger surface tension of water (see Table 6.2); in (d), the water droplet reaches its maximum spreading factor of 2.4 around 6.0 ms and diesel droplet achieves its maximum spreading factor of 3.1 around 11.0 ms. In a short period, after spreading as shown in (e), the water droplet begins receding under the effect of capillary force; however, it is difficult to visualize the receding in diesel droplet due to lower surface tension and higher viscosity of diesel fuel. Subsequently, the droplet becomes stable, corresponding to the equilibrium stage (not shown here).

Similarly, Fig. 6.8 (bottom) shows a series of *splashing* events for droplet impinging on a smooth plate. From left to right, there are (a) pre-impingement, (b) impingement, (c) splashing, (d) further splashing, and (e) primary deposited equilibrium. In (a) and (b), the initial droplet size of diesel (D_0 =2.87 mm) and water (D_0 =3.6 mm) are the same as mentioned in the non-splashing case; After interacting with the solid surface, droplets spread radially and splash at 1.0 ms in (c), and the stronger splashing is observed in diesel droplet in comparison to water due to higher surface tension of water. According to Yarin and Weiss (1995) and O'Rourke and Amsden (1996, 2000), the splashing threshold corresponds to the formation of a kinematic discontinuity. As well, the velocity discontinuity, located at the boundary between fluid moving outward from the splashing location and slower moving fluid on the surface, results in fluid to be ejected away from the surface. The secondary droplets are then generated; in (d), the diesel and water droplets further splash into a number of secondary droplets, and because of smaller surface tension in diesel case, more satellite droplets are formed in diesel case. On the other hand, oscillation is observed in water case because of the higher surface tension of water; around 40 ms after droplets impinging on the plate, as shown in (e), both diesel and water droplets tend to achieve the equilibrium stage while the spreading diameter in diesel is larger than that in water case.

6.4.1 *Splashing Criteria*

As discussed in Introduction section, the splashing threshold of $Ca\lambda^{\frac{3}{4}} = u > 17 \sim 18$ was suggested by Yarin and Weiss (1995). They studied a single-train droplet falling on a solid substrate with a thin film at a known impinging frequency (f). Figure 6.9 (top) provides the correlation between Capillary number (Ca) and dimensionless diffusion length (λ), and the black solid line represents the splashing criteria line obtained from Yarin and Weiss (1995). The data points shown in Fig. 6.9 (top) are our experimental results at various conditions (including variations of liquid viscosity, surface tension, smooth and roughened surfaces, heated plate), where the blue

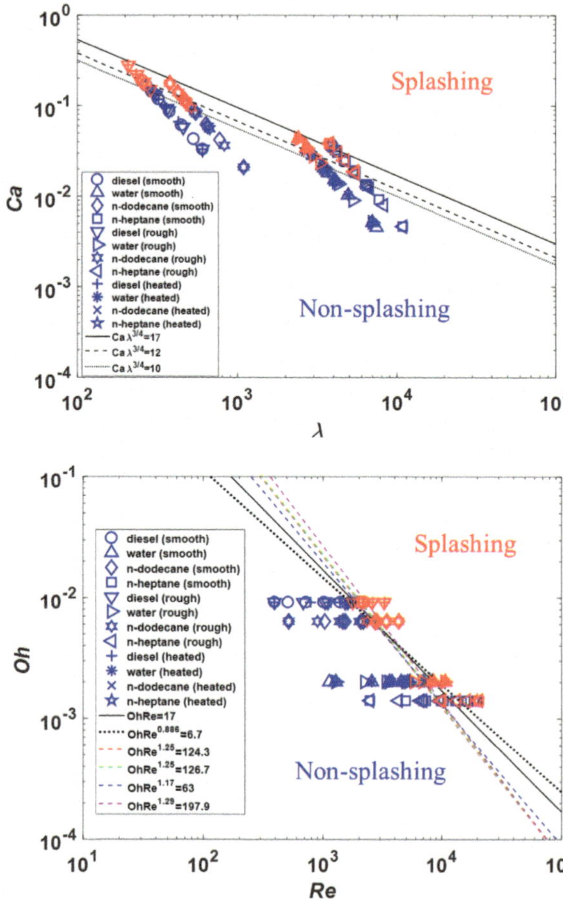

Fig. 6.9 Splashing criteria for various test conditions: Ca versus λ (top); *Oh* versus *Re* (bottom). (Zhao 2018)

points denote the non-splashing events while the red points signify the splashing events. The experimental results overall follow the same trend with the literature in predicting the non-splashing phenomena for water, diesel, and *n*-dodecane, but not for *n*-heptane. The data points from non-splashing cases with *n*-heptane fuel are observed to shift toward the splashing region. On the other hand, the data points representing splashing characteristics from other fuels cross the Yarin and Weiss's splashing criteria line (solid black line). As stated in previous, Yarin and Weiss's criterion may not work for many cases since the derived splashing threshold provides an explanation only for corona splash mechanism. Moreover, this correlation posed under an assumption of no direct interaction of droplet with the solid dry surface instead of a thin liquid film. Therefore, it may not be applied for droplet impingement directly on a dry surface.

The best correlation for the current experimental data is found to be between a dashed line showing $Ca\lambda^{\frac{3}{4}} = 12$ and a round dot line exhibiting $Ca\lambda^{\frac{3}{4}} = 10$

in Fig. 6.9 (top). It should be noted that the frequency (f) in the current work is assumed to be U_0/D_0 (Stanton and Rutland 1998) and λ can be further derived as $\lambda = Re^{1.5}/We$.

Another splashing criteria based on Oh and Re in Fig. 6.9 (bottom) have been also discussed previously; it was presented by Ma et al. (2017) by summarizing a larger number of researchers' experimental data at various test conditions. The black dashed line stands for the splashing correlation of $OhRe = 17$ from Ma et al. (2017), and the rest of four dashed lines exhibit the correlations of $OhRe^{1.25} = 124.3$, $OhRe^{1.25} = 126.7$, $OhRe^{1.17} = 63$, and $OhRe^{1.29} = 197.9$ from Geppert et al. (2016), Cossali et al. (1997), Vander Wal et al. (2006), and Bernard et al. (2017), respectively. Most blue symbols from our experiment are below these critical lines, while most red data points are above them. However, one of the exceptions occurs again in n-heptane case, rather than following the splashing criteria line of $OhRe = 17$, and n-heptane data points resides at $OhRe$ of 26. It is also observed that splashing on the roughened plate happens slightly below the $OhRe = 17$, because the probability of prompt splash increases as the amplitude of roughness increases (Rioboo et al. 2001).

A large number of experimental works are done on the droplet–wall interaction; however, due to the complexity of physics of droplet–wall interaction and the limitations of the experimental data, the splashing criteria are still necessary to be studied and improved. The best correlation with respect to the current experimental data and test conditions is found as follows:

$$OhRe^{0.886} = 6.7 \tag{6.28}$$

To further understand and examine the splashing correlation, for example, the splashing threshold from Yarin and Weiss (1995) determined by Ca and λ is discussed. As Eqs. (6.29) and (6.30), Ca represents the relative effect of viscous forces versus surface tension acting on an interface between a liquid and a gas; λ is known as the non-dimensional viscosity length.

$$Ca = \rho U_0 \nu/\sigma = We/Re \tag{6.29}$$

$$\lambda = \left(\frac{\nu}{f}\right)^{0.5} \sigma/(\rho \nu^2) \tag{6.30}$$

As the black solid line (splashing criteria line) shown in Fig. 6.10, at any splashing conditions, assuming the liquid density ρ, impact velocity U_0, and the droplet initial diameter D_0 are constants, Eqs. (6.29) and (6.30) become:

$$Ca \sim \nu/\sigma \tag{6.31}$$

$$\lambda \sim \sigma/\nu^{1.5} \tag{6.32}$$

At the same viscosity ν, Ca decreases and λ increases as the surface tension σ increases. This means that a relative longer viscosity length is required to overcome the surface tension force to ensure the splashing occurring, namely a larger surface

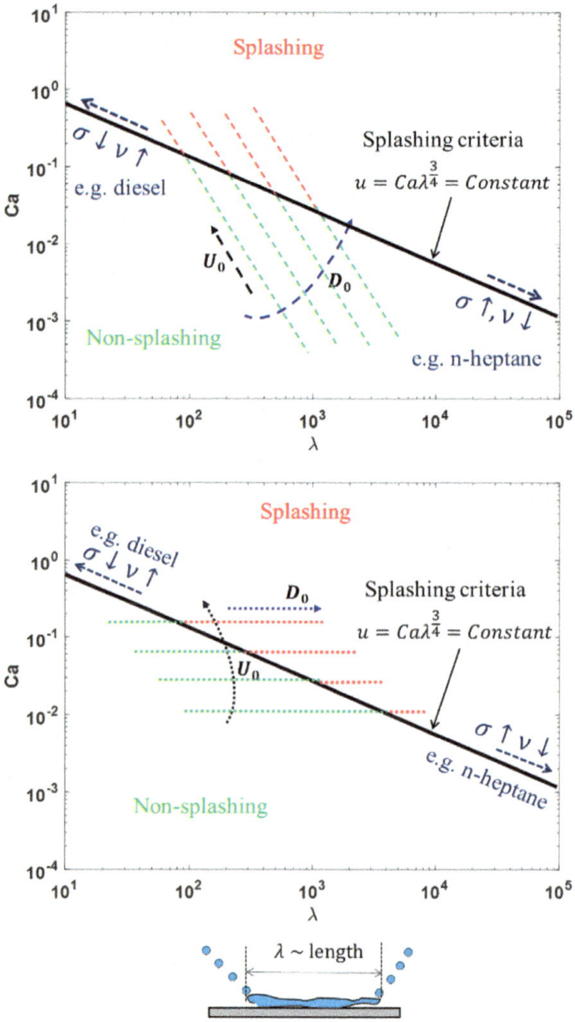

Fig. 6.10 Schematic of splashing criteria: red-green dashed line (D_0 = constant) (top); red-green dot line (U_0 = constant) (bottom) (Zhao 2018)

tension holds the droplet breakup until a certain viscosity length reaches and vice versa. It should be noted that viscosity is resistant to flow motion, while the surface tension is the force of attraction acting between the liquid molecules.

In the similar way, at the same surface tension σ, Ca increases and λ decreases as the viscosity v increases. This indicates that a relative shorter viscosity length is required to ensure the splashing occurrence, viz. a higher viscosity results in a shorter viscosity length after droplet spreading on the wall, resulting in the droplet breakup and vice versa.

Besides, it is interesting to point out that the data points in Fig. 6.9 are regularly seated in the figure with certain slopes. For example, four data sets are observed from

Fig. 6.9 (top) in terms of four different tested liquid fuels. Diesel and *n*-dodecane with similar liquid properties are shown in the left two sets while water and *n*-heptane are shown in the right two sets. In addition, Diesel and *n*-dodecane have relatively higher viscosity and lower surface tension than water and *n*-heptane. To describe and extend this phenomenon by a general way, as the red and blue lines shown in Fig. 6.10, with any given liquid fuel, the liquid properties remain same at a given condition, Eqs. (6.29) and (6.30) become:

$$Ca = C_1 U_0 \tag{6.33}$$

$$\lambda = C_2 D_0^{0.5} / U_0^{0.5} \tag{6.34}$$

where $C_1 = \rho\nu/\sigma$ and $C_2 = \sigma/\left(\rho\nu^{1.5}\right)$ are constants.

In Fig. 6.10 (top), as the impact velocity U_0 increases and the droplet initial diameter D_0 remains the same, Ca increases and λ decreases, causing higher chance of splashing occurrence and vice versa.

When the droplet initial diameter D_0 increases, Ca remains the same but λ increases. Therefore, the dashed line consisted by data points shifts toward the right as D_0 increases due to the change of λ as displayed in Fig. 6.10 (bottom). However, the slope of the line based on the data sets shows insignificant change with D_0. It was also found that this slope shows no substantial change when the same-size droplets with different liquid fuels were considered, as the experimental data points shown in Fig. 6.9.

To summarize the above analysis, the cross point of the splashing criteria line and the data sets with different liquid fuels and droplet initial diameter differentiates the non-splashing (blue) and splashing (red) characteristics. Since the current experiment tested, sub-millimeter-based droplets show the larger magnitude compared with typical droplets found in high-pressure sprays. Thus, the correlations and the concepts are summarized from Fig. 6.10 which might be proposed and extended to the actual sub-micrometer-based droplets splashing study. Other than the liquid properties and droplet size, from another point of view, only the spherical droplets are considered in the present work, but the droplet shape at pre- or post-impingement and further splashing are possible changed into an irregular shape. This behavior is also necessary to be taken into account for the universal splashing correlation development, which can be supported by Eulerian-based VOF simulations.

6.4.2 Post-impingement Evolution

In this section, the experimental results of a single diesel droplet impinging on a stainless steel plate will be presented. Three test cases were performed, one at the baseline surface temperature of 25 °C and two cases at elevated surface temperatures (150 and 260 °C) representative of the piston temperature in a modern diesel ICE (Miers et al. 2005). In addition, the experimental results at each condition are averaged from five repetitions, and after start of impingement (ASOI) time is presented for the post-impingement evolution.

6.4.2.1 Experimental Results

The experimental diesel droplet–wall impingement results at the baseline surface temperature condition (25 °C) are firstly presented. Figure 6.11 (black line) shows the temporal spreading factor and height ratio (left) and dynamic contact angle (right) at the baseline temperature condition. The impinged droplet achieves the maximum spreading factor about 8 ms. Then, the flattened droplet starts to recede under the capillary force and the spreading factor slightly reduces as a result of this recoiling. The spreading factor and height ratio remain same when the droplet becomes stable.

In Fig. 6.11 (right), the contact angle is around 150° at the onset of droplet–surface contact. Afterward, the contact angle reduces rapidly to about 100° and decreases

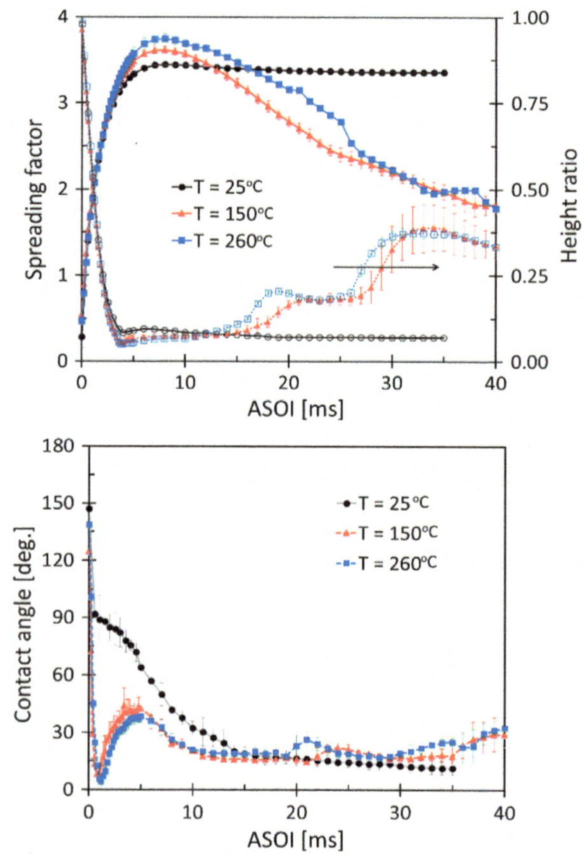

Fig. 6.11 Surface temperature effect on spreading factor and height ratio (top) and dynamic contact angle (bottom) at *We* = 207 (non-splashing conditions) (Markt et al. 2018)

throughout the rest of advancing phase. The receding phase initiates when the contact angle drops to 30° at approximately 10 ms and slowly decreases until 30 ms. During the receding phase, the droplet does not significantly recede as observed in Fig. 6.11. After 30 ms, the contact angle tends to be stable, signifying the start of the equilibrium stage.

6.4.2.2 Surface Temperature Effect

The experimental results of diesel droplet–wall impingement at two different surface temperatures, 150 and 260 °C, are performed in this section. These temperatures were selected as they are comparable with the piston surface temperature in diesel engines (Miers et al. 2005) and allow the study of surface temperature effect on post-impingement process. In the experiments, a 3-wire heat flux probe was used to measure surface, embedded and differential temperatures. The probe consists of two "J"-type thermocouples, one was installed at the plate surface and another was installed 2 mm under the surface thermocouple. The temperature profiles are not shown here for the sake of brevity, and one such temperature profile can be found in our previous work (Zhao 2018; Zhao et al. 2018).

Figure 6.11 (top) also shows the spreading factor and height ratio for diesel droplets impinging on the heated surface. Unlike the unheated case (25 °C), at both 150 and 260 °C, the droplet continues spreading until its maximum spreading diameter is achieved around 8 ms. In the same period, the height ratio shows little change until the receding phase begins and the droplet starts to oscillate slightly as it attempts to reshape. At the onset of receding, vertical elongation is observed as seen in Fig. 6.11 (top). Nevertheless, since the surface temperatures (150 and 260 °C) are below the Leidenfrost temperature of diesel (460 °C), the droplet fails to levitate above the surface (Zhao 2018). The spreading factor keeps decreasing in this stage while the height ratio shows a rapid increase at 20 ms and rapid reduction around 35 ms due to droplet oscillation. As a long time is required ($\gg$40 ms) for the droplet to stabilize, the equilibrium phase at higher surface temperature condition is not shown in Fig. 6.11.

Furthermore, surface temperature introduces additional complexity to the analysis of droplet–wall impingement phenomena by the change of liquid properties. Also, as heat transfer between solid–liquid and solid-surrounding gas occurs, droplet evaporation is possible. The viscosity and surface tension of diesel decrease with the increase of surface temperature; thus, the spreading diameter at 260 °C is larger than the one at 150 °C, until evaporation takes place around 25 ms. The height ratio at 260 °C is lower than the one at 150 °C before 13 ms. After that, the height ratio at both high surface temperatures shows fluctuations as observed in the high-speed images (not shown here). This phenomenon might be associated with heat transfer and droplet evaporation.

Since the surface temperature of 150 °C is smaller than the droplet saturation temperature (180–360 °C), the droplet evaporation is primarily driven by the vapor diffusion and the heat transfer occurs by conduction and free convection, while the surface temperature of 260 °C is within the range of the droplet saturation temperature. Therefore, the droplet evaporation, mostly driven by the heat transfer from the heated surface to droplet, may fall into the nucleate boiling regime (Zhao et al. 2018). In this regime, buoyancy starts to take place and moves up the formed vapor bubbles from the hot surface toward the liquid. It is possible that the droplet impingement with 260 °C surface temperature may experience this phenomenon; if so, the increase in height ratio may be due to rising vapor bubbles. Again, the viscous and surface tension forces reduce as the surface temperature increases; hence, the droplet recedes easily after reaching its maximum spreading distance compared to the lower surface temperature case.

In Fig. 6.11 (right), the dynamic contact angle at higher temperatures follows a similar trend with that at 25 °C. The contact angle is about 150° as long as the droplet contacts the surface, and it then reduces substantially to below 10°. Later, the contact angle quickly raises and starts to remain at a relatively stable value. The average advancing contact angle was found to be 74°, 34°, and 31° at 25, 150, and 260 °C, respectively; the average receding contact angle was 20°, 18°, and 21° at 25, 150, and 260 °C, respectively; and the average equilibrium contact angle was 12°, 25°, and 31°, 25, 150, and 260 °C, respectively.

6.5 Simulation Results

6.5.1 Single Droplet Impingement on an Unheated Surface

The droplet–wall interaction process is implemented with the VOF method in CONVERGE software (Richards et al. 2016). High-resolution interface capturing (HRIC) is activated for the current set of simulation to reconstruct the interface details. Since the air density and viscosity are much smaller than those of diesel fuel, the flow in the air has no significant effect on the flow in the droplet, and air flow works only around the droplet. Therefore, it is not necessary to provide a very large computational domain to signify an infinite domain. A 3-D computational domain ($18 \times 18 \times 8$ mm^3) was used to simulate the entire process of a single droplet impingement on an unheated solid surface. Initially, the fuel droplet with an initial velocity is located at a certain distance above the surface. The droplet travels downward toward the wall by the gravity force and reaches the surface with the impact velocity. In the simulation, the liquid phase is fuel and the gas phase is the surrounding air at atmospheric pressure. Open boundary conditions are applied at the top and side to simulate an infinite domain, and a no-slip condition is used for the impinged wall at the bottom (Zhao 2018; Zhao et al. 2018).

A specific contact angle θ of fluid at the wall is used as a boundary condition. The contact line velocity in the current work is simply approximated as the time derivative of the radius of the wetted area (r_{cl}) in Eq. (6.35), based on the studies by Šikalo et al. (2005) and Roisman et al. (2008),

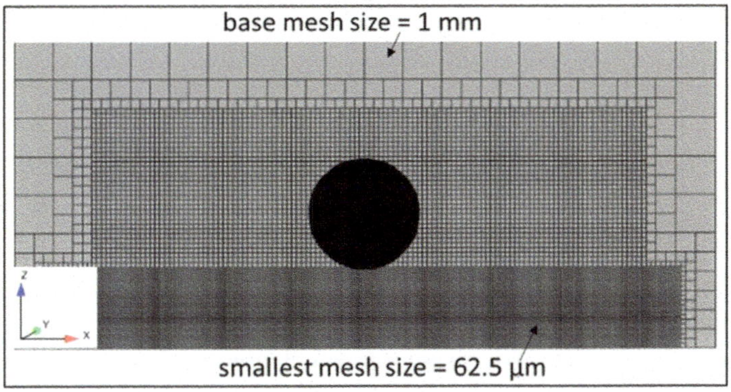

Fig. 6.12 Grid generation with the numerical 3-D diesel droplet (iso-surface at $\alpha = 0.5$) at 1.0 ms (Zhao 2018; Zhao et al. 2018)

$$U_{\mathrm{cl}} = \mathrm{d}r_{\mathrm{cl}}/\mathrm{d}t \tag{6.35}$$

After the contact line velocity is obtained, the advancing or receding phases are defined by the direction of contact line movement based on the dot product of the velocity vector with the unit-free surface normal.

Single diesel droplet impinging on the wall is selected for numerical study which serves for high-pressure diesel spray–wall interaction study in the future. The initial droplet–surface height is 52 mm, and incident droplet diameter is 2.87 mm. To reduce the computational cost, the initial location of droplet is set to 4 mm above the solid surface and the droplet shows the same diameter and velocity as in the experiment. The relevant parameters and liquid properties can be found in Tables 6.1 and 6.2. Constant contact angle with a value of $\theta = 13°$, following the experimentally measured equilibrium contact angle, was used in the simulations. The base mesh size was 1.0 mm in the simulation, and two levels of AMR based on void fraction were used. Further, three levels of fixed embedding were considered along with the droplet traveling path and four levels of fixed embedding were imposed near the impinged surface. Thus, the smallest grid size in the entire computational domain was 62.5 μm. Figure 6.12 shows the mesh generation in the vertical cross section with the diesel droplet at 1.0 ms.

The grid convergence study was performed for two minimum grid sizes. Simulations with minimum grid sizes of 62.5 μm were performed using three and four levels of embedding refinement for the droplet traveling path and near impinged surface regions, respectively. In addition, the embedding on the impinged surface is set to four times thicker in 62.5 μm case than that in 125 μm case, which has a substantial effect on the droplet shape after impinging on the plate. In terms of cell count, the 62.5 μm case resulted in a peak cell count of ~2.3 million, while 125 μm case resulted in a peak cell count of ~1.3 million, which requires at least twice more computationally demanding in 62.5 μm case. Based on the grid convergence, computational demand, and droplet shape considerations, 62.5 μm case grid size was selected as the reference minimum grid size for the current study.

The spreading factor and height ratio with the two grid sizes are plotted respectively in Fig. 6.13 and validated by experimental results. The simulation results with finer grid size (62.5 μm) overall match well with experimental data compared with the coarse grid size case. The spreading factor and height ratio show a stronger agreement with experimental results before ASOI of 5 ms. With the increase of spreading factor, the discrepancy between numerical and experimental results is observed, which might be by reason of contact angle applied in the simulation. The difference between the experimental and numerical spreading factor is about 3.0% at the maximum spreading diameter. Afterward, the simulation results in the receding stage around ASOI of 10 ms are quite comparable with experimental data. Further, due to the computational demanding of numerical study, the experiments also show a longer spreading stage than the simulation.

Figure 6.14 shows a sequence of high-speed images and the corresponding simulation results during the droplet impingement on the unheated surface. The simulation results generally provide a good agreement with the experimental data with respect to the droplet shape, impinging time, and spreading process. At ASOI of 8 ms, the numerical spreading diameter and the droplet structure show difference with the experimental result, which might be as a result of the surface tension. When the maximum spreading diameter is reached, the flattened droplet then starts receding under the capillary force and finally tends to be stable (not shown here).

Figure 6.15 presents pressure coefficient (C_p) and the induced flow field in a vertical plane through the center of the droplet during droplet–wall impingement. The pressure coefficient (C_p) is defined as (Malgarinos et al. 2014)

$$C_p = \frac{P - P_\infty}{\frac{1}{2}\rho_{\text{liq}} U_0^2} \tag{36}$$

where P is the pressure of the computational domain, P_∞ is the pressure on the far field, ρ_{liq} is the liquid fuel density, and U_0 is the impact velocity.

The same scale in terms of the initial droplet size as that in Fig. 6.14 is applied into Fig. 6.15. In Fig. 6.15, pressure increases up to 1.6 times of droplet initial kinetic energy at the initial stage of droplet impinging on the surface, because a dimple is formed based on the droplet impact (Maitra et al. 2014). In the same stage, the velocity magnitude is about 1.2 times of initial impact velocity near impinged surface region. The vortex on the top of the droplet is observed. At the remaining spreading stage from ASOI of 2 ms, C_p is 0.1 times lower than that in the initial spreading phase and it reaches its maximum value on the leading edge of spreading. The velocity magnitude in this remaining stage is also lower compared to that at the initial spreading phase, and it is around 0.6 times of impact velocity. The vortex is visible on the droplet rim in both left and right; it also changes the direction when the receding phase initiates. It appears that the vortex motion is associated with the motion of the droplet leading edge, where the size of vortex is proportional to the contact line velocity. This relation can be evidenced by the fact that the variation

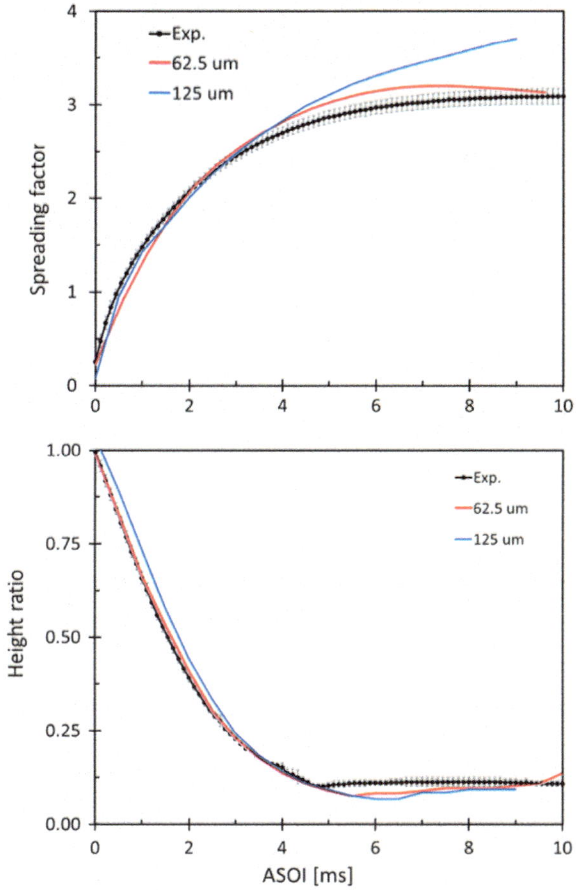

Fig. 6.13 Comparison of spreading factor and height ratio between experiment and simulation results (Zhao 2018; Zhao et al. 2018)

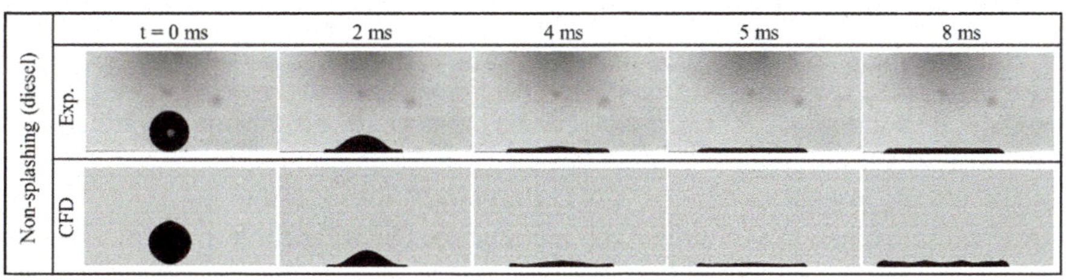

Fig. 6.14 A sequence of high-speed images (top) and the corresponding simulation results (bottom) (3-D iso-surface of droplet in black with $\alpha = 0.5$) (Zhao 2018; Zhao et al. 2018)

observed in the vortex magnitude is similar to that of contact line velocity during the spreading process.

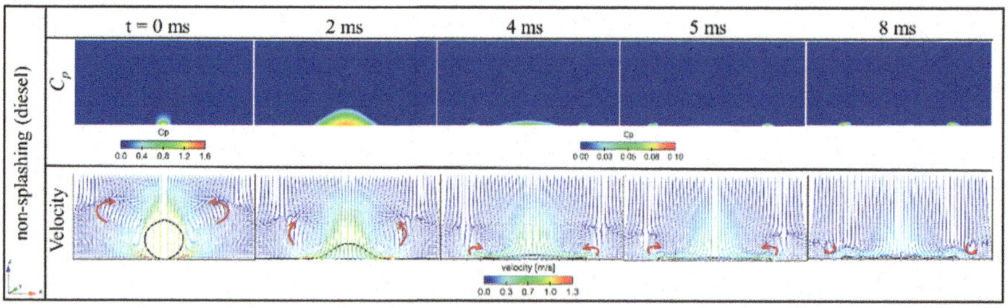

Fig. 6.15 Pressure coefficient C_p (top) and induced flow field profiles shown by velocity vector (bottom) on a vertical plane through the center of droplet. The first C_p legend corresponds to the first picture, while the second legend corresponds to the rest of four pictures (Zhao 2018; Zhao et al. 2018)

6.5.2 Multi-droplet Impingement on a Heated Surface

To validate the evaporation sub-model, a water droplet in the cross-stream of hot air is simulated in a 3-D domain and compared with the published results (Schlottke and Weigand 2008). Overall, the simulation based on the evaporation sub-model matches well with the results by Schlottke and Weigand (2008). After the evaporation sub-model is validated, the evaporation of spherical droplets' impingement on a heated surface and the droplet number and surface temperature effects on evaporation are studied by three cases.

Figure 6.16 shows the arrangement of multiple *n*-heptane droplets from both top and front views. Table 6.3 gives the simulation conditions. The total liquid mass remains the same in all cases; therefore, the droplet size varies at each case and multiple droplets always have the equal size. The initial distance between any two droplets in Case 2 and Case 3 is equal to the radius of this set of droplet. The initial distance between each droplet center and wall is the same for all cases. The droplets fall with an initial velocity of 0.8 m/s. The ambient temperature and pressure of the domain including the droplet are 298 K and 1 atm. The surface temperature is maintained at 483 K, which is 10 K larger than the Leidenfrost temperature of *n*-heptane (473 K). Hence, the evaporation of the droplets falls into the film boiling regime referred to Fig. 6.2. The contact angle between the droplet and the hot surface is set to 120° (Nikolopoulos et al. 2007a).

The computational domain is a 3-D domain (8 mm × 5 mm × 5 mm) with air inside at atmospheric pressure and temperature of 298 K. A non-uniform grid with the maximum size of 200 μm and minimum size of 50 μm in x, z directions and maximum size of 613 μm and minimum size of 7.5 μm in y direction is generated. A finer mesh is applied both in the center of the domain and near the hot surface region, where the heat conduction to the droplet is maximum and droplet shape after impact is also dependent on grid resolution. Figure 6.17 shows the grid generation of Case 1, and the similar grid distributions are used in Case 2 and Case 3.

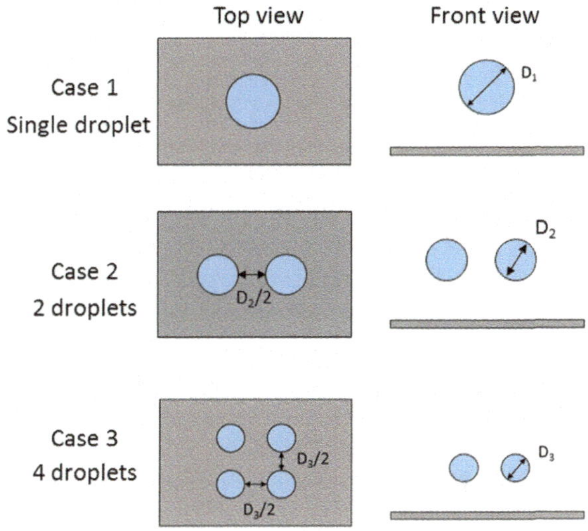

Fig. 6.16 Multiple droplets' arrangement of three cases (Potham et al. 2017; Zhao 2018)

Table 6.3 Simulation parameters of three cases

	Case 1	Case 2	Case 3
Fuel	*n*-heptane	*n*-heptane	*n*-heptane
Number of droplets	1	2	4
D_0(mm)	1.50	1.19	0.94
V_0(m/s)	0.8	0.8	0.8
We	41	33	26
Re	3750	2975	2360

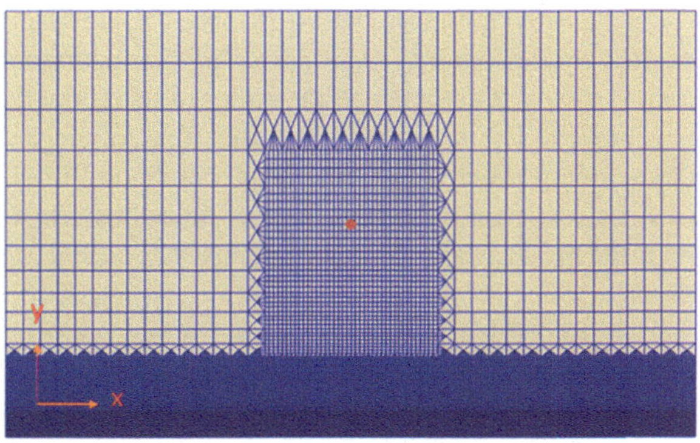

Fig. 6.17 Numerical grid distribution in Case 1 (Potham et al. 2017; Zhao 2018)

As all three cases correspond to film boiling regime, a vapor film can be observed between the droplet and the wall. This vapor layer prevents the droplets from getting into the physical contact with the hot surface. Heat transfer from the surface to droplets is driven by conduction through the vapor film.

Figure 6.18 presents the temporal variation of droplets' shape and vapor formation for Case 1, Case 2, and Case 3. A cut plane at $z = 0$ is presented for Case 1 and Case 2. A diagonal plane by passing the centers of droplets is used in Case 3. The white color signifies the liquid fuel phase, and the color with the legend indicates the vapor fuel phase in Fig. 6.18. Overall, the droplets' shape and vapor volume fraction follow the similar trend in all three cases. Also, the droplet shape and vapor volume fraction distributions of single droplet case match well with the results from Nikolopoulos et al. (2007a). As the droplets approach the hot surface around 3 ms, a portion of the liquid droplet vaporizes and a thin film is formed. This film prevents the physical contact between droplet and the hot surface. Droplets keep spreading after impingement till a thin neck region. It then starts to recede and rebound after 9 ms. At the rebound phase, droplets oscillate in shape from vertical elongation initially to near spherical shape in the later stage. The temperature is higher near the surface and decreases nearby the droplet. Moreover, from Case 2 and Case 3 (multi-droplet), it is also observed that droplets start to merge together after 3 ms as soon as they impinge on the wall. There is more vapor in the center of the domain at 5 ms, which might be caused by the pressure difference during multi-droplet impinging on wall and spread out to a larger region compared with single droplet. Finally, the multi-droplet merges into a single droplet about 8 ms and rebounds away from the hot surface.

In the current study, droplet levitation is calculated as the minimum of the vertical distances between the surfaces of the droplets to the hot surface. Droplet levitation has an effect on the surface temperature and vapor distributions around the droplet, and thus the evaporation rate. Figure 6.19 shows the variation of droplet levitation over time for three cases. Due to the different droplet size, initial droplet lift is also different in each case. Droplet lift approaches its minimum value in each case at about 2.5 ms when it gets closer to the surface. As the droplets spread and evaporate, the levitations increase due to the vapor distribution below the droplet. When the droplets reach their maximum spreading distances, the levitations reduce again since the force exerted by the vapor is overcome. The fluctuation of droplet levitation is observed until a steady state between the vapor mass below the droplet and the droplet mass is achieved. The levitation decreases to a certain extent and then increases during the receding stage. When the droplets rebound from the surface, droplet levitation in Case 1 (single droplet) is higher than those in multiple droplets. However, the opposite trend is shown when comparing Case 2 (two droplets) and Case 3 (four droplets). The reason is that droplets are arranged in two rows in z direction in Case 3, each droplet has two neighboring droplets opposing its spreading, resulting in the merged droplet rising earlier. The droplet levitation becomes similar in Case 2 and Case 3 after 14 ms.

The average droplet surface temperature is a key parameter associated with phase change of droplet. Theoretically, the average surface temperature of droplets with infinitesimally thin surface, undergoing phase change, must be saturation temperature. The saturation temperature of *n*-heptane is 371 K at atmospheric conditions; hence, the droplets start vaporizing when the droplets' surface temperature reaches the saturation temperature. Nevertheless, VOF simulations cause an interface smeared across few cells with finite thickness excluding the effect of grid resolution.

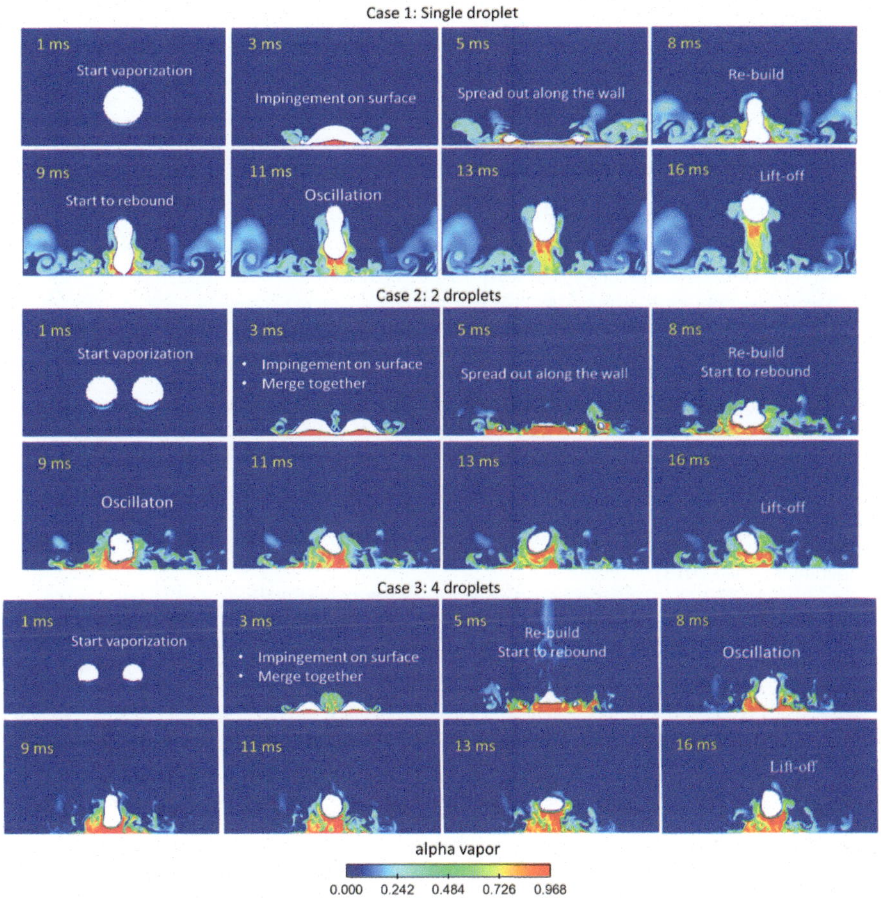

Fig. 6.18 Temporal evolution of liquid and vapor volume fractions (Potham et al. 2017; Zhao 2018)

Therefore, the average droplet surface temperature might be different with saturation temperature even with droplet phase change. Figure 6.20 presents the temporal average droplet surface temperatures in all the cases.

Initially, the droplets are at a room temperature of 298 K. The surface temperature increases steadily to the saturation temperature at 2.5 ms due to the droplets movement toward the hot surface. The droplet surface temperature significantly decreases after 8 ms in Case 1, because the droplet levitation increases at the same time as shown in Figs. 6.18 and 6.19. As the droplet travels away from the heated surface into a relatively cold domain, its surface temperature decreases. The reduction of droplets surface temperature is less distinct in Case 2 and Case 3. Due to less levitation occurred in Case 2 compared with that in Case 3 as shown in Figs. 6.18 and 6.19, the droplet surface temperature in Case 3 is lower than that in Case 2 after 7.5 ms.

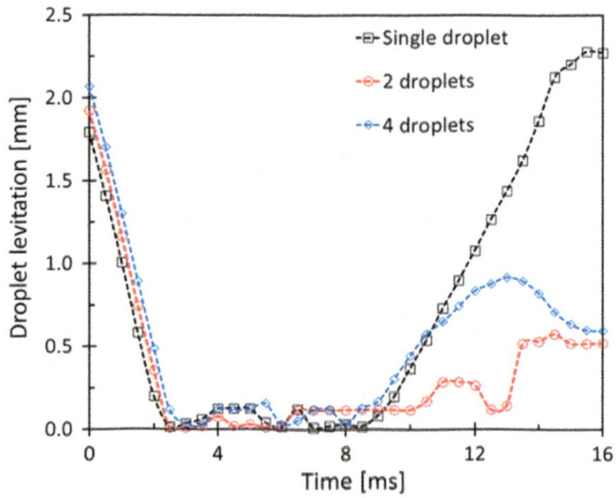

Fig. 6.19 Droplet lift-off height from the wall (Zhao 2018)

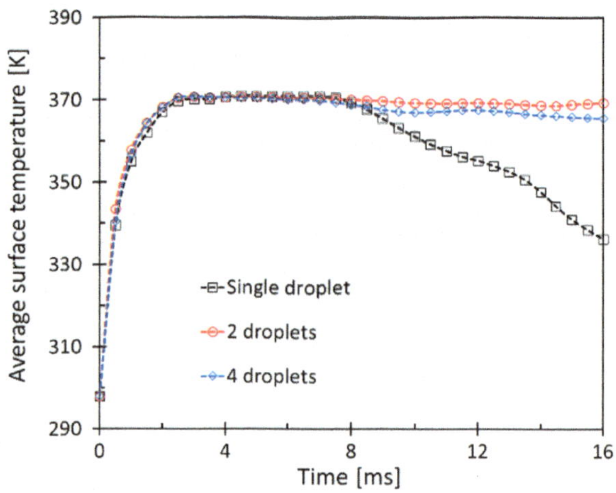

Fig. 6.20 Temporal variation of average surface temperature (Zhao 2018)

The liquid mass fraction inside the domain is normalized with the initial mass fraction. The liquid mass fraction versus time is plotted in Fig. 6.21. Initially, the liquid mass fraction is at 100% since droplets are at room temperature. The total liquid mass in the beginning is the same in three cases, and it decreases slightly due to mass diffusion from droplet surface to the surrounding air. During this period, the droplets' surface temperatures increase to the saturation temperature. Subsequently, there is a sudden reduction in liquid mass fraction caused by temperature and mass diffusion gradients. The evaporation rate in Case 3 is the highest; Case 2 is the next, finally Case 1. This is because of after droplets impacting on the hot surface, larger surface area occurred in Case 3 than Case 2 and Case 1. Larger surface area facilitates more heat transfer and thus higher evaporation rate. The evaporation rate considerably reduces after the droplets rebound. In Case 2 and Case 3, multi-droplet

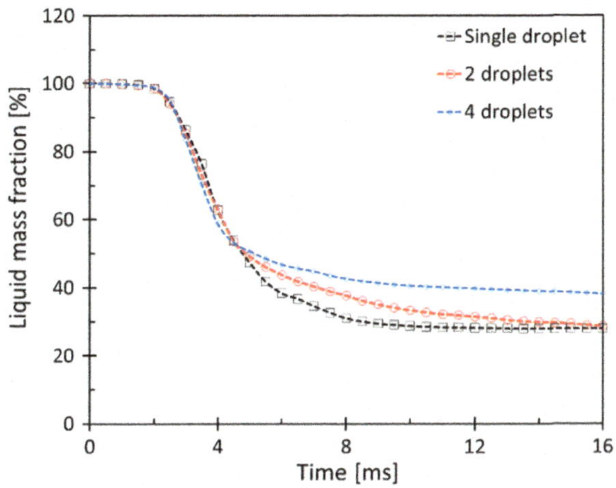

Fig. 6.21 Time dependency of liquid mass fraction (Zhao 2018)

merges at about 4.5 ms which leads to surface area decrease. Further, the evaporation rate decreases in Cases 2 and Case 3 compared to Case 1.

In addition, higher droplet levitation and less spreading cause less evaporation and higher liquid mass fraction in Case 3 compared to Case 2. After 5 ms, liquid mass fraction reaches its lowest value in Case 1 due to larger droplet spreading distance resulting in more evaporation. A very high droplet levitation ensures that evaporation is negligible in Case 1 after 10 ms, whereas a low evaporation rate causes continued decrease in liquid mass fraction in Case 2. Finally, liquid mass fractions are almost equal in Case 1 and Case 2 after 15 ms. Cumulative heat transfer to the droplet can be also calculated in terms of the evaporated liquid mass and the latent heat of vaporization. It is found that the maximum heat transfer shows in Case 1 and the minimum is in Case 3. However, the heat transfer in Case 2 continues to increase even beyond 9 ms and becomes almost the same with that of Case 1.

6.6 Further Discussion

For future experimental and numerical studies of droplet–wall impingement, various test conditions with multi-train droplets, including a sensitivity analysis of droplet size and ambient/injection conditions, as well as combustion, will be considered to further improve the correlation of splashing criteria. The evaporation solver will be improved to achieve a more accuracy of interface representation and numerical calculation. A more accurate modeling approach based on the physics and methodology obtained from droplet impingement with less parameter-tuning requirements for spray–wall impingement prediction will be developed.

6.7 Summary

In this chapter, a detailed analysis of the dynamic process of droplet–wall impingement with various conditions has been performed. The experimental work was implemented under the room temperature and atmospheric pressure; water, diesel, *n*-dodecane, and *n*-heptane were utilized as the tested fuels and injected at various *We* numbers. The splashing criteria during droplet impingement on the solid surface were investigated, and a new correlation in terms of the current experimental data was developed.

The current experimental study on the evolution of the dynamic process of droplet–wall interaction is one of the unique contributions to expand the database of relevant research, such as assisting the dynamic contact angle model development under Direct Numerical Simulations (DNS) or VOF methodologies. In the numerical work, VOF method was applied to characterize the single fuel droplet impingement on the solid surface and deliver a deep understanding of the dynamic impingement process in the CONVERGE™ framework. The main findings of this work are as follows:

- Overall, the experimental results of droplet impingement on the various ambient and surface conditions show a good agreement with the splashing and non-splashing phenomena observed in the published droplet–wall interaction models. A new splashing correlation with respect to *Oh* and *Re* based on the experimental data was found.
- The effect of surface temperature on droplet spreading factor, height ratio, and dynamic contact angle was studied. At higher surface temperature, the maximum spreading factor is larger than that at lower surface temperature case. Diesel droplet at different test conditions appears to oscillate slightly and attempt to reshape; the vertical elongation is observed during the receding phase.
- Simulation results match well with the experimental spreading factor and height ratio. The pressure coefficient C_p and the velocity magnitude are found to be larger at the initial stage of spreading. The vortex is observed on the top of droplet rim at the initial spreading phase. Later, it is visible around the droplet rim on both left and right with the cross-sectional view.
- The newly developed solver of evaporation sub-model was successfully implemented into the existing solver with VOF model. The new model was validated with the published results, and then the multi-droplet impingement on a hot surface was studied by using this evaporation sub-model. The simulations well predicted the droplet levitation characteristics as the surface temperature is higher than Leidenfrost point. As droplet number increases, the spreading distance decreases and it takes less time for droplets to recede. As well, droplet number in a perpendicular direction leads to reduction in droplet spreading and increase in lift-off. Higher lift-off results in lower average temperature.

Disclaimer and Funding Acknowledgement This research is supported by the Department of Energy (DOE), Office of Energy Efficiency and Renewable Energy (EERE), and the Department of Defense, Tank and Automotive Research, Development, and Engineering Center (TARDEC), under Award Number DE-EE0007292. The authors would like to thank Convergent Science for their support on simulation code. The authors also thank Dr. Lakshman Anumolu and Dr. Shaoping Quan of Convergence Science for their valuable recommendations.

References

Bai CX, Rusche H, Gosman AD (2002) Modeling of gasoline spray impingement. Atomization Sprays 12(1–3):1–27. https://doi.org/10.1615/AtomizSpr.v12.i123.10

Bernard R, Foltyn P, Geppert A, Lamanna G, Weigand B (2017) Generalized analysis of the deposition/splashing limit for one- and two-component droplet impacts upon thin films

Brant Foote G (1973) A numerical method for studying liquid drop behavior: simple oscillation. J Comput Phys 11(4):507–530. https://doi.org/10.1016/0021-9991(73)90135-6

Bussmann M, Mostaghimi J, Chandra S (1999) On a three-dimensional volume tracking model of droplet impact. Phys Fluids 11(6):1406–1417. https://doi.org/10.1063/1.870005

Cossali GE, Coghe A, Marengo M (1997) The impact of a single drop on a wetted solid surface. Exp Fluids 22(6):463–472. https://doi.org/10.1007/s003480050073

Fukai J, Zhao Z, Poulikakos D, Megaridis CM, Miyatake O (1993) Modeling of the deformation of a liquid droplet impinging upon a flat surface. Phys Fluids A 5(11):2588–2599. https://doi.org/10.1063/1.858724

Geppert A, Chatzianagnostou D, Meister C, Gomaa H, Lamanna G, Weigand B (2016) Classification of impact morphology and splashing/deposition limit for n-hexadecane. Atomization Sprays 26(10):983–1007. https://doi.org/10.1615/AtomizSpr.2015013352

Habchi C, Foucart H, Baritaud T (1999) Influence of the wall temperature on the mixture preparation in Di gasoline engines. Oil Gas Sci Technol—Rev IFP 54(2):211–222

Hu J, Xiong X, Xiao H, Wan K-t (2015) Effects of contact angle on the dynamics of water droplet impingement. Paper presented at the COMSOL, Boston

Kang BS, Lee DH (2000) On the dynamic behavior of a liquid droplet impacting upon an inclined heated surface. Exp Fluids 29(4):380–387. https://doi.org/10.1007/s003489900104

Leidenfrost JG (1966) On the fixation of water in diverse fire. Int J Heat Mass Transf 9(11):1153–1166. https://doi.org/10.1016/0017-9310(66)90111-6

Lewis SR, Anumolu L, Trujillo MF (2013) Numerical simulations of droplet train and free surface jet impingement. Int J Heat Fluid Flow 44:610–623. https://doi.org/10.1016/j.ijheatfluidflow.2013.09.001

Lindagren R, Denbratt I (2004) Influence of wall properties on the characteristics of a gasoline spray after wall impingement. https://doi.org/10.4271/2004-01-1951

Ma T, Feng L, Wang H, Liu H, Yao M (2017) A numerical study of spray/wall impingement based on droplet impact phenomenon. Int J Heat Mass Transf 112:401–412. https://doi.org/10.1016/j.ijheatmasstransfer.2017.04.110

Maitra T, Tiwari MK, Antonini C, Schoch P, Jung S, Eberle P, Poulikakos D (2014) On the nanoengineering of superhydrophobic and impalement resistant surface textures below the freezing temperature. Nano Lett 14(1):172–182. https://doi.org/10.1021/nl4037092

Malgarinos I, Nikolopoulos N, Marengo M, Antonini C, Gavaises M (2014) VOF simulations of the contact angle dynamics during the drop spreading: standard models and a new wetting force model. Adv Coll Interface Sci 212:1–20. https://doi.org/10.1016/j.cis.2014.07.004

Manzello S, Yang JC (2002) An experimental study of high Weber number impact of methoxynonafluorobutane $C_4F_9OCH_3$ (HFE-7100) and *n*-heptane droplets on a heated solid surface* 1, 45

Markt DP, Zhao L, Zhu X, Pathak A, Torelli R, Lee S-Y, Raessi M (2018) An experimental and computational study of a single diesel droplet impinging on a dry surface. Paper presented at the

ICLASS 2018, 14th triennial international conference on liquid atomization and spray systems, Chicago, IL, USA

Miers SA, Anderson CL, Blough JR, Inal MK (2005) Impingement identification in a high speed diesel engine using piston surface temperature measurements. https://doi.org/10.4271/2005-01-1909

Moreira ALN, Moita AS, Panão MR (2010) Advances and challenges in explaining fuel spray impingement: how much of single droplet impact research is useful? Prog Energy Combust Sci 36(5):554–580. https://doi.org/10.1016/j.pecs.2010.01.002

Mundo C, Sommerfeld M, Tropea C (1995) Droplet-wall collisions: experimental studies of the deformation and breakup process. Int J Multiph Flow 21(2):151–173. https://doi.org/10.1016/0301-9322(94)00069-V

Nikolopoulos N, Theodorakakos A, Bergeles G (2007a) A numerical investigation of the evaporation process of a liquid droplet impinging onto a hot substrate. Int J Heat Mass Transf 50(1):303–319. https://doi.org/10.1016/j.ijheatmasstransfer.2006.06.012

Nikolopoulos N, Theodorakakos A, Bergeles G (2007b) Three-dimensional numerical investigation of a droplet impinging normally onto a wall film. J Comput Phys 225(1):322–341. https://doi.org/10.1016/j.jcp.2006.12.002

Nukiyama S (1966) The maximum and minimum values of the heat Q transmitted from metal to boiling water under atmospheric pressure. Int J Heat Mass Transf 9(12):1419–1433. https://doi.org/10.1016/0017-9310(66)90138-4

O'Rourke PJ, Amsden AA (1996) A particle numerical model for wall film dynamics in port-injected engines. https://doi.org/10.4271/961961

O'Rourke PJ, Amsden AA (2000) A spray/wall interaction submodel for the KIVA-3 wall film model. http://dx.doi.org/10.4271/2000-01-0271

Otsu N (1979) A threshold selection method from gray-level histograms. IEEE Trans Syst Man Cybern 9(1):62–66

Potham SP, Zhao L, Lee S-Y (2017) Numerical study on evaporation of spherical droplets impinging on the wall using volume of fluid (VOF) model. https://doi.org/10.4271/2017-01-0852

Richards KJ, Senecal PK, Pomraning E (2016) CONVERGE manual (Version 2.3)

Rioboo R, Tropea C, Marengo M (2001) Outcomes from a drop impact on solid surfaces. Atomization Sprays 11(2):12. https://doi.org/10.1615/AtomizSpr.v11.i2.40

Roisman IV, Opfer L, Tropea C, Raessi M, Mostaghimi J, Chandra S (2008) Drop impact onto a dry surface: role of the dynamic contact angle. Colloids Surf A 322(1–3):183–191. https://doi.org/10.1016/j.colsurfa.2008.03.005

Schiaffino S, Sonin AA (1997) Molten droplet deposition and solidification at low Weber numbers. Phys Fluids 9(11):3172–3187. https://doi.org/10.1063/1.869434

Schlottke J, Weigand B (2008) Direct numerical simulation of evaporating droplets. J Comput Phys 227(10):5215–5237. https://doi.org/10.1016/j.jcp.2008.01.042

Schrader ME (1995) Young-Dupre revisited. Langmuir 11(9):3585–3589. https://doi.org/10.1021/la00009a049

Šikalo Š, Wilhelm H-D, Roisman IV, Jakirlić S, Tropea C (2005) Dynamic contact angle of spreading droplets: experiments and simulations. Phys Fluids 17(6):062103. https://doi.org/10.1063/1.1928828

Soriano G, Alvarado JL, Lin YP (2010) Experimental characterization of single and multiple droplet impingement on surfaces subject to constant heat flux conditions (49415), pp 707–715. https://doi.org/10.1115/ihtc14-22515

Stanton DW, Rutland CJ (1998) Multi-dimensional modeling of thin liquid films and spray-wall interactions resulting from impinging sprays. Int J Heat Mass Transf 41(20):3037–3054. https://doi.org/10.1016/S0017-9310(98)00054-4

Stow CD, Hadfield MG (1981) An experimental investigation of fluid flow resulting from the impact of a water drop with an unyielding dry surface. Proc R Soc Lond. A. Math Phys Sci 373(1755):419

Tamura Z, Tanasawa Y (1958) Evaporation and combustion of a drop contacting with a hot surface. Symp (Int) Combust 7(1):509–522. doi https://doi.org/10.1016/S0082-0784(58)80086-7

Tonini S, Cossali GE (2012) An analytical model of liquid drop evaporation in gaseous environment. Int J Therm Sci 57:45–53. https://doi.org/10.1016/j.ijthermalsci.2012.01.017

Trapaga G, Szekely J (1991) Mathematical modeling of the isothermal impingement of liquid droplets in spraying processes. Metall Mater Trans B 22(6):901–914. https://doi.org/10.1007/bf02651166

Vadillo DC, Soucemarianadin A, Delattre C, Roux DCD (2009) Dynamic contact angle effects onto the maximum drop impact spreading on solid surfaces. Phys Fluids 21(12):122002. https://doi.org/10.1063/1.3276259

Wal RLV, Berger GM, Mozes SD (2006) The splash/non-splash boundary upon a dry surface and thin fluid film. Exp Fluids 40(1):53–59. https://doi.org/10.1007/s00348-005-0045-1

Yang B, Ghandhi J (2007) Measurement of diesel spray impingement and fuel film characteristics using refractive index matching method. https://doi.org/10.4271/2007-01-0485

Yarin AL, Weiss DA (1995) Impact of drops on solid surfaces: self-similar capillary waves, and splashing as a new type of kinematic discontinuity. J Fluid Mech 283:141–173. https://doi.org/10.1017/S0022112095002266

Zhao L (2018) An experimental and computational study of fuel spray interaction: fundamentals and engine applications. (Doctor of philosophy), Michigan Technological University

Zhao L, Torelli R, Zhu X, Scarcelli R, Som S, Schmidt H, Lee S-Y (2017) An experimental and numerical study of diesel spray impingement on a flat plate. SAE Int J Fuels Lubr 10(2):407–422. https://doi.org/10.4271/2017-01-0854

Zhao L, Ahuja N, Zhu X, Zhao Z, Lee S-Y (2018) Splashing criterion and topological features of a single droplet impinging on the flat plate

Chapter 7
Modeling of Cavitation in Fuel Injectors with Single- and Two-Fluid Approaches

Kaushik Saha, Michele Battistoni, Sibendu Som and Xiangguo Li

Abstract In high-pressure fuel injection systems, cavitation is known to affect spray atomization processes. Modeling the cavitation phenomenon has become a necessity to ensure predictive quality and higher fidelity of the fuel spray simulations. Inside the fuel injectors, local pressures drop below the saturation pressure of fuels in regions of flow separations, such as inlet of holes and periphery of needles at low-lift conditions. Several cavitation models and multiphase modeling approaches have been employed in the literature to predict the extent of cavitation in the fuel injection systems. A review of these modeling approaches will be presented. Amongst the cavitation models, bubble-based and semi-empirical timescale-based ones are widely used. Mixture/single-fluid and Eulerian–Eulerian/two-fluid approaches have been adopted for fuel injection cavitation modeling. Two-fluid approach captures the interaction between the two phases, which is usually ignored in single-fluid approach. Comparative studies in the literature will be reviewed here to provide a comprehensive idea of the cavitation modeling approaches to the readers. The advantages and disadvantages of these models will be discussed in depth. Keeping in mind the conflicting requirements of accuracy and constraints of computational cost, recommendations will be provided for suitable cavitation modeling approaches.

K. Saha (✉)
Centre for Energy Studies, IIT Delhi, New Delhi, India
e-mail: Kaushik.Saha@ces.iitd.ac.in

M. Battistoni
Mechanical Engineering, University of Perugia, Perugia, Italy
e-mail: michele.battistoni@unipg.it

S. Som
Energy Systems Division, Argonne National Laboratory, Lemont, IL, USA
e-mail: ssom@anl.gov

X. Li
Mechanical and Mechatronics Engineering, University of Waterloo, Waterloo, ON, Canada
e-mail: xianguo.li@uwaterloo.ca

K. Saha et al. (eds.), *Two-Phase Flow for Automotive and Power Generation Sectors*, Energy, Environment, and Sustainability,
https://doi.org/10.1007/978-981-13-3256-2_7

Nomenclature

CFL	Courant Friedrichs Lewy number
D	Mass diffusivity (m^2/s)
f_v, f_l, f_g	Mass fraction of vapour, liquid and gas, respectively
k	Turbulent kinetic energy (m^2/s^2)
m_{vap}, m_{liq}	Mass of vapour and liquid phases in a cell (kg)
N'''	Bubble number density ($1/m^3$)
p	Local pressure (Pa)
p_{crit}	Critical pressure (Pa)
P_{inj}	Injection pressure (MPa)
P_{sat}, p_{sat}	Saturation pressure (Pa)
ΔP	Pressure differential, MPa
R	Bubble radius, m
R_P	Source term for cavitation modeling, kg/(m^3 s)
S_{11}	Strain rate (1/s)
t	Time (s)
T_{fuel}	Fuel temperature (K)
T_{sat}	Saturation temperature (K)
u	Local cell velocity (m/s)
u_i	Advected mean velocity (m/s)
u_j	Advecting mean velocity (m/s)
V	Volume of a cell (m^3)
x	Local cell vapour quality
$\bar{x}$	Local cell equilibrium quality
Y_m	Mass fraction of mth species

Greek

α	Void fraction (vapour + non-condensable gases)
α_q	Volume fraction of the qth phase
ϵ	Turbulence dissipation rate (m^2/s^3)
θ,θ_0	Equilibrium timescale and empirical time constant(s)
μ,μ_t	Dynamic and turbulent viscosity coefficient (kg/m s)
ρ, ρ_v, ρ_l, ρ_g	Density of mixture, vapour, liquid and gas (kg/m^3)

7.1 Introduction

Modern diesel injectors operate at high injection pressures (200 MPa or higher) with spray velocities on the order of few hundred m/s. In recent times, gasoline direct injectors also operate at higher than usual injection pressures (20 MPa or higher). Depending on fuel physical properties, needle lift, pressure differential and internal

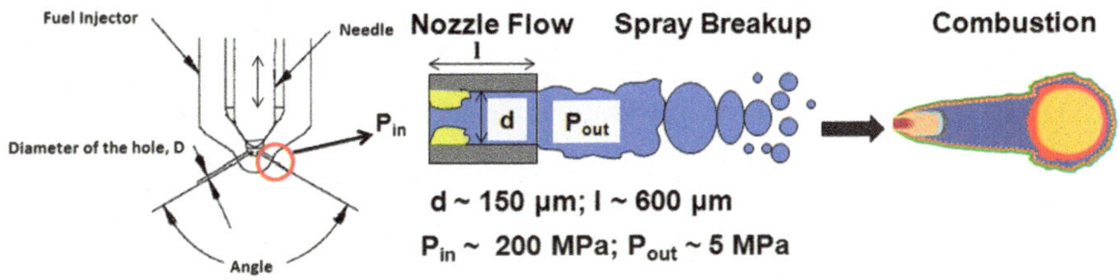

Fig. 7.1 Illustration of a cavitating nozzle and cavitation affecting spray breakup and combustion

injector geometry, local pressures inside the injector holes and in the sac region can drop below the saturation pressure of the concerned fuel resulting in vapour or void formation, a phenomenon called 'cavitation'. In the last two decades, cavitation has been known to affect the fuel injector erosion, the spray atomization process and consequently the combustion and pollutant formation processes (Baumgarten 2005; Payri et al. 2006; Som 2009). Figure 7.1 illustrates the effect of cavitation on spray combustion. The needle movement dictates the available passage for the fuel flow through the injector holes, affecting the flow features at the exits of the holes.

Investigation of cavitation in fuel injectors through experiments and numerical techniques has been carried out in the last few decades. An experimental investigation is extremely difficult in case of real-size commercial fuel injectors because of small dimensions of holes (~100 to 200 μm) and high fuel velocity (~400 to 600 m/s) inside the injector holes. Still, with some limitations on the operating conditions, real-size nozzles with optical access have been built and used for visualizations of cavitation phenomena (Afzal et al. 1999; Manin et al. 2018). Semi-empirical correlation [Nurick's correlation (Nurick 1976)] based on one-dimensional analysis predicting discharge coefficient (C_d) can only provide an estimate of the mass flow rate in case of round axisymmetric nozzles. Further information on the extent of cavitation and its effect on flow variables at the hole outlet cannot be obtained through such a correlation. Computational Fluid Dynamics (CFD), hence, has been a useful tool for analysing cavitation in fuel injectors for the last two decades.

Cavitation, a two-phase flow problem, should be solved using a multiphase modeling approach. The mixture multiphase approach is commonly adopted for cavitation modeling, since the approach is easy to implement and computationally cheap. Mixture approach results in only one set of conservation equations (mass, momentum, energy) and a single transport equation to estimate the instantaneous local composition of the two phases (liquid and vapour). Therefore, in a mixture approach, for any given computational cell the two phases are considered to be mixed homogeneously within that cell. The mixture multiphase approach without the consideration of relative velocity has been typically used for cavitation modeling. Although, in its generic form, the mixture approach does include the concept of slip or relative velocity (ANSYS 18). The reason proposed for neglecting the relative velocity is that the average flow velocity is considerably higher than the expected magnitude of the relative phase velocity (Saha 2014). Volume of fluid (VOF) is preferred when the two phases are considered immiscible and interface tracking is of prime importance. However, grid refinement requirement at the interfacial regions renders the VOF approach very expensive (Gorokhovski and Herrmann 2008). The full Eulerian–Eulerian approach takes into account the relative velocity between the phases, and hence, the interaction between the phases needs to be modelled more accurately. The Eulerian–Eulerian approach is usually more expensive than the mixture approach with a separate set of transport equations for both the phases (ANSYS 18). Eulerian–Lagrangian approach is rarely used for cavitation modeling because it is very expensive and not feasible to implement for multihole commercial fuel injectors. However, the Eulerian–Lagrangian approach, more commonly known as the discrete bubble model (DBM), is to date the most detailed approach in the field of numerical analysis of nozzle flow cavitation. The DBM approach involves bubble nucleation, bubble transport, collision, coalescence and breakup phenomena (Giannadakis et al. 2008). As a result, the DBM is computationally expensive and it is typically beyond the scope of the practical usefulness for fuel injection simulation today. Therefore, it is not included in the present review.

Prediction of the extent of cavitation is not only affected by the multiphase approach, but also due to cavitation model (source term modeling for phase change), turbulence modeling (different RANS based and LES approaches), geometry (sac region, hole shape: converging, straight, inlet shape: rounded or sharp) (Som 2009; Saha 2014; Schmidt and Corradini 2001). In terms of turbulence modeling, large eddy simulation (LES) has already been shown to be more reliable when flow transients are vital at low-needle lifts (needle opening and closing) (Battistoni et al. 2015). However, for a commercial fuel injector, even unsteady RANS (U-RANS) could be very expensive (Saha et al. 2016). Hence, LES is still not commonly adopted for cavitation simulations for commercial fuel injectors. Naturally, the quality of simulation prediction depends on the types of cavitation model and multiphase approach chosen for the study. Comparative assessments of cavitation models and multiphase approaches are rarely documented in the literature. The current chapter would like to address this less explored area and discuss some of the work reported in the literature.

The chapter is organized in the following manner: at the beginning, the governing equations involved in single- and two-fluid approaches will be described. In this chapter, single-fluid would indicate the mixture multiphase approach and the Eulerian–Eulerian multiphase approach will be addressed as two-fluid. The sub-models involved in the two-fluid approach for interaction between the two phases will be discussed. After that, different cavitation models, commonly used in the literature, will be described along with their origins, underlying assumptions and approximations, advantages and limitations. This will be followed by discussions on the published results for providing comparative assessments of the multiphase approaches coupled with various cavitation models.

7.2 Model Formulation

The model formulations available in the literature for solving the cavitating two-phase flow in a direct injection fuel injector will be reported here. The mass and momentum conservation equations for the single- and two-fluid approaches will be elaborated in this section. The source/sink terms in the transport equations for estimating the proportions of the constituent phases (liquid and vapour) are accounted for through cavitation model. Different cavitation models adopted in the literature are described in this section as well. The energy equation is often not solved in the cavitation modeling approach, under the assumption that cavitation phenomenon will be nearly isothermal (Saha 2014). However, in the case of modern fuel injectors there is still scope of inclusion of energy equation and examine its effect.

7.2.1 Multiphase Approaches

In case of the single-fluid approach, one set of conservation equations of mass and momentum is solved:

$$\frac{\partial \rho}{\partial t} + \frac{\partial u_j \rho}{\partial x_j} = 0 \tag{7.1}$$

$$\frac{\partial \rho u_i}{\partial t} + \frac{\partial \rho u_i u_j}{\partial x_j} = -\frac{\partial p}{\partial x_i} + \frac{\partial}{\partial x_j}\left[\mu_{\text{eff}}\left(\frac{\partial u_i}{\partial x_j} + \frac{\partial u_j}{\partial x_i} - \frac{2}{3}\frac{\partial u_i}{\partial x_i}\right)\right] \tag{7.2}$$

where the mixture density is calculated from

$$\frac{1}{\rho} = \frac{f_v}{\rho_v} + \frac{f_g}{\rho_g} + \frac{1 - f_v - f_g}{\rho_l} \tag{7.3}$$

where ρ, ρ_v, ρ_g and ρ_l are the densities of the mixture, the vapour phase, the non-condensable gases and the liquid phase, respectively. The vapour concentration is determined from the following transport equation:

$$\frac{\partial \rho f_v}{\partial t} + \frac{\partial u_j \rho f_v}{\partial x_j} = \frac{\partial}{\partial x_j}\left(\rho D \frac{\partial f}{\partial x_j}\right) + R_P \tag{7.4}$$

where R_P represents the phase change rate and will be elaborated in 'cavitation models'.

The effective viscosity in Eq. 7.2 is expressed as $\mu_{\text{eff}} = \mu + \mu_t$ where μ is the mixture (liquid and gaseous phases) molecular viscosity and μ_t is the turbulent viscosity which can be modelled. For example,

$$\mu_t = C_\mu \rho \frac{k^2}{\epsilon} \tag{7.5}$$

for $k - \epsilon$ models. In case of the two-fluid approach, transport equations are solved for both the phases. The pressure p is shared by both the phases. The conservation equations are as follows:

$$\frac{\partial \alpha_q \rho_q}{\partial t} + \frac{\partial \alpha_q u_{j,q} \rho_q}{\partial x_j} = R_P \tag{7.6}$$

$$\frac{\partial \alpha_q \rho_q u_{i,q}}{\partial t} + \frac{\partial \alpha_q \rho_q u_{i,q} u_{j,q}}{\partial x_j} = -\alpha_q \frac{\partial p}{\partial x_i} + \frac{\partial \tau_{ij,q}}{\partial x_j} + M_{qp} + u_{i,q} R_P \tag{7.7}$$

where

$$\tau_{ij,q} = \left[\alpha_q \mu_{\text{eff,q}} \left(\frac{\partial u_{i,q}}{\partial x_j} + \frac{\partial u_{j,q}}{\partial x_i} - \frac{2}{3} \frac{\partial u_{i,q}}{\partial x_i} \right) \right]$$

q stands for phase index, u_q is velocity of the qth phase and M_{qp} estimates the momentum exchange between the two phases. The closure equation becomes

$$\sum_{q=1}^{2} \alpha_q = 1 \tag{7.8}$$

The momentum exchange term can be evaluated in terms of interphase momentum exchange coefficient K_{qp} (ANSYS 18).

$$M_{qp} = K_{qp}(u_{i,q} - u_{i,p})$$

where $K_{qp} = \frac{\alpha_p \alpha_q \rho_q f}{\tau_q}$, f is the drag function dependent on the drag coefficient C_D and τ_q is the relaxation timescale.

7.2.2 *Cavitation Models*

There are various approaches to formulate a cavitation model. Most approaches are based on bubble dynamics and the approximation that pressure difference across the bubble interface is the main driving mechanism for growth and collapse (Brennen 1995). There is also a model that depends on the semi-empirical correlations and timescale estimation for phase change. The pros and cons of the cavitation models described here will also be discussed in this section. Typically, cavitation models involve the formulation of the source/sink terms to assess the growth and collapse of vapour cavities. Some of the cavitation models consider the presence of a very small amount (~ppm level) of non-condensable gases in the liquid fuel, and the relevant further details have been documented in the literature (Brennen 1995). Therefore, in general, the void fraction (volume fraction of gaseous species) in the two-phase flow is given by $\alpha = \alpha_g + \alpha_v$. In fuel injector holes, the liquid fuel is subjected to high pressure and high velocity. Hence, fuel compressibility becomes vital, which is not considered in all the cavitation models.

7.2.2.1 Schnerr and Sauer

In their cavitation model, Schnerr and Sauer (2001) neglected the presence of non-condensable gases. As a result, α denotes α_v only in Schnerr and Sauer model. They expressed α as,

$$\alpha = \frac{\frac{4}{3}\pi R^3 N'''}{1 + \frac{4}{3}\pi R^3 N'''} \tag{7.9}$$

One may arrive at the following, by combining liquid, vapour and overall mass conservation equations,

$$\frac{\partial(\rho_v \alpha)}{\partial t} + \frac{\partial(u_j \rho_v \alpha)}{\partial x_j} = \dot{m}_v = \frac{\rho_v \rho_l}{\rho}\frac{D\alpha}{Dt} \tag{7.10}$$

Equation 7.10 has been the base equation of multiple cavitation models proposed in the literature. The bubble number density N''' was considered to be constant (on the order of $10^{13}\frac{1}{\mathrm{m}^{-3}}$). The temporal derivative of α was coupled with the vapour and liquid mass conservation equations. Finally, using linear Rayleigh equation the source term was derived as (Schnerr and Sauer 2001)

$$R_P = \frac{3\alpha(1-\alpha)}{R}\frac{\rho_v \rho_l}{\rho}(-1)^j \sqrt{\frac{2\left(|p - p_{\mathrm{sat}}|\right)}{3\rho_l}} \tag{7.11}$$

where

$$p - p_{\mathrm{sat}} \begin{cases} > 0 & j = 1 \quad \text{for vapour condensation} \\ < 0 & j = 2 \quad \text{for vapour formation} \end{cases}$$

Schnerr-Sauer model has been combined with both the single- and the two-fluid approaches (ANSYS 18). The bubble number density N''' was defined in terms of per unit liquid volume rather than per unit control volume (containing the two-phase mixture). Such an approach may lead to undesirable outcomes, which will be elaborated in the Results and Discussion section.

7.2.2.2 Zwart-Gerber-Belamri

Zwart et al. (2004) assumed the bubble size to be constant in their cavitation model. They also did not consider the presence of non-condensable gases. If $p < p_{\mathrm{sat}}$ (vapourization),

$$R_P = F_{\text{vap}} \frac{3\alpha_{\text{nuc}}(1-\alpha)\rho_v}{R} \sqrt{\frac{2(p - p_{\text{sat}})}{3\rho_l}} \tag{7.12}$$

else (condensation),

$$R_P = F_{\text{cond}} \frac{3\alpha\rho_v}{R} \sqrt{\frac{2(p - p_{\text{sat}})}{3\rho_l}} \tag{7.13}$$

where the parameters used in this model are $\alpha = \alpha_v$, $F_{\text{vap}} = 50$; $F_{\text{cond}} = 0.01$, $R = 10^{-6}$ m is the bubble radius, $\alpha_{\text{nuc}} = 5 \times 10^{-4}$ is the nucleation site volume fraction. The vapourization equation was made different from the condensation. Their constant bubble radius assumption is not physically correct since bubble radius is supposed to change with the evolution of void fraction. Despite these limitations, their model has achieved fair amount of success because the parameters F_{vap} and F_{cond} were tuned to obtain reasonable predictions. Zwart-Gerber-Belamri model has also been implemented with both the single- and the two-fluid approaches (ANSYS 18).

7.2.2.3 Saha-Abu-Ramadan-Li

Saha-Abu-Ramadan-Li model (Saha et al. 2013; Saha and Li 2016) has also been based on Eq. 7.10. They considered the presence of non-condensable gases. Cavitation occurs when the local mean effective pressure is below the saturation pressure.

$$p_{\text{eff}} < p_{\text{sat}} \tag{7.14}$$

where the local mean effective pressure p_{eff} is considered to include the cumulative effect of local static pressure, stresses, turbulent pressure fluctuations.

$$p_{\text{eff}} = p - p' - (\mu + \mu_t) S_{11} \tag{7.15}$$

where p is the mean local static pressure, p' is the turbulent pressure fluctuation and can be taken as $p' = 0.47\rho k$. S_{11} corresponds to the highest eigen-value of the deviatoric strain tensor; to determine S_{11} first $det(S - \lambda I)$ is equated to zero, where S is the shear strain tensor and I is the identity matrix. Roots of λ are thus obtained, and the highest value of λ is used to estimate S_{11}. The phase change rate can be derived by combining the mass conservation equations for the mixture, liquid and vapour phases, respectively, as follows

$$R_P = \frac{3\alpha}{R} \frac{\rho_v \rho_l}{\rho} (-1)^j \sqrt{\frac{2(|p_{\text{eff}} - p_{\text{sat}}|)}{3\rho_l}} \tag{7.16}$$

where

$$p_{\text{eff}} - p_{\text{sat}} \begin{cases} > 0 & j = 1 \quad \text{for vapour condensation} \\ < 0 & j = 2 \quad \text{for vapour formation} \end{cases}$$

and R is the effective bubble radius, α is the void fraction $(\alpha_v + \alpha_g)$, and R can be expressed in terms of α and number density of bubbles $[N''' \sim 10^{12}\ (\mathrm{m}^{-3})]$. Therefore, the expression of α used by them was,

$$\alpha = \frac{4}{3}\pi R^3 N'''$$

Unlike the two previous cavitation models, this cavitation model considered compressibility of both the liquid and gaseous phases.

7.2.2.4 Alajbegovic et al., Battistoni et al.

Another commonly used bubble-based model is implemented in AVL FIRE CFD code (Alajbegovic et al. 1999; Battistoni and Grimaldi 2018; Battistoni et al. 2014a). It is coupled with two-fluid approach. The starting point of this model is based on Rayleigh bubble growth as well. The rate of increment of the bubble radius is provided as,

$$\dot{R} = \sqrt{\frac{2}{3}\left(\frac{p_{\text{sat}} - p_{\text{eff}}}{\rho_l} - R\ddot{R}\right)} \tag{7.17}$$

The p_{eff} accounts for the turbulent fluctuations affecting the local pressure deviating from the thermodynamic pressure.

$$p_{\text{eff}} = p - C_E \frac{2}{3}\rho_l k_l \tag{7.18}$$

where $C_E = 1.2$ and k_l is the turbulent kinetic energy in the liquid phase surrounding the vapour bubbles (Battistoni et al. 2014a). The rate of change of mass of a cavitation bubble is given by

$$\frac{\partial m_b}{\partial t} = \rho_v 4\pi R^2 \dot{R} \tag{7.19}$$

The modeling approach did not limit itself to mono-disperse bubble population approximation for representing the cavitating regions. The approach needed to estimate bubble size distribution to account for poly-disperse bubble distribution. Additional transport equations were solved for N''' and A''', which denote bubble number density and interface area density, respectively.

$$\frac{\partial N'''}{\partial t} + \nabla \cdot \left(N''' \vec{v}^{(0)}\right) = \sum R_j + R_{ph} \tag{7.20}$$

and

$$\frac{\partial A'''}{\partial t} + \nabla \cdot \left(A''' \vec{v}^{(2)}\right) = \sum \phi_j + \phi_{ph} \tag{7.21}$$

The zeroth and second moment average velocities of the bubble size distribution are indicated by $\vec{v}^{(0)}$ and $\vec{v}^{(2)}$, respectively. For the momentum interaction between the two phases, the model approach uses drag coefficient (C_D) and turbulent dispersion coefficient (C_{TD}). A value of $C_{TD} = 0.1$ is typically used. It is interesting to point out that additional treatment was included in the framework to account for the increment in the turbulent viscosity due to the relative velocity of the two phases by using Sato coefficient of $C_{\text{Sato}} = 0.6$.

7.2.2.5 Homogeneous Relaxation Model

Homogeneous relaxation model is a type of phase change model that is based on the empirical constants (Bilicki and Kestin 1990; Downar-Zapolski et al. 1996; Gopalakrishnan and Schmidt 2008). The HRM was formulated for solving flash boiling problems. However, it has been successfully implemented in commercial CFD codes like CONVERGE (Richards et al. 2018) for analysing cavitation problems in diesel injectors (Zhao et al. 2014; Battistoni et al. 2014a, b). HRM represents the phase transition by estimating the timescale to reach equilibrium two-phase composition. HRM falls in between the two extremes of thermodynamic two-phase models represented by the homogeneous equilibrium model (HEM) and the homogeneous frozen model (HFM) (Downar-Zapolski et al. 1996; Brennen 1995). Thus, HRM captures the essence of in-between real-world scenarios.

The source/sink terms R_P in the fuel species conservation equations are obtained through the HRM. At first, the rate at which the local vapour quality (x) approaches the equilibrium vapour quality is evaluated. The vapour quality is mathematically represented as $x = \frac{Y_v}{Y_v+Y_l} = \frac{m_{\text{vap}}}{m_{\text{vap}}+m_{\text{liq}}}$. The rate of change of local vapour quality, ($\frac{Dx}{Dt}$), provides the estimate of R_P. Through HRM we get,

$$\frac{Dx}{Dt} = \frac{\overline{x} - x}{\theta} \tag{7.22}$$

The timescale θ is calculated as

$$\theta = \theta_0 \alpha^{-0.54} \psi^{-1.76} \tag{7.23}$$

where

$$\theta_0 = 3.84 \times 10^{-7}; \ \psi = \frac{|p_{\text{sat}} - p|}{p_{\text{crit}} - p_{\text{sat}}}; \ \alpha = \alpha_v + \alpha_{N_2} + \alpha_{O_2}$$

Non-condensable gases are included in α. If x^0 is the current local vapour quality of a cell and x^1 is the corresponding value for the next time step, with a time step size of Δt, then x^1 can be computed as—$x^1 = \bar{x} - \left(\bar{x} - x^0\right) e^{-\Delta t/\theta}$. Therefore,

$$R_P = \frac{(x^1 - x^0)(m_{\text{vap}} + m_{\text{liq}})}{V \Delta t} = (x^1 - x^0)\rho \frac{(Y_v + Y_l)}{\Delta t} \tag{7.24}$$

where V is the volume of the cell and m_{vap} and m_{liq} are the vapour and liquid masses, respectively, in that cell.

7.3 Results and Discussion

In this section, published results from the literature will be discussed emphasizing on the effects of different cavitation models and the choice of multiphase approaches. The effects seen in both local and global perspectives will be presented and elaborated here.

The aim of the cavitation simulation in a fuel injector is to provide physically correct inputs for the spray simulation. Therefore, it is important to investigate the flowfield characteristics at the nozzle exit. Saha and Li (2016) delved into the effects of change in cavitation models as well as the multiphase approaches (Saha and Li 2016). RNG $k - \epsilon$ turbulence model was adopted in their study. In case of an axisymmetric nozzle with 50 MPa inlet pressure and 5 MPa back pressure, the liquid volume fractions predicted by three different cavitation models at the nozzle exit were plotted with respect to the normalized radius (r/r_0, r_0 being the radius of the outlet) and shown in Fig. 7.2. There are considerable differences in the two-phase compositions at the nozzle exit. Difference in the extent of predicted cavitation regions (c.f. Fig. 10 in Saha and Li 2016) leads to such variations in the liquid volume fraction profiles. However, all the models predict values of the same order and have the same qualitative trend with the maximum value at the centre and then decreasing towards the wall and then increasing again. At the nozzle exit, cavitating zones tend to drift away from the wall and hence the liquid concentration increases near the wall. The Saha-Abu-Ramadan-Li model tends to predict considerable vapour concentration in the middle of the nozzle compared to the other two models. Schnerr and Sauer model's definition of bubble number density is not physically correct, which is also the main difference in its formulation with respect to the Saha-Abu-Ramadan-Li model. Moreover, consideration of non-condensable gases assists in a realistic expansion of voids when the Saha-Abu-Ramadan-Li model is used. Zwart-Gerber-Belamri model does not consider the non-condensable gases. Moreover, that model uses tuned constants which cannot be expected to yield the desired prediction in a wide range of scenarios.

Saha and Li (2016) clearly showed the effect of changing the multiphase approach on cavitation through vapour volume fraction contours. Figure 7.3 presents the change in vapour volume fraction contours for the axisymmetric nozzle, when the multiphase approach is changed keeping the cavitation model the same.

It was seen that a time step size of 10^{-6} secs was sufficiently small to capture the cavitation characteristics. There were noticeable differences in the predictions using the single- and two-fluid approaches. Multiphase coupled algorithm and per phase turbulence modeling approach were adopted for the two-fluid simulations as they have been proved effective in the parametric study of different numerical schemes by Saha and Li (2016). ANSYS Fluent ANSYS (18) provides different options for

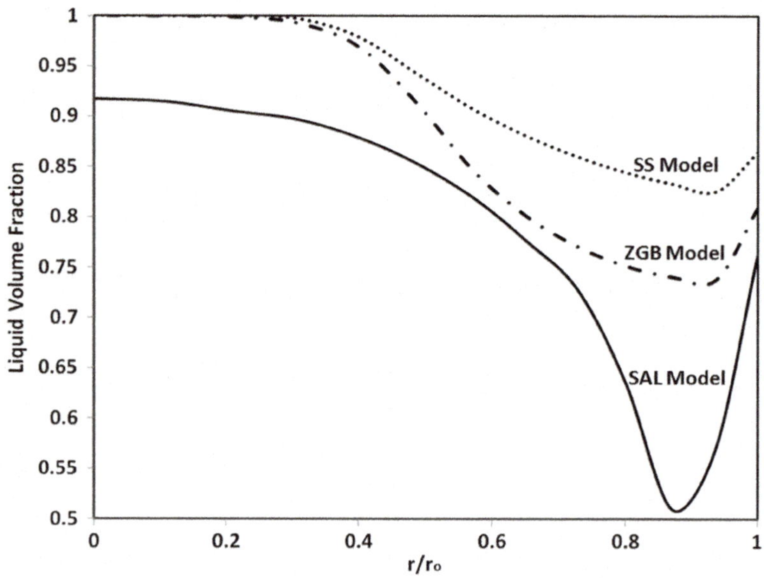

Fig. 7.2 Comparison of the predicted liquid volume fraction profiles at the nozzle exit of the axisymmetric nozzle using the three cavitation models (SS—Schnerr and Sauer; ZGB—Zwart-Gerber-Belamri; SAL—Saha-Abu-Ramadan-Li) coupled with the single-fluid approach subjected to the inlet pressure of 50 MPa and the outlet pressure of 5 MPa (from Saha and Li 2016, reprinted with permission from ASME)

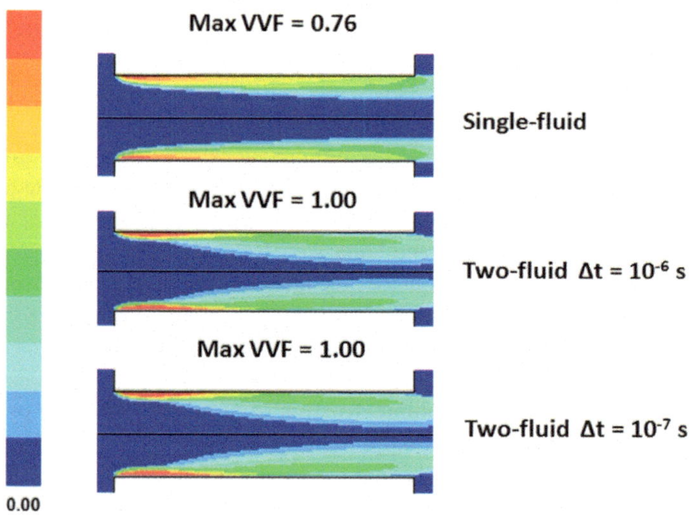

Fig. 7.3 Comparison of the vapour volume fraction contours for the axisymmetric nozzle for the Zwart-Gerber-Belamri model coupled with the single- and two-fluid approaches at the inlet pressure of 100MPa and fixed outlet pressure of 5 MPa. (The maximum vapour volume fraction is abbreviated as Max VVF. All the two-fluid results are at 0.5 ms.) (from Saha and Li 2016, reprinted with permission from ASME)

two-fluid approach in terms of turbulence modeling and pressure–velocity coupling. With multiphase coupled algorithm, there could be one set of turbulence transport equations with two sets of mass and momentum conservation equations. Another option is turbulence per phase where two sets of transport equations are solved for mass, momentum and turbulence. It was demonstrated by Saha and Li (2016).

Saha and Li (2016) also compared the model predictions with the experimental findings from Winklhofer et al. (2001). The experimental studies were carried out in a throttle nozzle with almost square cross-sections. Apart from measuring global variables, such as mass flow rates, the authors also estimated velocity profiles at a distance of 53 μm from the inlet section of the nozzle using fluorescence techniques. For 8.5 MPa of pressure differential across the nozzle (inlet pressure was 10 MPa), the model predictions (Schnerr and Sauer, Zwart-Gerber-Belamri: both single fluid and two fluid; Saha-Abu-Ramadan-Li: only single fluid) were compared with the measured values and presented in Fig. 7.4. Winklhofer et al. (2001) mentioned that uncertainties were involved in capturing velocity data in the near-wall regions. All the models were able to predict the qualitative trend of the velocity variation in the central regions and not in the near-wall regions. In all cases, there were over-predictions of the velocity which were expected as the mass flow rate predictions were also higher than the experimental value. The predictions from the Saha-Abu-Ramadan-Li model and the Schnerr and Sauer model with the single-fluid approach agree best with the experimental values at the core of the flow compared to the predictions of the other models. Both the Schnerr and Sauer and Zwart-Gerber-Belamri models coupled with the two-fluid approach show kinks which are not visible when coupled with the single-fluid approach. Therefore, cavitation models which do not perform well when coupled with single-fluid approach still have the potential to provide better prediction when coupled with two-fluid approach. The Saha-Abu-Ramadan-Li model prediction, only coupled with single-fluid approach, is visibly different in the near-wall region compared to the other cases, showing two kinks indicating well-developed cavitating regions and flow recirculation. The kinks near the wall indicate that the presence of vapour is accelerating the flow near the wall. The disparities in the prediction are due to the difference in the predicted concentration of vapour by the cavitation models coupled with the multiphase approach in the near-wall regions.

Battistoni et al. (2014a) compared predictions of cavitation models implemented in two different CFD codes—AVL FIRE and CONVERGE—which are commonly used in the internal combustion engine community. The AVL FIRE involves two-fluid approach coupled with bubble dynamics cavitation model. The CONVERGE code uses single-fluid approach coupled with homogeneous relaxation model. The standard $k - \epsilon$ model and standard wall function were used by both the codes. The X-ray experiments carried out by Duke et al. (2013) were utilized by Battistoni et al. in their work. X-ray-based studies are typically capable of providing quantitative information on voids which was not possible to retrieve from the data by Winklhofer et al. (2001). The nozzle was cylindrical, 2.5 mm long and 0.5 mm in diameter. The authors plotted the total void fraction (depth-integrated using the 3D data from simulations) along the length of the nozzle and compared them with the experimental data, as shown in Fig. 7.5.

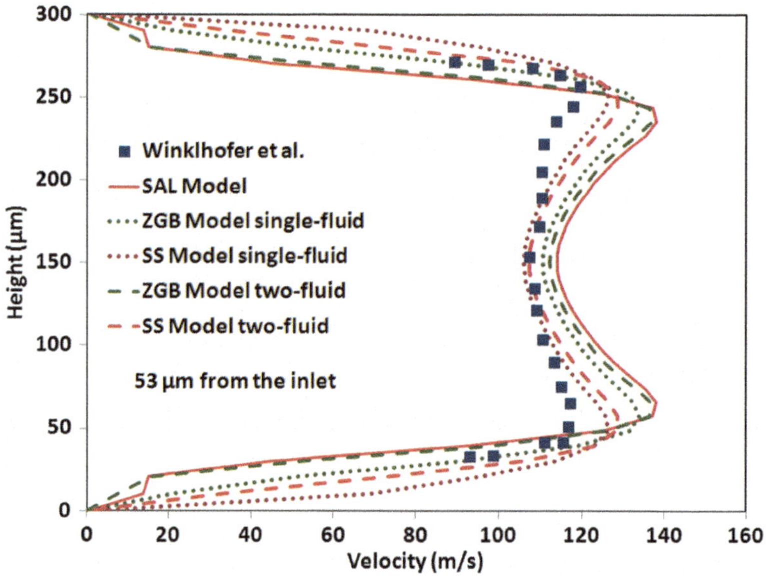

Fig. 7.4 Comparison of the predicted velocity profiles from the three (SAL: Saha-Abu-Ramadan-Li; ZGB: Zwart-Gerber-Belamri; SS: Schenrr and Sauer) cavitation models using the single- and two-fluid approaches, at the pressure differential of $\Delta P = 8.5$ MPa with the experimental values measured at a location, 53 μm from the nozzle inlet section (Winklhofer et al. 2001) (from Saha and Li 2016, reprinted with permission from ASME)

Fig. 7.5 Computed total volume fraction α (vapour+air) along the channel axis evaluated in one cell layer thick slices, compared to the experimental data (Duke et al. 2013) (from Battistoni et al. 2014a, reprinted with permission from ASME)

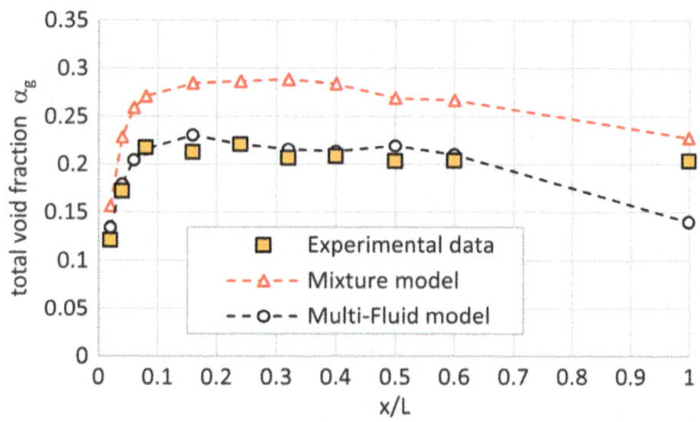

Right after the inlet edge, the void fraction goes up after attaining the peak value remains constant in the experiment. The models could predict the rise of void fraction but were unable to maintain the peak value. Nevertheless, the two-fluid approach of AVL FIRE did relatively a better job. The authors also highlighted the experimental uncertainties were not available in this data, which prevent to draw a comprehensive conclusion regarding a better cavitation model. The authors emphasized that the inclusion of non-condensable gases could be vital in void prediction since vapour can only form near the inlet and adjacent to the wall where the local pressure is

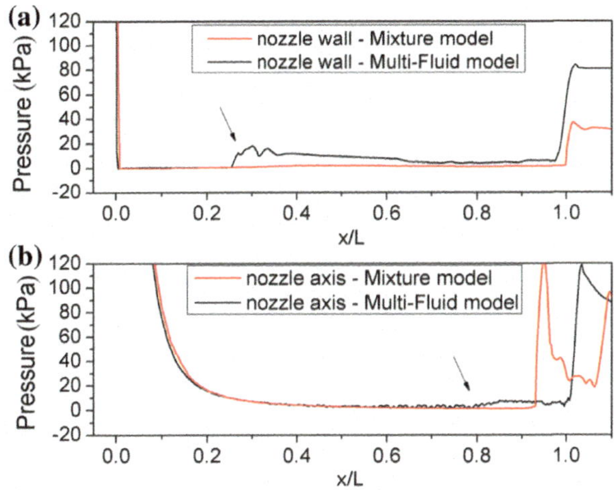

Fig. 7.6 Pressure profiles along the nozzle wall (**a**) and nozzle centreline (**b**) (from Battistoni et al. 2014a, reprinted with permission from ASME)

sufficiently low. In the central regions of the nozzle, the local pressure could not be low enough to trigger the cavitation. Hence, expansion of the dissolved gases could be the cause for void formation away from the nozzle wall.

The pressure field is generally a good indicator of the void distribution in the domain. Battistoni et al. (2014a) looked into the pressure variation along the nozzle wall and nozzle centreline, which are shown in Fig. 7.6. It is seen that both the codes are capable of predicting low-pressure values downstream of the inlet where cavitation is expected. However, in case of two-fluid approach in AVL FIRE the pressure recovery occurs at around $x/L = 0.25$ indicating flow re-attachment and disappearance of cavitation, which qualitatively matches with experimental visualization by Duke et al. (2013). On the other hand, with single fluid in CONVERGE the pressure recovery does not occur, implying the cavitating regions are extending downstream considerably more compared to the two-fluid approach in AVL FIRE. For both the approaches, compressibility was considered. As a result, the pressure in the cavitating regions never reached unphysical negative values. For both the centreline and nozzle wall cases, the pressure fluctuates before approaching the outlet pressure. The qualitative patterns are similar, but the locations of pressure drop and rise differ. This disparity in the evolution of pressure field is strongly dependent on the multiphase approach treatment. In this scenario, consideration of the two-phase interactions is possibly the reason behind the superior performance of the two-fluid approach.

7.4 Summary and Concluding Remarks

The current study reviews the different cavitation modeling approaches typically used for numerical analysis of the two-phase flows in the fuel injection systems. The governing equations of single- and two-fluid approaches are provided, followed by the description of some of the commonly adopted cavitation models. The pros and cons of the different cavitation models are discussed elaborately. The effects of changing the cavitation models as well as the multiphase approach are demonstrated using some of the published results in the literature. It is evident that for a given multiphase approach, different cavitation models may yield considerably different results. The cavitation models, unable to provide the desired level of predictions with single-fluid approach, can do better when coupled with two-fluid approach. However, if the cavitation model is devoid of tuned model constants and/or questionable approximations, reasonable predictions could be obtained even when coupled with single-fluid approach.

Acknowledgements UChicago Argonne, LLC, Operator of Argonne National Laboratory (Argonne), a U.S. Department of Energy Office of Science laboratory, is operated under Contract No. DE-AC02-06CH11357. The US Government retains for itself, and others acting on its behalf, a paid-up nonexclusive, irrevocable worldwide licence in said article to reproduce, prepare derivative works, distribute copies to the public, and perform publicly and display publicly, by or on behalf of the Government. This research was partially funded by the U.S. Department of Energy (DOE) Office of Vehicle Technologies, Office of Energy Efficiency and Renewable Energy under Contract No. DE-AC02-06CH11357. The authors wish to thank Gurpreet Singh and Leo Breton at DOE, for their support. The authors are grateful to ASME Publishing section for granting the permission for using published materials in this chapter.

References

Afzal H, Arcoumanis C, Gavaises M, Kampanis N (1999) Internal flow in diesel injector nozzles—modeling and experiments, S492/S2, IMechE

Alajbegovic A, Grogger HA, Phillip H (1999) Calculation of transient in nozzle using the two-fluid model. In: Proceedings of 12th annual conference on liquid atomization spray systems, IN, USA

ANSYS 18 Fluent Manual

Battistoni M, Grimaldi CN (2018) Analysis of transient cavitating flows in diesel injectors using diesel and biodiesel fuels. SAE Int J Fuels Lubricants 3(2):879–900

Battistoni M, Som S, Longman DE (2014a) Comparison of mixture and multifluid models for in-nozzle cavitation prediction. ASME J Eng Gas Turbine Power 136:(061506) 1–12

Battistoni M, Xue Q, Som S, Pomraning E (2014b) Effect of off-axis needle motion on internal nozzle and near exit flow in a multi-hole diesel injector. SAE Int J Fuels Lubricants 7(1):10–21

Battistoni M, Xue Q, Som S (2015) Large-Eddy Simulation (LES) of spray transients: start and end of injection. Oil Gas Sci Technol 71(1):4

Baumgarten C (2005) Mixture formation internal combustion engines. Springer, Berlin

Bilicki Z, Kestin J (1990) Physical aspects of the relaxation model in two-phase flow. Proc R Soc Lond A 428:379–397

Brennen CE (1995) Cavitation and bubble dynamics. Oxford University Press, Oxford

Downar-Zapolski P, Bilicki Z, Bolle L, Franco J (1996) The non-equilibrium relaxation model for one-dimensional flashing liquid flow. Int J Multiph Flow 22(3):473–483

Duke D, Kastengren A, Tilocco FZ, Powell C (2013) Study of cavitation phenomenon using different fuels in a transparent nozzle by hydraulic characterization and visualization. ILASS-Americas, Paper No. 8

Giannadakis E, Gavaises M, Arcoumanis C (2008) Modelling of cavitation in diesel injector nozzles. J Fluid Mech 616:153–193

Gopalakrishnan S, Schmidt D (2008) A computational study of flashing flow in fuel injector nozzles. SAE technical paper 2008-01-0141

Gorokhovski M, Herrmann M (2008) Modeling primary atomization. Annu Rev Fluid Mech 40(1):343–366

Manin J, Pickett L, Yasutomi K (2018) Transient cavitation in transparent diesel injectors. Paper ID 365 ICLASS

Nurick WH (1976) Orifice cavitation and its effect on spray mixing. ASME J Fluids Eng 98:681–687

Payri F, Arregle J, Lopez JJ, Hermens S (2006) Effect of cavitation on the nozzle outlet flow, spray and flame formation in a diesel engine. SAE technical paper 2006-01-1391. SAE World Congress

Richards KJ, Senecal PK, Pomraning E (2018) CONVERGE v2.4 Manual, Convergent Science, Madison, WI

Saha K (2014) Modelling of cavitation in nozzles for diesel injection applications. University of Waterloo, Waterloo

Saha K, Li X (2016) Assessment of cavitation models for flows in diesel injectors with single- and two-fluid approaches. ASME J Eng Gas Turbine Power 138:(011504) 1–11

Saha K, Abu-Ramadan E, Li X (2013) Modified single-fluid cavitation model for pure diesel and biodiesel fuels in direct injection fuel injectors. ASME J Eng Gas Turbine Power 135:(062801) 1–8

Saha K, Som S, Battistoni M, Li Y, Pomraning E, Senecal PK (2016) Numerical investigation of two-phase flow evolution of in- and near-nozzle regions of a gasoline direct injection engine during needle transients. SAE Int J Engines 9(2):1230–1240

Schmidt D, Corradini M (2001) The internal flow of diesel fuel injector nozzles: a review. Int J Engine Res 2(1):1–22

Schnerr GH, Sauer J (2001) Physical and numerical modeling of unsteady cavitation dynamics. In: Fourth international conference on multiphase flow

Som S (2009) Development and validation of spray models for investigating diesel engine combustion and emissions. University of Illinois at Chicago, Ph.D. thesis

Winklhofer E, Kull E, Kelz E, Morozov A (2001) Comprehensive hydraulic and flow field documentation in model throttle experiments under cavitation conditions. In: 17th international conference on liquid atomization and spray systems Europe, Zurich

Zhao H, Quan S, Dai M, Pomraning E, Senecal PK, Xue Q, Battistoni M, Som S (2014) Validation of a three-dimensional internal nozzle flow model including automatic mesh generation and cavitation effects. ASME J Eng Gas Turbine Power 136(9):(092603) 1–10

Zwart P, Gerber A, Belamri T (2004) A two-phase flow model for predicting cavitation dynamics. In: Fifth international conference on multiphase flow

Chapter 8
Characterization of Biodiesel Sprays

Chetankumar Patel, Joonsik Hwang, Avinash Kumar Agarwal and Choongsik Bae

Abstract Internal combustion engines are widely popular and useful in our life to meet different power requirements. Gaseous and particulate emissions emitted from these engines pose major environmental and health issues. Environment-friendly alternate fuels like biodiesel for diesel engine and alcohols for gasoline engines are gaining popularity steadily in the last decade due to faster depletion of conventional fuels reserves and adaptation of strict emission regulations worldwide. However, it is significantly important to review the spray characteristics of these alternative fuels because engine performance and emissions are largely dependent on air–fuel mixing process to a great extent. This chapter mainly focuses on different optical techniques used for spray characterisation. There are two types of spray characteristics, which are important in the context of internal combustion engines, namely macroscopic spray characteristics and microscopic spray characteristics. Macroscopic spray characteristics such as spray tip penetration and spray cone angle are generally characterised by Mie scattering, shadowgraphy and schlieren techniques, by using high-speed CCD camera. Microscopic characteristics such as spray droplet size distribution and droplet velocity distributions are generally measured using phase Doppler interferometry (PDI) technique.

Keywords Biodiesel · Spray characteristics · Macroscopic spray characteristics Microscopic spray characteristics · Spray top penetration · Droplet size distribution · Phase Doppler interferometry

---

C. Patel · A. K. Agarwal (✉)
Engine Research Laboratory, Department of Mechanical Engineering, Indian Institute of Technology Kanpur, Kanpur 208016, India
e-mail: akag@iitk.ac.in

J. Hwang · C. Bae
Engine Lab, Department of Mechanical and Aerospace Engineering, Korea Advanced Institute of Science and Technology (KAIST), Daejeon, South Korea

K. Saha et al. (eds.), *Two-Phase Flow for Automotive and Power Generation Sectors*, Energy, Environment, and Sustainability,
https://doi.org/10.1007/978-981-13-3256-2_8

Nomenclature

μ	Viscosity
ρ	Surface tension
ΔP	Pressure difference
f_d	Doppler frequency
δ	Fringed spacing
ms	Millisecond
mg	Milligram
μm	Micron

Abbreviation

CVCC	Constant volume combustion chamber
PDI	Phase Doppler interferometry
PIV	Particle image velocimetry
LIF	Laser-induced fluorescence
FPS	Frame per second
FIP	Fuel injection pressure
MPa	Mega Pascal
WCO	Waste cooking oil
TKE	Turbulence kinetic energy

8.1 Introduction

Diesel engines play a vital role in our life, especially in sectors like energy and transportation. However, their contribution to environmental pollution is raising eyebrows of environmentalist and policy-makers alike. Battery-operated vehicles are also emerging as threat to the automotive companies. However, some new alternate fuels are emerging as effective solutions to tackle this environmental degradation. Therefore, it is desirable to design biofuel-operated engines to achieve fuel efficiency and low emissions levels. Spray and combustion visualisation studies in an optical engine and constant volume combustion chamber (CVCC) are gaining momentum in last few decades for meeting stringent emission norms and utilization of alternate fuels. In this context, spray investigation of biofuels is of vast importance to design the alternate fuelled engines.

Biodiesel has emerged as a strong alternative to mineral diesel. However, it suffers from relatively higher density and viscosity compared to conventional diesel, which significantly affects the fuel injection system and fuel atomization process (Table 8.1). Fuel atomization plays a vital role in the combustion of fuels. Several

Table 8.1 Comparative physical properties of diesel and biodiesel (Hwang et al. 2017)

Property	Diesel	Karanja biodiesel
Carbon/hydrogen/oxygen (%)	86.27/13.73/0	78.25/12.03/9.72
Lower heating value (MJ/kg)	42.98	39.89
Density (@288 K) (kg/m^3)	820	886
Kinematic viscosity (@313 K) (mm^2/s)	2.2	5.66
Surface tension (@313 K) (N/m)	0.026	0.03
Flash point (K)	329	417

optical methods such as direct visualisation (spray imaging), phase Doppler interferometry (PDI), particle image velocimetry (PIV) and laser-induced fluorescence (LIF) are very useful for measurement of various spray atomization characteristics. These studies give valuable information on the fuel spray characteristics: macroscopic as well as microscopic.

8.2 Spray Characterization

Macroscopic spray studies provide information on the spray characteristics such as spray tip penetration and spray cone angle, while microscopic spray studies provide information about droplet size distribution and droplet velocity distribution of the sprays. Researchers conducted these studies in atmospheric as well as pressurised conditions in constant volume spray and combustion chambers. In recent years, IC engines have been extensively modified for optical access in order to carry out various optical diagnostics. This chapter discusses the measurement principles of these optical diagnostics, methodologies and some discussion on the results obtained by various researchers.

8.2.1 Macroscopic Spray Investigations

Macroscopic spray investigations involve two important characteristics, namely spray tip penetration and spray cone angle (Fig. 8.1). Spray tip penetration is the distance between the nozzle tip of the injector and the farthest point in the spray plume on the spray axis. Spray cone angle is the angle between the straight lines of the two farthermost points on the outline of the spray plume.

Several researchers conducted spray investigations in a constant volume spray chamber for various biodiesels. Patel et al. (2016a) conducted spray visualization experiment on a constant volume spray chamber at a FIP of 20 MPa and at ambient pressure of 0.1, 1 and 2 MPa. Figure 8.2 shows the schematic of experimental set-up

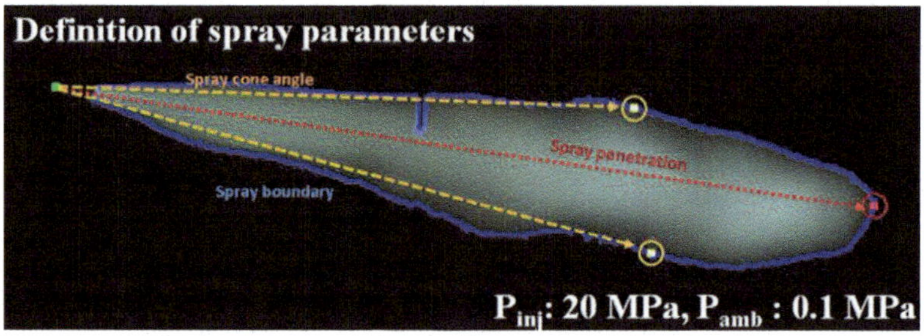

Fig. 8.1 Macroscopic spray parameter definitions (Patel et al. 2016a)

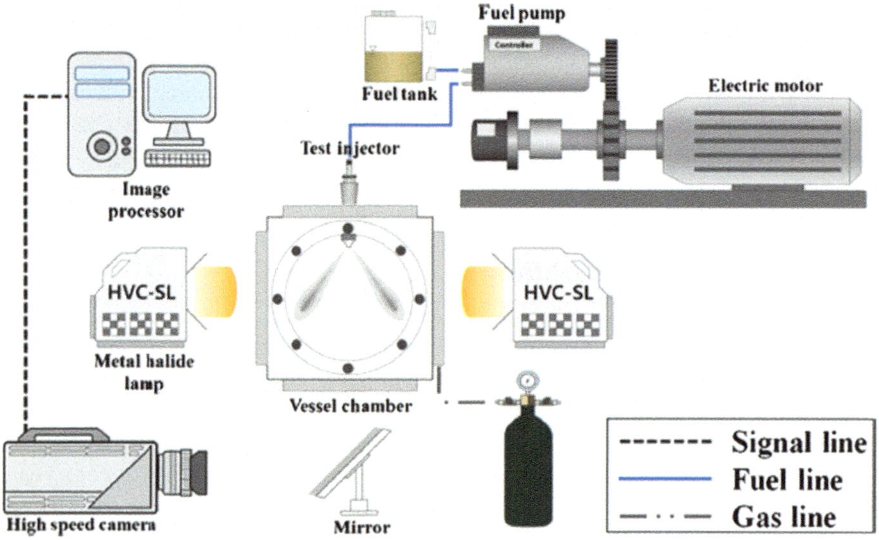

Fig. 8.2 Experimental set-up for spray visualisation (Patel et al. 2016a)

for visualisation of the spray. They conducted experiment using a mechanical fuel injector at an injection pressure of 20 MPa. Electric motor-driven fuel pump was utilized to supply the fuel to the injector of the fuel injection system. Spray visualization was conducted using high-speed CCD camera at 10,000 frames per second (FPS) using a 105-mm-focal length lens, an image processor and two illuminating lamps.

Image processing was done for the assessment of spray tip penetration and spray cone angle (Fig. 8.3). Captured images were processed using the following steps: (i) original image was converted into binary image; (ii) pixels of the background region were assigned a value of '0', while those in the spray region were assigned a value '1'. This helped in defining the boundary surrounding the spray region; (iii) these images were further enlightened by using several filters. Finally, these images were utilized for calculation of spray tip penetration and spray cone angle (Patel et al. 2016b).

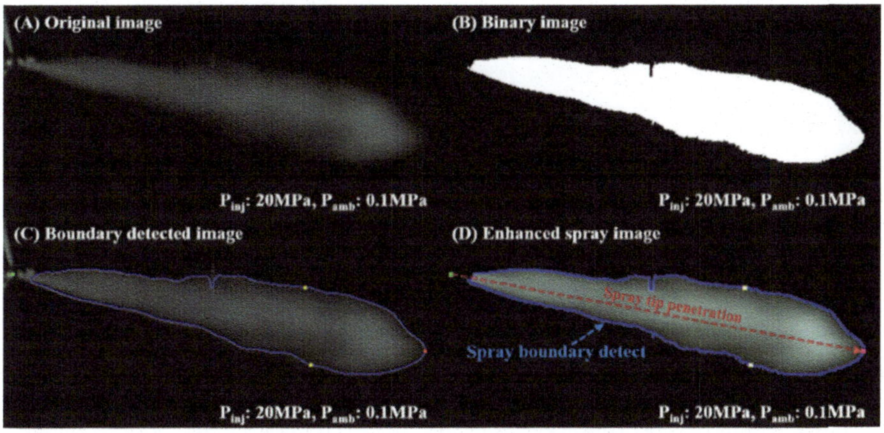

Fig. 8.3 Post-processing methodology for captured spray images using Matlab (Patel et al. 2016b)

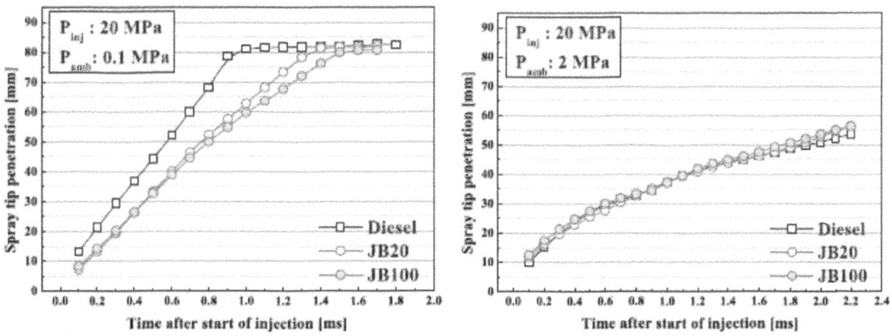

Fig. 8.4 Spray tip penetration of fuel sprays at different ambient pressures (Patel et al. 2016a)

Figure 8.4 shows the spray tip penetration and spray cone angle for biodiesels blends vis-à-vis baseline mineral diesel. Diesel sprays evolved relatively earlier compared to biodiesel and its blends. Spray tip penetration reached its maximum earlier for diesel under ambient pressure conditions. However, at higher ambient pressure conditions, there were no significant differences observed amongst these test fuels. Figure 8.5 shows the spray cone angle for these test fuels. It was observed that with increasing ambient pressure, spray cone angle increased, regardless of the test fuel used. Spray cone angle for biodiesel becomes relatively narrower after few ms of the injection at both ambient and higher ambient pressure conditions.

Suh et al. (2008) measured spray development for diesel and biodiesel blend (BD20) at a FIP of 100 MPa for single injection condition. The difference between the spray development of diesel and biodiesel was less, which suggested that biodiesel blend has less influence on the spray development.

Figure 8.6 shows the macroscopic spray images captured by Mie scattering technique. Spray characteristics in a CVCC were investigated by Hwang et al. (2016) at

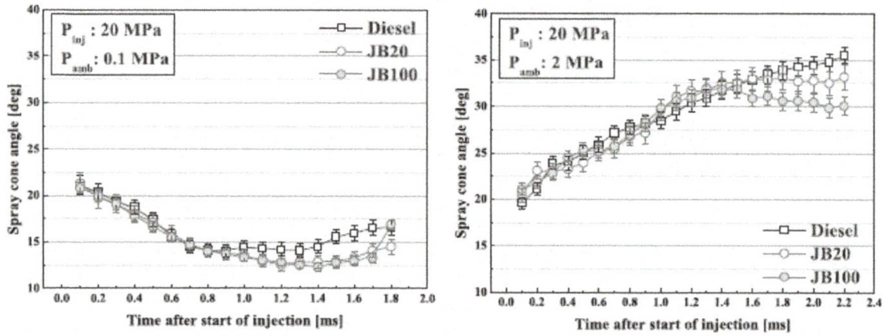

Fig. 8.5 Spray cone angle of fuel sprays at different ambient pressures (Patel et al. 2016a)

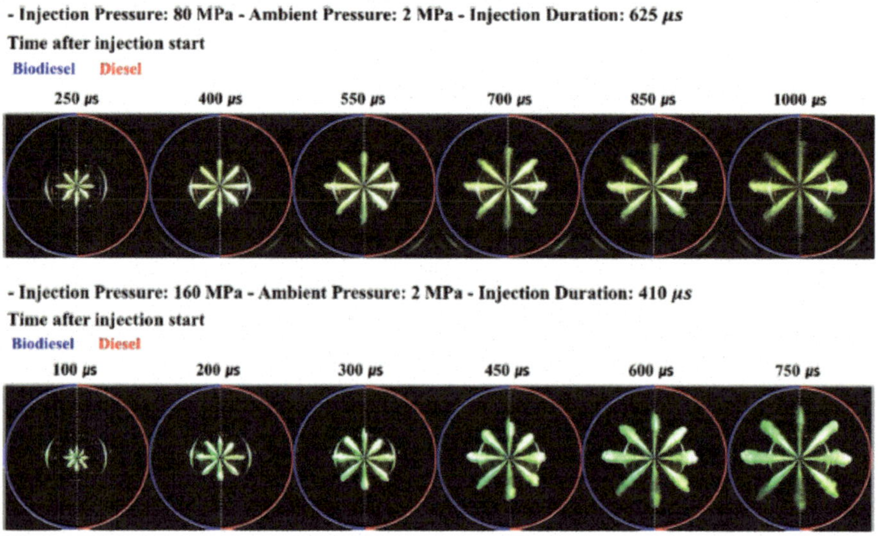

Fig. 8.6 Macroscopic spray images at fuel injection pressure of 80 and 160 MPa (Hwang et al. 2016)

0.1 MPa ambient pressure and 300 K temperature at a FIP of 80 and 160 MPa for waste cooking oil biodiesel and mineral diesel. They reported similar spray evolution for diesel and biodiesel (Fig. 8.6). WCO biodiesel exhibited longer injection delay and liquid penetration length but narrower spray angle compared to baseline mineral diesel, primarily due to higher viscosity, density and surface tension of biodiesel (Fig. 8.7).

Agarwal et al. (2013) reported that in general, biodiesel exhibited longer spray tip penetration and wider spray cone angle compared to baseline mineral diesel. They also reported improved atomization characteristics for lower biodiesel blends compared to baseline diesel. Biodiesel obtained from palm and waste cooking oil

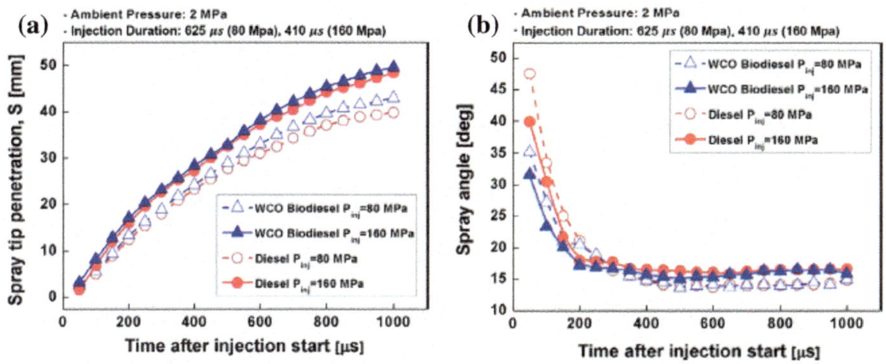

Fig. 8.7 **a** Spray tip penetration length, and **b** spray cone angle for diesel and WCO biodiesel (Hwang et al. 2016)

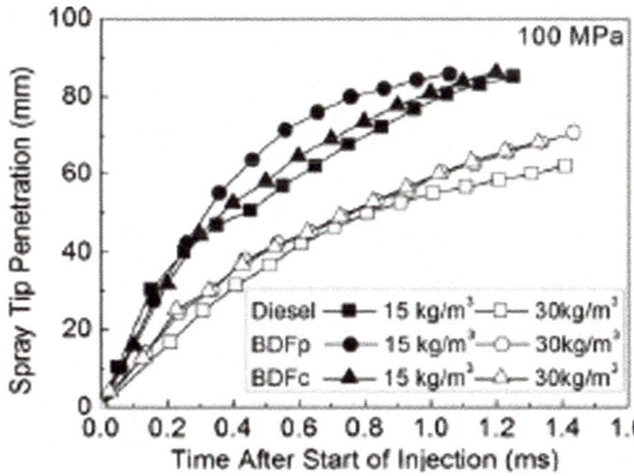

Fig. 8.8 Spray tip penetration at FIP of 100 MPa (Wang et al. 2010)

exhibited longer injection delay and longer spray penetration length at higher fuel injection pressures (Wang et al. 2010) (Fig. 8.8) primarily due to relatively higher surface tension as well as viscosity.

He et al. (2008) performed the spray characterization in the constant volume spray chamber. They reported that increase in injection duration resulted in increased spray tip penetration and wider cone angle in case of biodiesels. Park et al. (2011) reported that addition of diesel or bio-ethanol with biodiesel slightly reduces the spray tip penetration and had only little effect on the spray cone angle. Spray tip penetration increased, while spray cone angle reduced with increasing ratio of biodiesel in the blend (Gao et al. 2009; Zhao et al. 2008).

Som et al. (2010) conducted simulations to compare the effects of different fuel properties of biodiesel sprays. Figure 8.9 shows the cavitation contours inside the injector nozzle for diesel and biodiesel. They observed that cavitation started at the inlet of the nozzle orifice and continued till its exit in case of diesel, while cavitation

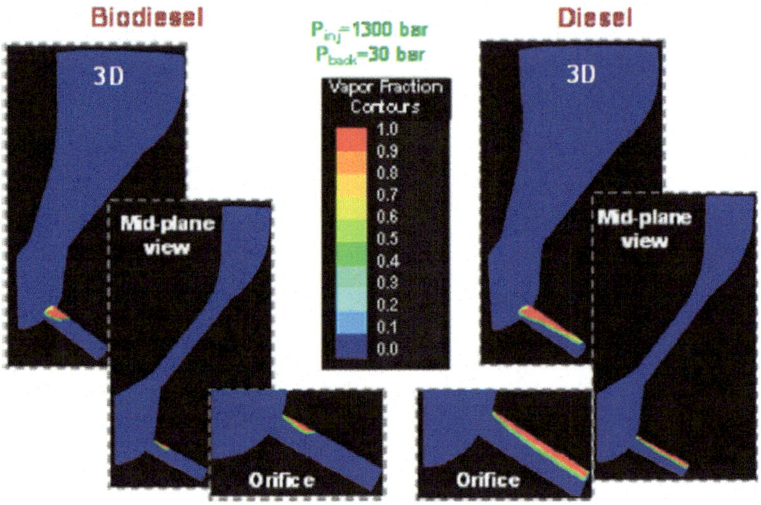

Fig. 8.9 Vapour fraction contours for diesel and biodiesel inside the injector (Som et al. 2010)

appeared at the inlet of the nozzle orifice in case of biodiesel, but it did not last till the exit of the nozzle. Biodiesels have relatively higher viscosity and density and relatively lower vapour pressure than baseline mineral diesel, which contributed to the differences in cavitation patterns (Battistoni and Grimaldi 2012; Saha et al. 2013). Turbulence was observed to be relatively lower in biodiesel at the exit of the injector nozzle. These parameters together resulted in higher spray tip penetration and lower spray cone angle for biodiesels (Som et al. 2010).

Yu et al. (2017) conducted simulation of the flow of diesel and biodiesel inside the nozzle orifice of the injector. Figure 8.10a shows the vapour phase fraction contours, and Fig. 8.10b shows the turbulence kinetic energy (TKE) contours in the injector nozzle for diesel and biodiesel at 50 MPa FIP and 0.1 MPa ambient pressure. They reported that internal cavitation appeared at the inlet of the orifice and extended till the exit of the nozzle, while in case of biodiesel, it appeared for a short distance in the orifice. This was due to relatively higher viscosity of biodiesel that suppressed the development of cavitation in the injector nozzle. Figure 8.10b also shows that biodiesel has relatively smaller high-intensity TKE zone compared to baseline mineral diesel. These results exhibited relatively lower vapour phase fraction and TKE in case of biodiesel in the nozzle orifice compared to baseline diesel because of relatively higher viscosity and higher surface tension of biodiesel. All these factors led to relatively longer spray penetration length and narrower spray cone angle for biodiesels compared to baseline mineral diesel.

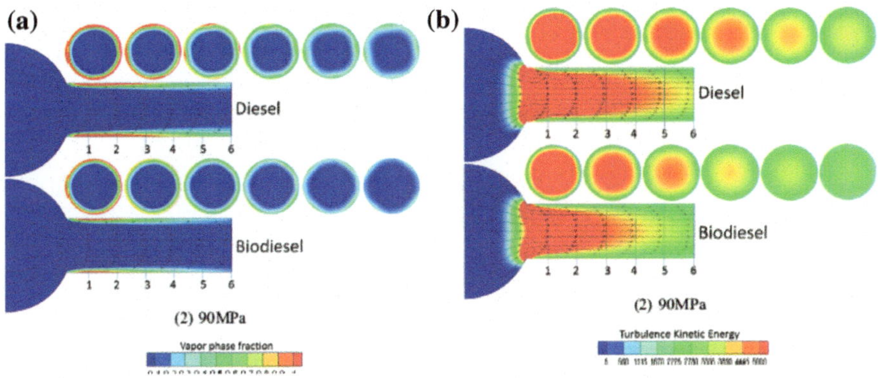

Fig. 8.10 Countours of **a** vapour phase fraction, and **b** turbulent kinetic energy (TKE) for diesel and biodiesel inside the injector nozzle (Yu et al. 2017)

8.2.2 Microscopic Spray Investigations

Spray development process after fuel injection from the nozzle is shown in Fig. 8.11. Spray plume can be divided into two sections. Initial part is called 'primary spray break-up region', where the break-up of liquid jet into ligaments and large droplets occurs. Conversion of large droplets into small droplets occurs in the 'secondary spray break-up region' (https://www.brighton.ac.uk/advanced-engineering/research-projects/les-sprays.aspx). Secondary break-up region is the most important region for microscopic spray investigations.

Diameter and velocity of the fuel droplets are important microscopic spray characteristics that need to be investigated after the break-up length in secondary spray break-up region. SMD and velocity measurements of diesel and biodiesel at 16 different positions were carried out by Hwang et al. (2017) at 0.1 MPa ambient pressure and 300 K temperature for biodiesel vis-à-vis baseline mineral diesel (Fig. 8.12). They moved the measurement location in horizontal axis by 1 mm and in vertical axis by 10 mm in the secondary break-up region of the spray plume. Figure 8.12 shows the phase Doppler interferometery (PDI) set-up for the measurement of droplet size and velocity distributions. This system consists of two transmitters and a receiver. The angle between each transmitter and receiver was set to be 30°. Transmitter 1 emits two laser beams (blue and green of wavelength 532 and 491 nm, respectively), while transmitter 2 emits one beam (yellow of 561 nm wavelength). Each laser beam was split into two separate beams of equal intensity, and a phase shift was introduced in the second beam using appropriate optics. All these six laser beams intersected at one point, which is called 'probe volume'. When the droplets pass through this probe volume, they generate Doppler burst signals (one for each wavelength) and alternate dark and bright interference fringes are formed in the probe volume. Whenever any droplet passes through the probe volume, it scatters the light, which is captured by

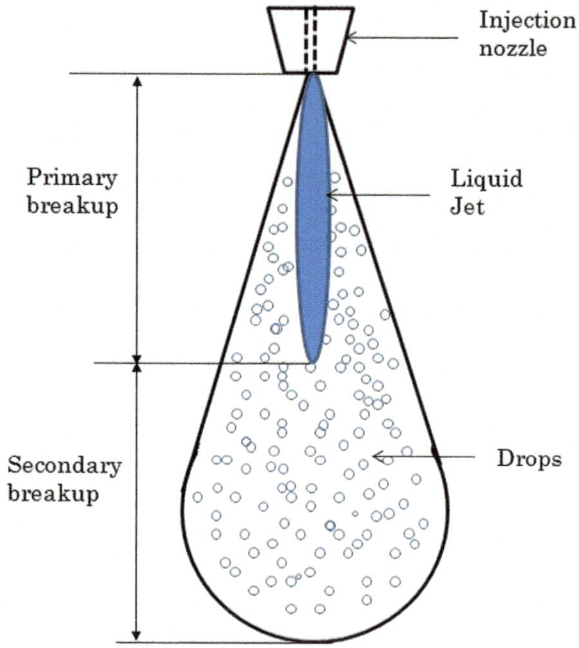

Fig. 8.11 Spray development process

the receiver. The frequency of the scattered light signal can be directly related to the velocity of the droplet and is given by the following relationship:

$$v = f_{\mathrm{d}}\delta \tag{8.1}$$

where f_{d} = Doppler frequency, and δ = fringed spacing

$$\text{Sauter Mean Diameter } D_{32} = \frac{\sum_i n_{c(i)} d_i^3}{\sum_i n_{c(i)} d_i^2} \tag{8.2}$$

where i = histogram bin number; nc = number of samples in each bin: corrected size count; d = diameter of the spherical droplet.

Surface volume divided by the surface area of droplet is called Sauter mean diameter (SMD). Droplet size and velocity play central role in fuel spray atomization process in any combustion device such as IC engine. Chi and Kim (1993) measured the droplet size distribution using Fraunhofer diffraction technique at different locations for varying FIP and ambient gas pressure for diesel. They reported a reduction in SMD with increasing FIP. They initially observed higher SMD of the spray droplets, which reached a constant value after 1–2.5 ms from the start of injection. Komada et al. (2013) reported that in a radial plane, droplets in diesel spray have relatively higher velocity in the centre and this velocity decreases with increasing radius at a FIP of 80 MPa. Wang et al. (2010) measured spray characteristics of palm and

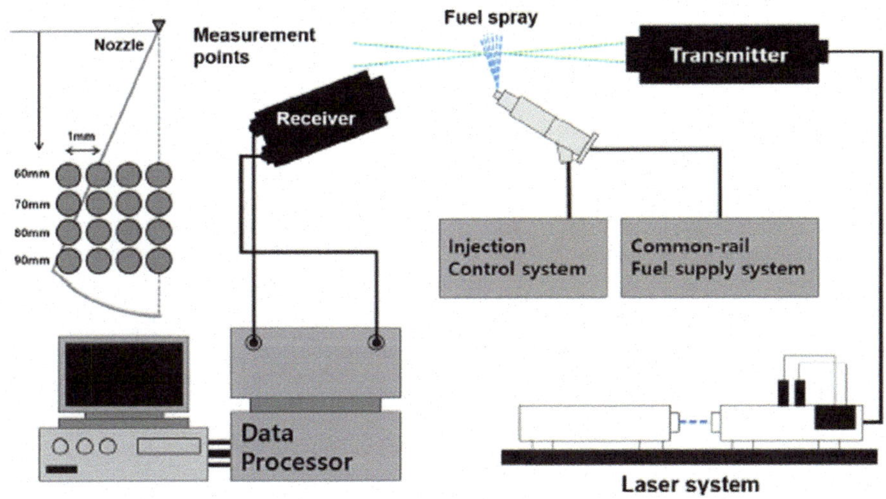

Fig. 8.12 Phase Doppler interferometry set-up (Hwang et al. 2017)

waste cooking oil biodiesels at FIPs of up to 300 MPa. They developed a correlation between SMD, fuel viscosity, surface tension and FIP. This relation is:

$$\text{SMD} = 6156\mu^{0.385}\sigma^{0.737}\rho_f^{0.737}\rho_a^{0.06}\Delta P^{-0.54} \tag{8.3}$$

Here, μ = Viscosity, ρ = Surface Tension, and ΔP = Pressure difference.

They found larger SMD of spray droplets from biodiesels due to their higher viscosity and surface tension analytically. SMD results at a FIP of 80 MPa under non-evaporating conditions after the break-up length are shown in Fig. 8.13 (Hwang et al. 2017). Droplet size distribution is represented in a circle for a given location in the fuel spray. SMD showed a decreasing trend in the axial direction, regardless of fuel type. SMD observed was relatively higher for biodiesel compared to baseline mineral diesel (Hwang et al. 2017). SMD of spray droplets in the axial direction along the centre line is represented in Fig. 8.14. SMD decreased with increasing FIP for all test fuels since higher FIP helps in improving the fuel spray atomization. Biodiesel showed higher SMD of spray droplets at all FIPs compared to baseline mineral diesel, due to its higher viscosity and surface tension (Hwang et al. 2017).

Patel et al. (2016c) reported relatively higher SMD for Jatropha biodiesel and blends compared to baseline mineral diesel at a FIP of 20 MPa under ambient conditions, 40 mm away from the injector nozzle exit, while the injected fuel quantity was varied from 12 to 32 mg/injection (Fig. 8.15). They also reported increased SMD with increasing fuel injection quantity. Velocity for biodiesel spray droplets was relatively lower compared to baseline mineral diesel (Fig. 8.16).

Lee et al. (2005) measured the SMD and velocity distribution of spray droplets using phase Doppler particle analyser (PDPA) (Fig. 8.17). They reported that SMD

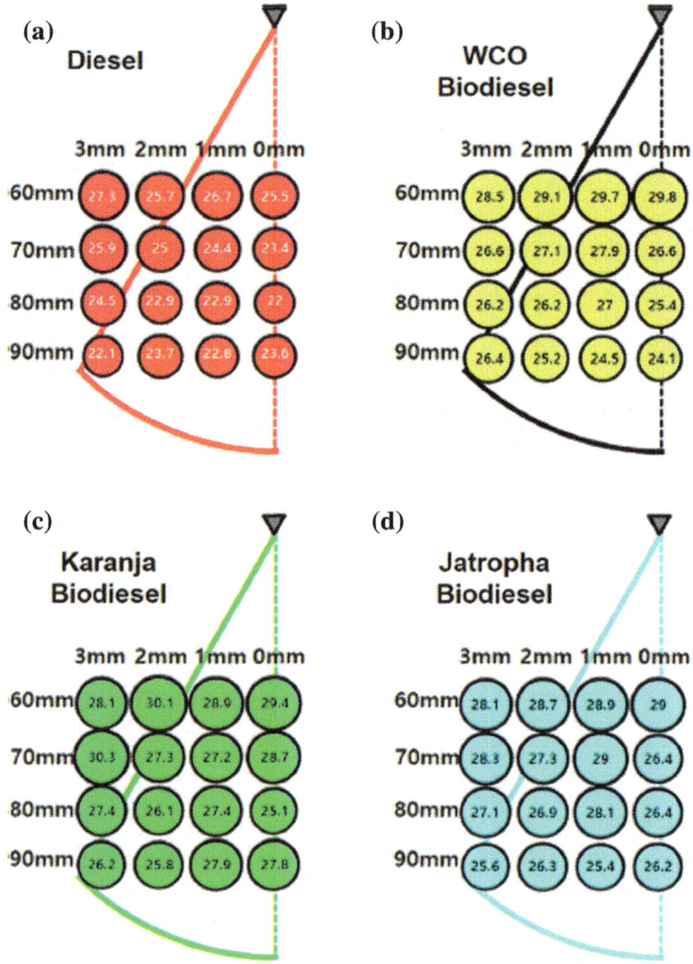

Fig. 8.13 Mean SMD distribution in the spray at a FIP of 80 MPa (Hwang et al. 2017)

and spray droplet velocity distribution of biodiesel blends were observed to be relatively higher compared to baseline mineral diesel. They indicated that higher surface tension led to lower Weber number, which increased the SMD distribution of the spray droplets compared to mineral diesel. Reduction in fuel injection velocity was due to higher surface tension, which increased friction between the nozzle surface and fuel flow.

Suh et al. (2008) conducted experiments for the measurement of droplet size and velocity distributions for diesel and biodiesel blends under single injection and pilot injection conditions, along the axial direction of the spray plume. With increasing axial distance, reduction in SMD in spray droplets was reported. They observed higher SMD of spray droplets and lower spray droplet velocity distribution

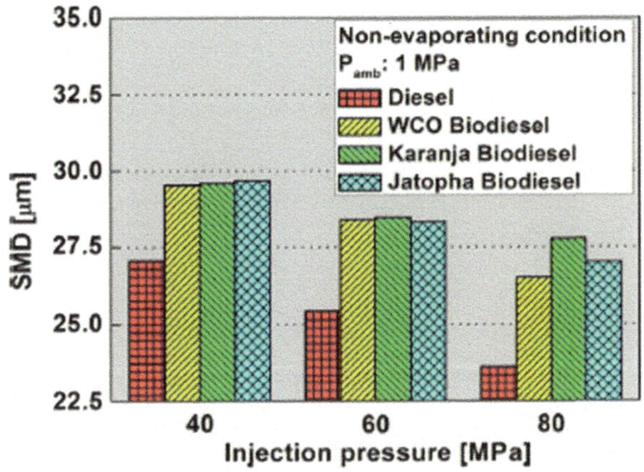

Fig. 8.14 Effects of FIP on SMD along spray centreline (Hwang et al. 2017)

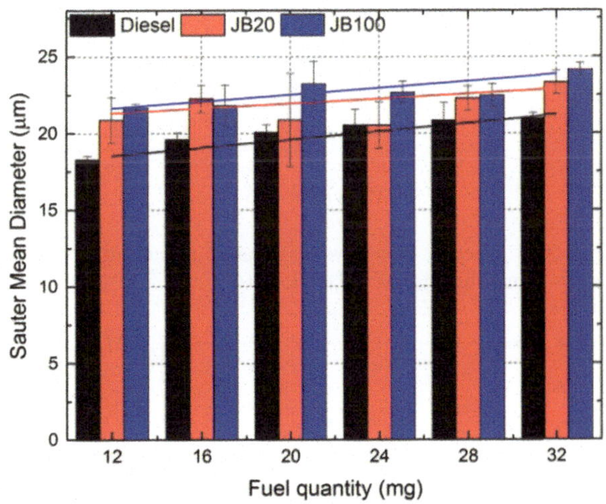

Fig. 8.15 SMD of spray droplets for different test fuels (Patel et al. 2016c)

of biodiesel spray compared to diesel spray. They also reported that spray atomization improves with pilot injection since the fuel injection velocity increases. Guan et al. (2015) conducted experiments using particle/droplet image analysis technique and reported relatively smaller SMD due to addition of di-n-butyl ether (DBE) in biodiesel. They found that centre of fuel spray has higher spray droplet number density compared to the peripheral zone. They also reported that biodiesels exhibited higher probability of larger droplets compared to baseline diesel; however, addition of up to 30% DBE resulted in increased probability of smaller droplets compared to biodiesel (Fig. 8.18) (Guan et al. 2015).

Gao et al. (2009) used biodiesel produced from three non-edible sources palm oil, used fried oil and Jatropha oil. They calculated SMD values for the spray droplets from the three test fuels and reported relatively higher SMD of biodiesel blends

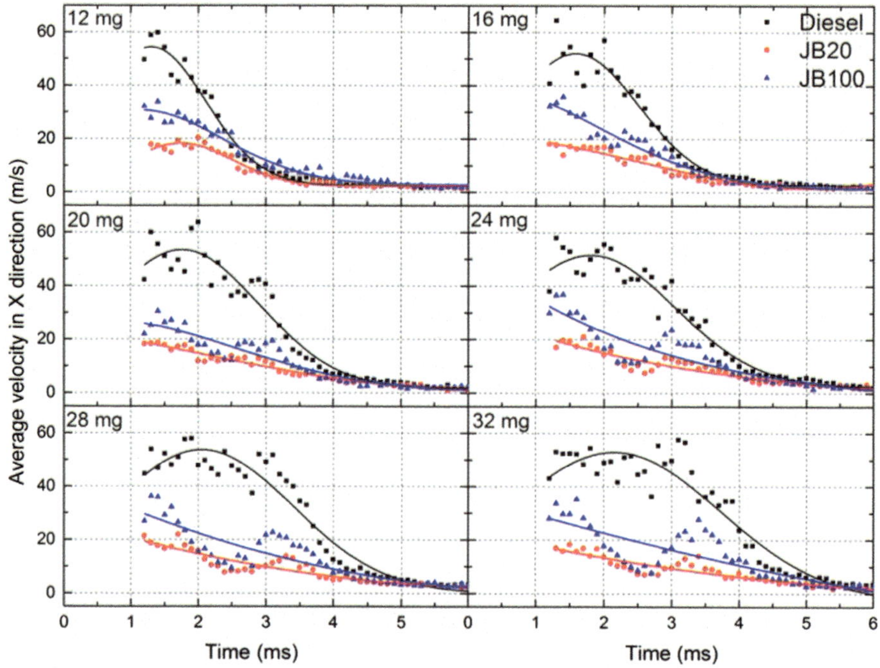

Fig. 8.16 Average droplet velocity variations in *X* direction for varying injected fuel quantity from 12 to 32 mg/injection (Patel et al. 2016c)

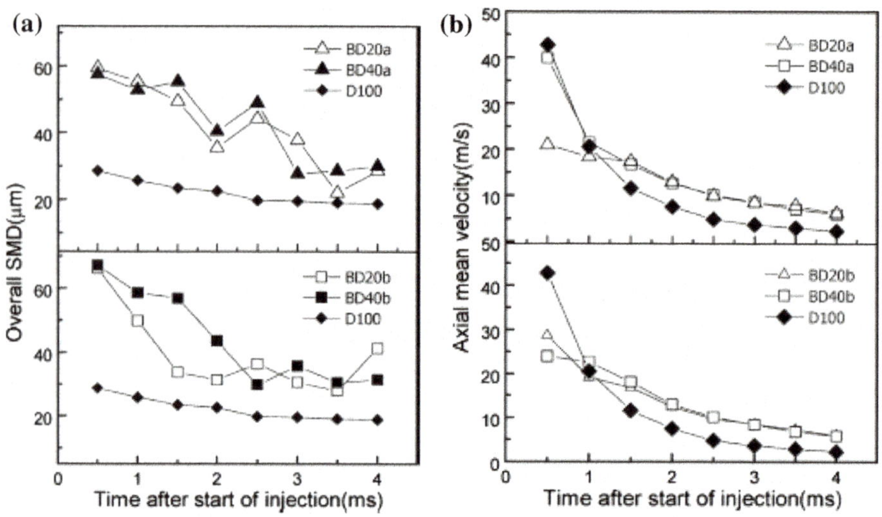

Fig. 8.17 **a** Droplet size distribution after the start of injection, and **b** axial mean velocity distribution after the start of injection for biodiesel blends (Lee et al. 2005)

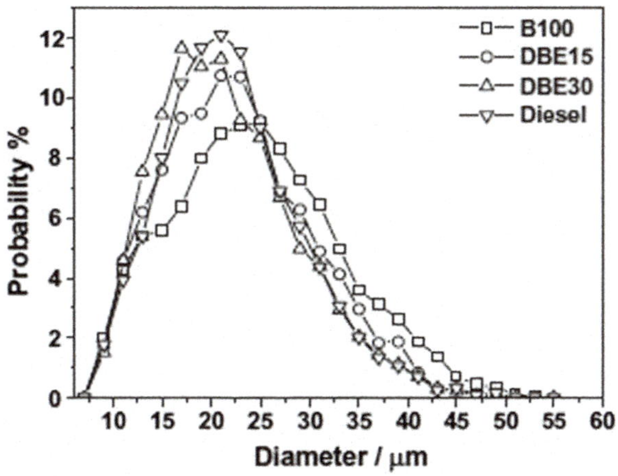

Fig. 8.18 Droplet size distribution for diesel, biodiesel and DBE blends (Guan et al. 2015)

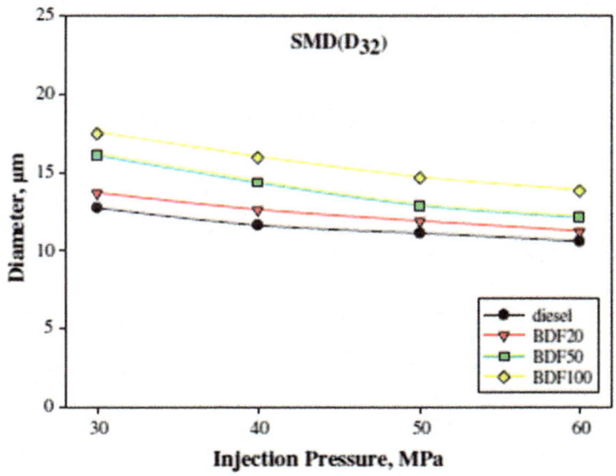

Fig. 8.19 Effect of biodiesel blending ratio and increasing FIP on SMD (Choi and Oh 2012)

compared to baseline diesel, primarily due to higher viscosity and surface tension of biodiesel compared to mineral diesel.

Choi and Oh (2012) used mineral diesel and biodiesel extracted from palm oil to analyse the effect of blending ratio and FIP on the fuel spray behaviour using a laser diffraction particle analyser. The results indicated that increasing FIP caused SMD of spray droplets to become smaller. In contrast, an increased biodiesel blending ratio caused SMD of the fuel spray droplets to become larger (Fig. 8.19).

These studies suggested that higher viscosity, density and surface tension of biodiesel play decisive role in increasing the droplet size and velocity distributions.

8.3 Conclusions

In this chapter, macroscopic and microscopic spray characteristics of biodiesels and blends are discussed. It emerges that physical properties of the fuel such as density, viscosity and surface tension play a major role in determining the spray characteristics. These properties affect the flow in the nozzle orifice of the injector and are responsible for relatively lower cavitation contour for biodiesels. This results in relatively longer spray tip penetration length and narrower spray angle for the biodiesel sprays under ambient conditions. However, it was observed that under higher ambient pressure conditions, the difference in macroscopic spray characteristics is not quite significant for different test fuels. Higher viscosity and surface tension significantly affect the droplet size and velocity distributions in case of biodiesel sprays. Majority of biodiesels showed relatively higher SMD and velocity distributions compared to baseline mineral diesel. Higher velocity distribution was observed due to relatively higher momentum after the injection of the biodiesel droplets from the injector nozzle. It can be therefore concluded that biodiesels exhibit relatively longer spray tip penetration, narrower spray cone angle, higher SMD and lower spray droplet velocity distribution compared to baseline mineral diesel under identical experimental conditions.

References

Agarwal AK, Chaudhary V, Agarwal A, Shukla PC (2013) Comparative study of macroscopic spray parameters and fuel atomization behaviour of straight vegetable oils (Jatropha), its biodiesel and blends. Thermal Sci 17:217–232

Battistoni M, Grimaldi CN (2012) Numerical analysis of injector flow and spray characteristics from diesel injectors using fossil and biodiesel fuels. Appl Energy 97:656–666

Chi Y, Kim E (1993) Measurement of droplet size distribution of transient diesel spray. SAE technical paper 931949

Choi S, Oh Y (2012) The spray characteristics of unrefined biodiesel. Renew Energy 42:136–139

Gao Y, Deng J, Li C, Dang F, Liao Z, Wu Z, Li L (2009) Experimental study of the spray characteristics of biodiesel based on inedible oil. Biotechnol Adv 27(5):616–624

Guan L, Tang C, Yang K, Mo J, Huang Z (2015) Effect of di-n-butyl ether blending with soybean-biodiesel on spray and atomization characteristics in a common-rail fuel injection system. Fuel 140:116–125

He C, Ge Y, Tan J, Han X (2008) Spray properties of alternative fuels: a comparative analysis of biodiesel and diesel. Int J Energy Res 32:1329–1338

https://www.brighton.ac.uk/advanced-engineering/research-projects/les-sprays.aspx

Hwang J, Bae C, Gupta T (2016) Application of waste cooking oil (WCO) biodiesel in a compression ignition engine. Fuel 176:20–31

Hwang J, Bae C, Patel C, Agarwal RA, Gupta T, Agarwal AK (2017) Investigations on air-fuel mixing and flame characteristics of biodiesel fuels for diesel engine application. Appl Energy 206:1203–1213

Komada K, Sakaguchi D, Tajima H, Ueki H, Ishida M (2013) Relation between tip penetration and droplet size of diesel spray. SAE technical paper 2013-01-1599. https://doi.org/10.4271/2013-01-1599

Lee CS, Park SW, Kown SI (2005) An experimental study on the atomization and combustion characteristics of biodiesel-blended fuels. Energy Fuels 19:2201–2208

Park S, Cha J, Lee C (2011) Spray and engine performance characteristics of biodiesel and its blends with diesel and ethanol fuels. Combust Sci Technol 183:802–822

Patel C, Lee S, Tiwari N, Agarwal AK, Lee CS, Park S (2016a) Spray characterization, combustion, noise and vibrations investigations of Jatropha biodiesel fuelled genset engine. Fuel 185:410–420

Patel C, Agarwal AK, Tiwari N, Lee S, Lee CS, Park S (2016b) Combustion, noise, vibrations and spray characterization for Karanja biodiesel fuelled engine. Appl Therm Eng 106:506–517

Patel C, Sharma N, Tiwari N, Agarwal AK (2016c) Effects of spray droplet size and velocity distributions on emissions from a single cylinder biofuel engine. SAE technical paper 2016-01-0994

Saha K, Abu-Ramadan E, Li X (2013) Modified single-fluid cavitation model for pure diesel and biodiesel fuels in direct injection fuel injectors. J Eng Gas Turbines Power 135(6):062801

Som S, Longman DE, Ramirez AI, Aggarwal SK (2010) A comparison of injector flow and spray characteristics of biodiesel with petro-diesel. Fuel 89:4014–4024

Suh HK, Roh HG, Lee CS (2008) Spray and combustion characteristics of biodiesel/diesel blended fuel in a direct injection common-rail diesel engine. J Eng Gas Turbines Power 130(3):032807

Wang X, Huang Z, Kuti OA, Zhang W, Nishida K (2010) Experimental and analytical study on biodiesel and diesel spray characteristics under ultra-high fuel injection pressure. Int J Heat Fluid Flow 31:659–666

Yu S, Yin B, Jia H, Wen S, Li X, Yu J (2017) Theoretical and experimental comparison of internal flow and spray characteristics between diesel and biodiesel. Fuel 208:20–29

Zhao XW, Han XK, He C, Tan JW (2008) Experimental study on spray characteristics of biodiesel oil. Chin Intern Combust Eng Eng 1:16–19

Part II
Multiphase Flow Application in the Industry

Chapter 9
LES and DNS of Multiphase Flows in Industrial Devices: Application of High-Performance Computing

Somnath Roy

Abstract High fidelity solutions of turbulent flow equations are obtained by large eddy simulation (LES) and direct numerical simulation (DNS). These techniques are devoted for resolving most of the energy-carrying scales in a turbulent flow. Grid resolution in LES or DNS is determined by the lengths of the finest scale of motion which is to be directly simulated. In multiphase flows, further refinement in the grid topology is required to capture the bubble or droplet front and also to resolve the small structures that are created in the wake zone of the bubble/droplet. Owing to the grid size and finer timescales, the computational complexities in LES or DNS are extremely high, and parallel computing resources are often deployed. The present chapter reviews computational efforts involved in turbulent multiphase flow simulation in industrial devices. Several high-performance computing (HPC) strategies like distributed computing using message passing interface (MPI), general purpose graphics processing unit (GPGPU) accelerated computing using CUDA and their hybridizations are also reviewed. Estimations of the computational requirement for simulation of large industrial devices are presented, and potential use of modern computational science and hardware are critically assessed.

Keywords Multiphase flow · Turbulence · Computational fluid dynamics
Large eddy simulation · Direct numerical simulation · High-performance computing · Parallelization · MPI · CUDA · GPGPU

Nomenclature

C_s^2 Smagorinsky constant
f_i ith component of body force
k Turbulent kinetic energy

S. Roy (✉)
Department of Mechanical Engineering, Indian
Institute of Technology Kharagpur, Kharagpur 721302, West Bengal, India
e-mail: Somnath.roy@mech.iitkgp.ac.in

K. Saha et al. (eds.), *Two-Phase Flow for Automative and Power Generation Sectors*, Energy, Environment, and Sustainability,
https://doi.org/10.1007/978-981-13-3256-2_9

p Pressure
Re Reynolds number
S_{ij} Large-scale strain rate tensor
t Time
u_i ith component of velocity
x_i ith coordinate
y^+ Wall coordinate
σ_{ij} Viscous stress term
τ_{ij}^t Sub-grid stress tensor
ε Turbulent dissipation rate
η Kolmogorov length scale
ν_t Sub-grid eddy viscosity

Acronyms

DNS Direct numerical simulation
GPGPU General purpose graphics processing unit
IBM Immersed boundary method
LBM Lattice Boltzman methods
LES Large eddy simulation
MPI Message passing interface
MRF Multiple reference frame
OpenMP Open multiprocessing
PBM Population balance model
PCI Peripheral component interconnect
RANS Reynolds-averaged Navier–Stokes
VOF Volume of fluids

9.1 Introduction

Multiphase flows are observed in a wide range of industrial applications involving energy sector, biochemicals, automobiles, petroleum, pulp and paper industries, etc. Complex flow phenomena with energy and momentum transfer at turbulent scales are associated with the operation of these devices. A two-phase flow is often observed to be of particulate nature, i.e., a dispersed particle-like phase is embedded in one continuous phase fluid. The particles show a range of length scales and their microscale dynamics interacts with the smaller scales of turbulent motion. Further, the particle dynamics show different flow features depending on the density ratio of continuous and particle phases. An accurate numerical modeling, therefore, should focus on resolving the smaller length scales. However, industrial devices like coal gasifier, slurry pumps, bubble separators also exhibit flow features at much larger scales.

Therefore, numerical simulation needs to predict flow behaviors at an extremely wide range of scales and the resulting requirement of computational expense can be restrictively high. The other types of multiphase flows consist of more than one dense dispersed system, where, multiple fluids are considered instead of fluid-particle systems. In these cases, the computational challenge is augmented by the complexities involved in interface tracking (Kaotka 1986) as the random turbulent velocities can significantly alter the interface topology and affect the levels of interfacial stresses. This is also to mention that the computational cost of tracking the particles or the interface is extremely high (Shin et al. 2017).

Design of industrial multiphase devices is motivated by the quest of the improving few important metrics like energy efficiency, reduced emission, production rate, longer life of the components, increase security. For a combustion device, in order to improve its efficiency, the fuel injection system and mixing parameters are to be improved. Similarly, the efficiency of the filtration and separation is to be augmented for reducing the content of unburned carbon and other hazardous particles through the emission nozzle. For an industrial mixing device, the mixing time is to be reduced with optimal penalty in power requirement in order to obtain better productivity. Marine applications are to be designed for better performance with reduced drag on them. For improved performance of oil/gas pipe networks, losses are to be minimized using understanding of multiphase flow profiles and separation zones. Cavitation in pumps and valves is to be well understood for predicting life of these devices. As a whole, it is well taken that detailed understanding of flow behavior is extremely necessary for better design and optimal specification of operating conditions of a large number of industrial devices. Also, the predictive and scheduled maintenances can be better planed once this understanding is well developed. Lab-scale experiments are often considered to be a route for developing the insights. However, scaling up has been reported as a much complex issue in multiphase experiments. Wang et al. (1993) reported that though velocities can be scaled from lab-scale prototype to the actual device, scaling could not be achieved for the droplet sizes. Similar scalability problems are reported for cavitation problems (Keller and Arndt 2000; Schnerr and Sauer 2001) where suitable non-dimensional parameters cannot be obtained. Pangarkar (2014) reviewed correlations based on mechanical similarity in multiphase reactors and concluded that the power of the Reynolds number varies for different devices and length scales. Considering this limitation of lab-scale experiments on multiphase flows, computational methods are to be critically designed for full-scale devices.

Computational fluid dynamics (CFD) techniques have been widely utilized for studying multiphase flows in industrial devices since last two decades. Few of the important works include Einberg et al. (2005), Gentric et al. (2005), Rohdin and Moshfegh (2007), Chu et al. (2009), Raynal et al. (2009). Most of these studies solved Reynolds-averaged Navier–Stokes equations with one or two equation-based eddy-viscosity models. Comparisons were made with experimental observations in terms of bulk parameters. However, turbulent statistics are either not compared with the field data or under/over predicted through these studies. It can be further observed that the considered devices have little geometric complexities. These are either a fixed domain or can be represented by simple rotational motion of the impeller blades and can be modeled efficiently by rotating/sliding mesh method using multiple reference frames (MRF) (Gentric et al. 2005). The mesh sizes considered are not more than a million, and hence, computational costs are not very high. This is to be noted that without further refinement, this mesh cannot sufficiently resolve small energy-carrying scales of turbulent flow, and hence, the turbulent stresses and their effects on smaller particles/interface topology cannot be predicted through these models.

Pierce and Moin (2004), Mahesh et al. (2006) present two pioneering works of numerical study of multiphase reacting flow through complex industrial geometries using LES. Both the studies have used near two million mesh points. Mahesh et al. (2006) used unstructured hexahedral volume elements and a Favre-averaging technique for filtering. A Lagrangian technique is used to model the fuel jet, and its breaking and atomization were also predicted. Satisfactory comparisons were also obtained in terms of turbulent kinetic energy statistics. Riber et al. (2009) evaluated different numerical models for prediction of particle dispersion in a particulate recirculating two-phase flow. They have been able to simulate motion of around 500,000 particles using near three million hexahedral fluid cells through multiprocessor distributed computing. Their implementation was shown to scale up to 64 processors. However, the geometric complexities were much less. We can compare their effort with the recent studies which have considered much complex industrial geometries like plowshare mixers or bladed tower mills (Cleary et al. 2017). In the later study, a much advanced computational framework called workspace (Cleary et al. 2015) has been used which scales well in large computing infrastructures. Yu et al. (2017) have also recently reported k-μ sub-grid model-based LES study of cavitation around highly skewed impellers using volume of fluid (VOF) method. They have used a well-resolved mesh which ensured $y+<50$ even near the propeller blades. Each of their simulation using openFoam took nearly 120 h in 128 processors. De Queiróz Lamas et al. (2017) presented commercial software (ANSYS CFX)-based predictions of multiphase flows through an industry-scale chute in which their finest mesh simulation has been performed over 11 million hexahedral mesh elements. Therefore, it can be observed that with the recent advances in computing power and multicore architectures, more accurate and better-resolved predictions of turbulent multiphase flows in complex industrial geometries have been obtained.

Microscale interaction between the different phases can only be predicted through DNS. DNS requires a mesh refinement of the order of the Kolmogorov scales and also needs the discretization schemes to be of very high order of accuracy. That essentially calls for spectral methods or use of higher order spatial stencils. Thus, the overall computational requirement for a DNS is much higher compared to LES or RANS models. Lakehal et al. (2002) reported a DNS study of interphase heat and mass transfer using VOF-based interface tracking for fundamental problems like collapse of a water column, rise of a single bubble. Lahey (2005) presented a two-fluid DNS study in a nuclear reactor and also evaluated the closure models of turbulent mass and momentum equations. However, Tryggvason et al. (2010) assessed that a billion-sized grid may be demanded by a complex geometry DNS. This has been a restrictive factor in present multiphase CFD research. Recently Balcázar et al. (2017)

studied bubble swarms using large computer architecture of 2048 processors and laid a roadmap of multiphase DNS in large geometries.

The present paper will try to address the issues and promises of performing high fidelity computational studies of multiphase flows in complex industrial geometries. In the next section (Sect. 9.2), we will briefly discuss the simulation techniques for multiphase turbulent flows and will explore the complexities related to LES and DNS in this regard. It will also try to assess the infrastructural requirement considering the associated computational complexities. The subsequent section (Sect. 9.3) will introduce high-performance computing (HPC) techniques and will review the existing HPC implementations of multiphase LES/DNS. Section 9.4 will discuss the potentials of modern HPC architecture in addressing the computational requirements. Section 9.5 will present the concluding notes.

9.2 Simulation Techniques for Turbulent Multiphase Flows

Eulerian–Eulierian approaches of turbulent multiphase flow simulation can be broadly classified as one-fluid (Tryggvason et al. 2006) or multi-fluid models (Crowe et al. 1996). In these approaches, efforts are taken to track the interface between different phases using methods like VOF, level set. The other class of approaches is based on Lagrangian particle tracking in a domain described by Eulerian flow (Picanno et al. 2015). In this particular section, we will like to focus on physics-based modeling approaches of multiphase turbulent flows. Other issues like interface tracking, handling of complex geometries will be discussed in Sect 9.4 in connection with application of HPC techniques.

9.2.1 Large Eddy Simulation (LES)

Large eddy simulation technique directly resolves the large-scale energy-carrying structures and model smaller scales of turbulent motion. The large structures are identified by filtering the Navier–Stokes equation at the grid level. The filtered Navier–Stokes equations for turbulent incompressible flow are given as:

$$\frac{\partial \overline{u_j}}{\partial x_j} = 0 \tag{9.1}$$

$$\frac{\partial \overline{u_i}}{\partial t} + \frac{\partial \overline{u_i u_j}}{\partial x_j} = -\frac{\partial \overline{p}}{\partial x_i} + \frac{\partial \overline{\sigma_{ij}}}{\partial x_j} - \frac{\partial \tau_{ij}^t}{\partial x_j} + \bar{f_i} \tag{9.2}$$

The filtering is typically done at the grid level which is at inertial subrange. The filtered equations capture all large energy containing eddies, and the sub-grid-scale momentum and energy fluxes are modeled through closure terms The fluctuations

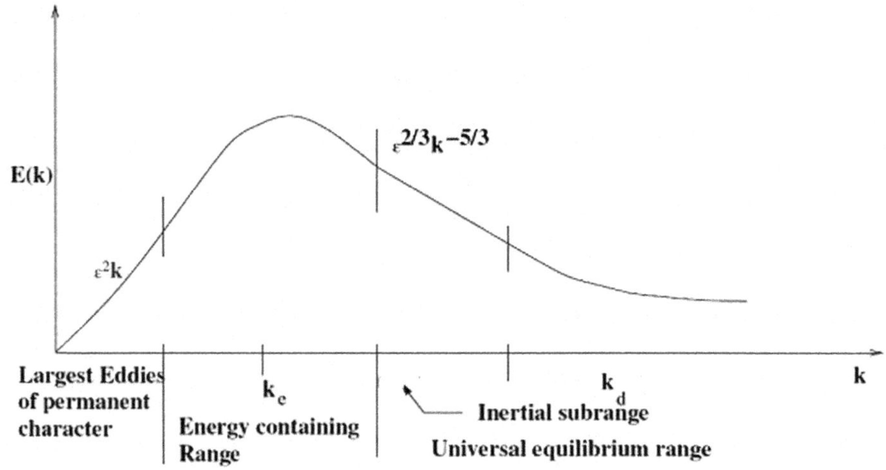

Fig. 9.1 Turbulent energy spectra showing the energy cascade at different scales

at sub-grid-scale are more universal in nature and are less affected by the boundary conditions (Kolmogorov 1941). A schematic showing the energy cascade at high Reynolds number flow is shown in Fig. 9.1.

Equation (9.2) introduces τ_{ij}^{t}, which is the sub-grid-scale-stress and obtained using an eddy-viscosity model as:

$$\tau_{ij}^{t}=\nu_t\overline{S}_{ij} \tag{9.3}$$

with $\nu_t = C_s^2\left|\overline{S}\right|$. The strain rate tensor is given by $\overline{S}_{ij} = \left(\frac{1}{2}\left(\frac{\partial u_i}{\partial x_j} + \frac{\partial u_j}{\partial xi}\right)\right)$, and $\left|\overline{S}\right|$ is the magnitude of the strain rate tensor. In the above equations, the over-bar represents the filtered quantities. The model constant C_s^2 can be obtained using Smagorinsky dynamic model (Smagorinsky 1963; Germano et al. 1991). For numerical stability, the model constant is limited to positive values only and smoothed locally. Detailed formulation of LES and sub-grid modeling can be found in Sagaut (2006) and Garnier et al. (2009).

Filtering and sub-grid-scale modeling for multiphase flows in Eulerian–Eulerian framework are more complex as closures are further needed for interphase mass and momentum transfer terms. The filtered governing equations with both one-fluid or multi-fluid assumption can be found in Labourasse et al. (2007). The filtered equations need closure terms for phase advection, interface mass and momentum transfer along with the sub-grid-scale turbulent stresses. It is also reported that a spatial filtering for properties and Favre-averaging of the velocities can help in obtaining more tractable set of governing equations.

LES has the challenge of incorporating wide ranges of length and timescales in the modeling. In a multiphase flow, dispersed phases introduce additional smaller scales of motion due to two factors: (a) Turbulent wakes are formed behind the particles,

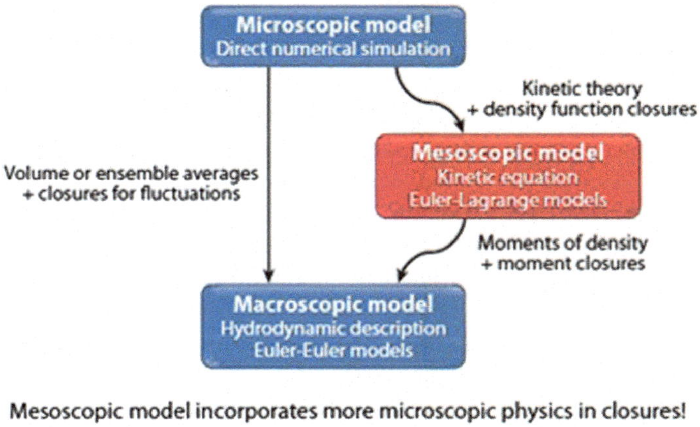

Fig. 9.2 Schematic of multiphase modeling approaches leading from DNS at microscopic level to Euler–Euler filtered equations for large eddy simulation (reproduced with permission from Fox 2012)

and (b) large-scale flow instabilities arise due to turbulent shear and density differences. Fox (2012) described that multiphase turbulent flow is actually a multi-scale multi-physics phenomena, where particle-wake dynamics at micro-level interacts with macroscopic flow physics. While a DNS can resolve all the scales up to the microscopic length, an LES will only resolve up to the inertial scales. The inertial-scale dynamics can be well represented by macroscopic Eulerian hydrodynamic models. However, Fox (2012) argued that information of the kinetics of microscopic particles may be lost while volume-averaging/filtering the microscopic activities to the inertial level. He, rather, suggested to introducing a mesoscopic model, which has to be constructed using density function closures on microscopic activities and then to apply filters and momentum and density closures for LES in macroscopic level. Schematic of these modeling approaches is shown in Fig. 9.2.

Fox and Verma (2003) demonstrated that filtering-based formulations are inconsistent with the underlying concepts of turbulent flows with dispersed particles. To address this, Pope (2010) proposed a self-conditioned LES model where the number density functions are closed at mesoscopic level using a conditional moment. Examples of large eddy simulation of turbulent multiphase flows are discussed in detail in Sect. 9.4.

9.2.2 Direct Numerical Simulation (DNS)

Direct numerical simulation (DNS) is the solution of the Navier–Stokes equations, resolving all the scales of motion while imposing the appropriate initial and boundary conditions. Each simulation produces a single realization of flow. In DNS, the mesh is fine enough to resolve the smallest scales of motion. DNS also needs numerical schemes with low dispersion and dissipation errors. As all the scales of turbulent motion are directly simulated, the simulation results are free from modeling errors. The smallest scales of turbulent fluctuations, which are even impossible to record through experimental measurements, can be well captured and visualized using DNS. Therefore, DNS can also be an efficient tool for designing turbulence models. Over the past few decades, DNS has been found useful in the study of transitional and turbulent flow physics. However, it has some limitations. Firstly, higher order discretization schemes are used in DNS in order to limit dispersion and dissipation errors; those schemes are not found to be very flexible in handling complex geometries and general boundary conditions. Secondly, DNS requires an extremely refined mesh, where the grid scale should be of the order of few Kolmogorov scales. Thus, the overall computing cost in DNS can be huge.

Tryggvason et al. (2010) discussed that DNS of multiphase flows is actually a multi-scale physics problem, where the individual defects near the interface boundaries of the droplets require a small-scale analysis and the modified constitutive relations from this analysis is used at a larger-scale Eulerian simulation to resolve the effects of the geometry and boundary conditions. Tryggvason et al. (2013) used this approach to study motion of hundreds of bubbles in turbulent channel flow and showed that multi-scale modeling can better predict the mass release from the bubbles.

Due to heavy computational requirement, DNS has been mostly restricted to simple flow configurations. However, some interesting insights into fundamental multiphase flow physics have been revealed through DNS studies, like effects of inter-particle interaction in heat transfer (Ström and Sasic 2015), breakup of swirl jet (Galbiati et al. 2016), modeling of atomization (Arienti et al. 2016), modulation of methane flames by external perturbation (de Souza et al. 2017). DNS has also been used for evaluation of various models used in LES and other higher-scale turbulent flow simulations (Tang et al. 2018; Vincent 2015).

9.2.3 Computational Complexities

To resolve all the scales of turbulent motion, the total number of grid points required is in the order of $Re^{9/4}$. The ratio of characteristic time to Kolmogorov timescale also varies in the order of $Re^{3/4}$. This implies that in order to compute the solution in a cube of unit characteristic length for a duration equal to characteristic time, the unsteady Navier–Stokes equation is to be solved for $O(Re^3)$ times. This high computational cost limits the application of DNS in low Reynolds numbers. It can be estimated that DNS of flow in simple geometry may require near billion meshes (Tryggvason et al. 2010) and a large number of computing cycles. Whereas, LES uses grid filters at inertial scale and the mesh requirement is less by hundred times. Wang et al. (2014) presented a scaling analysis for estimation of mesh sizes in turbulent wall bounder flow at friction Reynolds number, $Re_\tau = 180$. For this Reynolds number, Kolmogorov scale length in the wall coordinate is $\eta+\sim 1.5$. A channel is considered with respective aspect ratio of 4 and 1 in x- and z-directions and friction Reynolds number is defined based on the half channel height in y-direction. A mesh resolving the Kolmogorov scales thus needs $\Delta^{+} < 1.5$ and a minimum total number of mesh points of 6.7×10^{7}. Large eddy simulation on the same geometry may need less number of mesh points. For constant eddy-viscosity-based Smagorinsky model, they have used $\Delta^{+} = 5.64$ and reduced the mesh requirement to $256 \times 64 \times 64$.

However, this analysis is presented for a simple channel flow, and the mesh requirement will be higher by order of magnitude for industry-scale complex devices. Full-scale simulations are often essential requirements in multiphase CFD as scalability has been a long-debated issue. Capecelatro et al. (2016) observed that the predicted granular energy component of gravity-driven cluster induced turbulence increase linearly with the domain size, whereas the balance between them is nearly constant. Further, some other parameters like volume fraction fluctuation or particle settling velocity show different levels of dependence with the domain size. Therefore, a reduced scale simulation of turbulent multiphase devices may lead to erroneous predictions.

Roy and Acharya (2012) showed that even with a coarse grid with 70η spacing, LES of a lab-scale stirred tank reactor requires 4.2 million cells for grid independent solution. The computational complexities further enhance in a multiphase flow as the smaller dispersed particles (which are present in millions) are also to be modeled. The memory and computational clock requirement thus become higher than the limits of stand-alone computing servers. Therefore, parallel computing paradigms using HPC environments are to be developed. In the subsequent section, we will try to review the HPC techniques and their applicability for physics aware flow simulation in large-scale multiphase devices.

Prosperetti (2015) noticed that fully resolved simulations of turbulent particulate flow are restricted to simple low Reynolds number cases. He also observed that the near-wall turbulence behavior is very different for multiphase flows as compared to single-phase flows due to the fact that the particles near the wall do not show a no-slip behavior. Therefore, the closure models for multiphase turbulence must be more involved. Closure models are developed by averaging data obtained through fully resolved DNS. However, the large information due to thousands of particles and slow convergence rate of the averaging are the two major issues in developing these statistical correlations. Therefore, computational complexities are also involved in post-processing and analysis of DNS data, and advanced computational systems like parallel GPUs are to be deployed.

9.3 High-Performance Computing (HPC)

Modeling the fine structures of 3-D turbulent flow, even for simple geometries, requires a large number of grid points as well as computing cycles. Here, the conventional approaches of using high clock-speed, high RAM computers are found to be inefficient as the required number of floating point operations as well as memory access is enormously high. Overcoming the sequential bottleneck requires different approaches of computing. The most logical one is to use many processors which will work simultaneously on the same problem. This approach is popularly known as parallel computing. The main idea behind this HPC technique is basically breaking down a large computing task into a group of smaller tasks and executing them in parallel by a number of processors.

Before discussing the HPC implementations for multiphase CFD, it will be worthy to revisit the architectures of the parallel computing systems. There are two types of memory systems used in parallel computers: (1) distributed memory systems: where each processor carries its own onboard local memory which is not available to other processors in the network, and (2) shared memory systems: where a common memory area is made available to all the processors in the network. For a distributed memory system, data transfer takes place through inter-processor communication; and therefore, the performance tends to decrease rapidly with increase in number of processors. On the contrary, shared memory systems are expensive and often limited by the hardware restrictions. For shared memory architecture, data access by each processor may find a bottleneck by the bandwidth provided by the PCI bus. Based on the concurrent instructions streams and data streams, Flynn (1972) classified computer architectures. This classification is popularly known as Flynn's taxonomy. The classifications are: single instruction single data stream (SISD), single instruction multiple data streams (SIMD), multiple instruction single data stream (MISD), multiple instruction multiple data streams (MIMD). Among these architectures, SIMD and MIMD are most widely utilized for scientific computing. In SIMD architecture, a number of computing cores perform similar instructions over multiple data streams. This can be deployed in case of a shared memory multicore system (using openMP) or GPGPU (using CUDA). In MIMD architecture, each processor runs its own program on its own data. Typically, it has been observed that distributed memory multicore machines using MPI can give optimized performance in MIMD architecture. A typical comparison between shared memory and distributed memory architecture can be explained by Fig. 9.3. It can be shown that while shared memory processors can handle a large memory job together, they need a good synchronization between the concurrent processes. These concurrent processes are defined as "threads" in parallel multicore computing. In distributed memory architecture, the data as well as instructions are distributed into several machines which perform the tasks locally. However, they need to communicate via a network switch for inter-processor data transfer dynamically.

9.3.1 Multiprocessor Distributed Computing Using MPI

A domain decomposition-based parallelization scheme is usually used for CFD computations over large distributed memory parallel computing platforms (popularly known as clusters). In this method, the solution domain is divided into a number

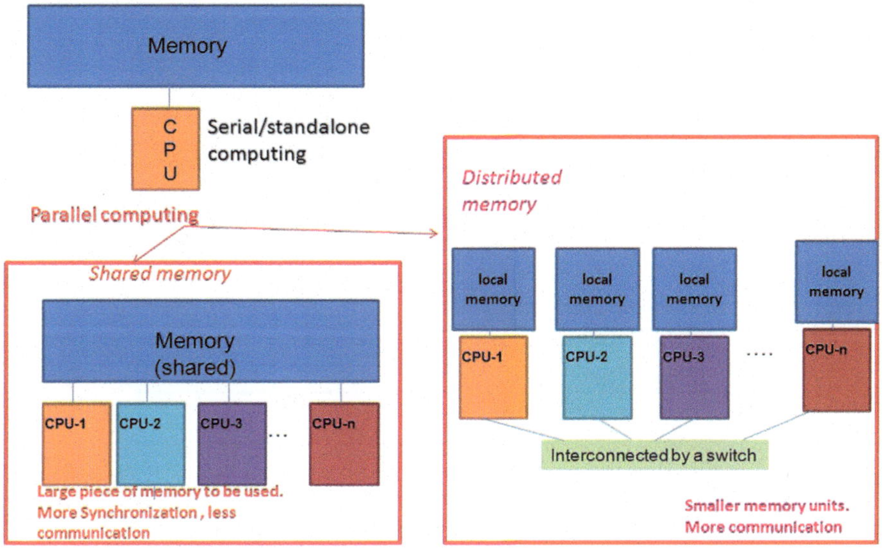

Fig. 9.3 Models of parallel computing architecture

of sub-domains and each sub-domain is assigned to a distinct processor in MIMD architecture. Each processor solves the discretized set of governing equations over the assigned sub-domain, subjected to the given boundary conditions. However, extra boundaries are created at the edges of each sub-domain, adjacent to the neighboring sub-domain, where the boundary condition needs to be defined (shown in Fig. 9.4) in order to solve the discretized governing equation for that particular sub-domain. Boundary conditions are assigned to that face of the sub-domain to maintain the continuity of the flow variables and fluxes. This can be achieved by allowing the data transfer among the different processors along the partition line of the sub-domains via standard MPI protocol (Gropp et al. 1999).

For curvilinear/block-structured curvilinear mesh, the solution domain is divided into an arbitrary number of hexagonal grid zones, Z, and the parallel solution strategy partitions these domains among an arbitrary number of processes, P, with the requirement that $P \leq Z$. Each process obtains a solution on its pre-assigned portion of the domain, subject to the boundary conditions for that part of the domain. Load balancing is an important step while assigning the block of zones to the processors. It is expected that applications using zones of approximately the same size will exhibit a higher level of parallel efficiency due to uniform communication and computation load and less latency. It is also important to see that neighboring zones are associated to same or neighboring processors to increase data localization and to reduce the volume of inter-processor data transfer. Data on the boundaries of each block are communicated to its adjacent blocks during every solution iteration. The parallel setup routines precompute the memory addresses of both the sources and destinations of all data transfer events, whether in-memory or inter-process. This is also to be

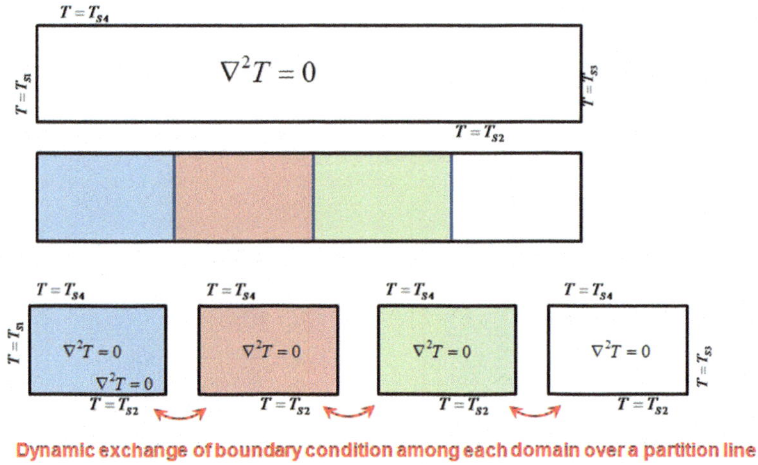

Fig. 9.4 Domain decomposition for Laplace equation of scalar, T, in a 2-D rectangular geometry

noted that the processors usually seat idle during data communication and increases latency of the application. Also to reduce latency, the computational effort must be divided among each processor such that each completes its work in about the same length of time. For a given number of computational zones, the inter-processor communications increase with the number of processes assigned to the problem. Also, if the sizes of the zones are made too small, the amount of data transferred, relative to the amount of CPU work required, will also increase. Assignment of grid zones to a user-defined number of processes is automatic and can be accomplished using graph partitioning software like METIS (Karypis and Kumar 1998). This strategy has been well discussed by Tyagi et al. (2007).

Usually the heaviest part of a CFD code consists of matrix solution routines. The size of the matrix to be solved by each processor can be limited to a smaller order by dividing the large problem into smaller sub-domain-based problems. Also, parallel solution of each sub-domain increases the overall speed. The matrix solvers are usually modified using additive/multiplicative Schwarz algorithm.

Alternatively OpneMP constructs can be deployed in a shared memory architecture, where parallel threads are created to compute for different rows of a matrix solver. OpenMP applications are mostly limited due to higher requirement of the RAM as the entire matrix is computed using a large contiguous memory. However, recent development of computer hardware has enabled CFD community to solve large problems using hybrid MPI/OpenMP constructs in heterogeneous clusters (de Wiart and Hillewaert 2015; Alvarez et al. 2018).

9.3.2 GPGPU-Accelerated Multicore Computing

Graphics processing units (GPUs) have emerged as highly parallel computing units over last few decades. A large number of streaming multiprocessors (SMs) are embedded in a GPU card; each of these multiprocessors contains computing cores with fixed number of registers. GPU has many more transistors devoted to data processing and less number of transistors for data control and cache. Each SM has a fast on-chip small-sized memory shared among its cores, whereas a larger off-chip memory is slower. This off-chip memory known as device memory is shared across different SMs. One of the most recent developments, Pascal P100 GPU cards, contains 56 SMs; each SM is having 64 FP32 computing cores, dedicated shared memory of only 64 kB and 4 MB L2 cache. The off-chip memory is 16 GB (Nvidia 2016). As suggested by their name, the GPUs were primarily designed for computer graphics and image processing (Nvidia 1999). As a later development, GPUs are modified to perform tasks similar to vector processors and deployed for scientific computing (Du et al. 2012). The general purpose computing on GPUs is termed as GPGPU computing where GPU is called as a device and CPU is called a host. A generic GPGPU architecture is shown in Fig. 9.5. A Kernel is the part of or a program itself that is executed on GPU device using large number of threads organized into thread blocks and grids. More details on the GPGPU programming model using threads and thread blocks can be found in the CUDA C Programming Guide (Nvidia 2011). CUDA allows us to program GPGPUs to carry out parallel computing at an extreme level and to launch "threads" on a multi-million scales. Thus, the mathematical solvers can be efficiently parallelized and sent to CUDA Kernel reducing the computation time drastically as compared to the serial counterpart. Double precision peak performance of NVIDIA P100 cards is reported to be around 5.3 TFlops (Nvidia 2016).

Though CPUs are much faster than GPGPUs for running a single set of instructions, they cannot execute multiple instructions concurrently. On the contrary, the GPGPUs, despite being much slower than CPUs, can process number of instructions of similar kind in parallel. However, the performance of GPU multiprocessors can be adversely augmented due to slow memory access and small onboard memory chips. Typically, parallel matrix solution algorithms are to be demonstrated with good "compute to global memory access ratio" for efficient GPU optimization (Dubois et al. 1986). With efficient memory management and kernel calls, efficient acceleration of matrix solvers has been reported in the works of Christen et al. (2007), Garland et al. (2008), Peña et al. (2014), Anzt et al. (2017), etc. In these studies, 7–30 times speedup is observed. It is noteworthy that the speedup has been increased over the years due to advancements of GPU hardware in terms of memory, bandwidth, and computing cores. Also, GPGPU computation is proved to be much energy efficient compared to multicore clusters.

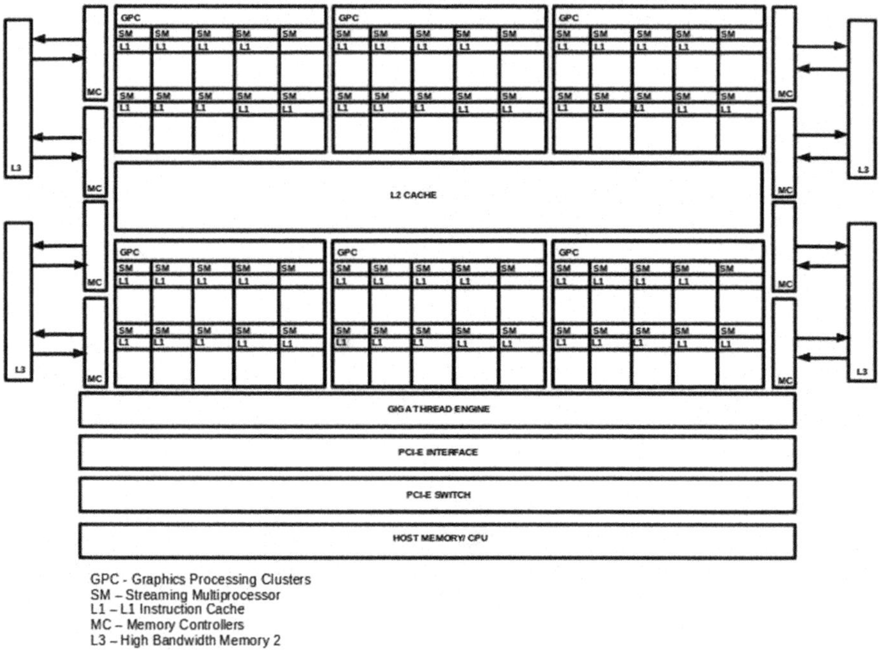

Fig. 9.5 Schematic of a GPGPU architecture showing streaming multiprocessors and memory register

A number of researchers have demonstrated GPGPU accelerators to obtain speedup of CFD solvers. Some of the important works are Cohen and Molemaker (2009), Jespersen (2010), Corrigan et al. (2012), Agrawal et al. (2015). This is noteworthy that most of the implementation came via augmenting simple CUDA constructs on legacy in-house codes.

9.3.3 Application of HPC Techniques in Computing Multiphase Turbulent Flows

Highly parallelized algorithms are used for simulation of multiphase flows by a number of researchers. Parallel computing methods are utilized for addressing computationally exhaustive issues like finer mesh requirements for DNS (Zeng et al. 2008; Wachs 2011), modeling of moving boundaries (Hartmann et al. 2006; Rodriguez et al. 2013), tracking of interphase front with higher geometric accuracy (Tryggvasn et al. 2001; Balcázar et al. 2014), and tracking of Lagrangian particles (Picano et al. 2015; Sweet et al. 2018). In this subsection, we will review the relevant literature.

Both DNS and LES require refined mesh to resolve finer scales of turbulent motions. For particulate flows, it is also important to resolve the mesh to capture the smallest scales of turbulent motion. Zeng et al. (2008) present a turbulent channel flow simulation where the diameter of the smallest particle is of the order of twice Kolmogorov length scale with the particle Reynolds number varying from 42 to 250. They have used spectral element methodology (SEM) over a 128^3 mesh, and the computations ran for 1.3 months in 256 processors. Ham et al. (2003) presented the multiphase LES simulation studies using arbitrary-shaped mesh in industrial combustor geometry with 18 injection holes. They have used an extremely refined mesh with 100 million control volumes. The detailed computer science indicators of their code have been reported along with a scalability up to 500 processors. This work, in particular, stands as one of the pioneering efforts of studying multiphase flow in industry-scale geometries. Much developments in LES of large multiphase devices have been reported in recent years due to development in computer architecture and CFD techniques. Olenik et al. (2015) studied dispersion of five million particles in a 2.5 MW IFRF coal combustor furnace and showed that LES predictions are much superior to RANS in terms of agreement with experimental data. Bauerheim et al. (2015) used near 35 million cells to study combustion instability in multi-point low NO_x annular combustor. Their work needed computing over a large number of time steps as they studied temporal behavior of the combustion flame and the vortex dynamics. Xie and Luo (2017) attempted a study of multi-scale physics in a two-phase stirred tank flow, where they have coupled CFD simulations with population balance model (PBM) for the dispersed droplets and reaction kinetics. They have used a commercial CFD software (ANSYS fluent) within whose framework ATRP kinetics and PBM models are coupled through user-defined functions. Production runs are obtained using 1.2 million cells in 64 CPUs. A physical time of 1000 s is simulated in 15 days runtime. Multiphase flow community has been efficiently using GPU-accelerated lattice Boltzmann model (LBM) for flow predictions. The advantages of LBM in a GPU architecture is that the issues of latency due to memory copying can be avoided as typical matrix solvers are not used. Kuznik et al. (2010) demonstrated a GPU-accelerated LBM simulation for 2048^2 mesh. Xiong et al. (2012) simulated near 120,000 particles with 75 μm diameter in 0.1 cm^3 domain using multiple GPUs. They have used a hybrid CPU-GPU architecture, where the inter-processor data transfer was carried by MPI or OpenMP calls. They reported that the scalability remains almost linear till 600 GPUs. Their work showcased a super scalable implementation of multiphase simulation using GPUs. DNS of droplet collision using LBM has been modeled in multiple GPUs, and 12 GTx titan GPUs are found to be 190 times faster than 12 i7 cores (Huang et al. 2018). Recent P100 GPUs are found to be twice faster than GTX titans. However, scalability is of reduced order in cases of Navier–Stokes solution due to memory management and iterative nature of large matrix solvers. Mayank et al. (2017), Sweet et al. (2018) reported GPU acceleration of two-phase Navier–Stokes LES solvers. The speedup is 7–14 times. The mesh requirement in a DNS is usually much higher than a LES. Wang et al. (2014) showed that to obtain comparable mean flow parameters, LES roughly requires 1/40 times mesh than DNS. They used a highly optimized code over eight K20 GPU cores.

Therefore, DNS of multiphase flows are restricted to simpler geometries and mostly focused on understanding local small-scale physics (e.g., Lebas et al. 2009, Shi et al. 2017). Beharfarid et al. (2013) used a smart approach of coupled DNS-RANS for simulation of fission gas discharge during loss of flow accident in a partially blocked sodium fast reactor. This basically used a multi-scale approach, where near 10 mm of the clad failure, DNS is used and RANS is used for the rest of the 0.5 m long geometry. The mesh requirement is limited to 1.5 million. The simulations ran for 1 million CPU hours in 2048 IBM blue gene cores using MPI. Sbrizzai et al. (2006) presented an early attempt of DNS of a full-scale stirred tank reactor with 10^4 particles with lowest diameter of the order of the Kolmogorov length scale. However, their calculation was achieved in a two-million mesh framework as they considered a rather lower level of the Reynolds number ($Re = 1636$).

One of the serious computational challenges for multiphase flow arises from the need of identification of the interface of the different phases. Especially for problems involving multiple bubbles coalescence, atomization or liquid jet instability, the interface undergoes drastic changes in topology and breaks/coalesces. Interface tracking algorithms are utilized to reconstruct smooth topologies while maintaining sharpness of the boundary. Algorithms like level set method, VOF are used for that. Herrmann (2003) discussed a domain decomposition-based scalable parallelization strategy for fast marching method solution of the parabolic equation for level set scalar transport. He showed that separate load balancing for both sides of the interface is essential for linear scalability of the implementation. He later extended this approach for unstructured mesh (Hermann 2008). However, this condition is difficult to achieve for large domain with a dispersed phase. Fu et al. (2015) reported up to 13x speed for CPU-GPU hybrid parallelization of level set-based multiphase front tracking solvers. Waters et al. (2017) used VOF method for prediction of spray breakup in a parallel multiprocessor environment.

Tryggvason et al. (2001, 2006) demonstrated a surface tracking methodology using marker points on the interface. These marker points are connected by a surface mesh. The multiphase flow equations are solved by one-fluid or multi-fluid model for different phases. The interface points move due to the local fluid velocities. Tryggvason and his group have extensively used this methodology for solving different applications of multiphase flow. The front advancement is followed by reconstruction and local smoothening of the interface, and the surface mesh is also dynamically refined to redefine the interfacial topology (Tryggvason et al. 2001). While the fluid domain is decomposed and the sub-domains are assigned to different processors, the interface nodes belonging to one sub-domain may move to another sub-domain and add up to the local memory belonging to a different processor. Bunner and Tryggvason (1999) introduced a separate parallelization strategy for the marker points. Their location information is stored in one particular node designated as the master. Based on the geometric location, this information is scattered from the master to other processors dynamically. For Lagrangian particle tracking in multiphase flows simulation, GPU acceleration has been well demonstrated as a simple and efficient strategy (Barker 2013; Fukushima et al. 2015). Picanno et al. (2015) used immersed boundary method to simulate motion of 10,000 particles in a turbulent channel flow with a DNS that uses more than 100 million cells. They have used a domain decomposition-based parallelization over 2048 cores of distributed memory cluster.

Immersed boundary method can also be used for modeling of complex and moving geometries used in industrial applications. A detailed review of this method can be found in Mittal and Iaccarino (2005). In this method, a structured fixed mesh may be employed to solve flow in any complex domain where the moving/deforming boundaries are modeled using velocity interpolation/force interpolation. Sun and Sakai (2016) demonstrated a fast solution method for multiphase flow in complex industrial devices using VOF and immersed boundary method. Tyagi et al. (2007) discuss a parallelization scheme for immersed boundary implementation in a baffled stirred tank. They have redistributed the grid blocks near the moving boundary for reducing latency during immersed boundary implementation and reported a tenfold increase in computational speed. Krishnan et al. (2017) developed a GPU-optimized immersed boundary solver for complex geometry flows. Very recently, Guitérrez et al. (2018) coupled moving mesh and immersed boundary method (IBM) to track bubbles using level set method in complex geometries.

9.4 Potential Use of Modern HPC Resources in Simulation of Industry-Scale Problems

In the previous sections, we have reviewed applications of HPC aware techniques for CFD simulation of multiphase flows. It has been identified that multiphase flows are actually a multi-scale phenomena and efficient modeling of different scales are required for accurate predictions in industry-scale geometries. As the drive for this predictions is better design of the components, the bulk flow parameters as well as microstructures of the flow are important in the simulations. Detailed discussion on potentials of DNS in industrial multiphase flows can be found in Reveillon et al. (2011). The wholesome computational requirement of this multi-scale simulation is extremely high as it will need to solve equations over a near billion mesh point for a long duration of physical time. While discussing the issues of dimensional scalability in multiphase CFD, Capecelatro et al. (2016) suggested that exa-scale computing can be a potential route to address computational challenges of accurate simulation of industry-scale flows. In one of the pioneering works on HPC-based CFD at large domain, Onodera et al. (2013) presented LES of single-phase wind flow simulation of 10 km $\times$ 10 km metropolitan area. They have handled a 0.8 billion grid points using multiple GPUs. Thus, multi-GPU and hybrid GPU-CPU infrastructures should be explored as the platforms for calculations of multiphase flow. Appleyard and Drikakis (2011) showed that an efficient memory management and memory reuse can help in acceleration of multiphase CFD with interface tracking in multi-GPU machines. Maruthi et al. (2017) showed that cutting edge P100 GPU can give 113 times speedup compared to a single CPU for DNS in turbine blade cascades. Pérez et al. (2018) used multi-level hybridization using MPI-OpenMP and CUDA for DNS of reacting flows with detailed chemistry. With advent of GPU direct technology, GPUs can communicate among them bypassing the CPU and interconnect switches and thus can reduce the latency (Otten et al. 2016).

Algorithmic developments are also vital for addressing larger problems in multiphase CFD. Few of the recent works in that direction can be briefly reviewed here. Shin et al. (2017) demonstrated 1024^3 mesh DNS of multiphase flow using massively parallel solvers and multigrid method. Juric et al. (2017) reported development of parallel CFD software for multiphase flows which can scale over 256,000 processors. Dyson (2018) accelerated the matrix solvers in OpenFOAM using GPUs and predicted spray dynamics. Ma et al. (2016) and Wu et al. (2018) have used data-driven models for prediction of multiphase flows. These models pose less computational complexity compared to deterministic CFD and also can be well accelerated in GPUs.

Multi-particle simulation in parallel architecture utilize either of the two different strategies: (i) A master node tracks all the particles and passes the information to the slave nodes, or (ii) each node is responsible for particles local to it, and they pass the relevant information to neighboring processors. Both these strategies introduce certain amount of latency. Bonnier et al. (2018) demonstrated an algorithm based on temporal locality of the particle data and redistributing the particles among the cores which improved the speedup of parallel particle tracking.

In 2016, US Department of Energy Office of Science and US Nuclear National Security Administration have jointly initiated the exa-scale computing project (ECP) to revolutionalize the computing power using cutting-edge technologies like hybrid parallelization, memory efficiency. This project aims at 50-fold increase in the application performances (Messina 2017). There have been similar initiatives in other parts of the globe (i.e., in European Union, Japan, China, and India). Under the aegis of these programs, worldwide scientific consortiums for exa-scale computing application on different domains are established. Brower et al. (2018) demonstrates a world spread exa-scale effort on lattice-QCD computations. In their 2030 vision document, NASA estimated the cost of computation for wall-modeled LES with billion grid points and asserted on the need of exa-scale computing (Slotnick et al. 2014). They have identified few grand challenges in aerospace CFD which will test the capabilities of the forthcoming HPC infrastructures. These problems involve aerodynamics and power generation in aircrafts. Their vision document has also identified the hurdles which have to be overcome by the present computational technology. These areas include higher power consumption, software extraction, programming environment, scalability of the code as well as the preprocessor and post-processor. Smith (2016) reports activities of a consortium formed by Department of Energy (US) Innovation Hub for full-scale reactor core simulation using exa-scale computing. Bolotnov (2015) asserted on the requirement of exa-scale HPC for fully resolved simulation of bubbly flow in reactor cores. Cela et al. (2016) reported a Brazil–Europe collaboration for energy research using exa-scale HPC. They have focused on wind energy and biomass-derived fuels. Apart from the use of advanced architecture and simulators, they have also asserted on fundamental CFD issues like improving the CFD microscale models and dynamic downscaling strategy for transferring the effects of mesoscale boundary conditions to the microscale solver.

Exa-scale computing initiatives will follow several challenges in terms of implementation at both software and hardware levels. Shalf et al. (2010) discuss the major challenges in establishing an exa-scale infrastructure. One serious difficulty will emerge due to huge power requirement by thousands of CPUs and GPUs involved in processing data at exaflops rate. Exa-scale architecture is typically categorized as a hybrid CPU-GPU architecture, which requires massive power for chip operation and cooling. To address this, better energy aware algorithms are to be developed. Focuses on improved thermal design of the chips will also complement the energy requirement by the processing cores (Kulkarni et al. 2018). The other challenges are interconnect latency, memory latency, resilience, software systems, data management, exa-scale algorithm development, etc. Ashraf et al. (2018) showed that MPI-OpenMP-CUDA tri-hybrid solvers can be designed with efficient algorithms for better performance–energy trade-off. As a step ahead, Eurolab-4-HPC team presents a long-term vision for development HPC infrastructure for 2023–2030 (Ungerer and Carpenter 2018). This vision map calls for unifying HPC with high-performance data analytics, use of nonvolatile memory, better programming paradigm, and not the least green and power saving applications. Statistical models are finding popularity in multiphase CFD community (Gibou et al. 2018; Zhuang et al. 2018). The vision for coupling of HPC with data analytics will help in strengthening these activities.

Considering the above developments and efforts of several consortia on exa-scale computing (Kumar et al. 2015; Chang et al. 2017), scientific community can hope of exploring new insights of multiphase turbulent flows in complex large devices in coming years.

9.5 Summary

This chapter has presented a review of LES and DNS studies for prediction of complex multiphase multi-scale flow in industrial devices. The developments in HPC techniques and their use in multiphase CFD are also discussed. It is observed that though LES has been performed for industrial problems using HPC platforms, DNS studies have mostly been restricted to simplified flow configurations. However, DNS has been instrumental in understanding some of the fundamental phenomena of multiphase flows. Thus, contemporary research on exa-scale computing focuses on addressing challenges like DNS of large configurations, simulation of reacting multiphase flows.

Acknowledgements Author acknowledges support from Mr. Apurva Raj, Research Scholar, Department of Aerospace Engineering, Indian Institute of Technology Kharagpur, in preparation of the figures.

References

Agrawal S, Kumar M, Roy S (2015) Demonstration of GPGPU-Accelerated Computational Fluid Dynamic Calculations. In: Intelligent Computing and Applications, pp 519–525. Springer, New Delhi

Alvarez X, Gorobets A, Trias FX, Borrell R, Oyarzun G (2018) HPC2—a fully-portable, algebra-based framework for heterogeneous computing. Application to CFD. Comput Fluids

Anzt H, Gates M, Dongarra J, Kreutzer M, Wellein G, Köhler M (2017) Preconditioned Krylov solvers on GPUs. Parallel Comput 68:32–44

Appleyard J, Drikakis D (2011) Higher-order CFD and interface tracking methods on highly-parallel MPI and GPU systems. Comput Fluids 46(1):101–105

Arienti M, Oefelein J, Doisneau F (2016) Modeling primary atomization of liquid fuels using a multiphase DNS/LES approach (No SAND2016-7913R). Sandia National Lab. (SNL-CA), Livermore, CA (United States)

Ashraf MU, Eassa FA, Albeshri AA, Algarni A (2018) Performance and power efficient massive parallel computational model for HPC heterogeneous Exascale systems. IEEE Access 6:23095–23107

Balcázar N, Jofre L, Lehmkuhl O, Castro J, Rigola J (2014) A finite-volume/level-set method for simulating two-phase flows on unstructured grids. Int J Multiph Flow 64:55–72

Balcázar N, Castro González J, Rigola Serrano J, Oliva Llena A (2017) DNS of the wall effect on the motion of bubble swarms. In: Procedia computer science, volume 108. International conference on computational science, ICCS, 12–14 June 2017, Zurich, Switzerland, pp 2008–2017. Elsevier

Barker D (2013) Development of a scalable real-time Lagrangian particle tracking system for volumetric flow field characterization. Doctoral dissertation, University of Illinois at Urbana-Champaign

Bauerheim M, Jaravel T, Esclapez L, Riber E, Gicquel LYM, Cuenot B … Rullaud M (2015) Multiphase flow les study of the fuel split effects on combustion instabilities in an ultra low-nox annular combustor. In ASME turbo expo 2015: turbine technical conference and exposition. American Society of Mechanical Engineers, pp V04BT04A069-V04BT04A069

Behafarid F, Shaver D, Bolotnov IA, Antal SP, Jansen KE, Podowski MZ (2013) Coupled DNS/RANS simulation of fission gas discharge during loss-of-flow accident in generation IV sodium fast reactor. Nucl Technol 181(1):44–55

Bolotnov IA (2015) Exascale applications for nuclear reactor thermal-hydraulics: fully resolved bubbly flow in realistic reactor core fuel assembly

Bonnier F, Emad N, Juvigny X (2018) Software architecture for parallel particle tracking with the distribution of large amount of data. In: Proceedings of the high performance computing symposium. Society for Computer Simulation International, p 12

Brower R, Christ N, DeTar C, Edwards R, Mackenzie P (2018) Lattice QCD application development within the US DOE Exascale computing project. In: EPJ web of conferences, vol 175. EDP Sciences, p 09010

Bunner B, Tryggvason G (1999) Direct numerical simulations of three-dimensional bubbly flows. Phys Fluids 11(8):1967–1969

Capecelatro J, Desjardins O, Fox RO (2016) Effect of domain size on fluid–particle statistics in homogeneous, gravity-driven, cluster-induced turbulence. J Fluids Eng 138(4):041301

Cela JM, Navaux PO, Coutinho AL, Mayo-García R (2016) Fostering collaboration in energy research and technological developments applying new Exascale HPC techniques. In: 2016 16th IEEE/ACM international symposium on cluster, cloud and grid computing (CCGrid). IEEE, pp 701–706

Chang CS, Greenwald M, Riley K, Antypas K, Coffey R, Dart E, et al (2017) Fusion energy sciences Exascale requirements review. An office of science review sponsored jointly by advanced scientific computing research and fusion energy sciences, 27–29 Jan 2016, Gaithersburg, Maryland. USDOE Office of Science (SC), Washington, DC (United States). Offices of Advanced Scientific Computing Research and Fusion Energy Sciences

Christen M, Schenk O, Burkhart H (2007) General-purpose sparse matrix building blocks using the NVIDIA CUDA technology platform. In: First workshop on general purpose processing on graphics processing units, p 32
Chu KW, Wang B, Yu AB, Vince A (2009) CFD-DEM modelling of multiphase flow in dense medium cyclones. Powder Technol 193(3):235–247
Cleary PW, Thomas D, Bolger M, Hetherton L, Rucinski C, Watkins D (2015) Using workspace to automate workflow processes for modelling and simulation in engineering. In: MODSIM2015, 21st international congress on modelling and simulation, modelling and simulation society of Australia and New Zealand, Gold Coast, pp 669–675
Cleary PW, Hilton JE, Sinnott MD (2017) Modelling of industrial particle and multiphase flows. Powder Technol 314:232–252
Cohen J, Molemaker MJ (2009) A fast double precision CFD code using CUDA. Parallel Comput Fluid Dyn Recent Adv Future Dir, pp 414–429
Corrigan A, Camelli F, Löhner R, Mut F (2012) Semi-automatic porting of a large-scale Fortran CFD code to GPUs. Int J Numer Meth Fluids 69(2):314–331
Crowe CT, Troutt TR, Chung JN (1996) Numerical models for two-phase turbulent flows. Annu Rev Fluid Mech 28(1):11–43
de Queiróz Lamas W, Bargos FF, Giacaglia GEO, Grandinetti FJ, de Moura L (2017) Numerical modelling and simulation of multiphase flow through an industrial discharge chute. Appl Therm Eng 125:937–950
de Souza TC, Bastiaans RJM, De Goey LPH, Geurts BJ (2017) Modulation of a methane Bunsen flame by upstream perturbations. J Turbul 18(4):316–337
de Wiart CC, Hillewaert K (2015) Development and validation of a massively parallel high-order solver for DNS and LES of industrial flows. In: IDIHOM: industrialization of high-order methods-a top-down approach, pp 251–292. Springer, Cham
Du P, Weber R, Luszczek P, Tomov S, Peterson G, Dongarra J (2012) From CUDA to OpenCL: towards a performance-portable solution for multi-platform GPU programming. Parallel Comput 38(8):391–407
Dubois M, Scheurich C, Briggs F (1986) Memory access buffering in multiprocessors. In: ACM SIGARCH computer architecture news, vol 14(2). IEEE Computer Society Press, pp 434–442
Dyson J (2018) GPU accelerated linear system solvers for OpenFOAM and their application to sprays. Doctoral dissertation, Brunel University London
Einberg G, Hagström K, Mustakallio P, Koskela H, Holmberg S (2005) CFD modelling of an industrial air diffuser—predicting velocity and temperature in the near zone. Build Environ 40(5):601–615
Flynn MJ (1972) Some computer organizations and their effectiveness. IEEE Trans Comput C–21(9):948–960. https://doi.org/10.1109/tc.1972.5009071
Fox RO (2012) Large-eddy-simulation tools for multiphase flows. Annu Rev Fluid Mech 44:47–76
Fox RO, Varma A (2003) Computational models for turbulent reacting flows. Cambridge University Press, Cambridge
Fu Z, Yakovlev S, Kirby RM, Whitaker RT (2015) Fast parallel solver for the level set equations on unstructured meshes. Concurrency Comput: Pract Exp 27(7):1639–1657
Fukushima N, Katayama M, Naka Y, Oobayashi T, Shimura M, Nada Y, et al (2015) Combustion regime classification of HCCI/PCCI combustion using Lagrangian fluid particle tracking. Proc Combust Inst 35(3):3009–3017
Galbiati C, Tonini S, Weigand B, Cossali G (2016) Direct numerical simulation of primary break-up in swirling liquid jet. In: 9th international conference on multiphase flow (ICMF 2016). AIDIC
Garland M, Le Grand S, Nickolls J, Anderson J, Hardwick J, Morton S, et al (2008) Parallel computing experiences with CUDA. IEEE Micro (4):13–27
Garnier E, Adams N, Sagaut P (2009) Large eddy simulation for compressible flows. Springer Science & Business Media
Gentric C, Mignon D, Bousquet J, Tanguy PA (2005) Comparison of mixing in two industrial gas–liquid reactors using CFD simulations. Chem Eng Sci 60(8–9):2253–2272

Germano M, Piomelli U, Moin P, Cabot WH (1991) A dynamic subgrid-scale eddy viscosity model. Phys Fluids A 3(7):1760–1765

Gibou F, Hyde D, Fedkiw R (2018) Sharp interface approaches and deep learning techniques for multiphase flows. J Comput Phys

Gropp W, Lusk E, Skjellum A (1999) Using MPI: portable parallel programming with the message-passing interface, 2nd edn. The MIT Press

Gutiérrez E, Favre F, Balcázar N, Amani A, Rigola J (2018) Numerical approach to study bubbles and drops evolving through complex geometries by using a level set–moving mesh–immersed boundary method. Chem Eng J 349:662–682

Ham F, Apte S, Iaccarino G, Wu X, Herrmann M (2003) Unstructured LES of reacting multiphase flows in realistic gas turbine combustors. Minnesota Univ Minneapolis

Hartmann H, Derksen JJ, Van den Akker HEA (2006) Numerical simulation of a dissolution process in a stirred tank reactor. Chem Eng Sci 61(9):3025–3032

Herrmann M (2003) A domain decomposition parallelization of the fast marching method. German Research Foundation (DFG) Bonn (Germany)

Herrmann M (2008) A balanced force refined level set grid method for two-phase flows on unstructured flow solver grids. J Comput Phys 227(4):2674–2706

Huang TC, Chang CY, Lin CA (2018) Simulation of droplet dynamic with high density ratio two-phase lattice Boltzmann model on multi-GPU cluster. Comput Fluids

Jespersen DC (2010) Acceleration of a CFD code with a GPU. Sci Program 18(3–4):193–201

Juric D, Chergui J, Shin S, Kahouadji L, Craster RV, Matar OK (2017) Innovative computing for industrially-relevant multiphase flows

Karypis G, Kumar V (1998) METIS, a software package for partitioning unstructured graphs, partitioning meshes, and computing fill-reducing orderings of sparse matrices

Kataoka I (1986) Local instant formulation of two-phase flow. Int J Multiph Flow 12(5):745–758

Keller A, Arndt RA (2000) Cavitation scale effects—a presentation of its visual appearance and empirically found relations. In: Proceedings of NCTC50 international conference on propeller cavitation, April 2000, Newcastle upon Tyne

Kolmogorov AN (1941) Dissipation of energy in locally isotropic turbulence. In: Dokl. Akad. Nauk SSSR, vol 32(1), pp 16–18

Krishnan A, Mesnard O, Barba LA (2017) cuIBM: a GPU-based immersed boundary method code. J Open Source Softw 2:15

Kulkarni D, Tang X, Ahuja S, Dischler R, Mahajan R (2018) Experimental study of two-phase cooling to enable large-scale system computing performance. In: 2018 17th IEEE intersociety conference on thermal and thermomechanical phenomena in electronic systems (ITherm). IEEE, pp 596–601

Kumar N, Sringarpure M, Banerjee T, Hackl J, Balachandar S, Lam H … Ranka S (2015) CMT-bone: a mini-app for compressible multiphase turbulence simulation software. In: 2015 IEEE international conference on cluster computing (Cluster). IEEE, pp 785–792

Kuznik F, Obrecht C, Rusaouen G, Roux JJ (2010) LBM based flow simulation using GPU computing processor. Comput Math Appl 59(7):2380–2392

Labourasse E, Lacanette D, Toutant A, Lubin P, Vincent S, Lebaigue O, … Sagaut P (2007) Towards large eddy simulation of isothermal two-phase flows: governing equations and a priori tests. Int J Multiph flow 33(1):1–39

Lahey RT Jr (2005) The simulation of multidimensional multiphase flows. Nucl Eng Des 235(10–12):1043–1060

Lakehal D, Meier M, Fulgosi M (2002) Interface tracking towards the direct simulation of heat and mass transfer in multiphase flows. Int J Heat Fluid Flow 23(3):242–257

Lebas R, Menard T, Beau PA, Berlemont A, Demoulin FX (2009) Numerical simulation of primary break-up and atomization: DNS and modelling study. Int J Multiph Flow 35(3):247–260

Ma M, Lu J, Tryggvason G (2016) Using statistical learning to close two-fluid multiphase flow equations for bubbly flows in vertical channels. Int J Multiph Flow 85:336–347

Mahesh K, Constantinescu G, Apte S, Iaccarino G, Ham F, Moin P (2006) Large-eddy simulation of reacting turbulent flows in complex geometries. J Appl Mech 73(3):374–381
Maruthi NH, Ranjan R, Narasimha R, Deshpande SM, Sharma B, Nanditale SRT (2017) GPU acceleration of a DNS code for gas turbine blade simulations. In: Computational science symposium, IISc, Bangalore
Mayank K, Banerjee R, Mangadoddy N (2017) Development of GPU parallel multiphase flow solver for turbulent slurry flows in cyclone
Messina P (2017) The exascale computing project. Comput Sci Eng 19(3):63–67
Mittal R, Iaccarino G (2005) Immersed boundary methods. Annu Rev Fluid Mech 37:239–261
Nvidia CUDA (2011) Nvidia cuda c programming guide. Nvidia Corporation 120(18):8
Nvidia T (2016) P100 white paper. NVIDIA Corporation
NVIDIA Launches the World's First Graphics Processing Unit: GeForce 256. Nvidia. 31 August 1999. Archived
Olenik G, Stein OT, Kronenburg A (2015) LES of swirl-stabilised pulverised coal combustion in IFRF furnace No. 1. Proc Combust Inst 35(3):2819–2828
Onodera N, Aoki T, Shimokawabe T, Kobayashi H (2013) Large-scale LES wind simulation using lattice Boltzmann method for a 10 km × 10 km area in metropolitan Tokyo. TSUBAME e-Sci J Global Sci Inf Comput Center 9:1–8
Otten M, Gong J, Mametjanov A, Vose A, Levesque J, Fischer P, Min M (2016) An MPI/OpenACC implementation of a high-order electromagnetics solver with GPU direct communication. Int J High Perform Comput Appl 30(3):320–334
Pangarkar VG (2014) Design of multiphase reactors. Wiley
Peña AJ, Reaño C, Silla F, Mayo R, Quintana-Ortí ES, Duato J (2014) A complete and efficient CUDA-sharing solution for HPC clusters. Parallel Comput 40(10):574–588
Pérez FEH, Mukhadiyev N, Xu X, Sow A, Lee B J, Sankaran R, Im HG (2018) Direct numerical simulations of reacting flows with detailed chemistry using many-core/GPU acceleration. Comput Fluids
Picano F, Breugem WP, Brandt L (2015) Turbulent channel flow of dense suspensions of neutrally buoyant spheres. J Fluid Mech 764:463–487
Pierce C, Moin P (2004) Progress-variable approach for large-eddy simulation of nonpremixed turbulent combustion. J Fluid Mech 504:73–97
Pope SB (2010) Self-conditioned fields for large-eddy simulations of turbulent flows. J Fluid Mech 652:139–169
Prosperetti A (2015) Life and death by boundary conditions. J Fluid Mech 768:1–4
Raynal L, Rayana FB, Royon-Lebeaud A (2009) Use of CFD for CO_2 absorbers optimum design: from local scale to large industrial scale. Energy Procedia 1(1):917–924
Reveillon J, Péra C, Bouali Z (2011) Examples of the potential of DNS for the understanding of reactive multiphase flows. Int J Spray Combust Dyn 3(1):63–92
Riber E, Moureau V, García M, Poinsot T, Simonin O (2009) Evaluation of numerical strategies for large eddy simulation of particulate two-phase recirculating flows. J Comput Phys 228(2):539–564
Rodriguez JM, Sahni O, Lahey RT Jr, Jansen KE (2013) A parallel adaptive mesh method for the numerical simulation of multiphase flows. Comput Fluids 87:115–131
Rohdin P, Moshfegh B (2007) Numerical predictions of indoor climate in large industrial premises. A comparison between different k–ε models supported by field measurements. Build Environ 42(11):3872–3882
Roy S, Acharya S (2012) Effect of impeller speed perturbation in a Rushton impeller stirred tank. J Fluids Eng 134(6):061104
Sagaut P (2006) Large eddy simulation for incompressible flows: an introduction. Springer Science & Business Media
Sbrizzai F, Lavezzo V, Verzicco R, Campolo M, Soldati A (2006) Direct numerical simulation of turbulent particle dispersion in an unbaffled stirred-tank reactor. Chem Eng Sci 61(9):2843–2851

Schnerr GH, Sauer J (2001) Physical and numerical modeling of unsteady cavitation dynamics. In: Fourth international conference on multiphase flow, vol 1. ICMF New Orleans

Shalf J, Dosanjh S, Morrison J (2010) Exascale computing technology challenges. In: International conference on high performance computing for computational science. Springer, Berlin, Heidelberg, pp 1–25

Shi J, Herø EH, Solsvik J, Jakobsen HA (2017) Experimental and numerical study on single droplet breakage in turbulent flow

Shin S, Chergui J, Juric D (2017) A solver for massively parallel direct numerical simulation of three-dimensional multiphase flows. J Mech Sci Technol 31(4):1739–1751

Slotnick J, Khodadoust A, Alonso J, Darmofal D, Gropp W, Lurie E, Mavriplis D (2014) CFD vision 2030 study: a path to revolutionary computational aerosciences

Smagorinsky J (1963) General circulation experiments with the primitive equations: I. The basic experiment. Monthly Weather Rev 91(3):99–164

Smith TM (2016) Full system scale simulations of reactor cores enabled by ExaScale computing (No. SAND2016-12354C). Sandia National Lab. (SNL-NM), Albuquerque, NM (United States)

Ström H, Sasic S (2015) Detailed simulations of the effect of particle deformation and particle-fluid heat transfer on particle-particle interactions in liquids. Procedia Eng 102:1563–1572

Sun X, Sakai M (2016) Numerical simulation of two-phase flows in complex geometries by using the volume-of-fluid/immersed-boundary method. Chem Eng Sci 139:221–240

Sweet J, Richter DH, Thain D (2018) GPU acceleration of Eulerian-Lagrangian particle-laden turbulent flow simulations. Int J Multiph Flow 99:437–445

Tang JC, Wang H, Bolla M, Wehrfritz A, Hawkes ER (2018) A DNS evaluation of mixing and evaporation models for TPDF modelling of nonpremixed spray flames. In: Proceedings of the combustion institute

Tryggvason G, Bunner B, Esmaeeli A, Juric D, Al-Rawahi N, Tauber W, … Jan YJ (2001) A front-tracking method for the computations of multiphase flow. J Comput Phys 169(2):708–759

Tryggvason G, Esmaeeli A, Lu J, Biswas S (2006) Direct numerical simulations of gas/liquid multiphase flows. Fluid Dyn Res 38(9):660–681

Tryggvason G, Thomas S, Lu J, Aboulhasanzadeh B (2010) Multiscale issues in DNS of multiphase flows. Acta Math Sci 30(2):551–562

Tryggvason G, Dabiri S, Aboulhasanzadeh B, Lu J (2013) Multiscale considerations in direct numerical simulations of multiphase flows. Phys Fluids 25(3):031302

Tyagi M, Roy S, Harvey Iii AD, Acharya S (2007) Simulation of laminar and turbulent impeller stirred tanks using immersed boundary method and large eddy simulation technique in multiblock curvilinear geometries. Chem Eng Sci 62(5):1351–1363

Ungerer T, Carpenter P (2018) Eurolab-4-HPC long-term vision on high-performance computing. arXiv preprint arXiv:1807.04521

Vincent S (2015) DNS of multiphase flows: energy conservation, vorticity, LES modeling and a priori filtering. In: Turbulence and interaction 2015 TI 2015

Wachs A (2011) Rising of 3D catalyst particles in a natural convection dominated flow by a parallel DNS method. Comput Chem Eng 35(11):2169–2185

Wang HY, McDonell VG, Sowa WA, Samuelsen GS (1993) Scaling of the two-phase flow downstream of a gas turbine combustor swirl cup: part i—mean quantities. J Eng Gas Turbines Power 115(3):453–460

Wang X, Shangguan Y, Onodera N, Kobayashi H, Aoki T (2014) Direct numerical simulation and large eddy simulation on a turbulent wall-bounded flow using lattice Boltzmann method and multiple GPUs. Math Probl Eng

Waters J, Carrington DB, Francois MM (2017) Modeling multiphase flow: spray breakup using volume of fluids in a dynamics LES FEM method. Numer Heat Transf Part B: Fundam 72(4):285–299

Wu JL, Xiao H, Paterson E (2018) Data-driven augmentation of turbulence models with physics-informed machine learning. arXiv preprint arXiv:1801.02762

Xie L, Luo ZH (2017) Multiscale computational fluid dynamics–population balance model coupled system of atom transfer radical suspension polymerization in stirred tank reactors. Ind Eng Chem Res 56(16):4690–4702

Xiong Q, Li B, Zhou G, Fang X, Xu J, Wang, J … Li J (2012) Large-scale DNS of gas–solid flows on Mole-8.5. Chem Eng Sci 71:422–430

Yu C, Wang Y, Huang C, Wu X, Du T (2017) Large eddy simulation of unsteady cavitating flow around a highly skewed propeller in nonuniform wake. J Fluids Eng 139(4):041302

Zeng L, Balachandar S, Fischer P, Najjar F (2008) Interactions of a stationary finite-sized particle with wall turbulence. J Fluid Mech 594:271–305

Zhuang Y, Liu D, Chen X, Ma J, Xiong J, Liang C (2018) Statistic model for predicting cluster movement in circulating fluidized bed (CFB) risers. J Taiwan Inst Chem Eng

Multiphase Flow its Application in Water Management and Harvesting in Fuel Cells

Tibin M. Thomas, Pallab Sinha Mahapatra, Raman Vedarajan and Ranjan Ganguly

Abstract Increased emission of carbon dioxide into the atmosphere from the fossil fuel-powered automobiles and power plants is one of the major sources of global warming. Using renewable and clean sources of energy as a fuel can control this. Among the other available alternatives, fuel cells have emerged as a promising source of clean energy due to their high efficiency, low/zero emission rate, modular design, and portability. Besides, fuel cells are capable of producing water as a by-product, making them an attractive option for potable water. Recent trends in the global automotive market show a strong trend of gravitating toward hydrogen fuel cell-powered automotive from the existing battery-operated automotive in the coming years. Owing to the high power density characteristics, polymer electrolyte membrane (PEMFC) fuel cell has been considered to be the most attractive one as the primary power source in fuel cell vehicle. Attaining slug-free drainage of water from the gas diffusion layers (GDLs) in PEMFC is one of the key challenges in their commercialization. Water management of hydrogen fuel cell can be optimized by extensive analysis of two-phase heat transfer phenomena like condensation and evaporation happening across GDL of the fuel cell. Excessive accumulation of water droplet on the GDL reduces the overall efficiency of the fuel cell. Thus, water removal from GDL is very important. Studies have shown that the quality of the water from a PEMFC meets the standard health requirements for drinking, indicating the importance of harvesting water from fuel cell exhausts as a sustainable drinking water source. The challenge in such water harvesting lies in achieving high condensation rate with minimum cooling energy penalty. This chapter reviews the fuel cell in general with a focus on multiphase phenomena and its use in water management in the GDL of fuel cells

T. M. Thomas · P. S. Mahapatra (✉)
Department of Mechanical Engineering, IIT Madras, Chennai 600036, India
e-mail: pallab@iitm.ac.in

R. Vedarajan
Centre for Fuel Cell Technology, International Advanced Research Centre for Powder Metallurgy, Phase I - 2nd Floor, IITM Research Park, Taramani, Chennai 600113, India

R. Ganguly
Department of Power Engineering, Jadavpur University, Kolkata 700098, India

K. Saha et al. (eds.), *Two-Phase Flow for Automotive and Power Generation Sectors*, Energy, Environment, and Sustainability,
https://doi.org/10.1007/978-981-13-3256-2_10

and harvesting drinking water from fuel cell exhaust. The background fundamentals are provided, and the state of art is discussed. Finally, the future perspective of water management and harvesting in fuel cells is provided in a larger backdrop of global energy–water nexus.

Keywords Water harvesting · PEM fuel cell · Water management
Self humidification

Nomenclature

$\dot{m}_{\text{air}}$	Mass flow rate of air [kg/s]
$\dot{m}^{w}_{\text{evap}}$	Evaporation rate of water [kg/s]
$\dot{m}^{w}_{\text{prod}}$	Water production rate [kg/s]
$\dot{m}_{\text{v}-\text{l}}$	Phase change mass transfer rate [kg/s]
A_{c}	Cross-sectional area of flow channel [m^2]
A_{m}	Effective membrane cross-sectional area [m^2]
A_{FC}	Active area of the fuel cell [m^2]
B	One half of the flow-field channel height [m]
c	Chord length [m]
$\text{C}_{\text{H}_2\text{O}}$	Water concentration in the membrane [kmol/m^3]
C_{mem}	Concentration of the membrane [kmol/m^3]
d	Pore diameter of GDL/CL [m]
D_i	Bulk diffusion constant [m^2/s]
D_i^{B}	Binary diffusion constant [m^2/s]
D_i^{eff}	The effective diffusion coefficient [m^2/s]
D_i^{Kn}	Knudsen diffusion constant [m^2/s]
D_{m}	Self-diffusion coefficient [m^2/s]
F	Faraday constant [C/kmol]
F_{D}	Force exerted due to drag [N]
F_{P}	Force exerted due to pressure [N]
F_{S}	Force exerted due to surface tension [N]
h	Height of the generated droplet [m]
i	Current density [Amp/m^2]
I_{ion}	Ionic current density [Amp/m^2]
$J^{\text{m}}_{\text{diff}}$	Mass flux due to diffusion [$\text{kg/m}^2\text{s}$]
$J^{\text{m}}_{\text{EOD}}$	Mass flux due to EOD [$\text{kg/m}^2\text{s}$]
k_{p}	Proton conductivity [s/m]
Kn	Knudsen number
$\delta^{\text{lattice}}_{\text{liq}}$	Average distance between liquid water lattices [m]
M_{air}	Molecular weight of air [kg/kmol]
$M_{\text{H}_2\text{O}}$	Molecular weight of water [kg/kmol]
m_{mem}	Mass of the dry membrane [kg]
N	Number of channels

n_d	EOD coefficient
$n_{SO_3^-}$	Number of moles of SO_3^- in the membrane [kmol]
P	Pressure [Pa]
P_{sat}	Saturation pressure at a temperature of T (K) [Pa]
P_{vap}	Vapor pressure at a temperature of T (K) [Pa]
Q_w	The amount of water generated from PEM fuel cell [L/s]
Q_w^{net}	Net amount of water collected from the fuel cell [L/s]
R	Universal gas constant [8.314 KJ/kmol K]
r	Radius of curvature of the spherical cap [m]
RH_{in}^c	Inlet relative humidity at cathode channel [%]
s_{lq}	Volume fraction of liquid water
T	Temperature [K]
U	Air velocity at the cathode inlet [m/s]
U_g^S, U_L^S	Superficial air/water velocity [m/s]
W	Power output of the fuel cell [kW]
x	Condenser efficiency

Greek Symbols

α	Ratio of net water flux to proton flux
l_g	Average distance between two successive collisions [m]
ε	Porosity
λ	Water content
μ	Viscosity of air [kg/ms]
ρ_{air}	Density of the air [kg/m^3]
ρ_{H_2O}	Density of water [kg/m^3]
ρ_{mem}	Density of the membrane [kg/m^3]
σ	Surface tension [N/m]
θ_A	Advancing contact angle [°]
θ_R	Receding contact angle [°]
θ	Static contact angle of the surface [°]
ζ	Flow stoichiometry of air

Abbreviation

Ca	Capillary number
CAH	Contact angle hysteresis
CL	Catalyst layer
EOD	Electro-osmotic drag
GDL	Gas diffusion layer
MCFC	Molten carbonate fuel cell
MCL	Maximum contaminants level
MEA	Membrane electrode assembly
MPL	Microporous layer
ORR	Oxygen reduction reaction

PEM	Polymer electrolyte membrane
PEMFC	Polymer electrolyte membrane fuel cell
PRR	Proton reduction reaction
PTFE	Poly-tetra-fluoro-ethylene
TDS	Total dissolved solids
USEPA	United States Environmental Protection Agency
WER	Water energy ratio

10.1 Introduction

Environment-related issues such as global warming, climate change, and air pollution have become more critical in the present decade because of the increased emission of carbon dioxide from the fossil-fuel-fired engines. An assessment of the various fuels over the production to consumption stages in a different type of automobiles is depicted in Fig. 10.1. In gasoline vehicles, greenhouse gases are emitted at the production and consumption stages. However, for electrical vehicles, greenhouse gases are emitted during the electricity generation in the power plant and zero emission at the consumption phase. Fuel cell vehicles are globally accepted as the most environment-friendly automobile with zero greenhouse gas emission at any stages of the fuel life cycle subjected to the hydrogen production method. Presently, the majority of the leading automobile manufacturers have a fully developed model of fuel-cell-powered vehicles. Majority of the automotive manufacturers are active in further research and development of their own fuel cell vehicle by collaboration with other leading manufacturers and government agencies for a sustainable environment. In the economic perspective, many hurdles need to be overcome to bring the fuel cell vehicles into the commercial market. The main challenges associated with the fuel-cell-powered vehicles are lack of infrastructure for fuel storage and involvement of rare-earth elements (like catalyst material, components of electric propulsion systems) in the fuel cell systems. All developed countries are spending a substantial amount for the infrastructure development of fuel refilling station as like gasoline stations reflect the global interest on zero-emission fuel cell vehicles in the future.

Fuel cells are the promising clean energy electrochemical devices to convert the chemical energy stored in the hydrogen fuel into electricity. The discovery of fuel cell is credited with Sir William Robert Grove in 1839 (Andújar and Segura 2009). Later under the leadership of Thomas Grubb and Leonard Niedrach, General Electric (GE) has developed fuel cell technology for Gemini space program of NASA in 1959 (Andújar and Segura 2009). The fuel cells are categorized into different types depending on the electrolyte used in the cell. The most important types of fuel cells are phosphoric acid fuel cells (PAFC), alkaline fuel cells (AFC), polymer electrolyte membrane fuel cells (PEMFC), direct methanol fuel cells (DMFC), solid oxide fuel cells (SOFC), and molten carbonate fuel cells (MCFC). According to the specific application, any suitable type of fuel cell can be selected. In stationary power generation applications, the widely used fuel cells are MCFC, SOFC, and PAFC. However,

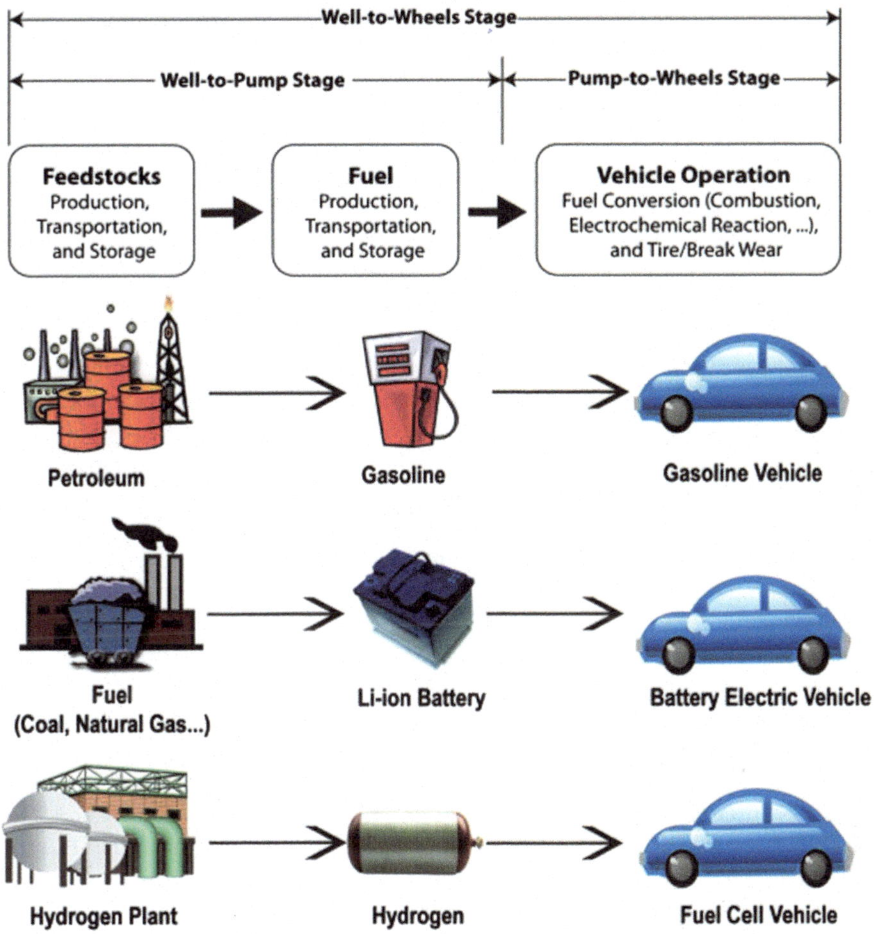

Fig. 10.1 Life cycle assessment of the various fuel in a different type of automobiles—gasoline vehicles, battery-powered electrical vehicle, fuel cell vehicle (reproduced with permission from Hwang 2013)

the suitable fuel cells for transportation and portable application are PEMFC, DMFC, and AFC. Out of all fuel cells, majority of the research and developments focus on PEMFC. PEM fuel cell is getting great attention since it is an ideal alternative power source for battery-powered electric vehicles. The important features of PEM fuel cells such as high power density, zero-emission characteristics, high efficiency, and low operating temperature are the main reasons for selecting them in automobile application. But the commercialization of this technology in the global automobile market becomes viable only after the further improvements in terms of durability, cold-start capability, cruising stage, and cost reduction.

Dwindling reserve of freshwater is a major threat to this planet despite the fact that around 71% of earth surface is covered with water. Due to the industrialization and environmental pollution, we are contaminating the available freshwater indiscriminately. This will affect the quality of potable water from the natural resources on earth, and drinking water is more likely to become dearer in future decades. It has been estimated that one-third of the global population is affected by the scarcity of freshwater. For conserving global water supplies, there is a need to explore alternative and sustainable technologies for the collection of freshwater. Studies show that the quality of the water produced from the PEM fuel cell is relatively pure and can be used for drinking water purposes with a proper purification system (Hristovski et al. 2009; Tibaquirá et al. 2011a, b). Therefore, the water generated from the fuel cell vehicle during operation is separately stored in a tank after purification and it can be used for the drinking purposes of the travelers.

Liquid water comes in the fuel cell through chemical reactions and condensation of water vapor. Water management of the fuel cell is one of the biggest challenges in the commercialization of the fuel cell. Liquid water and gases create a two-phase flow situation in the flow channel of the fuel cell. In case of excess water accumulation in the cathode side GDL, overall efficiency of the fuel cell decreases. Therefore, proper drainage of water is required from the GDL. Increasing the gas temperature (Wilkinson et al. 1998), applied vibration (Mughal and Li 2006), increasing pressure drop (Voss et al. 1995), proper flow path design (Anderson et al. 2010) generally helps in water management in the fuel cell. Both experimental and numerical investigations have been performed for the proper design of the flow channels. Multiphase analysis for the water management of the fuel cell is required to understand the underlying competition between the gravity, surface tension, and shear forces. Surface wettability also has a major role in water removal from the surfaces. The wettability of the surfaces can be modified by applying different coatings or by chemically changing the property of the surfaces. By applying PTFE coating, Owejan et al. (2007) reported that the water accumulation decreases inside the channels of PEMFC. Earlier studies show that the choice of suitable wettable surfaces for water removal is debatable. Both hydrophilic and hydrophobic surfaces have merits and demerits (Lu et al. 2011).

The objective of this chapter is to present the multiphase phenomena associated with the polymer electrolyte membrane fuel cell and its application in advanced water management and harvesting strategies. This chapter is structured as follows. The basic introduction and operating principle of the various components in the PEM fuel cell are presented in Sect. 10.2, and the water transport mechanism of each component is explained in Sect. 10.3. Section 10.4 illustrates the two-phase flow patterns in the flow-field channels and presents the overall overview of phase change heat transfer phenomena occurred during the fuel cell operation. Water management issues and strategies considered for the optimum design of fuel cell systems are summarized in Sect. 10.5. The feasibility of fuel-cell-generated water as a drinking water solution is reviewed in Sect. 10.6.

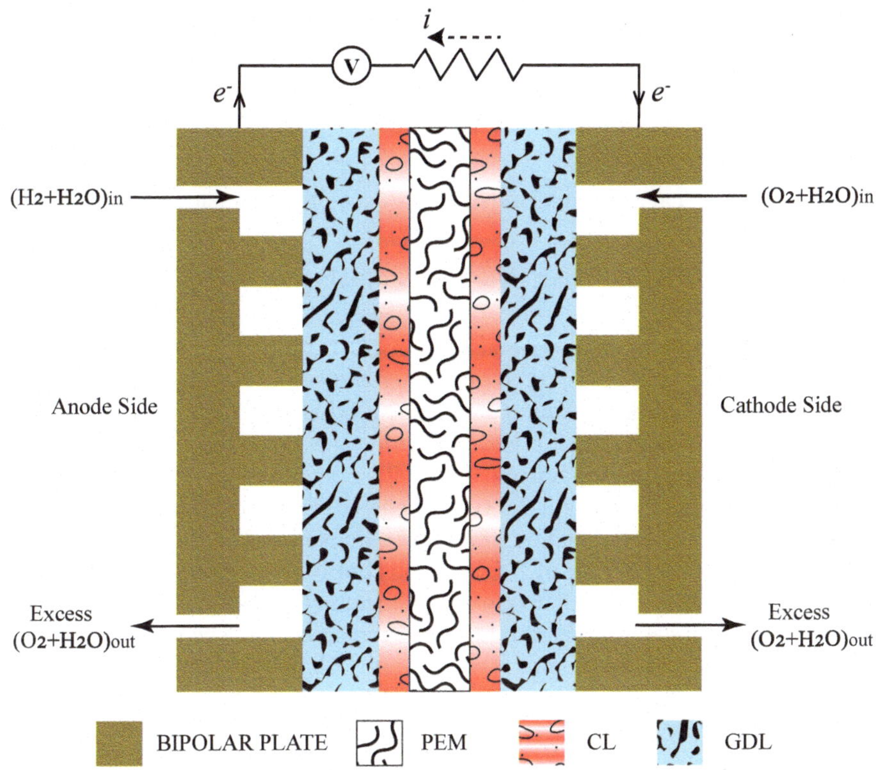

Fig. 10.2 Cross-sectional view of a single-cell PEMFC (membrane electrode assembly with two bipolar plates on both the sides)

10.2 Construction and Operating Principle of PEMFC

Fuel cells may be looked up as a promising zero-emission electrochemical devices capable of converting the chemical energy stored in the hydrogen to electricity and water. The basic design of PEMFC is composed of one membrane electrode assembly (MEA) and two bipolar plates placed on both sides of MEA as shown in Fig. 10.2. The typical MEA is made by binding five thinner porous structured layers. It consists of one polymer electrolyte membrane (PEM), catalyst layer (CL) attached to the sides of PEM, and two gas diffusion layer (GDL) attached above the CL. As illustrated in Fig. 10.2, PEMFC has cathode and anode side on either side of the membrane. In PEM fuel cell, hydrogen (H_2) is given as a fuel at the anode side and oxygen (O_2) is given as an oxidant at cathode side through the flow field of the bipolar plate.

At the anode side CL surface, hydrogen splits into electron and proton and the corresponding hydrogen reduction reaction (HOR) is expressed as,

$$H_2 \rightarrow 2H^+ + 2e^- \tag{10.1}$$

At the cathode side, water and electricity produced from oxygen reduction reaction (ORR) are given by,

$$\frac{1}{2}O_2 + 2H^+ + 2e^- \rightarrow H_2O + \text{Electrical Energy} + \text{Heat} \tag{10.2}$$

The overall chemical reaction in a hydrogen fuel cell is summarized as,

$$H_2 + \frac{1}{2}O_2 \rightarrow H_2O + \text{Electrical Energy} + \text{Heat} \tag{10.3}$$

Polymer electrolyte membrane acts as an electrolyte of the fuel cell. It allows the transport of conducting proton and water across the membrane and acts as a barrier for gases. Presently, Nafion membrane is widely used as a benchmark membrane in all kinds of PEMFC. It is developed in the 1970s, by the modification of Teflon (Grot 1976). The structure of Nafion membrane consists of flexible perfluorocarbon, hydrophobic poly-tetra-fluoro-ethylene (PTFE) backbone, and hydrophilic $H^+SO_3^-$ ions. Commercial Nafion membranes are Nafion-112, Nafion-115, Nafion-117, Nafion-1110 with a sizes of 2, 5, 7, 10 mil (1 mil=25.4 μm), respectively. Nafion membrane can work with a relatively higher operating temperature of around 190 °C without any chemical degradation. PEM membrane needs to be operated at a well-hydrated condition for better proton conductivity. The water content (λ) in a Nafion membrane is correlated as (Li 2005),

$$\lambda = \frac{m_{\text{mem}}}{n_{SO_3^-}\,\rho_{\text{mem}}} C_{H_2O} \tag{10.4}$$

where m_{mem} is the mass of the dry membrane, $n_{SO_3^-}$ is the number of moles of SO_3^- in the membrane, ρ_{mem} is the density of the membrane, and C_{H_2O} is the concentration of water in the membrane.

The catalyst layer is fabricated with platinum-loaded carbon particles and assembled between the GDL and PEM membrane. On the anode side, CL dissociates the hydrogen into electrons and protons and it allows the transport of proton through the porous structures and restricts the flow of electrons. On the cathode side CL, oxygen combines with electrons and protons and produces electrical energy and water. The porosity value of the CL is ranging between 0.2 and 0.4, and the thickness of the typical CL is in the order of 0.01 mm. The GDL is usually a carbon paper or carbon cloth with porous structures, and it acts as the electrical connection between CL and the bipolar plate. It distributes the reactant evenly along with the bipolar plate and removes water from GDL/CL, GDL–bipolar plate interfaces. The porosity value of the GDL is higher than that of CL and in the range of more than 0.5 (Jiao and Li 2011).

Graphite is generally used for the construction of bipolar plate because of its exceptional chemical stability features. Aluminum, titanium, stainless steel, nickel, polymer composites are also used as an alternative material for the bipolar plate.

Hence, a proper selection of bipolar plate material and design of flow field based on the application can enhance the performance of the fuel cells. The important functions of bipolar plates in fuel cells are

1. Even distribution of fuel (anode side) and oxidant (cathode side) in the flow-field channels.
2. To facilitate heat and water management.
3. To separate individual cells in a fuel cell stack.
4. To carry the generated electrical energy to an external circuit.

10.3 Water Transport Mechanism in a PEMFC

The transport mechanism of water in a PEM fuel cell is varying with the local operating conditions and different materials of each of its components. In normal operating conditions, water exists either in liquid or vapor state. There is also a possibility of solidification in PEMFC components which are frequently observed during the starting of automobiles in winter season (Oszcipok et al. 2005). The water transport mechanism becomes complex at this operating condition which is beyond the scope of this chapter. A detailed explanation of the water transport mechanism in the individual component of a PEM fuel cell at normal operating conditions is presented in the following subsections.

10.3.1 Polymer Electrolyte Membrane (PEM)

Optimized performance of the fuel cell is normally achieved by maintaining a fully hydrated membrane at all the operating conditions. A portion of the water content at the anode side is transported along with the proton transport. This proton transport depends on the temperature and amount of water content at the membrane anode side. The chemical structure of a Nafion membrane is composed of polymer side chain with sulfuric acid for the help of proton transport whereas hydrophobic poly-tetra-fluoro-ethylene back bone for the mechanical stability. High proton conductivity exhibits with the large concentration of SO_3^- ions. The proton transport from one polymer chain to the adjacent one is shown in Fig. 10.3. Each of the polymer chains attracts one H^+ ions. The void space between the two polymer chain is absorbed by water molecule, and this volume is increased with the increase in the anode side water content. The bond between H^+ and SO_3^- becomes weak in the presence of absorbed water in the void volume. Water molecules are strongly attracted to SO_3^-, and it helps in H^+ jumping from one active site to another. The vibration of polymer chain due to the attraction between neighboring polymer chains also promotes the proton transport in the membrane. A Nafion membrane at a well-hydrated condition contains nearly 20 water molecules for each SO_3^- ions. A relationship between proton conductivity (k_p) with water content (λ) and temperature (T) is given by Springer et al. (1991),

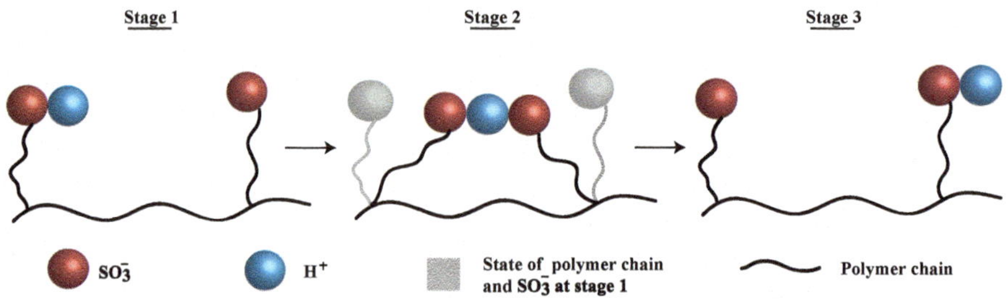

Fig. 10.3 Various stages of proton transport mechanism from one polymer chain to adjacent chain

$$k_p = (0.5139\lambda - 0.326)\exp\left(1268\left[\frac{1}{303.15} - \frac{1}{T}\right]\right) \tag{10.5}$$

In Nafion membrane, proton conductivity varies linearly with the amount of water content. It is observed that proton conductivity decreases when water content value becomes lower than 2. The optimum range of water content value for Nafion membrane is 2–5. A limiting case is analyzed when no water is available in the membrane resulting in a low proton conductivity due to the direct proton transport between two adjacent charged sites. The protons are transported over the membrane in the form of hydrogen–water ions like H_3O^+, $H_5O_2^+$, or something similar. Transport of water along with the proton transport in the pores of the membrane is termed as electro-osmotic drag (EOD). In the absence of pressure and concentration gradient, EOD coefficient is defined as the ratio between the number of moles of water to the number of moles of proton crossing through the electrolyte membrane. The mass flux ($J_{\mathrm{EOD}}^{\mathrm{m}}$) due to EOD is expressed as

$$J_{\mathrm{EOD}}^{\mathrm{m}} = -n_d \frac{I_{\mathrm{ion}}}{F} \tag{10.6}$$

where F is the Faraday constant, I_{ion} is ionic current density, and n_d is the EOD coefficient. Correlations have been developed from experimental measurements for the calculation of EOD coefficient. Two widely used correlations for the computation of EOD coefficient for numerical simulations are (Springer et al. 1991; Zawodzinski et al. 1995)

$$n_d = \frac{2.5\lambda}{22} \tag{10.7}$$

and

$$n_d = \begin{cases} 1 & ; \quad \lambda \leq 14 \\ 0.1875\lambda - 1.625 & ; \quad \text{otherwise} \end{cases} \tag{10.8}$$

In fuel cell, water is produced on the cathode side. So the larger value of water concentration at cathode side creates a concentration gradient which transports the water to the anode side by slow molecular diffusion through the void spaces in the membrane. The flux due to mass transfer ($J_{\text{diff}}^{\text{m}}$) of water over the membrane due to the molecular diffusion is given as

$$J_{\text{diff}}^{\text{m}} = -D_{\text{m}} \nabla C_{\text{mem}} \tag{10.9}$$

where D_{m} is the membrane self-diffusion coefficient and C_{mem} is the concentration of the membrane. The negative sign indicates that direction of mass transfer is always in the direction of decreasing concentration. The self-diffusion coefficient depends on the temperature and water content of the membrane. Zawodzinski et al. (1991) correlated the value of self-diffusion coefficient as,

$$D_{\text{m}} = \begin{cases} 2.692661843 \times 10^{-10} & ; \quad \lambda \le 2 \\ 10^{-10} \exp\left(2416\left[\frac{1}{303} - \frac{1}{T}\right]\right)\left(0.87[3-\lambda] + 2.95[\lambda - 2]\right) & ; \quad 2 < \lambda \le 3 \\ 10^{-10} \exp\left(2416\left[\frac{1}{303} - \frac{1}{T}\right]\right)\left(2.95[4-\lambda] + 1.642454[\lambda - 3]\right) & ; \quad 3 < \lambda \le 4 \\ 10^{-10} \exp\left(2416\left[\frac{1}{303} - \frac{1}{T}\right]\right)(2.563 - 0.33\lambda + 0.0264\lambda^2 - 0.000671\lambda^3) & ; \quad \lambda > 4 \end{cases} \tag{10.10}$$

Motupally et al. (2000) also correlated the self-diffusion value from the same experimental data by another method which is given by,

$$D_{\text{m}} = \begin{cases} 3.1 \times 10^{-7}\lambda\left(\exp[0.28\lambda - 1]\right)\exp\left(\frac{-2346}{T}\right) & ; \quad 0 < \lambda < 3 \\ 4.17 \times 10^{-8}\lambda\left(161\exp[-\lambda] + 1\right)\exp\left(\frac{-2346}{T}\right) & ; \quad 3 \le \lambda < 17 \end{cases} \tag{10.11}$$

The above two expressions are widely used in the numerical modeling of PEMFC although there is a significant difference between both. The maximum value of diffusion coefficient is observed when the amount of water content (λ) becomes nearly 3 in both correlations (Jiao and Li 2011).

10.3.2 Gas Diffusion Layer (GDL)

Gas diffusion layer is comprised of nanometer-scale porous and tortuous structures through which flow occurs. Based on the flow channel design, the water transport mechanism in a GDL can either be convection-dominated, diffusion-dominated, or mixed. Higher flow rates and high-pressure inlet conditions of the reactant gases help in the effective removal of water from the GDL surface. Presence of concentration gradient at the flow channel and reactant gases also ensures an optimum water removal at the cathode side. The flow characteristics in the GDL strongly depend on the flow channel design. The various flow channel designs (Wood et al. 1998) widely adopted in the PEMFC fabrication are detailed in Sect. 10.5.2.2.

The movement of the gaseous molecule in a porous/tortuous structure in a GDL or CL is restricted by pore walls resulting in a decrease in diffusion coefficient. The effective diffusion coefficient (D_i^{eff}) in GDL/CL is correlated by Bruggeman (1935) which is given as

$$D_i^{\text{eff}} = D_i \varepsilon^{1.5} \tag{10.12}$$

where D_i is the bulk diffusion constant of gaseous species i, and ε is the porosity value of GDL/CL. The force arising out of the surface tension also plays a major role in the transport of water in GDL/CL. The pressure difference (ΔP) across the interface between two phases due to the surface tension (σ) is related by the Young–Laplace equation

$$\Delta P = \frac{2\sigma}{r} = \frac{4\sigma \cos\theta}{d} \tag{10.13}$$

where r is the radius of curvature of the liquid phase (based on spherical cap assumption), d is the GDL/CL pore diameter, and θ is the static contact angle. The contact angle is defined as the angle between the solid–liquid and liquid–vapor contact line at triple line point. It is considered to be a measure of surface wettability. If θ is less than 90°, the surface is considered to be hydrophilic, while for θ greater than 90°, the surface is hydrophobic, as indicated in Fig. 10.4. The liquid transport by convection mode is advanced due to the higher flow velocity arising from the higher pressure differential across the interface. Pressure difference at the interface increases with the decrease in the GDL/CL pore diameter. The mobility of the water drop also depends on the surface wettability. Water mobility of a hydrophobic surface is higher than that offered by a hydrophilic due to the fact that hydrophobic surfaces, in general, exhibit a lower contact angle hysteresis (CAH) (de Gennes et al. 2004). Generally, GDL/CL is treated with PTFE to promote the surface hydrophobicity and, therefore, droplet mobility.

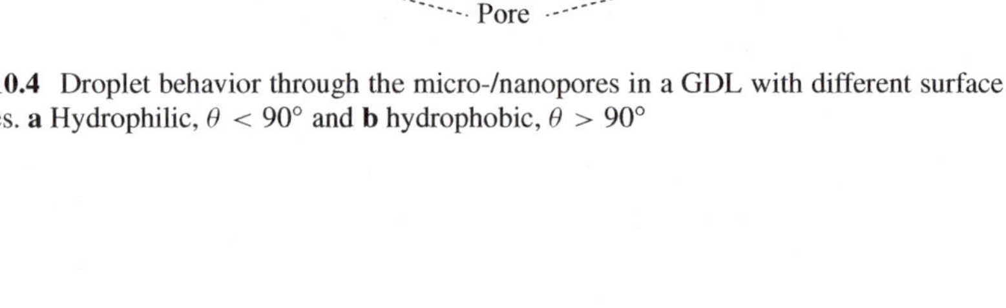

Fig. 10.4 Droplet behavior through the micro-/nanopores in a GDL with different surface wettabilities. **a** Hydrophilic, $\theta < 90°$ and **b** hydrophobic, $\theta > 90°$

10.3.3 Catalyst Layer (CL)

The most intricate form of water transport occurs in the porous structures of catalyst layer. The effective pore size of CL is much lower than GDL and in the order of few nm to 1000 nm (Uchida et al. 1996). The electro-oxidation reaction sites in the catalyst layer consist of embedded platinum and carbon particles with ionomer (polymer electrolyte) part. The function of embedded carbon and platinum particles is to promote the catalyst reaction, whereas ionomer is to transport the proton through PEM as explained in Sect. 10.3.1. Water is absorbed in the anode side interface (between PEM and CL) and released at the cathode side interface by ionomer. Water supplied at the cathode side is transported through the pores of the catalyst layer in either liquid or vapor state depending on the operating conditions. By achieving a proper balance in the porosity and number of reaction sites makes the membrane at a fully hydrated condition with good proton conductivity. The pores in the CL allow a small percentage of reactant gases to cross the membrane. These gases are transported through the membrane by diffusion and wasted without electrochemical reaction. This decreases the total efficiency of the cell. The self-humidifying membrane is capable to suppress the gas cross-over effectively and detailed in Sect. 10.5.2.4. Knudsen number associated with the water transport in liquid and vapor phases ($\mathrm{Kn}^{\mathrm{liq}}$ and Kn^{g}) through the nanopores of the CL is expressed as,

$$\mathrm{Kn}^{\mathrm{liq}} = \frac{\delta_{\mathrm{liq}}^{\mathrm{lattice}}}{d} \tag{10.14}$$

$$\mathrm{Kn}^{\mathrm{g}} = \frac{l_{\mathrm{g}}}{d} \tag{10.15}$$

where $\delta_{\mathrm{liq}}^{\mathrm{lattice}}$ is the average distance between the liquid water lattices, l_{g} is the average distance between two successive collision of gaseous molecules, and d is the pore diameter. Along with gas diffusion, catalyst layer exhibits Knudsen diffusion (collision of gas molecules with pores wall). The expressions for binary diffusion coefficient (collision between gaseous molecules, D_i^{B}) and Knudsen diffusion (D_i^{Kn}) are given by Wang (2004); Bird (2002)

$$D_i^{\mathrm{B}} = D_i^{\mathrm{B,ref}} \left(\frac{T}{T_{\mathrm{ref}}}\right)^{1.5} \left(\frac{P_{\mathrm{ref}}}{P}\right) \tag{10.16}$$

$$D_i^{\mathrm{Kn}} = \frac{1}{3}\left(\frac{8RT}{\pi M_i}\right)^{0.5} d \tag{10.17}$$

where $D_i^{\mathrm{B,ref}}$ is the binary diffusion coefficient at a pressure of P_{ref} and a temperature of T_{ref}, P and T are the local pressure and temperature, R is the universal gas constant, and M_i is the molecular weight of the gas species i. Knudsen diffusion coefficient is significant only in CL layer, so total diffusion coefficient on each component is summarized as

$$D_i^{\text{tot}} = \begin{cases} D_i^{\text{B}} & ; \quad \text{in GDL and flow channel} \\ \left(\frac{1}{D_i^{\text{B}}} + \frac{1}{D_i^{\text{Kn}}} \right)^{-1} & ; \quad \text{in CL} \end{cases} \tag{10.18}$$

At 1 bar, 80 °C operating conditions with a CL pore size of 10 nm, Knudsen number is calculated as 0.03 and 8.2 for the liquid and gaseous phases, respectively (Jiao and Li 2011). The continuum models are only applicable for the case of Knudsen number value of less than 10^{-3}. Thus, the macroscopic approach is not valid in the catalyst layer because of the higher value of Knudsen number. This problem makes the numerical modeling of the CL more complex. A hybrid model with molecular dynamics and continuum approach can be a better solution to the numerical modeling of the catalyst layer.

10.4 Overview of Multiphase Flows in a PEMFC

The flow of water and O_2 in the flow-field channels of PEM fuel cell is an extremely complex phenomenon as compared with the conventional multiphase channel flow. The main reason for such complexity is the introduction of water and consumption of reactant (O_2) at multiple positions along the flow channel through the porous GDL surface. This creates a spatiotemporal variation of water buildup on the GDL. Based on the surface properties, droplets of different shapes form on the GDL. After coalesce with the adjacent droplets, the droplet size becomes larger than the capillary length scale of water and detaches from the GDL. Detachment of these drops from GDL also occurs due to the imparted pressure from reactant gases. Water exists as either vapor or liquid state based on the local hot spots in the flow field arising from the nonuniform temperature distribution.

10.4.1 Mechanism of Droplet Generation and Detachment from GDL

Various groups have investigated the dynamics of the droplet on a GDL surface. Kumbur et al. (2006) have successfully developed a mathematical model to analyze the droplet behavior on a GDL, and the results were in good agreement with the conducted experiments. The static force, balance analysis of a single droplet sitting on a GDL is given by,

$$F_{\text{S}} + F_{\text{P}} + F_{\text{D}} = 0 \tag{10.19}$$

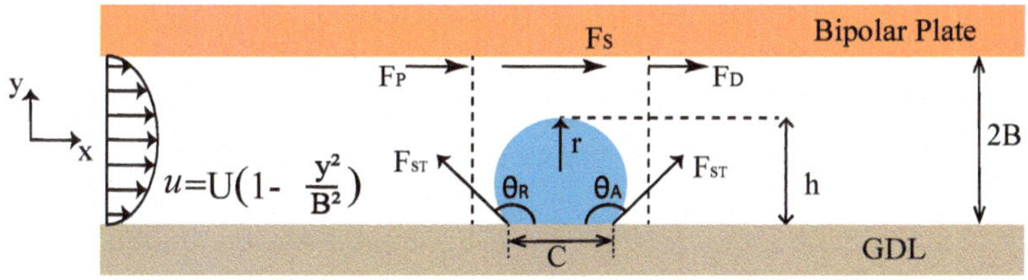

Fig. 10.5 Cross-sectional view of the flow field parallel to the membrane plane; the state of a growing droplet just before the shearing effect due to the drag force from the high-pressure oxidant supply

where F_S is the exerted shear force due to the wall, F_P is the pressure force exerted from the pressure gradient, and F_D is the drag force exerted on the droplet, and it is equal to the adhesion force of the GDL just before the detachment of the droplet. The droplet will detach from GDL when the drag force overtakes the adhesive force between pore wall and the droplet. The forces in Eq. 10.19 is modified by incorporating the relevant parameters such as flow velocity (U), channel height ($2B$), droplet height (h), contact angle hysteresis (Δ), and chord length (c) as shown in Fig. 10.5. The expressions for each forces acting on droplet from Kumbur et al. (2006) analysis are given by,

$$F_S = \frac{12\mu BUh^2}{(B - h/2)^2(1 - \cos\theta_A)^2} \tag{10.20}$$

$$F_P = \frac{24\mu B^2Uh^2}{(B - h/2)^3(1 - \cos\theta_A)^2} \tag{10.21}$$

$$F_D = -\frac{\pi}{2}c\sigma\left(\frac{[\sin(\Delta - \theta_A) - \sin\theta_A]}{(\Delta - \pi)} + \frac{[\sin(\Delta - \theta_A) - \sin\theta_A]}{(\Delta + \pi)}\right) \tag{10.22}$$

where Δ is the difference between advancing (θ_A) and receding (θ_R) contact angles, and μ is the viscosity of air. This analytical model concluded that for a constant width and constant drop size, lowering the channel height improves the rate of droplet removal. A hydrophilic GDL surface on the flow-field results in a film generation on the GDL surface. This film acts as a resistance to the smooth supply of reactant for the electrochemical reaction on CL surface, leading to a poor performance. Hence, a hydrophobic surface is preferred on the GDL surface. But the flow-field walls of the bipolar plate is to be hydrophilic for the faster water removal rate.

The reactant gas velocity causing the detachment phenomena on the GDL is inversely related to the droplet size (Ous and Arcoumanis 2007). The detachment of smaller droplets is difficult due to the large value of adhesive force between GDL pore wall and the droplet. Temperature also plays an indirect role on detachment phenomena, since the temperature is inversely proportional to the surface tension. As temperature increases, surface tension force of the droplets decreases leading to quicker droplet detachment. The effect of contact angle hysteresis is also important for droplet detachment. The reactant gas flow imparting a deformation on the droplet and causing the formation of two contact angle is called advancing and receding contact angles. The amount of deformation is a strong function of capillary number (Ca), which is the ratio of viscous to surface tension force. For larger Ca, the deformation increases with an increase in Ca (Shirani and Masoomi 2008). The effect of contact angle hysteresis on droplet detachment was studied by Fang et al. (2008) using multiple three-dimensional simulation with volume-of-fluid (VOF) approach in a microchannel of a width of 500 μm and depth of 45 μm. In their computational model, water was supplied from the bottom side of the channel and air was supplied perpendicular to the water flow direction at the inlet section of the microchannel. Parametric studies were conducted in this work by varying the air velocity ranging from 13.38 to 32 m/s and water velocity of 0.09 and 0.11 m/s. According to the results obtained from their numerical simulations, it shows that contact angle hysteresis helps in the droplet detachment and slug elongation process.

10.4.2 Flow Patterns

The characteristics of two-phase flow in a PEM fuel cell are analyzed from the flow patterns observed in the flow-field channels. The visualization of flow patterns can be done by different methods such as neutron radiography, magnetic resonance imaging, and nuclear magnetic resonance. These methods are too expensive, and these do not resolve all the droplets as an individual one. The easiest method to understand the flow behavior on a GDL is to construct the flow channels with an optically accessible material for the visualization.

Hussaini and Wang (2009) conducted an in situ fuel cell experiment with flow visualization. They used a transparent Lexan plate above the flow field, and high-resolution images were recorded for analyzing the complex two-phase flow patterns in the flow-field channels. Multiple experiments were conducted at a cell temperature of 80 °C with current densities of 0.2, 0.5, 0.8 A/cm^2 and gas humidification of 26, 42, and 66%. The different types of flow patterns observed in the flow channels of PEM fuel cell are shown in Fig. 10.6 (Hussaini and Wang 2009).

The characteristics of different flow regimes are briefly explained below.

1. **Single-phase flow**: Water droplets are not observed in this regimes because of the quick evaporation of the generated drops on the GDL surface.
2. **Droplet flow**: Water droplets adhere to the hydrophobic GDL surface due to the surface tension force. Flowing reactant gases cause the droplet shearing and distortion.
3. **Film flow**: The mass influx of the water at the further downstream of the flow channel increases with the generation of water. A liquid water film is formed at the hydrophilic channel walls, and it grows with the coalescence of the tiny

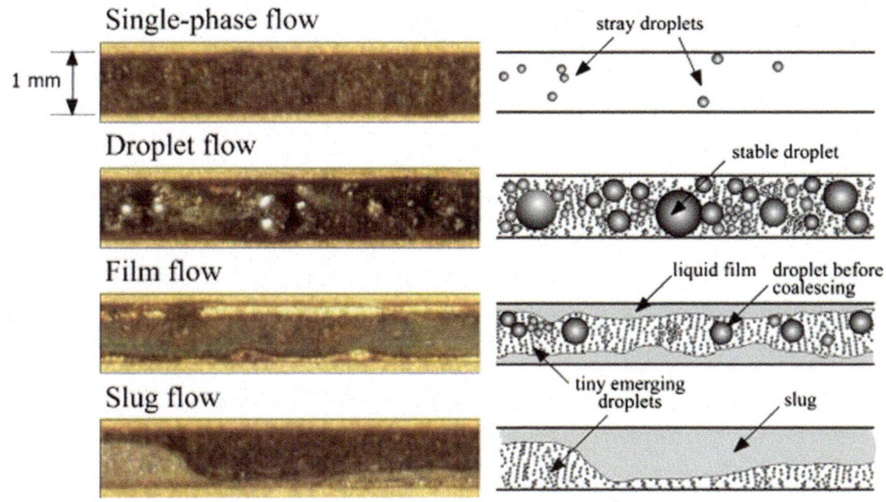

Fig. 10.6 Two-phase flow patterns in a rectangular-shaped flow field of PEM fuel cell (reproduced with permission from Hussaini and Wang 2009)

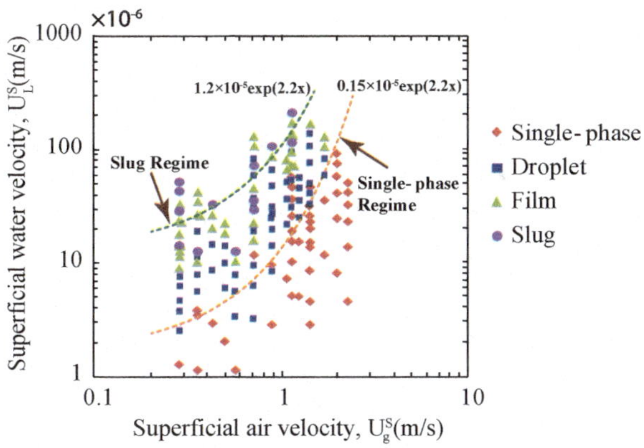

Fig. 10.7 The flow regime maps for different water and air velocities for a rectangular shaped flow field (reproduced with permission from Hussaini and Wang 2009)

droplets generated on the GDL surface. A wavy-shaped interface is formed due to the continuous influx and coalescence of tiny droplets with the adhered film.

4. **Slug flow**: Further film growth causes the formation of the multiple slow-moving slugs of different shapes and sizes. Larger slugs can even form liquid bridge between the walls and block the reactant gases flow, resulting in a poor performance of the cell.

The two-phase flow patterns in a PEMFC are strongly dependant on the operating conditions. Figure 10.7 shows the flow map of two-phase flow on a hydrophobic GDL from Hussaini and Wang (2009). Flow map of superficial phase velocities shows the conditions of occurrence of the four different regimes. From mass balance, the superficial gas (U_g^S) and liquid (U_L^S) velocities on a GDL are given by,

$$U_g^S = \frac{1}{N\rho_{air}A_c}\left[2.38\zeta\frac{M_{air}iA_M}{2F}\right] \tag{10.23}$$

$$U_L^S = \frac{1}{N\rho_{H_2O}A_c}[\dot{m}_{prod}^{w} - \dot{m}_{evap}^{w}] \tag{10.24}$$

where ζ is the flow stoichiometry of air, $\dot{m}_{prod}^{w}$ is rate of water produced through GDL, and $\dot{m}_{evap}^{w}$ is the water evaporation rate. These parameters are defined as,

$$\zeta = \frac{\dot{m}_{air}}{2.38M_{air}(iA_m/2F)} \tag{10.25}$$

$$\dot{m}_{prod}^{w} = (1+2\alpha)\frac{M_{H_2O}iA_m}{2F} \tag{10.26}$$

$$\dot{m}_{evap}^{w} = \frac{M_{H_2O}iA_m}{2F}\left[(2.38\zeta - 0.5)\frac{P_{sat}}{P - P_{sat}} - 2.38\zeta\frac{P_{sat}RH_{in}^{c}}{P - P_{sat}RH_{in}^{c}}\right] \tag{10.27}$$

where RH_{in}^{c} is the inlet relative humidity in the cathode side flow channel, N is the number of channels, A_c is the cross-sectional area of each channel, M_{air} and M_{H_2O} are the molecular weight of air and water, respectively, i is the current density, A_m is the effective MEA area, α is the ratio of net water flux with proton flux, $\dot{m}_{air}$ is the mass flow rate of air, P and P_{sat} are the fuel cell operating pressure and saturation pressure of water, ρ_{air} and ρ_{H_2O} are the density of air and water. The exponentially fitted curves on the flow map are splitting into different flow regimes. Generally, traditional channel flows have one inlet and one outlet, but for a PEM fuel cell channel comprised of multiple inlet and outlet channels. This makes the accurate computation of superficial velocity more challenging.

In another published literature, two-phase flow in the cathode flow-field channel is investigated with different superficial air velocities using high-speed camera (Lu et al. 2009). This study focuses on pressure drop characterization, flow maldistribution, and visualization of two-phase flow patterns associated with the PEM flow channels. They have conducted ex situ experiments with eight parallel channels according to the proposed flow-field design of Owejan et al. (2009). The dimensions of channels are length = 183 mm, width = 0.7 mm, depth = 0.4 mm and land width between the adjacent channels = 0.5 mm. To avoid mechanical shear on the GDL associated with straight channels, a weaving angle of 5° was considered in the channel design. To obtain a good optical clarity, gas channels were machined on Lexan plate and vapor polishing was carried out over the channel. The inlet pressure of air was adjusted with an air pressure valve, and the flow rate is controlled through rotameter. The flow rate of water in the chamber (constructed for ex situ experiments) was controlled through four syringe pumps, and it supplies liquid water through GDL pores. Pressure taps are placed on each channel, air inlet section for the steady-state pressure measurements.

This experimental setup simulates the operating environment of working PEM fuel cell. Major findings from their experiments relevant to water management of PEM fuel cells are (Lu et al. 2009),

1. Severe flow maldistribution can occur at an operating condition of low air velocities due to the holdup of larger water slugs on the GDL. Excessive water accumulation can occur even on the hydrophobic GDL surface with low air flow rate and high water injection rate.
2. The possible reasons for maldistribution on PEM fuel cell gas channels are water accumulation on GDL, nonuniform intrusion of GDL surface and unsuitable design of inlet manifold and gas channels.
3. Surface tension is the main resistive force to the slug motion. When a slug coalesces with other drops or film, the velocity increases sharply, and generally, the slug shape got distorted and transforms into film and eventually drained out through flow field.
4. Water accumulation is prevented from GDL surface at higher air stoichiometry of above 10. The generated water is removed by applied high shear force, and mist flow pattern can be observe in the gas channels. At higher air flow rate, evaporation phenomena also significantly contribute to the water removal mechanism.
5. Mist flow is considered to be as the efficient water removal mode from PEM flow field. But it requires higher pumping power due to the higher reactant flow rate. Generally, slug flow is observed in the flow channels, which is undesirable because of the severe performance degradation. The most preferred flow pattern is considered to be as film flow due to the low-pressure drop requirement and higher water removal capabilities.

10.4.3 Phase Change Phenomena

The generated water in the fuel cell exists either in liquid or vapor phase depending on the local temperature and pressure. The state of water also changes locally with the time of operation. Evaporation and condensation phenomena are the general phase change processes occurring in the GDL and flow field of a PEM fuel cell. The mode of phase change process strongly depends on the local saturation temperature and water vapor concentration. An optimum water removal situation from the fuel cell can be attained from the extensive analysis of condensation and evaporation phenomena. This section briefly introduces the phase change phenomena applicable during the fuel cell operation.

10.4.3.1 Condensation

In condensation, phase change occurs from the metastable vapor state to stable liquid phase. Condensation process is initiated from the nucleation sites of a surface. It may occur in the form of droplets (drop-wise condensation) or film (film-wise condensation). In a PEM fuel cell, the condensate droplets are generated on both sides of GDL along with the flow-field side of the bipolar plate surface (Basu et al. 2009). Multiple mechanisms are responsible that determine the rate of the liquid transport in the GDL when the droplets are formed on the GDL pores. These droplets can either detach from the GDL surface or it can remain on the GDL surface according to the magnitude of the adhesive force between the droplet and surface as explained in Sect. 10.4.1. If the generated droplets are remaining on the GDL surface it will grow and coalesce with the neighbor drops subsequently and result in a liquid film formation on the GDL surface.

The formation of the condensate film on the GDL surface is not desired since it blocks the pores on the GDL surfaces. It stops the oxidant supply to the catalyst layer for the required electrochemical reaction. Film formation on the gas diffusion layer can be avoided by blowing the oxidizer under pressure through the flow-field channels. A low energy surface coating (e.g., a hydrophobic coating) can promote the droplet detachment (Jiao and Li 2011). These coatings prevent the film formation on GDL surfaces passively. The generated condensate has to be removed periodically from the GDL for the good performance of the fuel cell. Condensation also occurs at the bipolar plate surface in the flow field. Rendering the bipolar plate side with hydrophilic coatings promotes the condensate removal due to the film formation on the surface. Based on the mode of condensation, it is concluded that a flow-field channel having drop-wise condensation on the GDL surface side with film condensation on the bipolar plate surface side is the best design for the optimum water removal in a PEM fuel cell (Cai et al. 2006). Wettability patterning techniques (Ghosh et al. 2014; Mahapatra et al. 2016) can also be used for faster condensate drainage.

10.4.3.2 Evaporation

Evaporation is the phase conversion from the liquid to the vapor at a temperature lower than the boiling temperature from a liquid surface. In a PEM fuel cell, evaporation mainly occurs from the GDL pores. Evaporation phenomena also occur from the small-sized condensed and detached droplets on the GDL surface. Evaporation phenomena inside the GDL pore are complex, and the rate of evaporation is faster under the flow channel side as compared with the evaporation rate under the rib side of the flow field (Inoue et al. 2011).

The evaporation process influences the heat management of PEM fuel cell. Evaporation helps the cooling process of the fuel cell components when it is working under the high-temperature condition. Lal et al. (2018) showed that speed of the oxidant gases does not have any influence on the evaporation rate. They also concluded from their study that any change in the evaporation rate is not proportional to the change in the carrier gas binary diffusion constant. The rate of evaporation is affected by the pore size and the local condition of the GDL.

Due to the optical inaccessibility, it is very difficult to analyze the phase change process in the PEM fuel cell components experimentally. Computational models are extremely effective for analyzing the phase change processes in a fuel cell. Different numerical models are reported in the literature to study the phase change process (He et al. 2000; Natarajan and Van Nguyen 2001). The most widely used formula to calculate the phase change rate during evaporation and condensation is given by (Wu et al. 2009),

$$\dot{m}_{\mathrm{v-l}} = \begin{cases} \gamma_{\mathrm{evap}} \varepsilon s_{\mathrm{lq}} \frac{(P_{\mathrm{vap}} - P_{\mathrm{sat}})}{RT} & ; \quad \text{if } P_{\mathrm{vap}} < P_{\mathrm{sat}} \text{ (evaporation)} \\ \gamma_{\mathrm{cond}} \varepsilon (1 - s_{\mathrm{lq}}) \frac{(P_{\mathrm{vap}} - P_{\mathrm{sat}})}{RT} & ; \quad \text{if } P_{\mathrm{vap}} > P_{\mathrm{sat}} \text{ (condensation)} \end{cases} \tag{10.28}$$

where $\dot{m}_{\mathrm{v-l}}$ is the phase change mass transfer rate between liquid and vapor phases, γ_{evap} and γ_{cond} are the overall phase change rates during evaporation and condensation, respectively, ε is the porosity, s_{lq} is the volume fraction of the liquid water, and p_{vap}, p_{sat} are the vapor and saturation pressure corresponding to a temperature T.

10.5 Water Management in a PEM Fuel Cell

Optimum performance of a fuel cell can be achieved by proper thermal and water management strategies. Thermal management involves preventing dehydration and overheating of the membrane by continuous removal of heat. At normal operating conditions, a good water management strategy plays a major role on the performance and durability as compared to the thermal management strategy of the fuel cell. This section presents the issues and strategies associated with the water management in a PEM fuel cell.

10.5.1 Issues on Fuel Cell Water Management

Two major issues associated with the water management in a fuel cell are membrane drying and water flooding. A guaranteed high-performing fuel cell situation can be achieved by maintaining a dynamic equilibrium between liquid water flooding and membrane drying.

10.5.1.1 Dehydration of the Membrane

Membrane dehydration generally occurs at the anode side of the membrane. The main reasons for membrane dehydration are insufficient inlet humidification at the anode side and higher electro-osmotic drag (EOD) rate due to the higher current density (Schmittinger and Vahidi 2008). At higher current densities, mass transfer due to electro-osmotic drag is more than the mass transfer due to back diffusion which will tend to dry out of the membrane at anode side even when the cathode side is well hydrated. The bond between H^+ and SO_3^- is cannot be dissociated (detailed in 10.3.1) at the dry condition of the membrane resulting in a low ionic conductivity (Hickner et al. 2006). Thus, low ionic conductivity decreases the number of active sites in the CL and makes it difficult to transport the protons to the membrane. The high ionic conductivity of the membrane is ensured by maintaining a good hydrated condition at the membrane by a larger water content electrolyte.

The higher rate of EOD resulted from the higher current density immediately pulls the water molecule from the anode side. In this case, back diffusion of water occurred by the concentration gradient from the cathode to anode is not sufficient to keep the PEM in a hydrated state. Moreover, the membrane pores will shrink at the dehydrated state and it can cause a slow back-diffusion rate. The dehydration causes the membrane to become brittle, and cracks form in the membrane. The generated cracks allow the H_2 and O_2 gases to flow across the membrane. This results in an uncontrolled chemical reaction between these gases leading to the formation of hot spots on the membrane (Huang et al. 2006). Formation of the cracks and hot spots perpetuates once initiated on the membrane and eventually destructs fuel cell membrane. It can be concluded that the life of a fuel cell becomes shorter if the fuel cells work with the dry operating conditions. So higher value of relative humidity at the inlet gases ensures fully hydrated condition of the membrane, and it also results in a good proton conductivity at the membrane.

10.5.1.2 Fuel Cell Flooding

The phenomena of excessive accumulation of water on the MEA are called as fuel cell flooding. The excess water blocks the pores in the GDL surfaces, and it reduces the transport rate of the reactants to the CL. The output of the fuel cell will significantly reduce when the flooding starts. In a flooded situation, the local pressure of the reactants can shoot up and the liquid water will flush out from some locations of the GDL. This temporarily removes the blockage on the GDL surface and allows transport of gases to the CL active sites. The periodic pressure buildup and water removal during the flooding state cause a continuous and unstable fluctuation in the performance of the fuel cell and degrade the life of the cell.

The performance of a fuel cell is predicted from the polarization curve. It is the curve drawn between current density and cell voltage. Figure 10.8 qualitatively shows the polarization curve for a typical PEM fuel cell with and without flooding. The curve 1 presents the optimum case where no flooding occurs, and curve 2–4 presents the polarization curve with different degrees of water flooding rate. It shows that a significant loss of performance occurs when flooding starts. Flooding in a fuel cell occurs if the water formation rate becomes larger than water removal rate.

Generally, flooding occurs at the cathode side of the MEA, but it can also observe in the anode side at some operating conditions. The main reasons for cathode flooding are (a) excess water formation by ORR with an increase in load or current density,

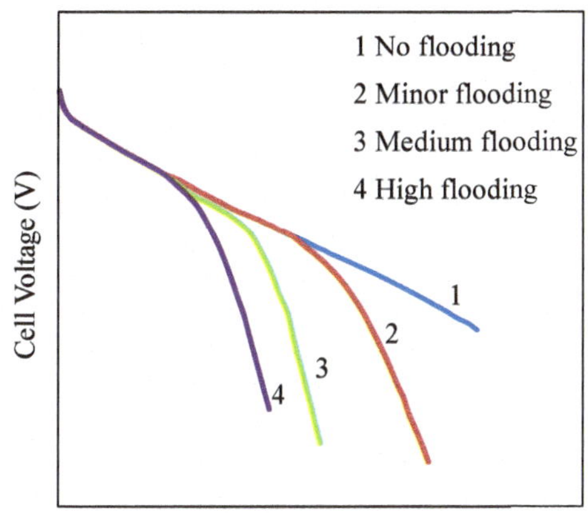

Fig. 10.8 Polarization curve of a typical PEM fuel cell with the different degrees of flooding: curve 1—no flooding, curve 2 to 4—increasing order of degree of flooding rate

(b) larger EOD rate due to the high electric field supply, and (c) over-humidification of reactant gases. The major portion of water is produced at the cathode side CL by oxygen reduction reaction. It is transported away from the CL to bipolar plate flow channels by water vapor diffusion, evaporation and capillary transport of liquid expelled through GDL. At higher current densities operation, the water production rate in a fuel cell is higher than the water removal rate. Thus, excess water blocks either flow-field channels and/or the GDL pores leading to a decrease in the active sites of the CL and flooding occurs.

The temporary blocks of the GDL resulted from the flooding causes the oxygen shortage at the catalyst layer. This phenomenon is called starvation. The load requirement in the fuel cell can cause the proton reduction reaction (PRR) in place of ORR on the membrane cathode side when a sufficient amount of oxygen is not present in the CL active sites. The electron-consuming process such as PRR due to the oxygen shortage is given by,

$$2H_3O^+ + 2e^- \rightarrow H_2 + 2H_2O \tag{10.29}$$

Flooding phenomena on the anode side are less frequent than cathode flooding. Anode flooding occurs when (a) a low current density cell operating in low temperature and low gas flow rate environment and (b) at low current density with smaller relative humidity at anode side as compared with cathode side. Flooding can occur even in the flow channel also. The fuel cell flooding causes (a) decrease to the performance of the cell, (b) increased pressure drop of the reactants due to the water accumulated at GDL pores, (c) uneven distribution of temperature and the current density of the membrane, and (d) degradation of the durability of the fuel cell.

A dynamic equilibrium between liquid flooding and membrane drying needs to be maintained in a fuel cell for the high performance. Fuel cell stacks are made by interconnecting each single cells in series. It is more challenging to maintain a precise equilibrium between flooding and membrane drying in a fuel cell module as compared to the single cell alone.

10.5.2 Strategies of Water Management

The optimum performance of a fuel cell mainly depends on the proper management of the generated water at the cathode side GDL/CL interface by electrochemical reaction. The investigations from the numerous researchers have concluded different strategies to attain a slug-free drainage of water from the GDL surface. This section presents the salient strategies on water management in the different components of a fuel cell.

10.5.2.1 Optimization of Operating Condition

The important operating conditions of PEM fuel cells include the inlet humidity of H_2 and O_2 gas streams, flow rate of inlet streams, fuel cell temperature and pressure, the current density of the cell, stoichiometric ratio of air–fuel supply. A sufficient inlet humidification is an important operating condition to avoid membrane drying. Flooding issues can be avoided by manipulating the main operating conditions such as temperature, relative humidity, and pressure of the inlet gaseous streams. A higher flow rate reactant gases remove the generated water from cathode side flow field and GDL either by evaporation or advection (Pasaogullari and Wang 2005). Generated water at the cathode can also be flushed out by increasing gas temperature and counter-flow operation of reactant gases (Wilkinson et al. 1998). An applied vibration (Mughal and Li 2006) can also help the removal of water by consuming extra power. An applied pressure drop between cathode and anode can transport water from cathode to anode via membrane, and this is called anode water removal (Voss et al. 1995). The main problem associated with the higher temperature and higher pressure drop operating condition is it aggravates the problems of rupture of the fuel cell membrane.

10.5.2.2 Flow-Field Design

The commonly used different types of flow-field designs (Wood et al. 1998) are illustrated in Fig. 10.9. In the case of conventional flow design, the transport mechanism is dominated by diffusion. The main features of conventional flow design are smaller inlet pressure and smaller inlet–outlet flow length. Both diffusion and convective modes of water transport are observed in serpentine design. It requires higher inlet pressure for the water transport due to the longer flow length. The highest pressure gradient occurs in interdigitated design, and the water transport mechanism is dominated by convective mode.

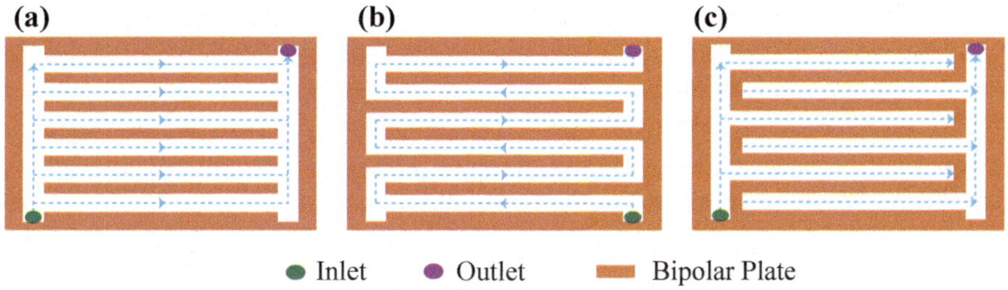

Fig. 10.9 Conventional flow channel designs in a commercial PEMFC, **a** conventional diffusion-dominated flow, **b** serpentine diffusion and conduction, and **c** interdigitated convection-dominated flow

An appropriate flow-field design helps in the easy removal of water as well as transport of O_2 gases through the GDL. An efficient flow-field design is one of the excellent strategies against the water flooding phenomena. As the transport mechanism in a conventional flow field (as shown in Fig. 10.9a) is diffusion-dominated, it results in a limited flow of reactant gases leading to the excessive accumulation of water on the GDL. In the case of the serpentine flow field (as shown in Fig. 10.9b), the liquid is transported by diffusion and convection modes. Serpentine flow-field design is characterized by longer channel length with a small cross-sectional area. The pressure drop at the pores of the GDL serpentine design is larger than that along the channel length. This difference in pressure drop drives the convective mode of water transport. This design also allows transport of sufficient reactant gases to the catalyst layer through GDL pores for the electrochemical reaction. Serpentine flow-field design is considered as the best among these three, and it is the widely adopted design for commercial applications. In interdigitated design (see Fig. 10.9c), the reactant is transported by convective mode. Water flooding phenomena in interdigitated design are avoided by the continuous removal of the generated water droplets on the GDL due to the shear force caused by the gas flow.

10.5.2.3 MEA Design

A thinner PEM membrane of thickness $\approx 10\,\mu m$ shows a good water management resulting from the higher back-diffusion rate. Even with a low anode side humidification thinner membrane will be at a better-hydrated condition. The drawback of the thinner membrane is its lower durability and higher reactant gas cross-over between anode and cathode. The membrane thickness is optimized to be 25–40 μm as a standard value for the fuel cell application (Freire and Gonzalez 2001). High-temperature ($T > 100\,^{\circ}C$) fuel cells are getting more interest in the past few years. At higher temperature, membrane water exists only as vapor state. So the water management can effectively be done in this operating condition by taking the benefit of single-phase analysis as compared with multiphase.

Gas diffusion layer has a critical role in the water management strategies in fuel cells. Engineering the GDL surface by treating it with a hydrophobic material like PTFE is the one way of promoting the removal rate of droplets. Increased loading of PTFE causes undesired effects on fuel cells like increased mass transport resistance, decreased electrical conductivity and porosity. The optimum range PTFE loading on GDL is 10–30% (Lim and Wang 2004). GDL should also continuously remove the excess heat generated from the fuel cell to prevent the formation of local hot spots in the electrode. The widely used substrate for the GDL is carbon paper, carbon cloth, and carbon non-woven. At a fully humidified condition, carbon cloth performs better performance as compared to carbon paper due to an efficient water removal rate (Ralph et al. 1997). But, at a dry operating condition carbon paper GDL surface shows a better performance as compared with carbon cloth (Quick et al. 2009).

Water management issue in a PEM cell can also be addressed by the modification of the CL microstructures. An optimal microstructure with sufficient transport path for water and gases in the CL is needed to avoid the chances of flooding issues on the cathode side CL. The optimum microstructure is designed by interconnecting the Nafion membrane with the carbon in the CL and partial filling of the void spaces with the nanosized hydrophobic particles. This design maintains a good electronic and ionic conductivity along with an optimal gas and liquid path across the CL structure.

Another method for improving the performance of a fuel cell is by assembling a microporous layer (MPL) between the CL and GDL interfaces. But the reason for the performance improvement is still not very clear and a subject for debate. Many experimental studies are conducted for understanding the physical explanation of MPL effect on improvement of performance and water management (Lu et al. 2010; Weber and Newman 2005). However, investigations have been concluded with different explanations. One widely accepted reason for the enhancement is that the optimum pore distribution in the MPL helps the water management issues (Qi and Kaufman 2002).

10.5.2.4 Self-humidified Membrane

As discussed in the earlier sections, proper humidification of the membrane is an essential requirement for the efficient proton transport. The performance of the fuel cell can be improved by ensuring a high degree of hydration levels on the membrane. So far, the discussion in this chapter was limited to the fuel cell system with external humidifiers. A membrane design with self-humidification capabilities is one of the novel methods to avoid the external humidifiers and to make the fuel cell systems more compact. The design concept of self-humidifying membrane is proposed by Watanabe et al. (1996). They have developed a self-humidifying Nafion membrane of 50 μm thickness with 0.07 mg/cm^2 weight percent of platinum (Pt) catalyst particles (diameter of 1–2 nm) and a few weight percent of titania or silica oxides.

The design concept of the self-humidified membrane is depicted in Fig. 10.10. The H_2 molecule from the anode side catalytically combines with the diffused O_2 molecules and produces water in the presence of Pt-dispersed particles in the mem-

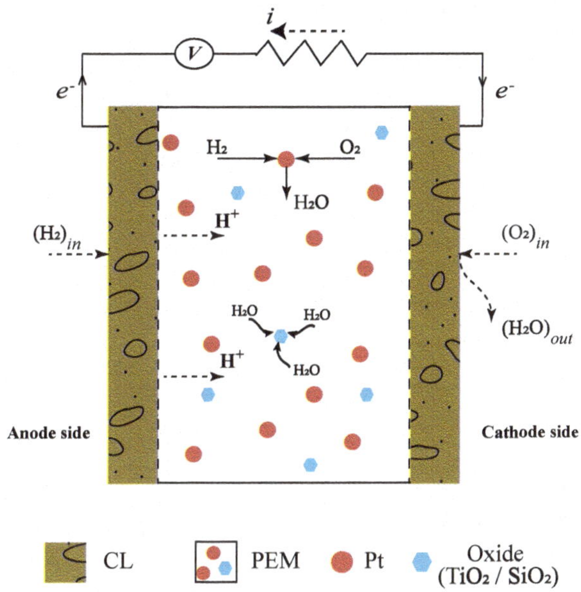

Fig. 10.10 Principle of self-humidified polymer electrolyte membrane (Nafion membrane with small weight percent of Pt catalytic particles and SiO_2/TiO_2 oxides)

brane. The oxide particle (TiO_2 or SiO_2) dispersed in the membrane adsorbs water and suppresses the gas cross-over in the membrane. Lowering the thickness of the membrane increases the gas cross-over rate but decreases the overall cell performance. This technique keeps the membrane at a fully hydrated state under any operating conditions. The main disadvantage of this technique is the increased cost of the membrane when Pt particles are added to it. To develop a low-cost self-humidifying membrane, researchers have looked for an alternative material to replace the Pt particles. Tungsten–carbide particles show the catalytic behavior as like Pt particles (Levy and Boudart 1973). Recently, Zheng et al. (2017) have successfully developed a durable self-humidifying composite membrane with a tungsten–carbide particle. The feasibility of these membranes has been investigated by in situ experiments. Their experimental results show a sufficient humidification, gas cross-over rate, and improved proton conductivity features with the tungsten–carbide composite membrane.

The design of self-humidified membrane is particularly advantageous for cruising vehicles where the operating condition changes abruptly. It is very difficult to control the amount of humidification on different loading conditions effectively by a control system. Whereas the self-humidified membrane has a capability to keep them at hydrated state continuously at any operating load. So it eliminates the complexity of the humidifier control system in different operating conditions. Overall cell system design is effectively simplified with internal humidification. Fuel-cell-powered vehicles with the self-humidified membrane is a good strategy for their commercialization.

10.5.2.5 Other Fuel Cell Design Strategies

Effect of surface wettabilies of the flow channel in two-phase flow dynamics was investigated by Lu et al. (2011) through ex situ experiments. Engineering the GDL surface with wettability patterns is one of the advanced water management techniques in a PEM fuel cell. The concept of hybrid GDL surface was introduced by Utaka et al. (2011). Figure 10.11 shows the concept of patterned wettability GDL surface with alternate wettable and non-wettable (hydrophobic) coatings. Owing to the higher surface energy characteristics of hydrophilic paths, the water generated at the GDL/CL interface is absorbed in the hydrophilic track. The sufficient reactant gas supply for the electrochemical reaction is ensured through the hydrophobic tracks. A higher oxygen diffusivity is reported with patterned GDL as compared with uniform wettability (hydrophobic) GDL surface (Utaka et al. 2011). Recently, a novel method

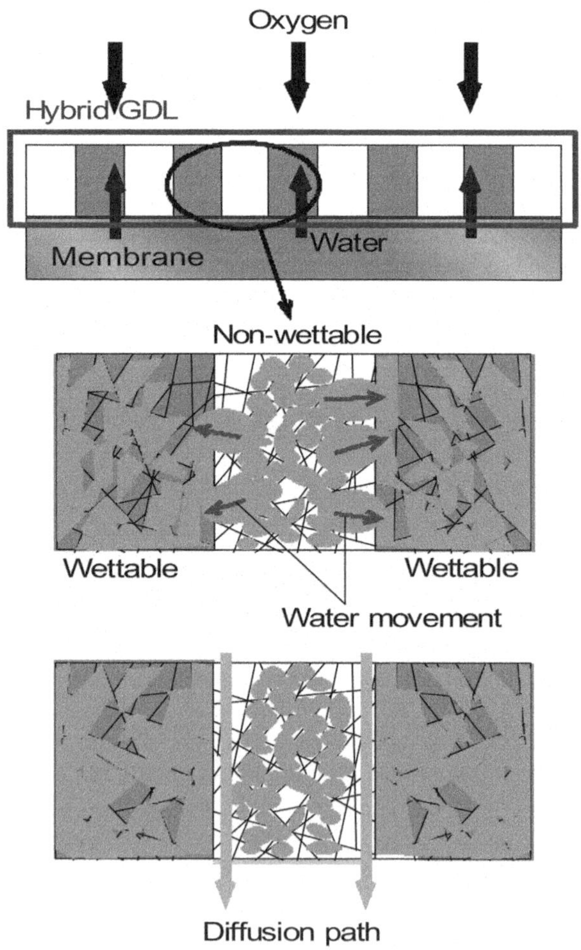

Fig. 10.11 Hybrid GDL surface with alternate hydrophobic–hydrophilic tracks. The motion of water from the hydrophobic track to hydrophilic at the underneath of GDL surface is shown in the middle image. Diffusion path of reactant gases is shown in bottom picture (reproduced with permission from Utaka et al. 2011)

for preparing patterned GDL surfaces based on radiation grafting is demonstrated by Forner-Cuenca et al. (2015). In situ experiments of PEM fuel cell with patterned GDL show a significant enhancement in the power density as compared with the non-patterned GDL surface (Forner-Cuenca et al. 2016). The water transport in the GDL is effectively managed by optimizing the track widths, the area ratio between hydrophobic and hydrophilic track, contact angle, etc.

Electro-osmotic (EO) pumping technique is an another efficient water management way in a PEM cell by placing the porous glass EO pumps with an integrated reservoir system at the cathode channel wall (Buie et al. 2006). The generated water on the GDL/CL interface forces out from the porous GDL by hydrophobic forces, and it coalesces with multiple drops. These water droplets are wicked into the porous hydrophilic structures of EO pump and transported to the integrated reservoir once the glass structures are sufficiently filled with liquid water. At certain operating conditions, EO improves the stability and performance of the fuel cell along with the efficient water removal mechanism.

In addition to the discussed strategies, many other techniques are also developed for achieving a dynamic equilibrium for the generated water in the fuel cell. Ge et al. (2005) have developed a new strategy by mounting a hydrophilic wick structures to the cathode flow-field channels of a serpentine design. This method improves the water removal by absorption mechanism of the wicking material. Other methods regarding flow-field design strategies are (a) by using a porous bipolar plate (Yi et al. 2004), (b) by modifying serpentine design with baffle blocks (Jang et al. 2006), and (c) triangular/stepped/trapezoidal cross-sectional flow field (Metz et al. 2008; Min 2009; Gopalan and Kandlikar 2014).

10.6 Water Harvesting from PEM Fuel Cell

The amount of electricity production is interconnected with the water consumption rate. Water is consumed in the majority of the power generation techniques like thermal, nuclear, solar, and natural gas thermo-electric power plants. Water is one of the precious commodities in the environment for the survival of living beings. Fuel cells are the promising energy production devices capable to generate clean water and electricity with a negligible amount of water consumption. The generated water from the fuel cell is pure, but the foreign particle present in the H_2 and O_2, contaminants in the flow channels, degradation of the membrane and fuel cell components, etc., are contaminating the fuel cell generated water. This section reviews the potential of PEM fuel cell as an alternative potable water solution.

Water harvesting from fuel cells is a novel method for the potable water solution of astronauts in space. Early studies in this area were conducted in the Gemini space program of NASA for checking the quality of the generated water from the fuel cell in space. An alkaline fuel cell stack with organic electrode was used in this space shuttle. Unfortunately, the presence of formaldehyde, *p*-benzaldehyde sulfonic acid, and sulfo-benzoic acid was observed in the generated water (Collier et al. 2006).

Even after the purification process of filtering and carbon sorption, the targeted quality of potable water could not be achieved in the Gemini mission. In the Apollo space program, a fuel cell with sintered nickel electrode was used instead of the organic electrode in Gemini mission. These fuel cells resolved the water quality issues associated with the presence of impurities observed in the Gemini space program. But the presence of bis (penta-methylene-di-thiocarbamate) Ni(II) was observed in the water quality testing when the samples were collected from the water storage tank of the spaceship (Sauer and Calley 2018). The NASA report on the potable water system of Apollo mentioned the value of total dissolved solids (TDS) as 0.73 mg/L and pH as 5.4 (Sauer and Calley 1973). A comprehensive technology was developed during the Apollo program for the future water requirement in spacecraft.

Later, the quality of the generated water from the US Space Shuttle and Mir space stations was analyzed by Orta et al. (1998) using ion chromatography and capillary electrophoresis method. The quality of water is analyzed from the generated water sample collected before and after the mission. Presence of organic ions was rarely detected during the quality measurements. They reported that the quality of water from both space shuttle followed the drinking water standards with a presence of negligible amount of cations and anion at $\mu g/L$ concentrations. All the above studies verified the feasibility of the generated water from the fuel cell as a drinking source in space. These studies portrayed fuel cells as a promising technique for the simultaneous generation of power and potable water in the outer space.

10.6.1 Water Yield Calculation from PEM Fuel Cell

Water is generated as a by-product of a fuel cell by the simple electrochemical reaction as expressed in Eq. 10.3. The measure of water generated from the PEMFC is mathematically expressed by Buie et al. (2006),

$$Q_w = \frac{i A_{FC} M_{H_2O}}{2F\rho_{H_2O}} \tag{10.30}$$

where i is the current density, A_{FC} is the active area of the fuel cell, M_{H_2O} and ρ_{H_2O} are molecular weight and density of water, respectively, and F is the Faraday constant. The generated water in a fuel cell typically exists as both liquid and vapor states depending on the operating temperature, operating pressure, current density, amount of humidification, etc. A condenser unit should be required in the water collection system for converting the vapor form of water to liquid. So the net amount of water can be collected from a fuel cell system which is reduced by a fraction depending on the efficiency of the condenser unit. Considering a condenser efficiency of $x\%$, the net amount of water collected (Q_w^{net}) from the fuel cell is expressed as,

$$Q_{\mathrm{w}}^{\mathrm{net}} = Q_{\mathrm{w}} \times \frac{x}{100} \quad (10.31)$$

Hristovski et al. (2009) have extensively modeled the water yield from a commercially available fuel cell stack (46 polymer membranes) of Ballard Power Systems Inc by assuming 100% water collection capacity. Their analysis was carried out by operating the fuel cell at a current of 52 A and a power output of 1.63 kW. The water capturing rate from this stack is found to be 19 L/day, which is more than the average US households drinking water consumption rate of ≈5.2 L/day (USEPA 2001).

To relate the amount of water generated to the power output (W) of a fuel cell, a parameter can define as water energy ratio (WER) and it is expressed as,

$$\mathrm{WER} = \frac{Q_{\mathrm{w}}^{\mathrm{net}}}{W} \quad (10.32)$$

Unlike the negative value of WER for the conventional power generation techniques, fuel cells have a positive value of WER. Another study is conducted by Tibaquirá et al. (2011b) to calculate the water yield from a laboratory-scale PEM fuel cell with different types of Nafion membranes. According to their results, WER is changing with the current density and the type of the membrane. The approximate value of the average electrical power requirement per household in the USA is ≈31 kWh (US Energy Information Administration 2005) as per the survey conducted by US Energy Information Administration. So the average WER required for a typical household is approximately 0.17 L/kWh. Water energy ratio of ≈1 L/kWh is obtained in the Tibaquirá et al. (2011b) work when the laboratory-scale PEM fuel cell is operated at a current density of 0.7 A/cm^2. According to their result, the WER from the fuel cell is nearly six times more than the required WER for a household. From all these aspects, we can conclude that fuel cell is the promising technology to meet the average electricity and potable water requirement of a household.

10.6.2 Water Quality Measurements from PEM Fuel Cell

It is notable that fuel cell can generate water as a by-product during their operation but the quality of this water when it is used for drinking purposes has always been debatable. In a pioneering work, Hristovski et al. (2009) investigated the quality of water produced from a PEM fuel cell. This work compared the quality of the water produced from the six distinctive PEMFC situated in the various states of USA. It did not evaluate the effect of the operation condition of the fuel cell with water quality, but it verified the feasibility of the drinking water production from the PEMFC. The water quality test results obtained from various PEMFC showed that quality of water from the fuel cells was satisfactory. The contaminant levels in the majority of water samples were found to be lower than the maximum contaminant level (MCL) values of standard potable water as indicated by the United States Environmental Protection

Agency (USEPA) (1996). But in some samples, the concentration levels of lead and antimony were higher than MCL. Diffusion of fuel cell materials with the generated water might be the possible reason for the presence of these contaminants. These issues on water quality can be overcome by modifying the piping materials, fuel cell materials, etc.

In another published literature, the water quality comparison on various types of the commercial fuel cell was investigated with a molten carbonate fuel cell (MCFC) and polymer electrolyte membrane fuel cell (PEMFC) by Tibaquirá et al. (2011a). They have used 1 kW of PEM fuel cell (ReliOn, I-1000) and 300 kW MCFC (Fuel Cell Energy, DFC-300) for the water quality comparison. Water samples from both fuel cells were collected, and concentrations of major salts, metals, nonmetals, organic compounds, total oxide carbon, etc., pH, and conductivity were analyzed in accordance with the USEPA standards. According to their results, the average pH value is 7.0 ± 0.7 for PEMFC and 7.2 ± 0.4 for MCFC which falls in the acceptable range of 6.5 and 8.5 as per USEPA potable water standards. The average conductivity is found to be $5.9 \pm 1.8\ \mu s/cm$ for PEMFC and $306 \pm 281\ \mu s/cm$ for MCFC. The conductivity of MCFC shows a very high value as compared with PEMFC, but both the values are lower than the standard potable water. The concentration of aluminum and nickel in a PEMFC is higher than MCL. In the case of MCFC, nitrite and manganese concentration is higher than the MCL. Based on the analytical data, they concluded that the nature of water from these fuel cells is relatively pure, and the collected water from the fuel cell can be made as potable with a proper water treatment process.

Tibaquirá et al. (2011b) also developed a laboratory-scale experimental setup of PEM fuel cell in Arizona State University, USA, to study the influence of operating conditions on water quality as shown in Fig. 10.12. Water recovery system in this setup consists of a condenser and storage tank. Electrical heaters are used in the experimental setup for maintaining the working temperature of the fuel cell between 30 and 100 °C. They have conducted multiple experiments with different operating temperatures on five different set of fuel cell systems such as Nafion-111 with DI water humidification, Nafion-212 with DI water humidification, Nafion-212 with pipe water humidification, self-humidified Nafion-212 with graphite made bipolar plate, and self-humidified Nafion-212 with stainless steel made bipolar plate. Humidifiers are not used in the experimental setup as in Fig. 10.12, when the fuel cell is operated with a self-humidified membrane. The presence of contaminants in the fuel cell water is coming from the fuel cell materials, whereas the membrane composition is not contaminating the generated water. According to their results, they inferred that water generated from the PEM fuel cell needs a purification system to make it potable. The metal concentration of the fuel cell water reflects the lifespan of the fuel cell components. Therefore, the robustness of a fuel cell can qualitatively analyze from the metal concentration values of the generated water.

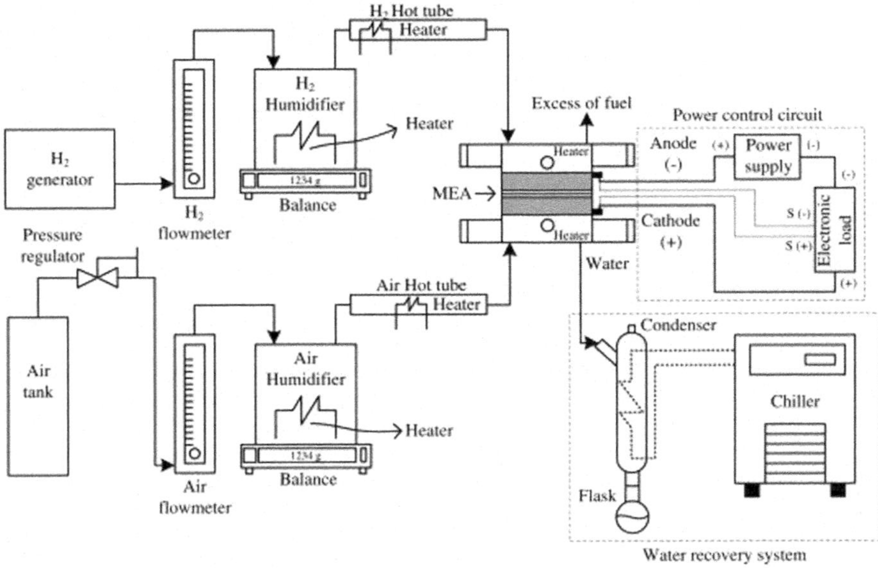

Fig. 10.12 A laboratory-scale experimental setup developed by Tibaquira et al. to recover water from a PEM fuel cell. Humidifiers are omitted for self-humidifying membrane (reproduced with permission from Tibaquira et al. 2011b)

10.7 Summary

This chapter reviews the potential of PEM fuel cells as a primary power source in electric vehicles and the feasibility of exhausted water as a potable water source. Presence of rare-earth elements like platinum as an essential element in the major components of the PEM fuel cells is one of the key challenges for their commercialization. The development of fuel cell material with alternate composite materials can replace platinum particles and can effectively reduce the total cost of fuel cells. Water transport mechanism in a PEM fuel cell comprises of multiple complex phenomena like multiphase flows (condensation, evaporation, boiling, melting, etc.), molecular diffusion, electro-osmotic forces, catalytic reaction, electrochemical reaction, flow through porous media, etc. A detailed understanding of the water transport mechanism in different components of fuel cell can help in the design of high-performing fuel cells by developing advanced water management strategies. Gas diffusion layer acts as a porous opening for the supply of reactant gases and the removal of product water. An optimum dynamic balance of reactant supply and product water removal can be achieved by analyzing the multiphase flows and proper modifications in the cathode side of the GDL surfaces. This chapter also discusses some recent advancements in the perspective of fuel cell materials and water management techniques.

References

Anderson R, Zhang L, Ding Y, Blanco M, Bi X, Wilkinson DP (2010) A critical review of two-phase flow in gas flow channels of proton exchange membrane fuel cells. J Power Sources 195(15):4531–4553

Andújar JM, Segura F (2009) Fuel cells: history and updating. A walk along two centuries. Renew Sustain Energy Rev 13(9):2309–2322

Basu S, Wang C-Y, Chen K-S (2009) Phase change in a polymer electrolyte fuel cell. J Electrochem Soc 156(6):B748–B756

Bird RB (2002) Transport phenomena. Appl Mech Rev 55(1):R1–R4

Bruggeman DAG (1935) The calculation of various physical constants of heterogeneous substances. I. The dielectric constants and conductivities of mixtures composed of isotropic substances. Ann Phys 416:636–791

Buie CR, Posner JD, Fabian T, Cha S-W, Kim D, Prinz FB, Eaton JK, Santiago JG (2006) Water management in proton exchange membrane fuel cells using integrated electro-osmotic pumping. J Power Sources 161(1):191–202

Cai YH, Hu J, Ma HP, Yi BL, Zhang HM (2006) Effects of hydrophilic/hydrophobic properties on the water behavior in the micro-channels of a proton exchange membrane fuel cell. J Power Sources 161(2):843–848

Collier A, Wang H, Yuan XZ, Zhang J, Wilkinson DP (2006) Degradation of polymer electrolyte membranes. Int J Hydrogen Energy 31(13):1838–1854

de Gennes P-G, Brochard-Wyart F, Quere D (2004) Capillarity and wetting phenomena: drops, bubbles, pearls waves, vol 291

Fang C, Hidrovo C, Wang F-u, Eaton J, Goodson K (2008) 3-d numerical simulation of contact angle hysteresis for microscale two phase flow. Int J Multiph Flow 34(7):690–705

Forner-Cuenca A, Biesdorf J, Gubler L, Kristiansen PM, Schmidt TJ, Boillat P (2015) Engineered water highways in fuel cells: radiation grafting of gas diffusion layers. Adv Mater 27(41):6317–6322

Forner-Cuenca A, Biesdorf J, Manzi-Orezzoli V, Gubler L, Schmidt TJ, Boillat P (2016) Advanced water management in pefcs: diffusion layers with patterned wettability iii. operando characterization with neutron imaging. J Electrochem Soc 163(13):F1389–F1398

Freire TJP, Gonzalez ER (2001) Effect of membrane characteristics and humidification conditions on the impedance response of polymer electrolyte fuel cells. J Electroanal Chem 503(1–2):57–68

Ge S, Li X, Hsing I-M (2005) Internally humidified polymer electrolyte fuel cells using water absorbing sponge. Electroch Acta 50(9):1909–1916

Ghosh A, Beaini S, Zhang BJ, Ganguly R, Megaridis CM (2014) Enhancing dropwise condensation through bioinspired wettability patterning. Langmuir 30(43):13103–13115

Gopalan P, Kandlikar SG (2014) Effect of channel materials and trapezoidal corner angles on emerging droplet behavior in proton exchange membrane fuel cell gas channels. J Power Sources 248:230–238

Grot WG (1976) Heat-treated fluorocarbon sulfonylamine cation permselectivity. US Patent 3969285, 13 July 1976

He W, Yi JS, Van Nguyen T (2000) Two-phase flow model of the cathode of PEM fuel cells using interdigitated flow fields. AIChE J 46(10):2053–2064

Hickner MA, Fujimoto CH, Cornelius CJ (2006) Transport in sulfonated poly(phenylene)s: proton conductivity, permeability, and the state of water. Polymer 47(11):4238–4244

Hristovski KD, Dhanasekaran B, Tibaquirá JE, Posner JD, Westerhoff PK (2009) Producing drinking water from hydrogen fuel cells. J Water Supply Res Technol AQUA 58(5):327–335

Huang X, Solasi R, Zou Y, Feshler M, Reifsnider K, Condit D, Burlatsky S, Madden T (2006) Mechanical endurance of polymer electrolyte membrane and PEM fuel cell durability. J Polym Sci Part B: Polym Phys 44(16):2346–2357

Hussaini IS, Wang C-Y (2009) Visualization and quantification of cathode channel flooding in PEM fuel cells. J Power Sources 187(2):444–451

Hwang J-J (2013) Sustainability study of hydrogen pathways for fuel cell vehicle applications. Renew Sustain Energy Rev 19:220–229

Inoue G, Ishibe N, Matsukuma Y, Minemoto M (2011) Simulation of liquid water evaporation in GDL for PEMFC under gas purge condition. In: ASME-JSME-KSME 2011 joint fluids engineering conference. American Society of Mechanical Engineers, pp 3585–3591

Jang J-H, Yan W-M, Li H-Y, Chou Y-C (2006) Humidity of reactant fuel on the cell performance of PEM fuel cell with baffle-blocked flow field designs. J Power Sources 159(1):468–477

Jiao K, Li X (2011) Water transport in polymer electrolyte membrane fuel cells. Prog Energy Combust Sci 37(3):221–291

Kumbur EC, Sharp KV, Mench MM (2006) Liquid droplet behavior and instability in a polymer electrolyte fuel cell flow channel. J Power Sources 161(1):333–345

Lal S, Lamibrac A, Eller J, Büchi FN (2018) Determination of water wvaporation rates in gas diffusion layers of fuel cells. J Electrochem Soc 165(9):F652–F661

Levy RB, Boudart M (1973) Platinum-like behavior of tungsten carbide in surface catalysis. Science 181(4099):547–549

Li X (2005) Principles of fuel cells. CRC Press, New York

Lim C, Wang CY (2004) Effects of hydrophobic polymer content in GDL on power performance of a PEM fuel cell. Electrochim Acta 49(24):4149–4156

Lu Z, Kandlikar SG, Rath C, Grimm M, Domigan W, White AD, Hardbarger M, Owejan JP, Trabold TA (2009) Water management studies in PEM fuel cells, Part II: ex situ investigation of flow maldistribution, pressure drop and two-phase flow pattern in gas channels. Int J Hydrogen Energy 34(8):3445–3456

Lu Z, Daino MM, Rath C, Kandlikar SG (2010) Water management studies in PEM fuel cells, Part III: dynamic breakthrough and intermittent drainage characteristics from GDLs with and without MPLs. Int J Hydrogen Energy 35(9):4222–4233

Lu Z, Rath C, Zhang G, Kandlikar SG (2011) Water management studies in PEM fuel cells, Part IV: effects of channel surface wettability, geometry and orientation on the two-phase flow in parallel gas channels. Int J Hydrogen Energy 36(16):9864–9875

Mahapatra PS, Ghosh A, Ganguly R, Megaridis CM (2016) Key design and operating parameters for enhancing dropwise condensation through wettability patterning. Int J Heat Mass Transf 92:877–883

Metz T, Paust N, Müller C, Zengerle R, Koltay P (2008) Passive water removal in fuel cells by capillary droplet actuation. Sens Actuators A: Phys 143(1):49–57

Min C-H (2009) Performance of a proton exchange membrane fuel cell with a stepped flow field design. J Power Sources 186(2):370–376

Motupally S, Becker AJ, Weidner JW (2000) Diffusion of water in nafion 115 membranes. J Electrochem Soc 147(9):3171–3177

Mughal A, Li X (2006) Experimental diagnostics of PEM fuel cells. Int J Environ Stud 63(4):377–389

Natarajan D, Van Nguyen T (2001) A two-dimensional, two-phase, multicomponent, transient model for the cathode of a proton exchange membrane fuel cell using conventional gas distributors. J Electrochem Soc 148(12):A1324–A1335

Orta D, Mudgett PD, Ding L, Drybread M, Schultz JR, Sauer RL (1998) Analysis of water from the space shuttle and mir space station by ion chromatography and capillary electrophoresis. J Chromatogr A 804(1–2):295–304

Oszcipok M, Riemann D, Kronenwett U, Kreideweis M, Zedda M (2005) Statistic analysis of operational influences on the cold start behaviour of PEM fuel cells. J Power Sources 145(2):407–415

Ous T, Arcoumanis C (2007) Visualisation of water droplets during the operation of PEM fuel cells. J Power Sources 173(1):137–148

Owejan JP, Trabold TA, Jacobson DL, Arif M, Kandlikar SG (2007) Effects of flow field and diffusion layer properties on water accumulation in a PEM fuel cell. In: ASME 2007 5th international

conference on nanochannels, microchannels, and minichannels. American Society of Mechanical Engineers, pp 311–320
Owejan JP, Gagliardo JJ, Sergi JM, Kandlikar SK, Trabold TA (2009) Water management studies in PEM fuel cells, Part I: fuel cell design and in situ water distributions. Int J Hydrogen Energy 34(8):3436–3444
Pasaogullari U, Wang C-Y (2005) Two-phase modeling and flooding prediction of polymer electrolyte fuel cells. J Electrochem Soci 152(2):A380–A390
Qi Z, Kaufman A (2002) Improvement of water management by a microporous sublayer for PEM fuel cells. J Power Sources 109(1):38–46
Quick C, Ritzinger D, Lehnert W, Hartnig C (2009) Characterization of water transport in gas diffusion media. J Power Sources 190(1):110–120
Ralph TR, Hards GA, Keating JE, Campbell SA, Wilkinson DP, Davis M, St-Pierre J, Johnson MC (1997) Low cost electrodes for proton exchange membrane fuel cells performance in single cells and ballard stacks. J Electrochem Soc 144(11):3845–3857
Sauer RL, Calley DJ (1973) Apollo experience report: potable water system
Sauer RL, Calley DJ Biomedical results of apollo-potable water supply. Report no. Technical report, LC-75-600030
Schmittinger W, Vahidi A (2008) A review of the main parameters influencing long-term performance and durability of PEM fuel cells. J Power Sources 180(1):1–14
Shirani E, Masoomi S (2008) Deformation of a droplet in a channel flow. J Fuel Cell Sci Technol 5(4):041008
Springer TE, Zawodzinski TA, Gottesfeld S (1991) Polymer electrolyte fuel cell model. J Electrochem Soc 138(8):2334–2342
Tibaquirá JE, Hristovski KD, Westerhoff P, Posner JD (2011a) Recovery and quality of water produced by commercial fuel cells. Int J Hydrogen Energy 36(6):4022–4028
Tibaquirá JE, Hristovski KD, Westerhoff PK, Posner JD (2011b) Water quality and yield from polymer electrolyte membrane fuel cells. Int J Hydrogen Energy 36(20):13022–13031
Uchida M, Fukuoka Y, Sugawara Y, Eda N, Ohta A (1996) Effects of microstructure of carbon support in the catalyst layer on the performance of polymer-electrolyte fuel cells. J Electrochem Soc 143(7):2245–2252
US Energy Information Administration (2005) Residential energy consumption survey: energy consumption and expenditures tables. www.eia.doe.gov
USEPA (1996) Setting standards for safe drinking water. Environmental Protection Agency, U.S
USEPA (2001) National primary drinking water regulations; arsenic and clarifications to compliance and new source contaminants monitoring; final rule. Fed Reg 66(14):6975
Utaka Y, Hirose I, Tasaki Y (2011) Characteristics of oxygen diffusivity and water distribution by x-ray radiography in microporous media in alternate porous layers of different wettability for moisture control in gas diffusion layer of PEFC. Int J Hydrogen Energy 36(15):9128–9138
Voss HH, Wilkinson DP, Pickup PG, Johnson MC, Basura V (1995) Anode water removal: a water management and diagnostic technique for solid polymer fuel cells. Electrochim Acta 40(3):321–328
Wang C-Y (2004) Fundamental models for fuel cell engineering. Chem Rev 104(10):4727–4766
Watanabe M, Uchida H, Seki Y, Emori M, Stonehart P (1996) Self-humidifying polymer electrolyte membranes for fuel cells. J Electrochem Soc 143(12):3847–3852
Weber AZ, Newman J (2005) Effects of microporous layers in polymer electrolyte fuel cells. J Electrochem Soc 152(4):A677–A688
Wilkinson DP, Voss HH, Fletcher NJ, Johnson MC, Pow EG (1998) Electrochemical fuel cell stack with concurrent flow of coolant and oxidant streams and countercurrent flow of fuel and oxidant streams. US Patent 5773160, 30 June 1998
Wood DL III, Yi Jung S, Nguyen TV (1998) Effect of direct liquid water injection and interdigitated flow field on the performance of proton exchange membrane fuel cells. Electrochim Acta 43(24):3795–3809

Wu H, Li X, Berg P (2009) On the modeling of water transport in polymer electrolyte membrane fuel cells. Electrochim Acta 54(27):6913–6927

Yi JS, Yang JD, King C (2004) Water management along the flow channels of PEM fuel cells. AIChE J 50(10):2594–2603

Zawodzinski TA Jr, Neeman M, Sillerud LO, Gottesfeld S (1991) Determination of water diffusion coefficients in perfluorosulfonate ionomeric membranes. J Phys Chem 95(15):6040–6044

Zawodzinski TA, Davey J, Valerio J, Gottesfeld S (1995) The water content dependence of electro-osmotic drag in proton-conducting polymer electrolytes. Electrochim Acta 40(3):297–302

Zheng W, Wang L, Deng F, Giles SA, Prasad AK, Advani SG, Yan Y, Vlachos DG (2017) Durable and self-hydrating tungsten carbide-based composite polymer electrolyte membrane fuel cells. Nat Commun 8(1):418

Chapter 11
Ferrofluids for Propulsion

Uddalok Sen and Souvick Chatterjee

Abstract Ferrofluids are colloidal suspensions of single-domain magnetic nanoparticles in a nonmagnetic carrier fluid. Despite the presence of both solid (nanoparticle) and fluid (carrier fluid) phases, the suspension behaves as a fluid in the presence of a magnetic field. The superparamagnetic nature of the suspended particles allows for the suspension to be manipulated by a magnetic field 'from a distance,' i.e., without any mechanical actuation system. Although ferrofluids were first developed in the 1960s for drawing liquid fuel against gravity in rocket propulsion systems, the application never materialized commercially in a large scale. However, ferrofluids have been extensively used for other applications such as ferrofluidic seals, loudspeakers, magnetic drug targeting, thermomagnetic convection, magnetic shape-shifting optical mirrors, and energy harvesting. Recently, it was shown that ferrofluids can be used in the micropropulsion sector. In the presence of a magnetic field, a ferrofluid surface naturally deforms to form sharp peaks, the phenomenon being known as Rosensweig instability. For ionic ferrofluids, an amplification of the electric field is present at the tip of these peaks, which leads to electrospray emission. The reaction force due to the emitted droplets results in forward propulsion. In the present chapter, a summary of the basics of ferrofluids and their manipulation techniques is presented, followed by a brief review of ferrofluid-based propulsion techniques.

11.1 Introduction

The need for very accurate positioning of small satellites has seen the birth and growth of electric propulsion techniques (Lev et al. 2017). These techniques can provide thrusts in the range of tens of micro-Newtons in an extremely controllable

U. Sen · S. Chatterjee (✉)
Department of Mechanical and Industrial Engineering,
University of Illinois at Chicago, Chicago, IL 60607, USA
e-mail: souvickchat@gmail.com

U. Sen
e-mail: uddalok.sen.us@gmail.com

K. Saha et al. (eds.), *Two-Phase Flow for Automotive and Power Generation Sectors*, Energy, Environment, and Sustainability,
https://doi.org/10.1007/978-981-13-3256-2_11

fashion, thus making them attractive for geostationary communication satellites, low earth orbit mega constellations, and earth observation satellites (Lev et al. 2017). Colloid thrusters, where droplets emanate from a colloidal fluid in the presence of an external field, are one of the popular methods of electric propulsion. An electrospray, where an ionic liquid undergoes jetting and subsequent breakup into droplets in the presence of an electric field, is a popular method of thrust generation in such colloid thrusters (Gamero-Castano and Hruby 2001).

An ionic liquid surface, in the presence of an electric field, deforms to form a conical meniscus (Taylor 1964; Taylor and McEwan 1965), known in the literature as a Taylor cone. The electric field imposes Maxwell stresses which pull the liquid (Yarin et al. 2014), while the surface tension of the liquid tends to minimize the surface area. An interplay between these two counteracting stresses results in the conical shape of the liquid-free surface. When a liquid flow is fed into this cone simultaneously, a steady jet ensues which further breaks up into droplets (Zeleny 1914). The 1960s and 1970s saw a boom in the research on electrospray for space propulsion applications (Huberman and Rosen 1974), but the interest gradually declined. The primary reasons identified (Gamero-Castano and Hruby 2001) for this initial decline included a decrease in space technology research in the post-Apollo era, the high (>10 kV) operating voltages required, and difficulties associated with scaling up. However, the recent rekindling of the interest in colloid thruster technology can be attributed to (Gamero-Castano and Hruby 2001) the suitability of the low-output thrust of a single electrospray for smaller spacecraft, the need for precise and fine thrust variations, and the knowledge of producing sprays of higher specific impulse at a lower onset voltage (15 kV). Although the thrust from a single electrospray is of the order of micro-Newtons, it can be increased significantly by having multiple emitters in an array (Velásquez-García et al. 2006).

The magnetic analog of the Taylor cone is the normal field instability (Rosensweig instability) (Cowley and Rosensweig 1967), observed when a magnetic field is applied to a magnetizable fluid (Afkhami et al. 2010). These fluid peaks are regularly spaced and static, and have been studied extensively (Boudouvis et al. 1987; Gailītis 1977; Rosensweig 1987). In terms of magnetizable fluids, ferrofluids are the most obvious choice. Ferrofluids are colloidal suspensions of single-domain magnetic nanoparticles (5–15 nm in diameter) in a nonmagnetic liquid phase (either water or hydrocarbon oil) (Odenbach 2004). The particles usually consist of ferrite, magnetite, or maghemite alloyed with nickel, cobalt, magnesium, or zinc (Puri and Ganguly 2014), and are usually prepared by chemical coprecipitation of ferrous salts or by mechanical grinding. An adsorbed surfactant layer is usually present in these particles to prevent agglomeration due to van der Waals forces and dipole–dipole interaction. The particles exhibit superparamagnetic nature in this size range; i.e., the particle magnetization curves, although being comparable to ferri- or ferromagnetic particles, show no hysteresis (Sen et al. 2017). A ferrofluid is nonmagnetic in the absence of a magnetic field, but an externally imposed magnetic field aligns the magnetic moments of the superparamagnetic nanoparticles, thus making the fluid magnetic. This feature has been harnessed for manipulating ferrofluids ‘from a distance’ by either active or passive means. Microfluidics (Nguyen 2012) has been a broad application area of ferrofluids, but they have been used for other applications such as thermomagnetic convection (Ganguly et al. 2004) and energy harvesting (Bibo et al. 2012).

Recent studies (Jackson et al. 2017; King et al. 2014; Mkrtchyan et al. 2013) show that the free surface of an ionic ferrofluid, in the presence of both magnetic and electric fields, can deform to produce a jet, which subsequently disintegrates into droplets. Several studies (Stone et al. 1999; Tyatyushkin and Velarde 2001; Zakinyan et al. 2012) exist that have investigated the deformation of ferrofluid droplets in the presence of both electric and magnetic fields. The present chapter highlights some of the popular manipulation techniques of ferrofluids, followed by a brief review of the ferrofluid-based propulsion techniques. The chapter concludes by summarizing the key findings and briefly highlighting the future scope of research in this area.

11.2 Ferrofluid Manipulation Techniques

The behavior of ferrofluid drops in magnetic fields has been extensively studied numerically (Rowghanian et al. 2016; Liu et al. 2011), experimentally (Tan and Nguen 2011; Yan et al. 2015; Tan et al. 2010; Long et al. 2009; Chen et al. 2009), and analytically (Brancher and Zouaoui 1987; Zhu et al. 2011; Lee et al. 2012). The present section highlights two representative studies that demonstrate some of the popular formation and breakup techniques of ferrofluid droplets.

11.2.1 Formation of Ferrofluid Droplets

In a recent study, Sen et al. (Sen et al. 2017) demonstrated the use of a microfluidic T-junction to generate ferrofluid droplets. A numerical simulation was performed employing the volume of fluid (VOF) method, where a magnetic dipole was placed underneath the T-junction geometry whose main channel had a Tween 80–water solution as the continuous phase and an oil-based ferrofluid (EFH3, Ferrotec) entered through the side channel as the discrete phase. The magnetic Kelvin body force acting on the ferrofluid was accounted for via a user-defined function. The dispersed phase flow rate was kept constant, while the flow rate of the continuous phase (in terms of its capillary number, Ca) and the dipole strength, m, was varied during the course of the study. The position of the magnetic dipole relative to the side channel was also varied.

The ferrofluid flow in the absence of any magnetic field is shown via snapshots in Fig. 11.1 for $Ca = 0.025$. The forces acting on the ferrofluid stream once it starts entering the main channel are primarily due to viscous, pressure, and surface stresses. As observed from the snapshots in Fig. 11.1(ii) through (vii), the ferrofluid thread increases in volume from ‘a’ to ‘b’, while the remaining is attached at the neck. A change in curvature of the neck (from convex to concave), as the ferrofluid

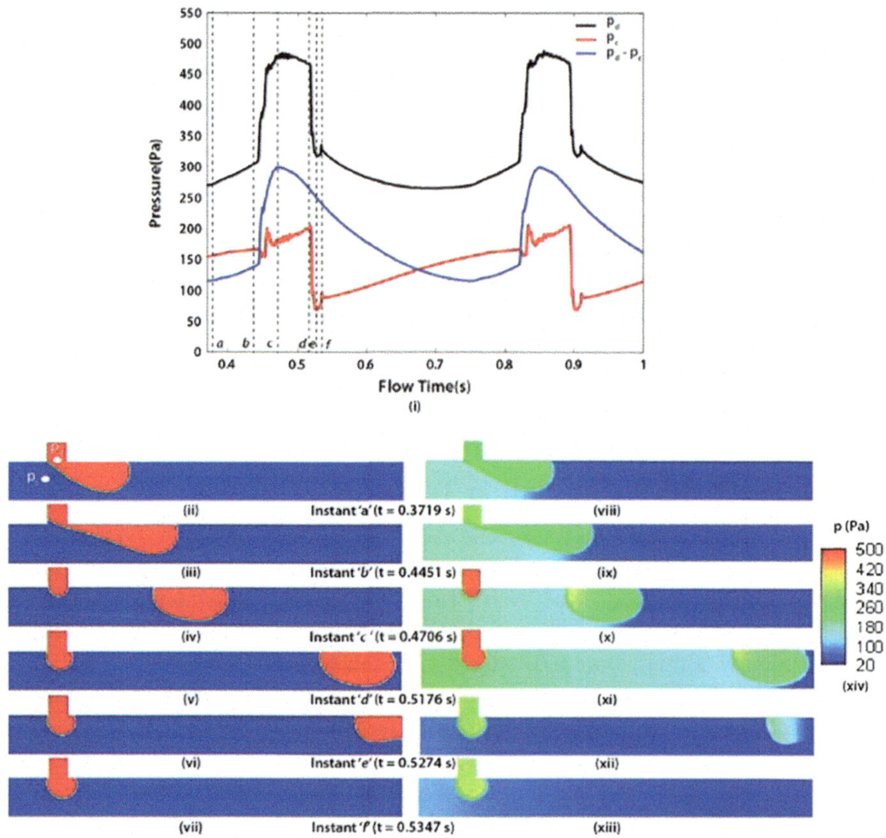

Fig. 11.1 (i) Variation of continuous and dispersed phase pressures with flow time at $Ca = 0.025$ and $m = 0$; (ii)–(vii) droplet shapes at the salient time instants shown in (i); locations of the pressure probes are shown in inset (ii); (viii)–(xiii) pressure contours at the salient time instants. The figure has been reproduced, with permission, from Sen et al. (2017)

progresses further into the main channel, causes the pressure p_d [measured at the location marked in Fig. 11.1(ii)] to rise [as seen in Fig. 11.1(i)]. p_c [measured at the location marked in Fig. 11.1(ii)] also increases [as seen in Fig. 11.1(i)] during this time as the ferrofluid volume inside the main channel blocks a certain fraction of the flow area of the continuous phase. The pressure rise is maximum between snapshots 'b' and 'c', until the point where the ferrofluid neck ruptures due to the squeezing and shearing action of the continuous phase. The rupture causes the formation of a droplet, which then adjusts its shape so as to have the minimum surface area. The departure of the detached droplet from the flow domain (instant 'e') results in an abrupt decrease of the flow resistance, thus causing the abrupt drops of p_c and p_d observed in Fig. 11.1(i).

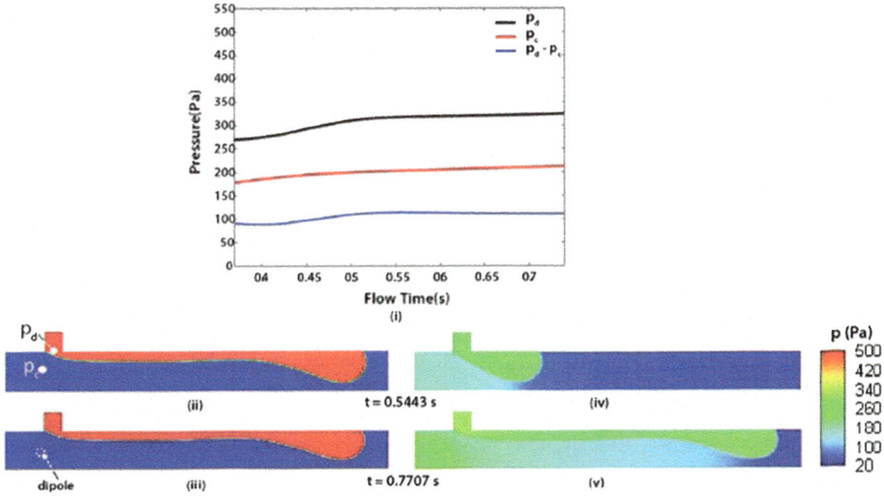

Fig. 11.2 (i) Variation of continuous and dispersed phase pressures with flow time at $Ca = 0.025$ and $m = 22.5 \times 10^{-6}$ A m^2 placed upstream; shapes of the dispersed phase at (ii) $t = 0.5443$ s and (iii) $t = 0.7707$ s. Top views also show locations of the pressure probes [inset (ii)] and location of the dipole under the microchannel [inset (iii)]; (iv), (v) pressure contours at the same salient time instants. The figure has been reproduced, with permission, from Sen et al. (2017)

When a magnetic dipole of strength 22.5×10^{-6} A m^2 is placed upstream of the side channel [as shown in Fig. 11.2(ii)], a significant change is observed in the ferrofluid flow behavior for the same $Ca = 0.025$. Droplet formation was not observed; instead, the ferrofluid flowed through the main channel as a continuous fluid stream, characteristic of jetting or parallel flow. Jetting is usually associated with higher capillary numbers where the strong shearing action of the continuous phase causes severe stretching of the dispersed phase. The magnetic Kelvin body force due to the presence of the dipole in this case is responsible for an extra stress on the ferrofluid volume in addition to the pressure, interfacial, and shear forces. Since the dipole is placed upstream of the location of the side channel, the ferrofluid experiences an attractive force opposite to the direction of motion of the continuous phase, resulting in a reduction of the ferrofluid flow velocity. The increased relative velocity results in an increased shear stress, which tips the system beyond the dripping-jetting transition point. The constant nature of the pressure plot [Fig. 11.2(i)] is also representative of a typical parallel flow regime, where the difference between p_d and p_c is due to the Laplace pressure difference caused by the unchanged curvature of the ferrofluid–water interface.

When a dipole of the same strength is placed downstream of the side channel, the flow behavior is distinctly different from that shown in Fig. 11.2. In this case, as the ferrofluid is entering the main channel, the dipole is to its downstream, thus attracting it in the forward direction. But once it crosses the location of the dipole, the direction of the force is reversed. The velocity of the ferrofluid is reversed, thus

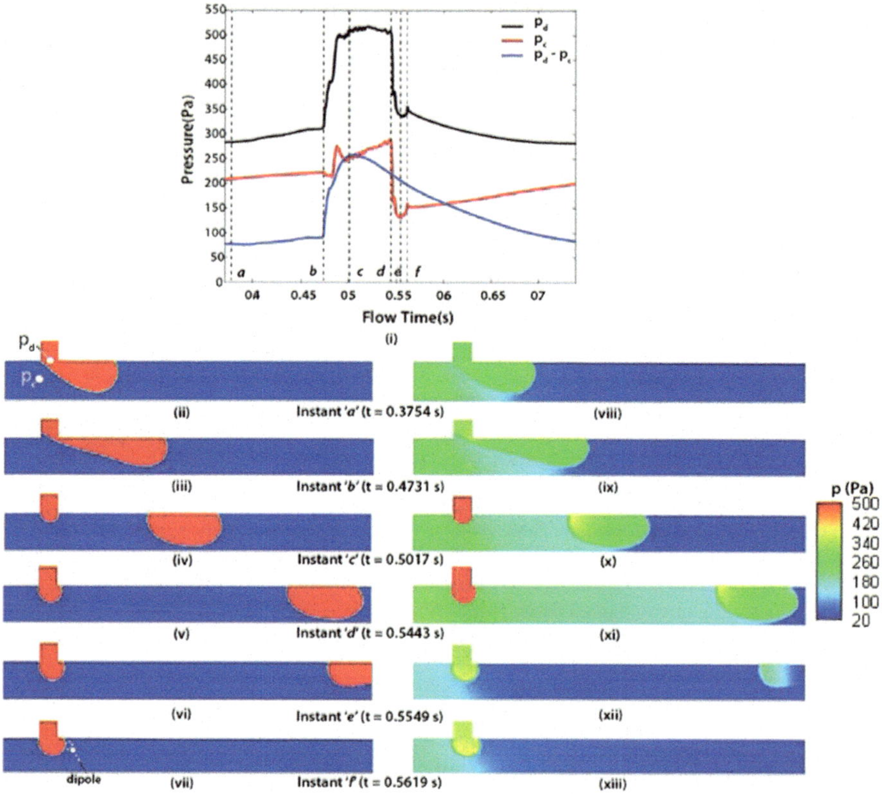

Fig. 11.3 (i) Variation of continuous and dispersed phase pressures with flow time at $Ca = 0.025$ and $m = 22.5 \times 10^{-6}$ A m^2 placed downstream; (ii)–(vii) droplet shapes at the salient time instants shown in (i); locations of the pressure probes are shown in inset (ii); (vii) location of the dipole under the microchannel; (viii)–(xiii) pressure contours at the same salient time instant. The figure has been reproduced, with permission, from Sen et al. (2017)

increasing the relative velocity, and consequently the shear between the dispersed and the continuous phases. However, this increase is not as high as the previous case when the dipole was upstream of the side channel. Also, while the leading part of the ferrofluid is experiencing a force in the opposite direction, the trailing part is still attracted toward the forward direction. This results in the stresses to not cross the dripping-jetting transition point. The droplet detachment phenomenon (Fig. 11.3) is similar to the nonmagnetic case. However, the detached droplet is attracted against the direction of flow by the magnetic dipole (which is downstream of the side channel location but upstream of the droplet), thus increasing its residence time within the channel.

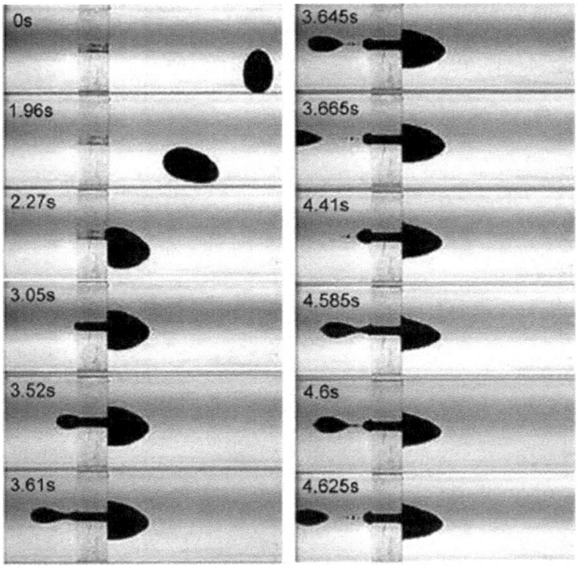

Fig. 11.4 Sequential images of the drop motion for the representative case of $d = 0.5$ mm, $H_c = 510$ Oe, and $l = 10$ mm at different times. l denotes the distance of the orifice from the electromagnet coil. The figure has been reproduced, with permission, from Chen et al. (2009)

11.2.2 Breakup of Ferrofluid Droplets

The breakup of a ferrofluid drop when it is squeezed through an orifice in the presence of a magnetic field was studied experimentally by Chen et al. (Chen et al. 2009). A ferrofluid drop (17 L volume) is placed inside a tube (filled with distilled water–glycerol mixture) that is fitted with an acrylic orifice of diameter, d. An electromagnet coil is placed on the other side of the orifice to that of the drop. When the coil is turned on, the magnetic field produced by the electromagnet attracts the drop toward it, resulting in a squeezing action through the orifice.

A representative case of ferrofluid drop breakup for orifice diameter, $d = 0.5$ mm, and local field strength, $H_c = 510$ Oe, is shown in Fig. 11.4. At $t = 0$, gravity and the weaker interfacial tension between the ferrofluid and the glycerol–water mixture are responsible for distorting the droplet. Upon activating the magnetic field, the drop moves toward the orifice until it comes into contact with the orifice plate. The confinement caused by the orifice increases the viscous drag, while the surface stresses also increase due to the change in curvature of the free surface. However, the magnetic force is still strong enough to pull the ferrofluid through the orifice. A thread of ferrofluid is stretched through the orifice, while the rest of the drop remains jammed upstream of the orifice location. The rounded leading edge of the elongated ferrofluid thread keeps progressing forward (toward the left in Fig. 11.4) which leads to necking and finally ruptures to form a new ferrofluid droplet. Post breakup, the remaining ferrofluid thread recoils due to surface tension. If the magnetic field is still prevalent after the first breakup, the remaining ferrofluid drop undergoes breakup in the same manner as the previous one to yield another ferrofluid droplet.

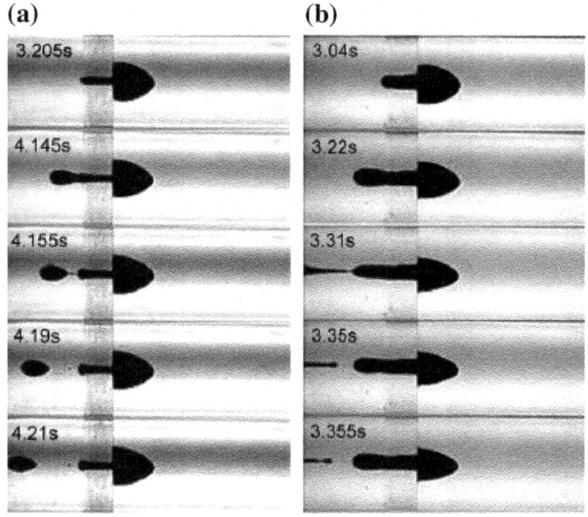

Fig. 11.5 Sequential images of drop motion for different diameters of $H_c = 510$ Oe and $l = 10$ mm: **a** $d = 0.3$ mm and **b** $d = 0.8$ mm. l denotes the distance of the orifice from the electromagnet coil. The figure has been reproduced, with permission, from Chen et al. (2009)

The effect of varying the diameter of the orifice at a constant local field strength (H_c) of 510 Oe is shown in Fig. 11.5a, b. With increasing confinement, the viscous drag is expected to increase. Moreover, a smaller orifice results in a larger curvature of the fluid thread, which leads to a higher resistance due to surface tension. Hence, it is more challenging for the same ferrofluid drop to traverse through a smaller orifice as compared to a larger one. For $d = 0.8$ mm (Fig. 11.5b), the ferrofluid leading edge is much more stretched as compared to the case where $d = 0.3$ mm and after less time as well.

Such unique abilities to manipulate ferrofluid are bound to attract attention from different spheres of innovation, and the propulsion community has also responded. The following section describes the relevance of a particular type of liquid for propulsion of miniaturized satellites and how ferrofluids can serve as a strong component in this context.

11.3 Ferrofluid-Based Propulsion Systems

One of the most striking things in the context of ferrofluids for propulsion is the history associated with the invention of ferrofluids. In the absence of gravity, it was a big challenge to pump fuel efficiently into a rocket engine. Papell (1963) at NASA came up with the idea of attaching magnetic properties to an otherwise nonmagnetic fuel so that it can be controlled under zero gravity by powerful magnets or pumped by switching magnetic fields using electromagnets. The idea was abandoned as solid fuel propellants came to reality.

After years of demonstrating multiple liquid manipulation techniques, demonstrated briefly in the previous section, ferrofluids again have the potential to find use in outer space propulsion operations. This innovation came to the limelight because of the increased usage of electrospray for spacecraft propulsion.

Recent years have seen a steep ascent in the usage of electric propulsion for transport of nanosatellites in space. These satellites perform a bunch of functions ranging from geostationary communication, CubeSat (mass less than 1.33 kg), interplanetary missions, earth observation, etc. Apart from agencies like NASA and the Department of Defense, many private companies are showing interest in entering into this market segment. According to SpaceWorks Enterprises (Buchen 2014), the number of microsatellites (1–50 kg) launched almost tripled in 2013 as compared to 2012 and is projected to increase 10× by 2020. CubeSat is a typical miniaturized satellite used for space research, and electric propulsion is a key mode of a thruster for these satellites. The commercial viability of this technology is proven by the planned mission Lunar IceCube by NASA to estimate the size and composition of water deposits on the Moon. Electric propulsion will be used for this project through an electric RF ion engine system, which produces 1.1 mN thrust and 2800 s specific impulse (defined as total impulse or change in momentum per unit propellant consumed) from 50 W total input power. Traditionally, electrospray is generated using microfabricated capillary needle electrodes. At the application of a critical voltage, the liquid forms a Taylor cone, the apex of which provides emission. A microsatellite can be propelled using an array of such cones. A new technology eliminates the need of an array of capillary electrodes by creating multiple parallel jetting instabilities through the application of simultaneous electric and magnetic fields. This requires the use of ionic liquid ferrofluids (ILFFs), which are created by suspending magnetic nanoparticles in a molten salt carrier liquid.

The Taylor cone instability of ionic liquid correlates to the Rosensweig instability for magnetic fluids, which replaces the support structure of capillary electrodes for traditional electrospray. Hence, the first step in this study is the creation of the magnetic ionic fluid with enough magnetic strength to demonstrate the Rosensweig instability. Although there are several efforts for creation of these two-phase fluids, a successful one was demonstrated by Jain et al. (2011). They created an ILFF based on ethylammonium nitrate (EAN); but it was noted to be hygroscopic and more viscous, which limits its use for propulsion applications. Hence, the other ILFF was developed by King et al. (2014). The latter was based on EMIM-NTf2 and was prepared by mixing a dispersion of sterically stabilized Sirtex magnetic nanoparticles in 50:50 (by weight) water and ethanol with EMIM-NTf2, followed by ultrasonification for 2 min. Water and ethanol were removed by rotary evaporation and overnight nitrogen purging. The classic *M–H* curve (magnetization flux field strength) shows zero hysteresis, which is characteristic of superparamagnetism. A set of permanent magnets provided enough field strength to display the Rosensweig instability. Once the instability is achieved using a magnetic field, an additional electric field generates electric stresses which, after a threshold, leads to electrospray emission.

King et al. (2014) showed a comparative study of the EAN and EMIM-NTf2-based ILFFs by monitoring the emission current. Their custom ILFF (based on EMIM-

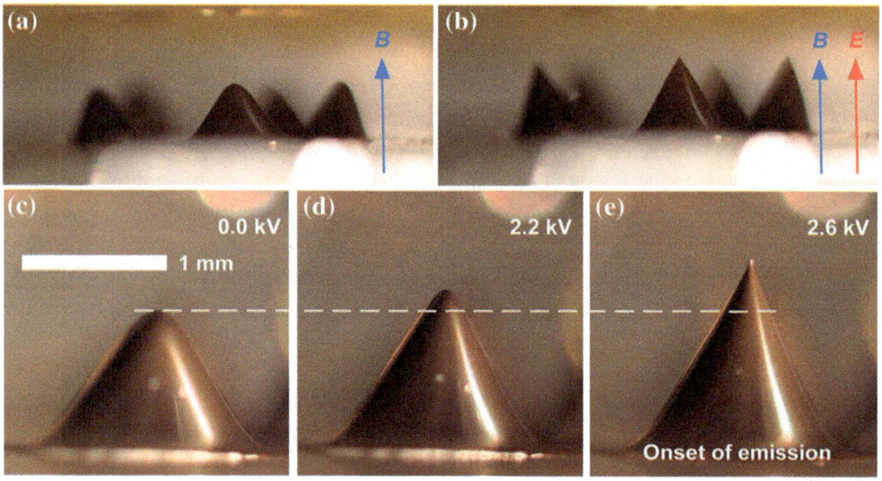

Fig. 11.6 Effect of superimposition of electric field on sessile ferrofluid droplet. **a** Profile image of the five tipped arrays created in ILFF under only a magnetic field. **b** Suitably strong electric field causing the tips to sharpen into current-emitting cones. **c–e** Magnetic features were incrementally strained and sharpened by the electric field, showing an abrupt onset of emission that in this case occurred at 2.6 kV to produce 2 μA of positively charged emission. The figure has been reproduced, with permission, from King et al. (2014)

NTf2) showed an abrupt onset of emission 2.6 kV to produce 2 μA of positively charged emission. The experiments showed five tips in the Rosensweig surface pattern as shown in Fig. 11.6, each of which has the potential to act as an individual electrospray emitter. Hence, for space propulsion applications, an increased surface area of ferrofluid can, in principle, create multiple such magnetic field-induced emitters (peaks), thus eliminating the need for microfabricated capillary electrodes. Their theoretical analysis also showed a lower critical electrical field in the presence of a magnetic field to cause the jetting instability. Also, up to 16 emission sites are hypothesized to be created using ILFF as the ionic liquid, which is a massive boost for propulsion of miniaturized satellites.

More recently, Jackson et al. (2017) showed a transient computation model to simulate morphology of a sessile ILFF droplet when subjected to electrical and magnetic forces. They discussed roles of the individual perturbing fields (electric and magnetic) in relation to the jetting instability. Their simulations showed excellent agreement with the experiments in terms of droplet height up to 85% of the critical onset voltage, after which the model slightly overpredicted the droplet deformation, as shown in Fig. 11.7 for two different magnetic fields. Stress measurements at the meniscus apex for a single simulated droplet showed magnetic stresses to dominate at lower applied voltages but rapidly overtaken by electrical stresses at emission onset. Their observance of a strong correlation between spray onset potential and increasing magnetic field strength clearly highlights the importance of the latter in this technology.

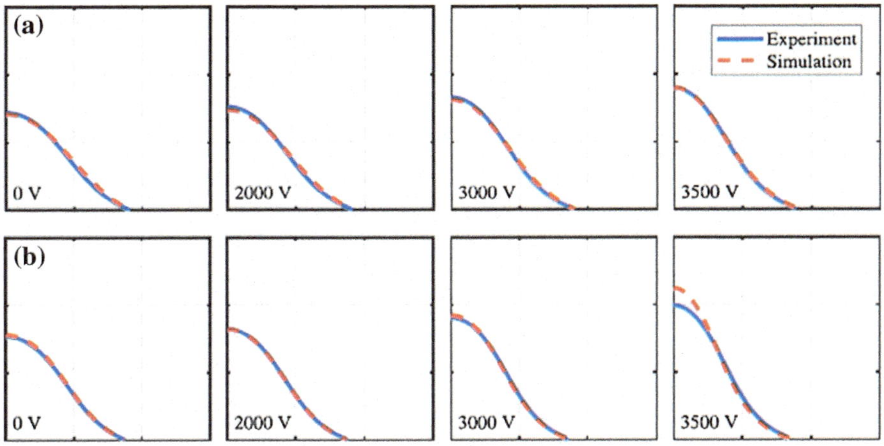

Fig. 11.7 Simulated and imaged contours for **a** 225 G and **b** 264G magnetic fields. The figure has been reproduced, with permission, from Jackson et al. (2017)

The most recent work in this segment by the same group was presented at the AIAA 2018 Joint Propulsion Conference, where Jackson and King (2018) showed the performance estimation of single-tip ionic liquid ferrofluids. A key performance metric is the thrust per emitter, which dictates feasibility of this source as electrospray emitter for spacecraft propulsion. The efficiency, η, of a thruster, which in turn is given by the ratio of power of electrospray jet (P_j) and electrical power input (P_{in}) such that:

$$\eta = \frac{P_j}{P_{in}} = \frac{T^2}{2\dot{m}P_{in}}$$

where T is the resultant thrust and $\dot{m}$ is the propellant mass flow rate. The other crucial parameter in this context is specific impulse of electrospray, which is determined by the exhaustive velocity of the thruster, given by total exit velocity of all ions that are emitted:

$$v_e = \sum v_i = \sum \sqrt{\frac{2q_i V}{m_i}}$$

where v_i is the exit velocity of a single ion, q_i is the charge of the ion, V is the acceleration voltage, and m_i the is mass of the ion. Both of these parameters depend on the mode of operation of the emitter: pure ion mode or droplet mode. In the former, charge-to-mass ratio is high leading to higher specific impulse and reduced thrust. Droplet mode leads to lower charge-to-mass ratio, lower specific impulse, and higher thrust. An in-between mode of operation of the emitter provides a broader range of operation.

Since thrust could not be measured directly, they approximated the parameter using the average mass-to-charge ratio of the emitted fluid. Their experiments also showed quantitative aspects in terms of emission test such that after a span of 45–180 min, the emission stabilized and transformed into a constant voltage mode with sustained emission for a span greater than 10 h, which resulted into self-extenuation of the source. A cyclical process of restarting and extenuation would continue till depletion of fluid or reaching of the maximum range of voltage source. They conducted three different tests for varying magnetic fields, which showed consistent results. An average specific impulse of 1385 s and an average thrust of 0.38 μN were estimated from a single emitter. This estimation has multiple assumptions such as

- All particles are accelerated to the full extraction potential.
- Single value accurately describes mass-to-charge ratio. In the configuration described, which is a mixed ion–droplet regime, the mass-to-charge ratio distribution ranges several orders of magnitude and is a future scope of this research.

However, incorporating the overprediction of thrust and specific impulse, the numbers are significantly close to the NASA Lunar IceCube electrospray-propelled nanosatellite. The ease of scaling up ILFF-based electrospray also provides a reasonable strength to the future of this emerging technology in the innovation ecosystem.

11.4 Conclusions and Future Scope

Ferrofluids have been an integral component of research at the intersection of magnetism and fluid mechanics for a significant period. The fundamental aspects of preparation of this type of fluid and manipulation techniques have been quite extensively studied both numerically and experimentally. The current chapter briefly summarizes some of the key studies on manipulation techniques of this special type of fluid. The existence of two phases in a typical sessile ferrofluid droplet makes numerical studies using the classical volume of fluid approach relevant and has been discussed in detail in this chapter. The ability of controlling morphology of the droplet using an external magnetic field makes it unique and has been elaborately studied in the literature. This chapter also highlights the relevant quantitative parameters involved in this physical phenomenon. Highlighting the fundamental novelty of ferrofluid, the chapter further discusses the importance of this fluid in light of electrospray-based propulsion.

After being conceived in the 1900s, the concept of electric propulsion saw momentum at the end of the last century, leading to maturity of the technology with market-wide acceptance for multiple functions onboard different types of satellites. With increasing demand, there is an increased chance of growth of different electric propulsion technologies to cater specific needs, which prospers potential usage of other technologies in the pipeline that includes electrospray and hence ferrofluid-based electrospray.

SpaceWorks's projection dataset shows an average growth of 23.8% over 2014–2020 in use of CubeSat with 2000–2750 micro-/nanosatellites requiring a launch. With increased interest in academia, eruption of commercial companies, and start-up activities, the micro-/nanosatellite market is growing significantly and future launches suggest this trend will continue. Hence, innovation will continue to thrive in this space with newer and efficient propulsion technologies.

As mentioned earlier, for propulsion applications using ionic liquid ferrofluid, an array of emitter is ideal. However, there is a crucial challenge to overcome in this segment. A small difference in heights of different peaks in an array creates different onset voltages resulting in different emission intensities. Creation of custom magnets and proper tuning of magnetic field gradient can produce uniform arrays, but those require further exploratory research. Successful dynamic simulations of the single-peak normal field instability of ferrofluid in the presence of electric and magnetic fields opened up possible studies of the effect of existing broad parameter range in this context without spending excessive time and money. Other research avenues include prevention of plasma formation, composition of emitted fluid to achieve the desired viscosity, magnetic, and ionic properties. The promising performance metric justifies further efforts to refine the technology in these grounds and is expected to show tangible positive outcomes in the near future.

References

Afkhami S, Tyler AJ, Renardy Y, Renardy M, St Pierre TG, Woodward RC, Riffle JS (2010) Deformation of a hydrophobic ferrofluid droplet suspended in a viscous medium under uniform magnetic fields. J Fluid Mech 663:358–384

Bibo A, Masana R, King A, Li G, Daqaq MF (2012) Electromagnetic ferrofluid-based energy harvester. Phys Lett A 376:2163–2166

Boudouvis AG, Puchalla JJ, Scriven LE, Rosensweig RE (1987) Normal field instability and patterns in pools of ferrofluid. J Magn Magn Mater 65:307–310

Brancher JP, Zouaoui A (1987) Equilibrium of a magnetic liquid drop. J Magn Magn Mater 65:311–314

Buchen E (2014) SpaceWorks' 2014 nano/microsatellite market assessment. In: 28th Annual AIAA/USU conference on small satellites

Chen C-Y, Chen C-H, Lee W-F (2009) Experiments on breakups of a mag- netic fluid drop through a micro-orifice. J Magn Magn Mater 321:3520–3525

Cowley MD, Rosensweig RE (1967) The interfacial stability of a ferromagnetic fluid. J Fluid Mech 30(4):671–688

Gailītis A (1977) Formation of the hexagonal pattern on the surface of a ferromagnetic fluid in an applied magnetic fluid. J Fluid Mech 82(3):401–413

Gamero-Castano M, Hruby V (2001) Electrospray as a source of nanoparticles for efficient colloidal thrusters. J Propul Power 17(5):977–987

Ganguly R, Sen S, Puri IK (2004) Thermomagnetic convection in a square enclosure using a line dipole. Phys Fluids 16(7):2228–2236

Huberman MN, Rosen SG (1974) Advanced high-thrust colloid sources. J Spacecraft 11(7):475–480

Jackson BA, King LB (2018) Derived performance metrics and angular efficiency of an ionic ferrofluid electrospray propulsion concept. In: 2018 Joint propulsion conference

Jackson BA, Terhune KJ, King LB (2017) Ionic liquid ferrofluid inter- face deformation and spray onset under electric and magnetic stresses. Phys Fluids 29:064105

Jain N, Zhang X, Hawkett BS, Warr GG (2011) Stable and water-tolerant ionic liquid ferrofluids. ACS Appl Mater Interfaces 3(3):662–667

King LB, Meyer E, Hopkins MA, Hawkett BS, Jain N (2014) Self-assembling array of magnetoelectrostatic jets from the surface of a superparamagnetic ionic liquid. Langmuir 30:14143–14150

Lee C-P, Yang S-T, Wei Z-H (2012) Field dependent shape variation of mag- netic fluid droplets on magnetic dots. J Magn Magn Mater 324:4133–4135

Lev DR, Emsellem GD, Hallock AK (2017) The rise of the electric age for satellite propulsion. New Space 5(1):1–11

Liu J, Yap YF, Nguyen N-T (2011) Numerical study of the formation process of ferrofluid droplets. Phys Fluids 23:072008

Long Z, Shetty AM, Solomon MJ, Larson RG (2009) Fundamentals of magnet-actuated droplet manipulation on an open hydrophobic surface. Lab Chip 9:1567–1575

Mkrtchyan L, Zakinyan A, Dikansky Y (2013) Electrocapillary instability of magnetic fluid peak. Langmuir 29:9098–9103

Nguyen N-T (2012) Micro-magnetofluidics: interactions between magnetism and fluid flow on the microscale. Microfluid. Nanofluid. 12:1–16

Odenbach S (2004) Recent progress in magnetic fluid research. J Phys Condens Matter 16:R1135–R1150

Papell S (1963) Low viscosity magnetic fluid obtained by the colloidal suspension of magnetic particles

Puri IK, Ganguly R (2014) Particle transport in therapeutic magnetic fields. Annu Rev Fluid Mech 46:407–440

Rosensweig RE (1987) Magnetic fluids. Annu Rev Fluid Mech 19:437–463

Rowghanian P, Meinhart CD, Campàs O (2016) Dynamics of ferrofluid drop deformations under spatially uniform magnetic fields. J Fluid Mech 802:245–262

Sen U, Chatterjee S, Sen S, Tiwari MK, Mukhopadhyay A, Ganguly R (2017) Dynamics of magnetic modulation of ferrofluid droplets for digital microfluidic applications. J Magn Magn Mater 421:165–176

Stone HA, Lister JR, Brenner MP (1999) Drops with conical ends in electric and magnetic fields. Proc R Soc Lond A 455:329–347

Tan SH, Nguen N-T (2011) Generation and manipulation of monodispersed ferrofluid emulsions: the effect of a uniform magnetic field in flow-focusing and T-junction configurations. Phys Rev E 84:036317

Tan S-H, Nguyen N-T, Yobas L, Kang TG (2010) Formation and manipula- tion of ferrofluid droplets at a microfluidic T-junction. J Micromech Microeng 20:045004

Taylor G (1964) Disintegration of water drops in an electric field. Proc R Soc Lond A 280(1382):383–397

Taylor GI, McEwan AD (1965) The stability of a horizontal fluid interface in a vertical electric field. J Fluid Mech 22(1):1–15

Tyatyushkin AN, Velarde MG (2001) On the interfacial deformation of a magnetic liquid drop under the simultaneous action of electric and magnetic fields. J Colloid Interface Sci 235:46–58

Velásquez-García LF, Akinwande AI, Martinez-Sanchez M (2006) A planar array of micro-fabricated electrospray emitters for thruster applications. J Microelectromech Syst 15(5):1272–1280

Yan Q, Xuan S, Ruan X, Wu J, Gong X (2015) Magnetically controllable generation of ferrofluid droplets. Microfluid Nanofluid 19:1377–1384

Yarin AL, Pourdeyhimi B, Ramakrishna B (2014) Fundamentals and applications of micro and nanofibers. Cambridge University Press, Cambridge

Zakinyan A, Tkacheva E, Dikansky Y (2012) Dynamics of a dielectric droplet suspended in a magnetic fluid in electric and magnetic fields. J Electrostat 70:225–232
Zeleny J (1914) The electrical discharge from liquid points, and a hydrostatic method of measuring the electric intensity at their surfaces. Phys Rev 3(2):69–91
Zhu G-P, Nguyen N-T, Ramanujan RV, Huang X-Y (2011) Nonlinear deformation of a ferrofluid droplet in a uniform magnetic fluid. Langmuir 27:14834–14841

Part III
Boiling and Condensation Phenomena

Chapter 12
Coolability of Heat-Generating Porous Debris Beds in Severe Accident Situations

Aranyak Chakravarty, Priyankan Datta, Koushik Ghosh, Swarnendu Sen and Achintya Mukhopadhyay

Abstract Molten fuel–coolant interactions in postulated severe accident scenario of nuclear reactors lead to the formation of a porous debris bed. Substantial heat generation takes place within such debris beds as a result of radioactive decay, and this needs to be continuously removed in order to maintain the temperature of the debris material within acceptable limits. This is achieved by boiling heat transfer using cooling water. Any failure in this regard can lead to re-melting of the material in an extreme situation and lead to further catastrophic consequences. In this context, it becomes imperative to have an assessment of the limit beyond which the debris cannot be maintained in a coolable condition. This limit is typically identified by the occurrence of *dryout,* i.e. water vapour accumulation within the debris bed. This chapter attempts to highlight the underlying mechanism and the pertinent factors contributing to dryout occurrence in typical debris beds. Various experimental studies and numerical modelling carried out in this regard are thoroughly reviewed. Augmentation of the dryout limit using available techniques is discussed in detail. A numerical model that has been developed for analysing multiphase flow and the associated heat and mass transfer in such porous debris beds are also presented in this chapter along with some salient results.

Nomenclature

a_i Interfacial area density, m^{-1}
c_p Specific heat capacity, J/kg.K
D_p Particle diameter, m

A. Chakravarty
School of Nuclear Studies and Applications, Jadavpur University, Kolkata 700032, India

A. Chakravarty · P. Datta · K. Ghosh · S. Sen (✉) · A. Mukhopadhyay
Department of Mechanical Engineering, Jadavpur University, Kolkata 700032, India
e-mail: sen.swarnendu@gmail.com

K. Ghosh
e-mail: kghoshjdvu@gmail.com

K. Saha et al. (eds.), *Two-Phase Flow for Automotive and Power Generation Sectors*, Energy, Environment, and Sustainability,
https://doi.org/10.1007/978-981-13-3256-2_12

$\boldsymbol{F}$ Solid–fluid drag force, kg.m^{-2}.s^{-2}
$\boldsymbol{g}$ Acceleration due to gravity, m/s^2
h Enthalpy, J/kg
K Permeability, m^2
K_r Relative permeability
m Mass transfer rate, kg.s^{-1}
p Pressure, Pa
q Volumetric heat transfer rate, W.m^{-3}
R Interfacial momentum exchange coefficient, kg.m^{-3}.s^{-1}
T Temperature, K
$\boldsymbol{V}$ Velocity, m.s^{-1}

Greek Letters

α Volume fraction
ε_f Porosity
η Passability, m
η_r Relative passability
λ Thermal conductivity, W.m^{-1}.K^{-1}
μ Viscosity, kg.m^{-1}.s^{-1}
ρ Density, kg.m^{-3}
ψ Sphericity

Subscripts, Superscripts, etc.

f Fluid phase
i Liquid–vapour interface
j Primary phase index
k Dispersed phase index
l Liquid phase
LC Liquid continuous regime
s Solid phase
sat Saturation value
v Vapour phase
VC Vapour continuous regime
$'''$ Volumetric quantities

12.1 Introduction

Management guidelines for situations involving severe accidents in nuclear reactors require prompt shutdown of a reactor following an accident, thereby ensuring termination of the nuclear fission reaction. However, a substantial amount of decay heat continues to be generated within the reactor core even after reactor shutdown. This is estimated to be about 6% of the reactor thermal power just after reactor shutdown. The decay heat generated undergoes an exponential decrease to approximately 1% of the reactor thermal power 1 h after reactor shutdown. Such a large amount of decay heat generated needs to be continuously removed in order to prevent temperature rise within the reactor core. Unchecked temperature rise within the core can potentially cause meltdown of the core. This is the major difficulty faced in accident management of nuclear reactors and also serves as the basic premise of safety in nuclear reactors.

A postulated severe accident in a nuclear reactor is characterised by a lack of sufficient heat removal systems from the core due to the failure of normal and emergency cooling systems. Continuous heat generation within the reactor core, coupled with a lack of cooling systems, results in tremendous heating of the core and causes the residual cooling water within the core to evaporate. This exposes the decay heat-generating fuel rods to water vapour. As a consequence, heat removal from the rods decreases leading to rapid temperature rise and in an extreme situation may result in meltdown of the fuel rods.

In case the fuel rods melt down, a melt pool is formed within the reactor core comprising the fuel and support structure materials. This is typically referred to as *corium*, and it relocates to the reactor pressure vessel (RPV) due to gravitational action in the form of melt jets or drops (see Fig. 12.1). Two situations may arise during this relocation depending on whether the reactor pressure vessel contains any residual water. If the RPV does not contain any residual water, the molten material comes in direct contact with the RPV wall and accumulates in the form of a molten pool in the RPV lower head (see Fig. 12.1). In contrast to this scenario, the superheated molten mass comes in contact with subcooled water if the RPV contains any residual water. During the descent through this water pool, the molten material jets undergo hydrodynamic interactions with liquid water and are disintegrated into smaller droplets (Mahapatra et al. 2018). This action is typically termed as *hydrodynamic fragmentation*. Film boiling of water starts simultaneously since the molten jet temperature exceeds the Leidenfrost temperature of water. This results in the formation of a meta-stable mixture comprising of water, water vapour and the fragmented material. Under certain conditions, the vapour film enveloping the fragments may undergo destabilisation and this may bring the superheated molten mass in direct contact with water. The thermal stress generated as a result of this interaction creates local pressurisation, and this can cause further disintegration of the fragmented molten mass. This is typically termed as *thermal fragmentation*. The fragmented material is eventually quenched, and it settles down in the RPV as a mass of decay heat-generating debris (see Fig. 12.1). The above-described phenomena are typically

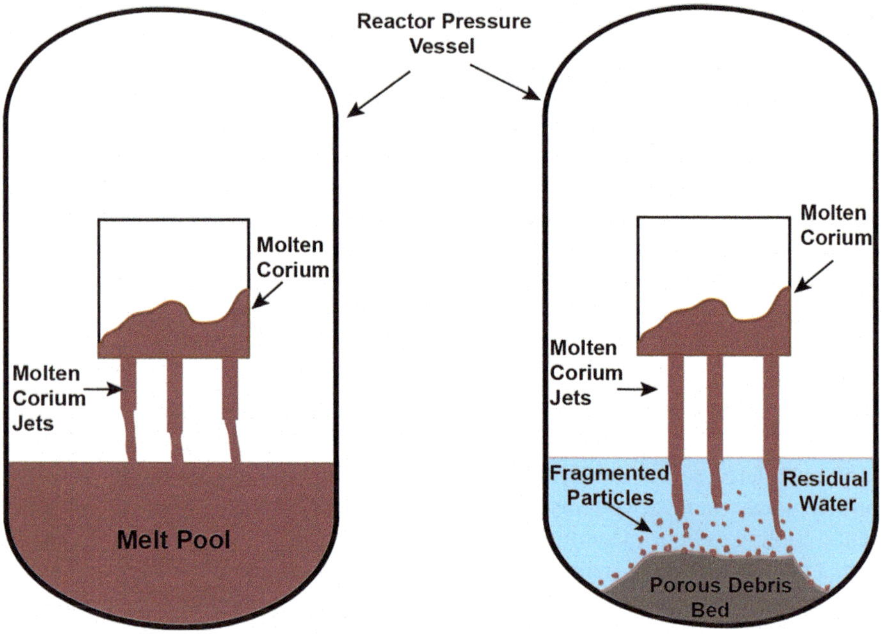

Fig. 12.1 Schematic representation of the formation of melt pool and debris bed in the RPV due to melt relocation

termed as the *in-vessel phase* in the accident progression sequence since the entire phenomena remain confined within the RPV.

An estimation of the decay heat generated within the debris yields a specific power between 100 and 300 W/kg and an equivalent power density between 0.5 and 1.5 MW/m^3. This obviously is dependent on the reactor type and composition of corium. Proper removal of such a large magnitude of decay heat generation can only be achieved by boiling heat transfer using cooling water. A stable and coolable condition can be achieved only if the evaporated water can be replaced continually—either by external injection of additional cooling water or by condensation and recirculation of the evaporated water. Also, it is necessary that the vapour generated within the debris is able to exfiltrate the debris and the cooling water is able to infiltrate into the debris. Failure to maintain the mass of debris in a coolable condition will lead to localised dryout and result in a subsequent rise of solid-phase temperature within the debris. This creates a possibility where the debris might be subjected to re-melting and again result in the formation of a molten pool in the RPV lower head. Thermal interaction occurs between the melt pool and the RPV wall, and in an extreme situation, this may result in failure of the RPV. Subsequently, the molten mass relocates to the reactor containment.

In the reactor containment, the molten mass is subjected to similar actions as in the in-vessel scenario. Failure to terminate the accident progression at this stage, how-

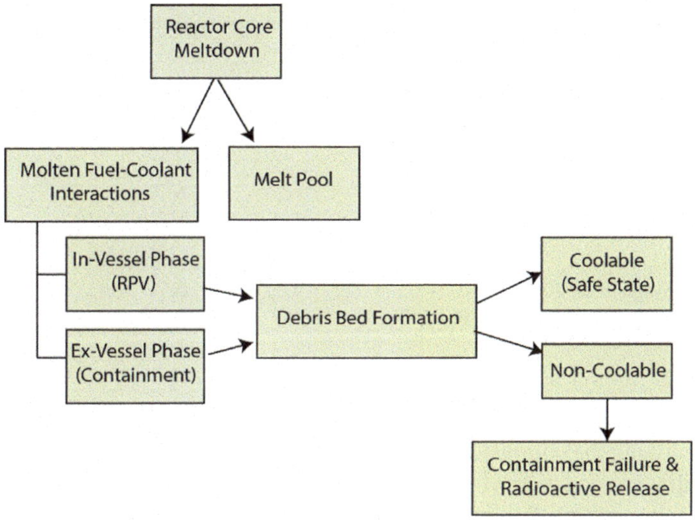

Fig. 12.2 Schematic of accident progression sequence considering molten fuel–coolant interactions

ever, can lead to steam explosion and the subsequent radioactive release to the environment. This phase of the accident sequence is termed as the *ex-vessel phase* since the entire phenomena occur outside the RPV. The accident progression sequence, described above, is schematically represented in Fig. 12.2.

It can, thus, be understood that ensuring adequate heat removal from the heat-generating debris and maintaining it in a coolable state is imperative for terminating the accident progression sequence. In this context, it becomes necessary to assess the limit beyond which the debris cannot be maintained in a coolable state. This limit is typically defined by the *dryout* condition and is represented in the literature in terms of the corresponding heating power (*dryout power*) or the heat flux that can be removed through the upper surface of the bed (*dryout heat flux*).

12.2 Debris Bed Characteristics

The composition and structure of a debris bed are primarily determined by the preceding interactions between the molten corium and the coolant. This is typically referred to by the term molten fuel–coolant interactions (MFCI). The phenomena of MFCI are itself dependent on various factors such as the size of the core breach, mass flux of the relocating molten corium, temperature of the molten corium and water pool. Therefore, it becomes essential to have a proper evaluation of the MFCI phenomena and its outcomes in order to carry out a comprehensive study on debris coolability.

Table 12.1 Debris bed composition observed in different experimental programmes

Experimental programme	Particle size range (mm)	Average porosity
KROTOS (Huhtiniemi and Magallon 2001)	<2	N/A
FARO (Magallon 2006)	<6	0.5–0.6
TROI (Song et al. 2003)	<6.35	N/A
DEFOR (Karbojian et al. 2009)	<10	0.46–0.7
COMECO (Singh et al. 2015)	<10 (without decay heat) <50 (with decay heat)	0.51

Several experimental investigations have been conducted in the past with a focus on characterising the composition of the debris. Experiments in the FARO and KROTOS facilities studied debris formation by considering a wide range of various parameters (Magallon 2006). Parameters considered include mass of the molten material, water pool depth, diameter of the melt jet and system pressure. Significant fragmentation of the melt jet was observed to occur irrespective of the operating conditions. The DEFOR experiments observed the effect of water pool subcooling on the fragmentation process. Similar investigations on debris formation have been carried out in TROI facility (Song et al. 2003) as well as the MISTEE-Jet facility (Manickam et al. 2016).

Observations from the above-mentioned experimental investigations reveal that the debris formed has an irregular and heterogeneous porous composition. This allows fluid movement through the interconnected void between the constituent particles. Significant uncertainty is, however, evident from reported experimental data on bed porosity and size of the constituent particles. Average porosity of the debris has been reported to vary between 0.25 and 0.7, while the particle size has been observed to range from very fine fragments (~100 μm) to large chunks (~50 mm). Table 12.1 summarises the observations reported from various experimental programmes on debris bed formation. These data have been used in recreating typical debris beds for the purpose of investigation into coolability of the debris beds.

A separate source of uncertainty arises with respect to the debris bed structure. It is expected that the debris bed formed due to MFCI will have a heap-like structure similar to that observed during pouring of granular material on a surface. Karbojian et al. (2009) observed that a heap-like debris bed is formed when the melt jet is completely fragmented into very fine particles (Fig. 12.3). Lin et al. (2017) characterised the structure of a debris bed in terms of the particle size. A flat-topped cylindrical bed structure was observed for very fine particles (<0.25 mm) while it evolved from a concave-shaped to a convex-shaped conical heap as the particle size was progressively increased to ~2.5 mm and beyond. This is represented in Fig. 12.4. The formation of heap-like beds has also been confirmed from a small-scale in-house experiment carried out on MFCI (Fig. 12.5).

Fig. 12.3 A typical debris bed formed as a result of MFCI in DEFOR experiment (used with permission from Karbojian et al. 2009)

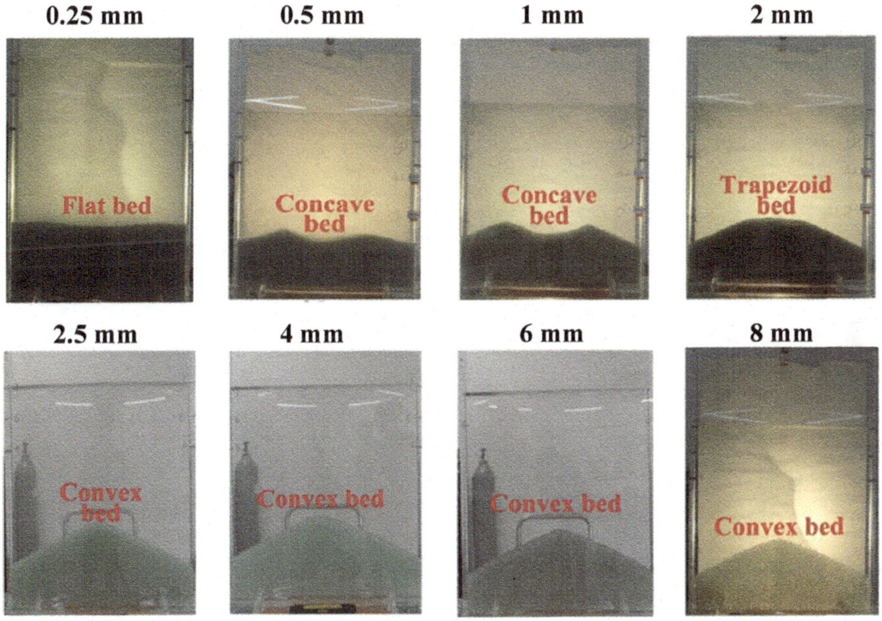

Fig. 12.4 Change in debris bed structure with particle size (used with permission from Lin et al. 2017)

12.3 Prior Investigations on Debris Coolability

12.3.1 Experimental Programmes

Observations from various experimental studies have enabled identification of the root cause of dryout occurrence in debris beds. Heat removal from the decay heat-generating solid particles is achieved by flooding the debris bed with cooling water.

Fig. 12.5 Debris structure observed in in-house experiments

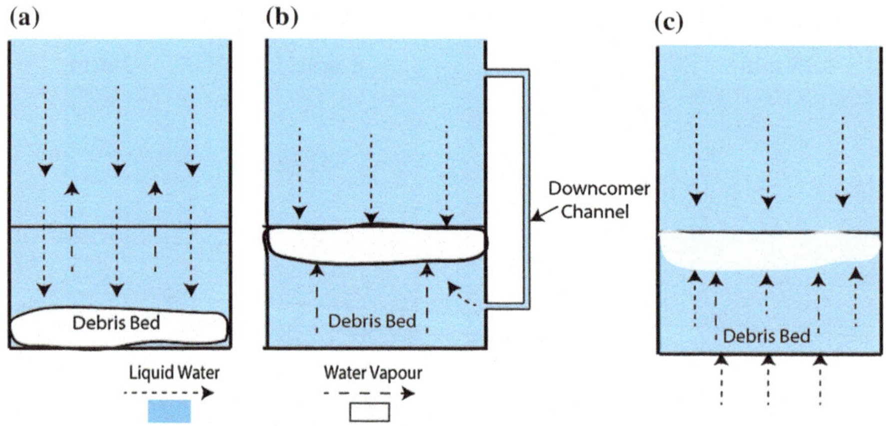

Fig. 12.6 Schematic representation of **a** top flooding and counter-current flow of water and vapour, **b** lateral flooding and **c** bottom flooding in one-dimensional debris beds

A high power density within the debris bed results in evaporation and boiling of the flooding water and leads to substantial vapour generation. The vapour generated, along with the heated flooding water, moves upwards due to buoyancy and ideally leaves the bed through the top surface. The vapour and heated water leaving the bed are replaced by additional cooling water which penetrates into the bed through its surfaces. A counter-current flow situation is, thus, established in the upper regions of the bed (see Fig. 12.6). The downward-moving cooling water faces resistance from the upward-moving vapour, and beyond a certain vapour mass flux, the cooling water would be unable to further infiltrate into the bed. The vapour, therefore, starts accumulating within the bed leading to subsequent dryout of the bed. This led to the conclusion that the establishment of the counter-current fluid flow mechanism within the bed is the major reason for dryout of the bed.

Dryout occurrence in debris beds is expected to be substantially influenced by the system pressure. One of the earliest reported works on the effects of system pressure on debris dryout is by Squarer et al. (1982). Reed et al. (1986) reported the impact of system pressure on dryout, as observed in the DCC experiments, by varying the system pressure in the range of 1–170 atmospheres for various particle sizes. The dependence of dryout phenomena on pressure was also studied by Miyazaki et al. (1986), in the DEBRIS experiments (Schäfer et al. 2006) and in the STYX experiments (Lindholm et al. 2006). In all these investigations, the DHF was observed to substantially increase with system pressure up to a 10 bar beyond which the DHF actually decreased with further increase in system pressure.

Dryout occurrence in debris beds has also been observed to be significantly influenced by the composition of the debris bed. Squarer et al. (1982) studied the effects of particle size and its distribution within the debris bed on DHF. The DCC experiments (Reed et al. 1986) considered the particle size to vary between 75 μm and 11.2 mm. Cha and Chung (1986) characterised the DHF as a function of water mass flux for a particle size varying between 1.5 and 4.0 mm in a forced convective situation over an inductively heated debris bed. The effect of particle size on dryout has also been studied in the STYX (Lindholm et al. 2006) and DEBRIS (Schäfer et al. 2006) experimental programmes. Results obtained from the experimental investigations indicate that the DHF becomes higher as the particle size is progressively increased. Porosity of the debris bed has also been observed to have a substantial impact on DHF (Ma and Dinh 2010).

The debris bed is expected to be non-homogeneous, and the particles constituting the bed are expected to be highly irregular in a realistic situation. This can be visualised from the photographs of experimentally obtained debris in Figs. 12.3, 12.4 and 12.5. The effect of a varying bed composition was experimentally investigated by Tung and Dhir (1987). They performed quenching and dryout experiments in an inductively heated particle bed with varying bed permeability in both axial and radial directions. The impact of heterogeneous composition of a debris bed was investigated in the DEBRIS experimental programme (Schäfer et al. 2006). A stratified bed composition was considered in the STYX (Lindholm et al. 2006) and POMECO (Nayak et al. 2006) experimental programmes. The DHF was observed to be substantially reduced in case of an axially stratified bed, with finer particles in the top layer, in the STYX facility (Lindholm et al. 2006). Morphology of the constituent particles has also been observed to have a significant impact on the dryout limit (Ma and Dinh 2010). Micro-inhomogeneity present in the constituent solid particles results in a localised high porosity zone, and this has also been observed to enhance the DHF of the bed by up to 50% when compared to a bed without such inhomogeneity (Ma and Dinh 2010).

The dryout limit of a debris bed can be expected to be augmented with appropriate modification of the fluid flow mechanism within the debris bed. Extensive investigations have also been carried out in this regard. Liquid–vapour flow within the porous bed can be effectively modified from counter-current mode (in a purely *top-flooding* situation) to co-current mode if additional coolant is injected from the bottom of the bed (*bottom flooding*) or across the lateral surface of the bed (*lateral flooding*). This is schematically represented in Fig. 12.6 and is expected to significantly enhance the dryout limit of the bed.

One of the earliest investigations on multidimensional flooding effects on coolability of debris beds is done by Wang and Dhir (1988). They studied the quenching characteristics of a heated particulate bed under bottom-flooding condition. Atkhen and Berthoud (2006) studied the effects of bottom flooding in the SILFIDE experimental facility. Observations reveal that coolant injection from the bottom of the debris bed, in addition to top flooding, is substantially more efficient compared to top-flooding mechanism only. Similar observations from experiments carried out in the DEBRIS facility have been reported by Schäfer et al. (2006) and Rashid et al. (2011). Ma and Dinh (2010) reported that the DHF is augmented by 40% in bottom-flooding situations when compared to top-flooding conditions. Emphasis on bottom coolant injection as a possible method of augmenting the dryout limit can be found in several other works (Bang and Kim 2010; Miscevic et al. 2006). The concept of bottom flooding of debris bed has been extended to the development of a core catcher device (Widmann et al. 2006).

The concept of lateral flooding of debris beds was experimentally investigated in the POMECO facility (Nayak et al. 2006) by utilising the concept of downcomers for recirculating the condensing vapour. The DHF was observed to be substantially augmented with the application of downcomers. The effect of lateral flooding has also been investigated by Takasuo et al. (2011) in the STYX experimental facility considering an irregular debris bed. It was reported that lateral flooding increased the DHF by 22–25% in case of homogeneous beds. However, in case of stratified beds, only a marginal increase was observed in the DHF with the use of lateral flooding. Similar observations of DHF enhancement in the DEBRIS experimental facility with the use of lateral flooding have been reported by Rashid et al. (2012).

As previously discussed, the quenched debris as a result of MFCI settles down in the RPV as a heap-like mass. However, it becomes difficult to predict beforehand the exact shape of the debris bed due to the uncertain nature of the parameters affecting it. Therefore, it becomes necessary to carry out investigations into the effects of bed structure on dryout occurrence. Surprisingly, all the experimental investigations discussed above only considered one-dimensional beds (see Fig. 12.7) and, as such, are not able to appreciate the multidimensional effects associated with heap-like beds (see Fig. 12.8). This has led to additional experimental investigations into the effects of bed geometry in the past few years. Takasuo et al. (2012) compared the dryout characteristics of conical and top-flooded cylindrical debris bed configurations in the COOLOCE experimental facility. A better coolability was observed in case of the conical bed when the two beds had the same height. However, if the two beds had the same volume with equal bed diameter, the conical bed was observed to have a lower dryout power density. This is due to a higher bed height in case of the conical bed which results in greater heat flux in upper parts of the bed. Thakre et al. (2014) compared the dryout power density of a triangular and a cylindrical bed in their study. The dryout power density in triangular beds was observed to increase by 69% and was attributed to the effects of multidimensional coolant infiltration in the triangular bed. Takasuo (2016) considered the following different bed configurations for analysis in the COOLOCE facility—conical, top-flooded cylinder, fully flooded cylinder, laterally flooded cylinder, cone on a cylindrical base and a truncated cone (see Fig. 12.9). The DHF increased by about 47–73% in bed configurations with multidimensional flooding effects.

Fig. 12.7 Top view of the particle bed used in DEBRIS experiments (used with permission from Schäfer et al. 2006)

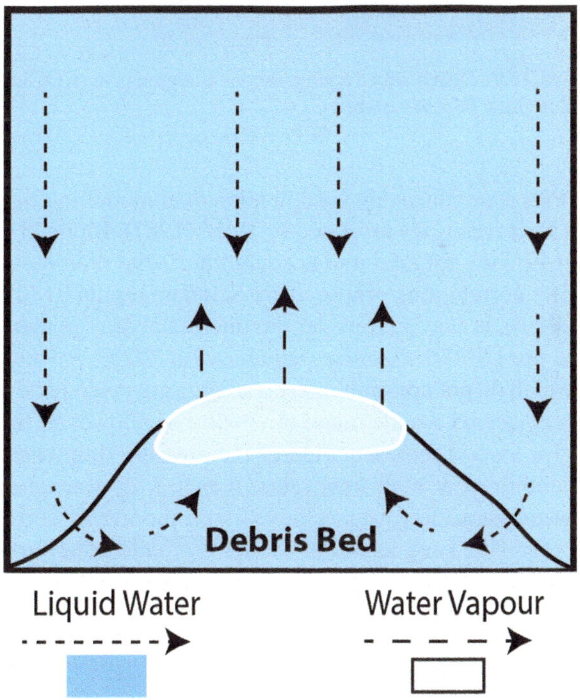

Fig. 12.8 Inherent multidimensional flooding associated with heap-like beds

Experimental investigations have also focused on developing correlations for the prediction of frictional pressure drop in two-phase flow situation through porous media. Frictional pressure drop determines the onset of counter-current flow limit for the liquid–vapour flow which contributes to dryout occurrence. These correla-

Fig. 12.9 Different bed configurations analysed in the COOLOCE experiments (used with permission from Takasuo 2016)

tions are extensively used in numerical modelling of the debris coolability problem. The correlations proposed by Reed (1982), Lipinski (1984) and Hu and Theofanous (1991) do not take into account interfacial drag between liquid and vapour phases. The correlations proposed by Schulenberg and Müller (1987) and Tung and Dhir (1988), however, consider the interfacial drag in their models. Experimental studies in the DEBRIS facility (Schäfer et al. 2006) have established that the correlations which do not consider interfacial drag are inadequate to predict the trends of pressure drop across a wide range of conditions. However, the Reed model was observed to give a good representation of the pressure drop characteristics for bottom-flooding conditions at high flow rates. A better representation of the qualitative behaviour of pressure drop was achieved with the use of correlations considering interfacial drag. The Tung and Dhir model has undergone further modifications with respect to consideration of smaller particles (Schmidt 2007) and the annular flow regime (Schmidt 2007; Taherzadeh and Saidi 2015; Li et al. 2018).

12.3.2 Numerical Studies

It is evident from the experimental studies on debris formation that a typical debris bed is composed of solid particles with interconnected voids. This gives it a porous structure, and as a result, the numerical models developed for porous media can be utilised in modelling flow through a debris bed. The earliest attempts at numerical predictions of dryout resulted in the development of empirical correlations and simplified models for DHF estimation in one-dimensional beds. These include the correlations proposed by Theofanous and Saito (1981), Lipinski (1984) and Schrock et al. (1986).

These correlations, however, are not suitable for DHF prediction in multidimensional situations. Multidimensional numerical models must be developed to predict the dryout occurrence in such situations. Such numerical models must take into account the existence of three distinct phases—solid particles, liquid water and water vapour—and must also consider the hydrodynamic and thermal interactions between the phases. The complex transport processes of two-phase flow and boiling heat transfer must also be considered in such models. Several attempts have been made at the development of such multidimensional numerical models which is capable of assessing debris coolability.

The debris coolability module in the severe accident analysis code SAMPSON provides a tool for predicting the safety margin following an accident by analysis of three-dimensional natural convection in a debris bed with simultaneous consideration of spreading, melting and solidification (Hidaka and Ujita 2001). In addition to this, the code is also equipped to evaluating wall failure by determining temperature distribution of the vessel walls.

Berthoud (2006) attempted to model the dryout of a debris bed with MC3D by implementing proper modifications to account for the complex processes associated with boiling heat transfer in porous media, including the presence of non-condensable gases. The modified code—MC3D-REPO—was validated with analytical solutions of various simplified problems as well as with two different experiments. Although the code could properly tackle the multidimensionality of the problem, its application became limited due to its assumption of thermal equilibrium in porous media and, as such, could not be applied for modelling post-dryout heat transfer from debris beds. Raverdy et al. (2017) reported the development made in MC3D with respect to tackling of the thermal non-equilibrium aspect of the problem.

Fichot et al. (2006) presented a multidimensional numerical model, considering thermal non-equilibrium, for two-phase flow in debris beds that is implemented in ICARE/CATHARE. Reasonable agreement was obtained in the prediction of DHFs, as available in the literature, in one-dimensional situations. Results of one-dimensional reflooding indicate the necessity of using a thermal non-equilibrium model and further indicate that the effects of channelling within the debris bed should be taken into account for more accurate modelling. The two-dimensional results presented highlight the influence of porous medium characteristics of a debris bed. As expected, water circulation is observed to improve when considering multidimensional flow in the bed. The dryout heat flux predicted in such situations is larger than that predicted by 1D modelling. The multidimensional flow situation establishes a flow pattern where steam can leave the debris bed in preferential channels, and there is reduced impact of counter-current flow on dryout occurrence. An improved model of heat transfer was implemented in ICARE/CATHARE by Bachrata (2012) and utilised to model reflooding situations in debris beds.

The numerical code WABE-2D (Bürger et al. 2006) was developed at IKE, University of Stuttgart, to simulate boil-offs and quenching of debris bed pertaining to debris coolability. The multidimensional effects along with top and bottom injection of coolants were discussed with proper constitutive laws for drag and interfacial friction. Experimental data from SILFIDE facility were used for the validation of the code. A different code developed at the same institute is MEWA. Takasuo et al. (2011) analysed the effects of lateral flooding in irregular debris bed using MEWA. Rahman (2013) utilised the MEWA code to perform extensive analysis on quenching as well as coolability of debris beds. Huang and Ma (2018) utilised the MEWA code to study the dryout phenomena in multidimensional heap-like beds. Numerical models have also been implemented in the framework of PORFLO (Takasuo 2015) and THERMOUS (Taherzadeh and Saidi 2015) programmes.

All the above-mentioned models and codes, however, have been sparingly used in addressing coolability of multidimensional heap-like beds. Also, no reported studies exist which have utilised commercially available computational fluid dynamics (CFD) tools for modelling the problem of debris coolability. The following section gives a brief description of a CFD model that has been developed by the authors and implemented in the framework of a commercial CFD platform (ANSYS Fluent) for analysis of debris coolability. Validations of the developed model and some salient results are also presented.

12.4 CFD Modelling of Multiphase Flow and Dryout Phenomena in Heap-like Debris Beds

Realistic numerical modelling of dryout phenomena in debris beds involves solution of the mass, momentum and energy transport equations in the debris bed as well as the surrounding clear fluid region for the liquid and the vapour phases. As can be understood from the preceding discussions, the agglomeration of the fragmented particles gives a porous nature to the debris bed, and hence, fluid movement becomes possible through the void between the particles. The well-established drag laws of porous media can, therefore, be utilised in modelling momentum transport through the debris bed. The local thermal non-equilibrium (LTNE) model is adopted in modelling energy transport through the porous debris bed in order to account for the large temperature difference existing within the heat-generating solid particles and the fluid phases, especially in dryout conditions.

The major assumptions made while developing the numerical model are summarised below.

1. The effect of capillary pressure is not considered in the model. As such, the same static pressure is shared by all the constituent phases, i.e. $p_l = p_v = p$.
2. The thermo-physical properties of all the phases are constant except the density of liquid and vapour phases which are modelled using the Boussinesq approximation.

3. Uniform heat generation takes place in the solid particles only.
4. The porous medium constituting the debris bed is homogeneous and isotropic.
5. The solid particles comprising the porous debris are perfectly spherical, i.e. $\psi = 1$ in the expression for permeability.

12.4.1 Governing Equations

The governing equations of mass, momentum and energy transport within the clear fluid region and the porous debris bed are stated below. The LTNE approach is followed for modelling the energy transport in debris bed.

Clear Fluid Region:

$$\frac{\partial}{\partial t}(\alpha_j \rho_j) + \nabla.(\alpha_j \rho_j \boldsymbol{V}_j) = m_{kj}''' \tag{12.1}$$

$$\frac{\partial}{\partial t}(\alpha_j \rho_j \boldsymbol{V}_j) + \nabla.(\alpha_j \rho_j \boldsymbol{V}_j \boldsymbol{V}_j) = -\nabla(\alpha_j p) + \mu_j \nabla^2 \boldsymbol{V}_j + m_{kj}''' \boldsymbol{V}_{kj} + R_{kj}(\boldsymbol{V}_k - \boldsymbol{V}_j) + \alpha_j \rho_j \boldsymbol{g} \tag{12.2}$$

$$\frac{\partial}{\partial t}(\alpha_j \rho_j h_j) + \nabla.(\alpha_j \rho_j \boldsymbol{V}_j h_j) = \alpha_j \lambda_j \Delta^2 T_j - q_{ji}''' + m_{kj}''' h_{ji} \tag{12.3}$$

Porous Debris Bed:

$$\frac{\partial}{\partial t}(\alpha_j \rho_j) + \nabla.(\alpha_j \rho_j \boldsymbol{V}_j) = m_{kj}''' \tag{12.4}$$

$$\frac{\partial}{\partial t}(\alpha_j \rho_j \boldsymbol{V}_j) + \nabla.\left(\frac{\alpha_j \rho_j \boldsymbol{V}_j \boldsymbol{V}_j}{\varepsilon_f}\right) = -\nabla(\varepsilon_f \alpha_j p) + \mu_j \nabla^2 \boldsymbol{V}_j$$

$$+\frac{1}{\varepsilon_f}\left(m_{kj}''' \boldsymbol{V}_{kj} + R_{kj}(\boldsymbol{V}_k - \boldsymbol{V}_j)\right) + \varepsilon_f \alpha_j \rho_j \boldsymbol{g} + \boldsymbol{F}_{sj} \tag{12.5}$$

$$\frac{\partial}{\partial t}(\varepsilon_f \alpha_j \rho_j h_j) + \nabla.(\alpha_j \rho_j \boldsymbol{V}_j h_j) = \alpha_j \varepsilon_f \lambda_j \nabla^2 T_j + q_{sj}''' - q_{ji}''' + m_{kj}''' h_{ji} \tag{12.6}$$

$$\frac{\partial}{\partial t}((1-\varepsilon_f)\rho_s c_{p,s} T_s) = (1-\varepsilon_f)\lambda_s \nabla^2 T_s + q_s''' - q_{sl}''' - q_{sv}''' - q_{si}''' \tag{12.7}$$

In the above equations, the subscripts j and k represent the primary and secondary fluid phases, respectively. The subscripts s, l and v refer to the solid, liquid and vapour phases, respectively. The liquid–vapour interface is denoted by the subscript i.

Equations 12.1 and 12.4 represent the mass transport, while Eqs. 12.2 and 12.5 represent the momentum transport in the clear fluid region and the debris bed, respectively. The term $\boldsymbol{F}_{sj}$ in Eq. 12.5 represents the momentum sink due to flow resistance offered by the porous composition of the debris bed. The general form of $\boldsymbol{F}_{sj}$ is expressed in a multiphase situation as

$$\boldsymbol{F}_{sj} = -\varepsilon_f \alpha_j \left(\frac{\mu_i}{K K_{r,j}} \boldsymbol{V}_j + \frac{\rho_i}{\eta \eta_{r,j}} |\boldsymbol{V}_j| \boldsymbol{V}_j \right) \tag{12.8}$$

where K and η represent the permeability and passability of the porous medium constituting the debris bed, respectively. These are defined as

$$K = \frac{\psi^2 D_p^2 \varepsilon_f^3}{150(1 - \varepsilon_f)^2} \tag{12.9}$$

$$\eta = \frac{\psi D_p \varepsilon_f^3}{1.75(1 - \varepsilon_f)} \tag{12.10}$$

Interfacial drag between the fluid phases is accounted for in the term R_{kj}. The term $m'''_{kj} \boldsymbol{V}_{kj}$ in the momentum transport equations represents the momentum exchange between the phases due to phase change where $\boldsymbol{V}_{kj}$ is the interphase velocity. The interphase velocities are determined as follows

$$\boldsymbol{V}_{kj} = \begin{cases} \boldsymbol{V}_k, if m'''_{kj} > 0 \\ \boldsymbol{V}_j, if m'''_{kj} < 0 \end{cases} \tag{12.11}$$

Energy transport in the fluid phases is represented by Eqs. 12.3 and 12.6 for the clear fluid region and the debris bed, respectively. Equation 12.7 represents the energy transport in the solid phase comprising the debris bed where q'''_s takes into account the volumetric heat generation in the solid particles.

12.4.2 Closure Relations

It becomes necessary to demarcate the fluid flow within the domain into different regimes in a multiphase flow situation. In this regard, the fluid flow regime is subdivided into three different regimes based on the volume fraction of the fluid phases—liquid continuous (or bubbly) regime, transition regime and vapour continuous (or droplet) regime. The regime demarcation adopted is summarised in Table 12.2.

Proper implementation of the numerical model also requires utilisation of appropriate closure relations for modelling various terms in the governing equations. These include the terms dealing with thermal and momentum exchange, mass transfer and turbulence. The various correlations used in this numerical model are listed in Table 12.3.

Table 12.2 Criterion adopted for demarcating the flow regimes

Flow regime	α_l	α_v
Liquid continuous (LC)	$\alpha_l \geq 0.7$	$\alpha_v \leq 0.3$
Transition	$0.01 < \alpha_l < 0.7$	$0.99 > \alpha_v > 0.3$
Vapour continuous (VC)	$\alpha_l \leq 0.01$	$\alpha_v \geq 0.99$

Table 12.3 Correlations utilised for modelling various closure terms

Closure term		Correlation	Remarks
Interfacial drag	Clear fluid region (R_{kj})	Schiller and Naumann (1935)	Implementation based on flow regime demarcation
	Debris bed $(R_{kj}, K_{r,j}, \eta_{r,j})$	Reed et al. (1986), Schulenberg and Muller (1987) and Schmidt (2007)	
Heat transfer	Convection from solid to liquid ($q'''_{s,l}$)	Ranz and Marshall (1952)	$\alpha_l \geq 0.7$ (or $\alpha_v \leq 0.3$)
	Boiling ($q'''_{s,i}$)	Nucleate boiling—Rhosenow (1952), film boiling—Bromley (1950)	$T_s > T_{sat}$
	Convection from solid to vapour ($q'''_{s,v}$)	Ranz and Marshall (1952)	$\alpha_v \geq 0.99$ (or $\alpha_l \leq 0.01$)
	Interfacial liquid–vapour heat transfer ($q'''_{j,i}$)	Ranz and Marshall (1952)	$0 < \alpha_v (\text{or}\, \alpha_l) < 1$
Mass transfer (m'''_{lv})		$\frac{q'''_{s,i}+q'''_{v,i}+q'''_{l,i}}{h_{v,sat}-h_{l,sat}}$	Only boiling and interfacial heat transfer considered
Turbulence		k–ε mixture model	Not considered within the debris bed due to very low fluid velocity

12.4.3 Implementation of the CFD Model in ANSYS Fluent

The numerical model detailed in the previous sections is implemented in the framework of the computational fluid dynamics (CFD) platform ANSYS Fluent. The porous media model available in ANSYS Fluent is adopted to take into account the effects of the porous debris bed. The Eulerian multiphase model of ANSYS Fluent is utilised in handling the transport equations for the fluid phases in multiphase situations (ANSYS 2012a).

Table 12.4 UDF modules utilised in implementation of the model (ANSYS 2012b)

Quantities	UDF Module
Interfacial momentum exchange coefficient	DEFINE_EXCHANGE_PROPERTY
Relative permeability and relative passability	DEFINE_PROFILE
Heat transfer terms	DEFINE_SOURCE
Mass transfer	DEFINE_MASS_TRANSFER
Transient term in solid energy transport equation	DEFINE_UDS_UNSTEADY
Diffusive term in solid energy transport equation	DEFINE_DIFFUSIVITY

The energy transport equation for the solid particles, when the LTNE approach is employed, is solved by defining the solid-phase temperature as a user-defined scalar (UDS) variable and solving an additional transport equation for the UDS. Various terms of the UDS transport equation are implemented by means of different user-defined functions (ANSYS 2012b).

Solution of the governing equations also requires proper implementation of the various closure relations. These include the interfacial momentum exchange coefficients in the clear fluid region as well as the debris bed, relative permeability and relative passability models for the debris bed, and the various heat transfer mechanisms. The implementation of these terms and coupling of the terms with the respective transport equations are achieved with extensive utilisation of user-defined functions (UDF) in ANSYS Fluent. Table 12.4 lists all the UDF modules utilised in the implementation (ANSYS 2012b).

12.4.4 Model Validations

In order to test the validity of the developed numerical model, the numerical predictions are compared with available experimental data or numerical data (in case experimental data are not available) for different situations.

Figure 12.10a–d represents the comparative results obtained for single-phase situations. Figure 12.10a compares the numerically calculated dimensionless temperature variation along the *x*-direction at different locations of a differentially heated porous enclosure undergoing natural convection with experimental data of Beckermann et al. (1987). Figure 12.10b shows a comparative plot of numerically obtained local Nusselt number along the cold wall of an enclosure containing heat-generating porous media in a natural convective situation. Both these comparisons are made using the LTE model. The validity of the LTNE model is assessed for a single-phase natural convective flow in a square enclosure filled with heat-generating porous media. A comparison of the average Nusselt number at the cold enclosure walls indicates good agreement with available numerical data of Baytas (2003) (see

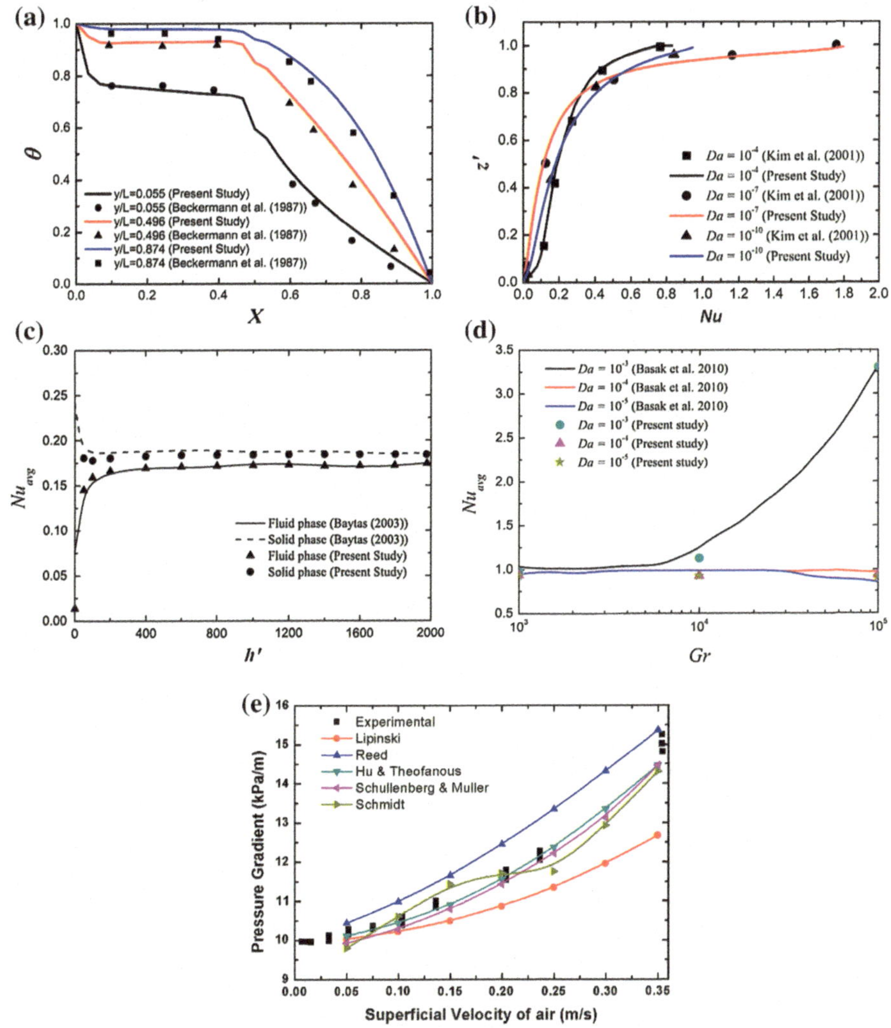

Fig. 12.10 Validation of the developed numerical model with **a** experimental data for single-phase LTE model, **b** numerical data for single-phase LTE model with internal heat generation, **c** numerical data for single-phase LTNE model with internal heat generation, **d** numerical data for single-phase LTE model in mixed convective situation, **e** experimental data of Li et al. (2015) for porous media drag models in two-phase flow situation (**a**–**d** used with permission Chakravarty et al. 2016, 2017, 2018)

Fig. 12.10c). The numerical model has also been validated for a single-phase mixed convective situation as represented by the comparative plot (Fig. 12.10d) of average Nusselt number with respect to Grash of number and Darcy number.

In a multiphase flow situation, the numerical model has been assessed for the two most challenging aspects encountered in numerical modelling of heat transfer from

Table 12.5 Comparison of experimental and numerical prediction of dryout

Bed configuration	Dryout power density (kW/m^3)		Deviation (%)
	Experimental (Takasuo 2016)	Numerical	
Cylindrical (top flooded)	1331.59	1370.75	2.94
Cylindrical (fully flooded)	2617.49	2610.96	0.24
Conical	2349.87	2415.14	2.77
Truncated conical	2602.0	2588.78	0.51

the debris bed, viz. pressure drop in porous media saturated with a two-phase mixture and heat transfer from the debris bed. Figure 12.10e represents the comparison of numerically obtained pressure gradient at different superficial air velocities against the experimental data of Li et al. (2015). The experiments were carried out in the DEBECO test facility with water and air at isothermal conditions. It can be observed that all the porous drag models are able to predict the pressure gradient with reasonable accuracy. Among these, the Schulenberg and Muller drag model is observed to have the best prediction. This drag model is, therefore, utilised for comparison of the heat transfer correlations utilised in the numerical model. The heat transfer aspect is validated in terms of prediction of the minimum dryout power density for different bed configurations against the experimental results of Takasuo (2016) obtained in the COOLOCE test facility. The results are tabulated in Table 12.5. It can be observed that a very close prediction of the minimum dryout power density is achieved for all the bed configurations.

In a nutshell, it can be stated from the above-stated comparisons that the developed numerical model is able to appreciably predict the fluid flow and heat transfer characteristics in porous media and clear fluid for single-phase and multiphase flow situations. It can, thus, be utilised in carrying out studies on debris coolability.

12.4.5 Salient Results and Discussion

The validated CFD model has been utilised in assessing coolability of typical debris beds—situations have been analysed with and without the consideration of phase change of the working fluid. Coolability analysis without considering phase change has been carried out for two different situations—natural convection and mixed convection. The salient results from these analyses are discussed in the following sections.

12.4.5.1 Analyses Without Considering Phase Change

The single-phase natural convection problem assumes a heat-generating debris bed located centrally on the base of a cylindrical enclosure (see Fig. 12.11). The problem is solved following both LTE and LTNE approaches. The side wall of the enclosure is assumed to be in a cold isothermal state, and all other walls are assumed to be adiabatic in the analysis carried out following the LTE approach (Fig. 12.11a). In contrast, in the analysis following the LTNE approach, the top wall is also assumed to be in a cold isothermal state and only the bottom wall is assumed to be in an adiabatic condition (Fig. 12.11b).

Results (Figs. 12.12 and 12.13) indicate that heat transfer takes place from the heat-generating debris bed to the cooler working fluid. This establishes a buoyancy-driven counter-clockwise fluid circulation within the enclosure. Fluid circulation results in the heated fluid first coming in contact with the top wall where thermal interaction takes place between the fluid and the wall depending upon the imposed boundary condition. In case of the adiabatic wall (LTE approach), heat transfer does not occur at the top wall, and as a result, the entire thermal energy transfer takes place at the side wall of the enclosure (Fig. 12.12). In the LTNE approach, however, the major amount of heat transfer takes place at the top wall due to its cold isothermal state, and hence, only the residual energy is transferred to the side wall (Fig. 12.13).

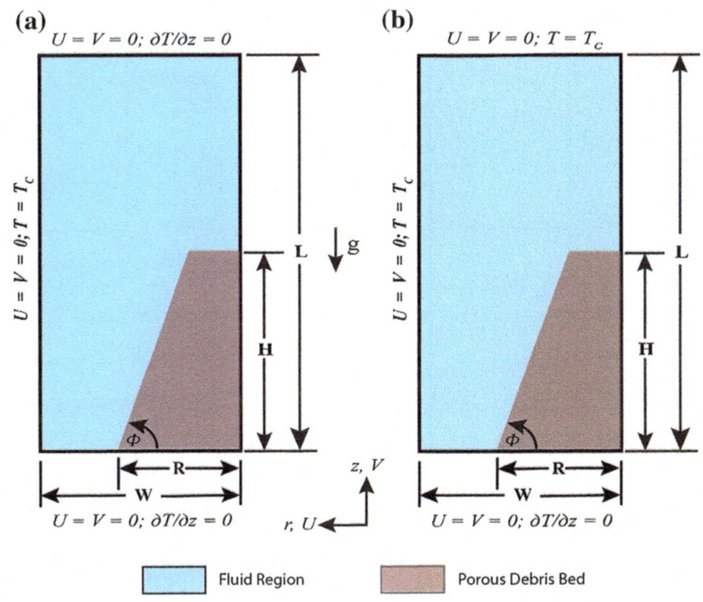

Fig. 12.11 Schematic representation of the problem geometry utilised for analysing single-phase natural convection using **a** LTE approach and **b** LTNE approach (redrawn with permission from Chakravarty et al. 2016, 2017)

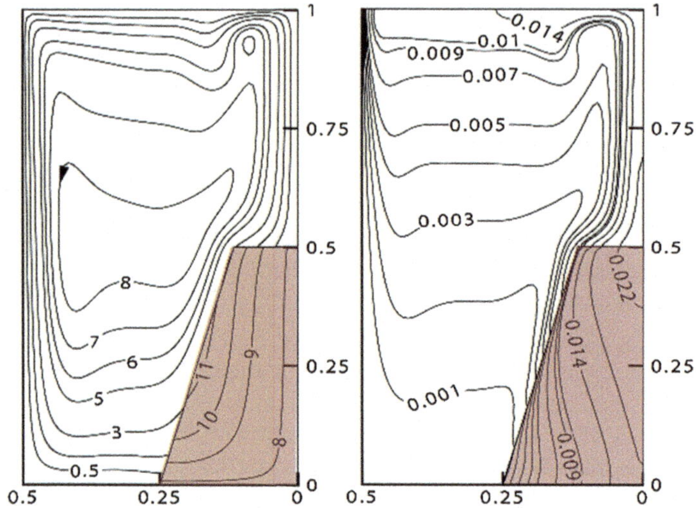

Fig. 12.12 Stream function (left) and isotherm (right) contour within the domain using the LTE approach (used with permission from Chakravarty et al. 2016)

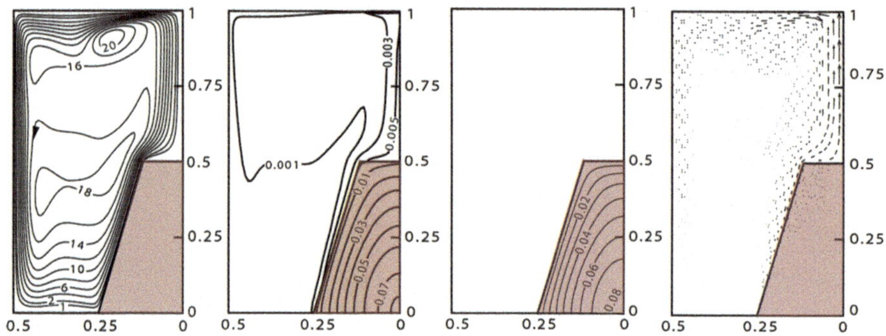

Fig. 12.13 Stream function, fluid isotherm, solid isotherm and energy flux vectors within the domain using the LTNE approach (used with permission from Chakravarty et al. 2017)

The effects of bed permeability, bed heat generation, thermal conductivity ratio and bed geometry on the natural convective fluid flow and heat transfer characteristics of the system are analysed in this investigation. All these parameters are observed to significantly influence the fluid flow mechanism and, consequently, the heat transfer characteristics as well. Figure 12.14 represents the heat transfer characteristics obtained with the LTE approach and LTNE approaches, in terms of dimensionless average Nusselt number(Nu_{avg}), with variations in bed heat generation (in terms of dimensionless Rayleigh number (Ra)) and bed permeability (in terms of dimensionless Darcy number (Da)). The effects of bed geometry and thermal

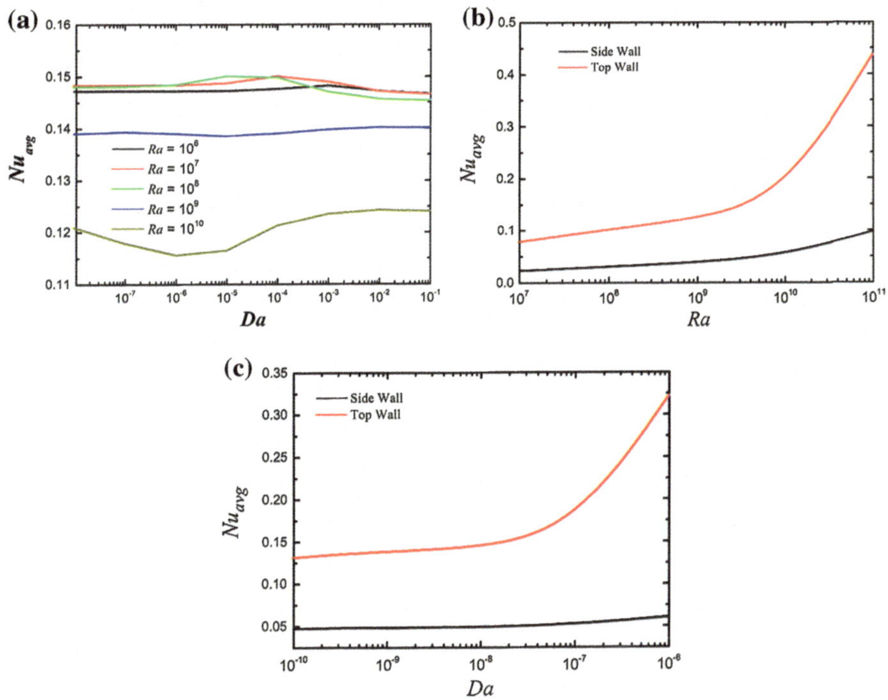

Fig. 12.14 Variations of **a** Nu_{avg} with Da for different Ra using LTE approach, **b** Nu_{avg} for top wall and side wall with Ra at $Da = 10^{-7}$ using LTNE approach and **c** Nu_{avg} for top wall and side wall with Da at $Ra = 10^{10}$ using LTNE approach (used with permission from Chakravarty et al. 2016, 2017)

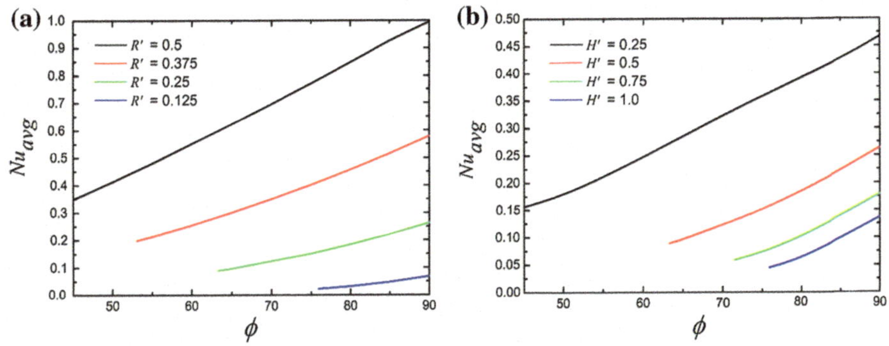

Fig. 12.15 Variation of Nu_{avg} with bed angle (ϕ) at $Ra = 10^8$, $Da = 10^{-4}$ for **a** different bed radii at $H' = 0.5$ and **b** different bed height at $R' = 0.25$ (used with permission from Chakravarty et al. 2016)

conductivity ratio on the heat transfer characteristics are evident from Fig. 12.15 and Table 12.6, respectively.

Table 12.6 Variation in Nu_{avg} with changing thermal conductivity ratio (λ)

λ	Nu_{avg}
1.0	0.14988
5.0	0.69978
10.0	1.39957

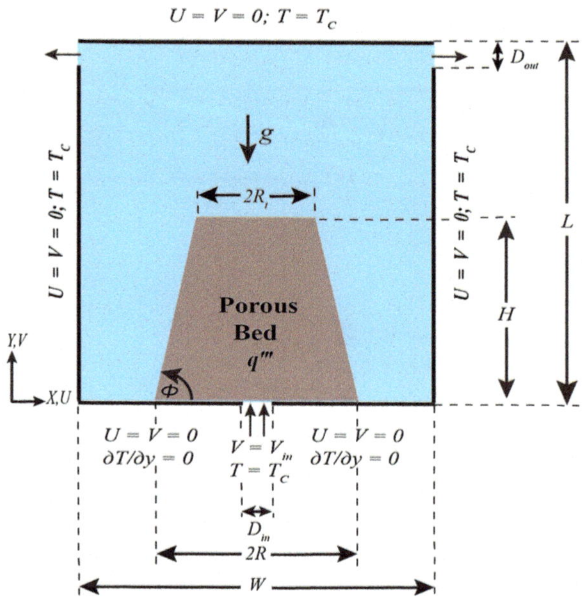

Fig. 12.16 Schematic representation of the problem geometry utilised for analysing single-phase mixed convection (used with permission from Chakravarty et al. 2018)

Significant information is obtained from this analysis regarding the dominant heat transfer mechanism. The dominant mode of heat transfer from the debris bed to the clear fluid region is observed to be dependent on the associated heat generation rate as well as the bed permeability. At a fixed heat generation rate, convective heat transfer is observed to dominate in a highly permeable bed. However, conductive heat transfer is observed to have an increasing influence as the permeability decreases. This change is observed to occur at a specific bed permeability for a given bed heat generation rate and is expressed in terms of the dimensionless number *RaDa*. In the LTE approach, this change is observed to occur at $RaDa = 10^3$ when $Ra < 10^8$ and at $RaDa = 10^4$ when $Ra > 10^8$. In contrast, this magnitude of *RaDa* is observed to be 100 in the analysis using LTNE approach.

The motivation for the mixed convection analysis is drawn from the concept of bottom injection of coolant as a means of augmenting coolability of debris beds. Such fluid injection creates a mixed convective flow situation within the cavity containing the debris bed. Analysis is carried out for this situation following the LTE approach considering single-phase laminar flow. Figure 12.16 represents the problem geometry analysed in this study. The bounding walls of the cavity are considered to be in an isothermally cooled state.

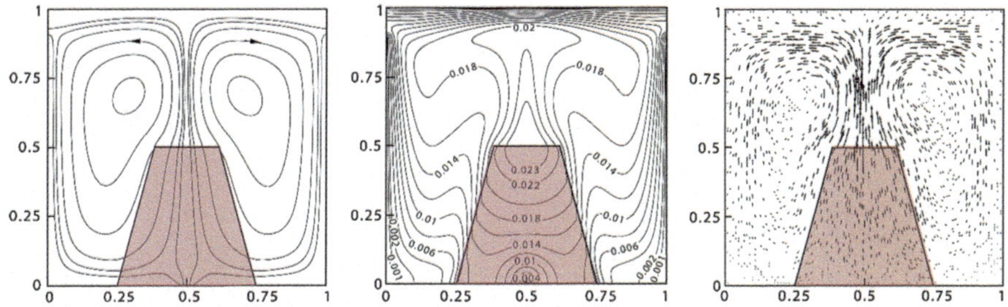

Fig. 12.17 Contours of stream function (left), isotherm (centre) and energy flux vectors (right) within the domain (used with permission from Chakravarty et al. 2018)

It is observed that fluid circulation remains symmetric within the enclosure and heat transfer at the top wall significantly exceeds that at the side walls, irrespective of the flow situation. The symmetric nature of fluid flow is evident from the contours of stream function and isotherms, as well as the energy flux vectors (Fig. 12.17). Another common feature observed is the location of the maximum temperature zone in the upper region of the bed which is primarily due to the effect of cold fluid injection from the bottom of the bed. This is in contrast to that observed in a purely natural convective situation where the maximum temperature rise is observed to occur in the inner regions of the bed (see Figs. 12.12 and 12.14).

Analysis shows that the fluid flow within the system is governed mainly by the combined effects of the following factors—permeability of the debris bed, natural convection as a result of internal heat generation within the debris bed and inertial flow due to bottom injection of cold fluid. Thermal energy transfer takes place from the debris bed to the cold walls and also to the system outlet by the coupled effects of the buoyancy-induced and inertial flow mechanisms. The relative dominancy of these flow mechanisms is denoted by the dimensionless Richardson number (Ri). Fluid flow mechanism changes from a strongly inertial flow at very low Ri to a buoyancy-driven dominated flow at very high Ri. Heat transfer characteristics of the system are represented in terms of Nu_{avg} at the cold walls for different situations. Stronger fluid injection at low Ri situations enables greater heat transfer from the debris bed. This results in a lower bed temperature rise. Consequently, a weaker temperature gradient is established within the cavity resulting in a smaller steady-state magnitude of Nu_{avg} (see Fig. 12.18). The injection strength of cold fluid decreases as Ri gradually increased, and as such, relatively less heat transfer takes place from the bed. This results in a very high temperature rise within the bed leading to a larger temperature gradient within the enclosure and a consequent increase in Nu_{avg}, as evident from Fig. 12.18.

An interesting observation made from this analysis shows that the fluid flow spreads laterally in the vicinity of the inlet channel as the permeability of the bed is reduced. This prevents the cooler fluid from reaching the upper regions of the bed. As a result, sufficient heat removal does not occur from this region and the maximum rise in temperature is observed in this region.

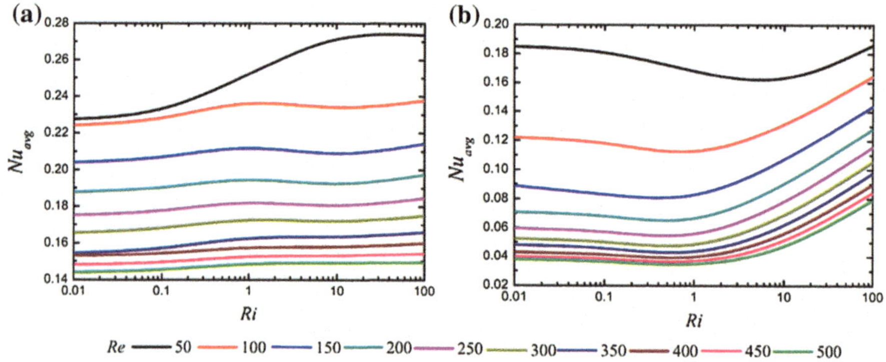

Fig. 12.18 Nu_{avg} variation with Richardson number (*Ri*) at different Reynolds number (*Re*) at a constant Darcy number (*Da*) at **a** top wall and **b** side walls (used with permission from Chakravarty et al. 2018)

12.4.5.2 Analysis Considering Phase Change

A critical issue faced in the debris coolability analysis considering phase change of cooling water is to devise an appropriate numerical method for identification of dryout in the debris beds. This is achieved by determining the minimum liquid volume fraction and the maximum solid-phase temperature within the debris bed at each time instant. The identification is done with the help of user-defined functions. It is concluded that dryout has occurred in the debris bed if the minimum liquid saturation becomes zero, and this condition is sustained throughout the rest of the time period, and the corresponding value of maximum solid temperature also indicates a sustained rise (of at least 5 K) above the steady-state temperature obtained using the immediately lower power level. This method of dryout identification is represented in Fig. 12.19 for the validation case (see Table 12.5) with the conical bed configuration (Fig. 12.20).

This method of dryout determination, however, does not give sufficient information with respect to the spatial location of the dryout zone. Spatial distribution of α_l and T_s within the domain at different time instances is utilised in order to obtain the spatial location of the dryout zone. Figure 12.21 represents liquid volume fraction and solid-phase temperature distributions for the truncated conical bed configuration (see Fig. 12.20) in a dryout situation. A counter-clockwise circulation of the liquid is established within the domain with the cold liquid entering the domain from the left side of the top boundary. The downward-moving cold liquid comes into contact with the upward-moving heated liquid as well as the vapour generated due to phase change, near the top surface and the upper portion of the lateral surface of the bed. This creates a counter-current flow situation for the fluids across the top surface. The flow situation becomes chaotic across the lateral surface with fluid flow taking place in different directions. As a result, the vapour generated starts to accumulate in these regions, which lowers the heat removal from the bed and ultimately results in dryout.

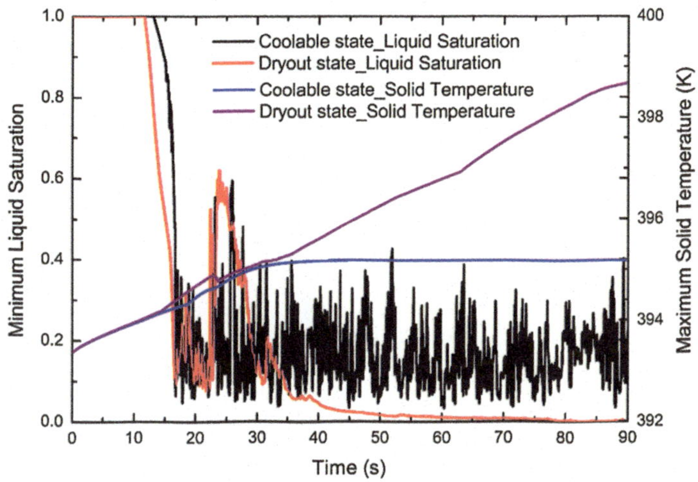

Fig. 12.19 Transient history of minimum liquid saturation and maximum solid temperature as indicator of dryout occurrence in the conical bed configuration with Schulenberg and Müller drag model

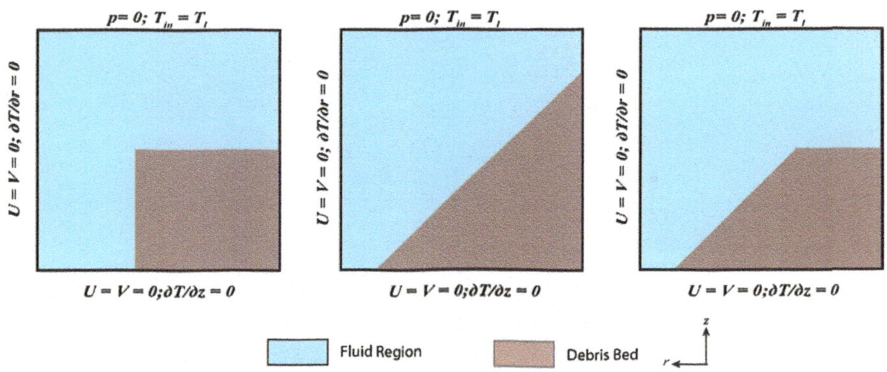

Fig. 12.20 Different bed configurations analysed considering phase change

The location of dryout is clearly evident from the liquid saturation as well as the solid temperature distributions. It can also be observed that the increase in temperature of the solid particles remains localised to the dryout region only. Temperature in the rest of the debris bed remains near the saturation value indicating sufficient cooling in these regions. Figures 12.22 and 12.23 represent the solid temperature distribution in dryout condition for the cylindrical and conical bed configurations, respectively.

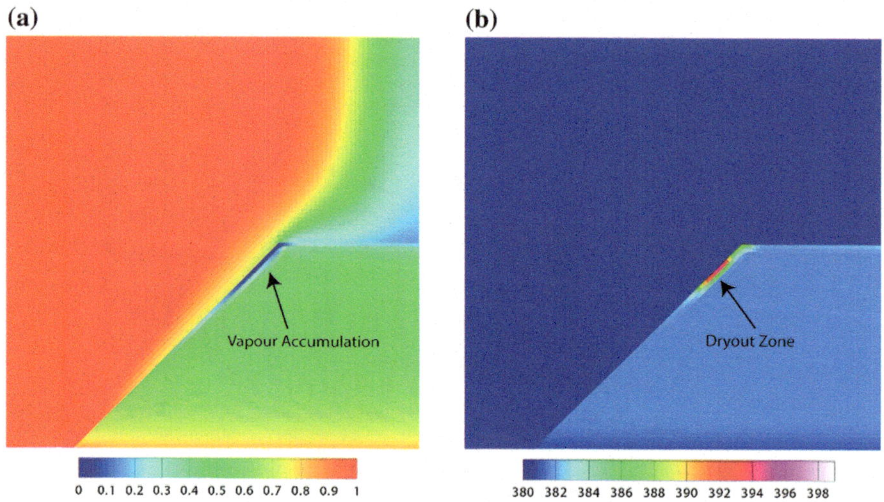

Fig. 12.21 **a** Liquid volume fraction and **b** solid-phase temperature (in K) distributions in dryout condition for the truncated conical bed configuration

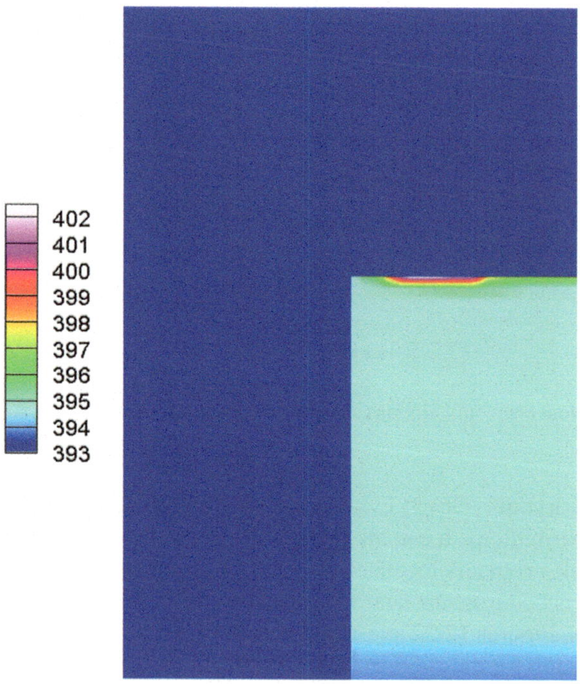

Fig. 12.22 Solid temperature distribution (in K) at dryout condition for the cylindrical bed configuration

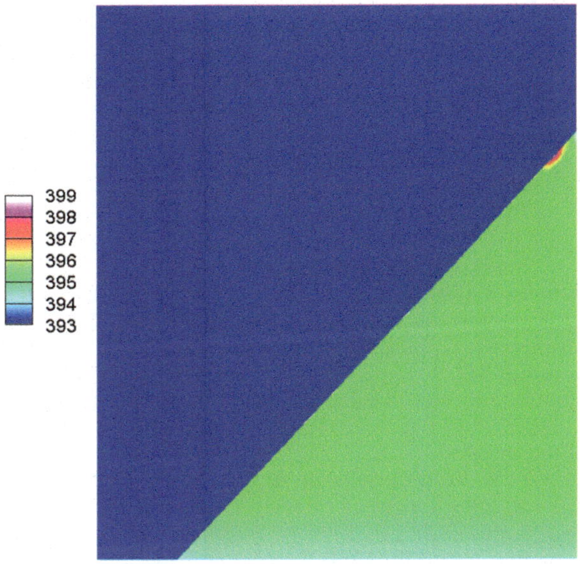

Fig. 12.23 Solid temperature distribution (in K) at dryout condition for the conical bed configuration

12.5 Conclusions

This chapter discusses in detail the phenomena of debris dryout in the context of severe accident in nuclear reactors. Dryout occurs in the debris beds due to insufficient heat removal leading to vapour accumulation and consequent temperature rise within the debris. This presents a possible scenario where the debris might undergo further re-melting and cause further accident progression. Therefore, it becomes necessary to assess the dryout limit of typical debris beds.

A computational fluid dynamics model for modelling the associated multiphase flow and boiling heat transfer is also presented in this chapter. This model is implemented within the framework of the commercial CFD platform ANSYS Fluent. Validations of the CFD model with existing experimental and numerical data are presented, and some salient results obtained using the implemented CFD model are also discussed. It can be inferred from these that the implemented CFD model is well equipped to handle single-phase and multiphase flows in clear fluid regions and heat-generating porous debris beds. Appreciable accuracy in dryout prediction of typical heap-shaped debris beds is also achieved using this CFD model.

Acknowledgements The authors are grateful to AREVA SA for providing fellowship to the first author.

References

ANSYS Inc. (2012a) ANSYS FLUENT theory guide

ANSYS Inc. (2012b) ANSYS FLUENT UDF manual

Atkhen K, Berthoud G (2006) SILFIDE experiment: Coolability in a volumetrically heated debris bed. Nucl Eng Des 236:2126–2134

Bachrata A (2012) Modeling of core flooding in a highly degraded reactor. Ph.D. thesis, University of Toulouse, France

Bang KH, Kim JM (2010) Enhancement of dryout heat flux in a debris bed by forced coolant flow from below. Nucl Eng Tech 42(3):297–304

Baytaş AC (2003) Thermal non-equilibrium natural convection in a square enclosure filled with a heat-generating solid phase, non-darcy porous medium. Int J Energy Res 27:975–988

Beckermann C, Ramadhyani S, Viskanta R (1987) Natural convection flow and heat transfer between a fluid layer and a porous layer inside a rectangular enclosure. J Heat Transf 109:363–370

Berthoud G (2006) Models and validation of particulate debris coolability with the code MC3D-REPO. Nucl Eng Des 236:2135–2143

Bromley LA (1950) Heat transfer in stable film boiling. Chem Eng Prog 46:221–227

Bürger M, Buck M, Schmidt W, Widmann W (2006) Validation and application of the WABE code: investigations of constitutive laws and 2D effects on debris coolability. Nucl Eng Des 236:2164–2188

Cha JH, Chung MK (1986) Forced flow dryout heat flux in heat generating debris bed. J Korean Nuclear Soc 18(4):273–280

Chakravarty A, Datta P, Ghosh K, Sen S, Mukhopadhyay A (2016) Numerical analysis of a heat-generating, truncated conical porous bed in a fluid-filled enclosure. Energy 106:646–661

Chakravarty A, Datta P, Ghosh K, Sen S, Mukhopadhyay A (2017) Thermal non-equilibrium heat transfer and entropy generation due to natural convection in an enclosure with a truncated conical, heat-generating porous bed. Transp Porous Med 116:353–377

Chakravarty A, Datta P, Ghosh K, Sen S, Mukhopadhyay A (2018) Mixed convective heat transfer in an enclosure containing a heat-generating porous bed under the influence of bottom injection. Int J Heat Mass Transf 117:645–657

Fichot F, Duval F, Trégourès N, Béchaud C, Quintard M (2006) The impact of thermal non-equilibrium and large-scale 2D/3D effects on debris bed reflooding and coolability. Nucl Eng Des 236:2144–2163

Hidaka M, Ujita H (2001) Verification for flow analysis capability in the model of three-dimensional natural convection with simultaneous spreading, melting and solidification for the debris coolability analysis module in the severe accident analysis code 'SAMPSON'. J Nucl Sci Technol 38(9):745–756

Hu K, Theofanous TG (1991) On the measurement of dryout in volumetrically heated coarse particle beds. Int J Multiph Flow 17:519–532

Huang Z, Ma W (2018) Validation and application of the MEWA code to analysis of debris bed coolability. Nucl Eng Des 327:22–37

Huhtiniemi I, Magallon D (2001) Insights into steam explosions with corium melts in KROTOS. Nucl Eng Des 204:391–400

Karbojian A, Ma WM, Kudinov P, Dinh T (2009) A scoping study of debris bed formation in the DEFOR test facility. Nucl Eng Des 239:1653–1659

Li L, Kong L, Zou X, Wang H (2015) Pressure drops of single/two-phase flows through porous beds with multi-sizes spheres and sands particles. Ann Nucl Energy 85:290–295

Li L, Zou X, Wang H, Zhang S, Wang K (2018) Investigations on two-phase flow resistances and its model modifications in a packed bed. Int J Multiphase Flow 101:24–34

Lin S, Cheng S, Jiang G, Pan Z, Lin H, Wang S, Wang l, Zhang X, Wang B (2017) A two-dimensional experimental investigation on debris bed formation behaviour. Prog Nucl En 96:118–132

Lindholm I, Holmström S, Miettinen J, Lestinen V, Hyvärinen J, Pankakoski P, Sjövall H (2006) Dryout heat flux experiments with deep heterogeneous particle bed. Nucl Eng Des 236:2060–2074

Lipinski RJ (1984) A coolability model for post-accident nuclear reactor debris. Nucl Technol 65:53–66

Ma WM, Dinh TN (2010) The effects of debris bed's prototypical characteristics on corium coolability in a LWR severe accident. Nucl Eng Des 240:598–608

Magallon D (2006) Characteristics of corium debris bed generated in large-scale fuel-coolant interaction experiments. Nucl Eng Des 236:1998–2009

Mahapatra PS, Datta P, Chakravarty A, Ghosh K, Manna NK, Mukhopadhyay A, Sen S (2018) Molten drop to coolant heat transfer during premixing of fuel coolant interaction. In: Basu S, Agarwal A, Mukhopadhyay A, Patel C (eds) Applications paradigms of droplet and spray transport: paradigms and applications. Springer, Singapore, pp 201–235

Manickam L, Kudinov P, Ma W, Bechta S, Grishchenko D (2016) On the influence of water subcooling and melt jet parameters on debris formation. Nucl Eng Des 309:265–276

Miscevic M, Rahli O, Tadrist L, Topin F (2006) Experiments on flows, boiling and heat transfer in porous media: Emphasis on bottom injection. Nucl Eng Des 236:2084–2103

Miyazaki K, Murai K, Ohama T, Yamaoka N, Inoue S (1986) Pressure dependence of the particle bed dryout heat flux. Nucl Eng Des 95:271–273

Nayak AK, Sehgal BR, Stepanyan AV (2006) An experimental study on quenching of a radially stratified heated porous bed. Nucl Eng Des 236:2189–2198

Rahman S (2013) Coolability of corium debris under severe accident conditions in light water reactors. Ph.D. thesis, Institute of Nuclear Technology and Energy Systems (IKE), University of Stuttgart, Germany

Ranz WE, Marshall WM (1952) Evaporation from drops. Chem Eng Prog 48:141–146

Rashid M, Kulenovic R, Laurien E, Nayak AK (2011) Experimental results on the coolability of a debris bed with multidimensional cooling effects. Nucl Eng Des 241:4537–4543

Rashid M, Kulenovic R, Laurien E (2012) Experimental results on the coolability of a debris bed with down comer configurations. Nucl Eng Des 249:104–110

Raverdy B, Meignen R, Piar L, Picchi S, Janin T (2017) Capabilities of MC3D to investigate the coolability of corium debris beds. Nucl Eng Des 319:48–60

Reed AW (1982) The effect of channeling on the dryout of heated particulate beds immersed in a liquid pool. Ph.D. thesis, Massachusetts Institute of Technology, Cambridge, USA

Reed AW, Bergernon ED, Boldt KR, Schmidt TR (1986) Coolability of UO2 debris beds in pressurized water pools: DCC-1 and DCC-2 experiment results. Nucl Eng Des 97(1):81–88

Rhosenow W (1952) A method of correlating heat transfer data for surface boiling of liquids. Trans. ASME 74:969–976

Schäfer P, Groll M, Kulenovic R (2006) Basic investigations on debris cooling. Nucl Eng Des 236:2104–2116

Schiller L, Naumann Z (1935) A drag coefficient correlation. Z Ver Deutsch Ing 77:318–320

Schmidt W (2007) Interfacial drag of two-phase flow in porous media. Int J Multiph Flow 33:638–657

Schrock VE, Wang CH, Revankar S, Wei LH, Lee SY (1986) Flooding in particle beds and its role in dryout heat flux prediction. In: Proceedings of the sixth meeting on debris coolability, Los Angeles, California, EPRI NP-4455, Palo Alto, California

Schulenberg T, Müller U (1987) An improved model for two-phase flow through beds of coarse particles. Int J Multiphase Flow 13:87–97

Singh N, Kulkarni PP, Nayak AK (2015) Experimental investigation on melt coolability under bottom flooding with and without decay heat simulation. Nucl Eng Des 285:48–57

Song JH, Hong SW, Kim JH, Chang YJ, Shin YS, Min BT, Kim HD (2003) Insights from the recent steam explosion experiments in TROI. J Nucl Sci Technol 40(10):783–795

Squarer D, Pieczynski AT, Hochreiter LE (1982) Effect of debris bed pressure, particle size and distribution on degraded nuclear reactor core coolability. Nucl Sci Eng 80(1):2–13

Taherzadeh M, Saidi MS (2015) Modeling of two-phase flow in porous media with heat generation. Int J Multiph Flow 69:115–127

Takasuo E (2015) Coolability of porous core debris beds: Effects of bed geometry and multi-dimensional flooding. PhD Thesis, VTT Technical Research Centre of Finland
Takasuo E (2016) An experimental study of the coolability of debris beds with geometry variations. Ann Nucl Energy 92:251–261
Takasuo E, Holmström S, Kinnunen T, Pankakoski P, Hosio E, Lindholm I (2011) The effect of lateral flooding on the coolability of irregular core debris beds. Nucl Eng Des 241:1196–1205
Takasuo E, Holmström S, Kinnunen T, Pankakoski P (2012) The COOLOCE experiments investigating the dryout power in debris beds of heap-like and cylindrical geometries. Nucl Eng Des 250:687–700
Thakre S, Li L, Ma WM (2014) An experimental study on coolability of a particulate bed with radial stratification or triangular shape. Nucl Eng Des 276:54–63
Theofanous TG, Saito M (1981) An assessment of class 9 (core-melt) accidents for PWR dry-containment systems. Nucl Eng Des 66:301–332
Tung VX, Dhir VK (1987) Quenching of debris bed having variable permeability in axial and radial directions. Nucl Eng Des 99:275–284
Tung VX, Dhir VK (1988) A hydrodynamic model for two-phase flow through porous media. Int J Multiph Flow 14:47–65
Wang CH, Dhir VK (1988) An experimental investigation of multidimensional quenching of a simulated core debris bed. Nucl Eng Des 110:61–72
Widmann W, Bürger M, Lohnert G, Alsmeyer H, Tromm W (2006) Experimental and theoretical investigations on the COMET concept for ex-vessel core melt retention. Nucl Eng Des 236:2304–2327

Chapter 13
Direct Contact Condensation of Steam in Subcooled Water

Priyankan Datta, Aranyak Chakravarty, Koushik Ghosh, Achintya Mukhopadhyay and Swarnendu Sen

Abstract Direct contact condensation (DCC) of steam in subcooled water is a phenomenon which is experienced in many applications such as thermal, chemical and nuclear engineering, particularly in energy generation devices since it enables immense energy transfer via the two-phase interface. However, under certain situations, steam water direct contact can lead to rapid condensation and result in fast (of the order of acoustic time scale) pressure transients which could have serious implications on structural integrity and safety, especially in nuclear power plants. Therefore, understanding of the underlying physics and characteristics of DCC phenomenon has a paramount importance. DCC is a complex thermo-hydraulic event in which the phase change process is governed by the interplay between several thermo-mechanical factors (e.g. local heat transfer coefficient, interfacial area density, turbulence intensity in the liquid phase) across the phasic interface. In this chapter, different situations of the DCC events, their characteristics and underlying mechanisms are discussed in detail. A detailed review of the earlier works which includes both system-level and interface scale modelling of the phenomena is also addressed in this chapter. An emphasis is given on the DCC events which are always associated with the large amplitude pressure spikes such as chugging and condensation-induced water hammer.

13.1 Introduction

Direct contact condensation (DCC) of steam in subcooled water occurs in various industrial applications such as nuclear, thermal and chemical engineering (Sideman and Maron 1982) as it is an effective way of the enhancement of heat transfer via the two-phase interface. DCC is a very complex thermo-hydraulic phenomenon since the dynamic interplay between condensation rate and turbulence intensity in the vicinity of the phasic interface control this event. The heat transfer during DCC is primarily

P. Datta · A. Chakravarty · K. Ghosh · A. Mukhopadhyay (✉) · S. Sen
Department of Mechanical Engineering, Jadavpur University, Kolkata 700032, India
e-mail: achintya.mukhopadhyay@jadavpuruniversity.in

K. Saha et al. (eds.), *Two-Phase Flow for Automotive and Power Generation Sectors*, Energy, Environment, and Sustainability,
https://doi.org/10.1007/978-981-13-3256-2_13

determined by the interfacial area between the two phases. This steam–water contact area is further governed by the localised eddies which in turn are generated due to momentum transfer to the water during phase change.

The underlying physics of the steam–water DCC events is still not very clear, and hence, a comprehensive understanding is required, specially, in the context of nuclear industry as under certain conditions DCC may lead to violent condensation and results in the formation of fast and dangerous pressure transients. The pressure surges generated due to DCC have sufficient potential to cause serious damage to the structural integrity of the nuclear reactor safety systems which may trigger an accidental situation. Thus, direct contact condensation research has got attention among the researchers since last three decades.

In this chapter, steam–water direct contact condensation process is discussed in detail. To understand the phenomenon in depth, a discussion will be carried out about different mechanisms as well as the characteristics of the DCC events. A review of the earlier works which includes both the system-level modelling (using different system codes such as RELAP, ATHLET) and the interface scale modelling (using volume of fluid approach) of the phenomena is addressed in this chapter. An emphasis is given on the DCC events which are always associated with the large amplitude pressure spikes (chugging and condensation-induced water hammer). Finally, the conclusions are outlined in Sect. 14.3.

13.2 DCC Mechanisms

From the experimental observations, Liang and Griffith (1994) reported that DCC may occur in the following two situations.

13.2.1 Blowing of Steam into Subcooled Water Pool

Injection of steam in subcooled water (either through vertical vent or horizontal vent line) is almost unavoidable in the light-water nuclear reactors during transient and in the accidental conditions (e.g. loss-of-coolant accident, loss-of-flow accident). This type of DCC event can be classified primarily into three different condensation modes, namely jetting, bubbling and chugging, depending on the steam mass flux and water subcooling. A detailed discussion on different condensation modes during steam blowdown is discussed in the subsequent sections.

13.2.1.1 Condensation Modes Near-Sonic Range

In case of higher steam flow rate, typically jetting condensation mode occurs during which steam enters into the subcooled water pool as a jet. The jet shape in this

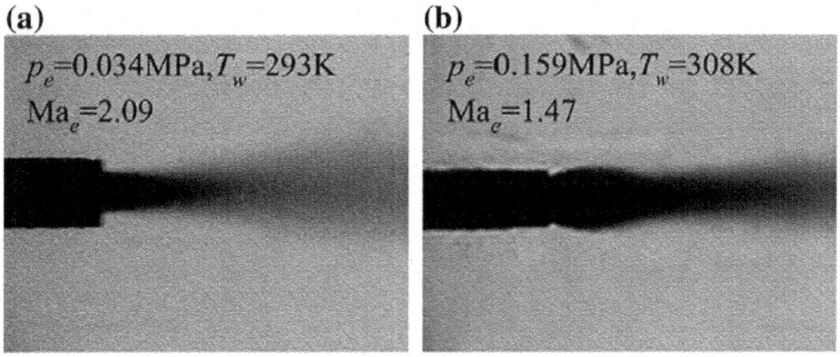

Fig. 13.1 Conical **a** and **b** ellipsoidal shape formation during supersonic steam jet condensation [used with permission from Wu et al. (2007)]

situation is primarily governed by the vapour phase momentum. The experimental observations (Xu et al. 2018; Wu et al. 2007, 2009; Kim et al. 2001, 2004; Chun et al. 1996; Simpson and Chan 1982) showed that as the steam flow rate approaches near-sonic or supersonic, stable jet shapes are formed. The jet shapes become either conical or ellipsoidal depending upon the degree of water subcooling. Chun et al. (1996) from his vapour injection experiment claimed that the conical shape is expected during high degree of subcooling. However, the jet shape changes to ellipsoidal or even divergent type with the decrease in subcooling. Figure 13.1 shows the conical and ellipsoidal shape formation during supersonic steam jet condensation.

13.2.1.2 Condensation Modes Within the Subsonic Range

Steam condensation phenomenon at lower flow rate becomes more complex since the interplay between interfacial condensation rate and vapour momentum ultimately determines the condensation modes. In this scenario, all the three condensation modes, i.e. jetting, bubbling and chugging, can occur depending on the steam mass flux and degree of water subcooling. However, in the subsonic regime more irregular and fuzzy jet surface is expected as compared to the sonic regime (Liang 1991). The steam jet characteristics at lower flow rate (within the subsonic range) are thoroughly investigated by Chan and Lee (1982). From the experimental observations, they proposed a condensation regime map as a function of steam mass flux (within a range between 1 and 175 kg/m^{2}s which corresponds to Mach number range between 0.1 and 0.5) and pool temperature (within 40–90 °C) as shown in Fig. 13.2.

(i) *Ellipsoidal jet regime (zone 1)*

This regime occurs at high steam mass flux and higher pool temperature. In this regime, due to the high vapour momentum and lower subcooling, steam exits from the vent and an ellipsoidal steam region is formed. This ellipsoidal steam zone grows further and encapsulates the vent pipe exit. Chan and Lee (1982) observed that during

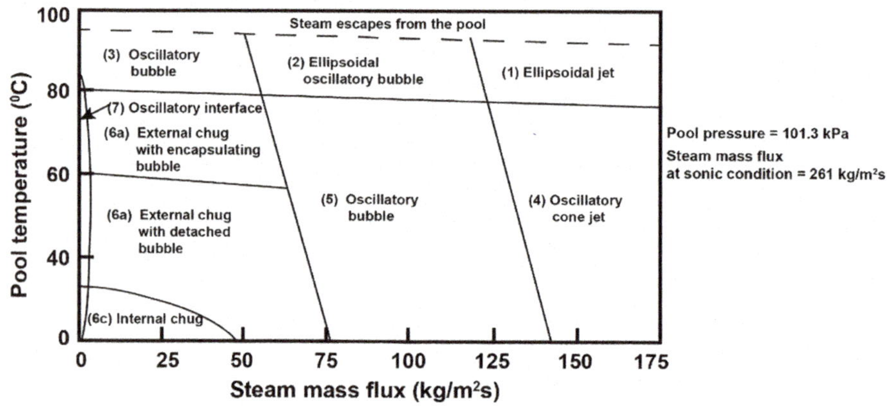

Fig. 13.2 Condensation regime map during steam injection into subcooled water [redrawn with permission from Chan and Lee (1982)]

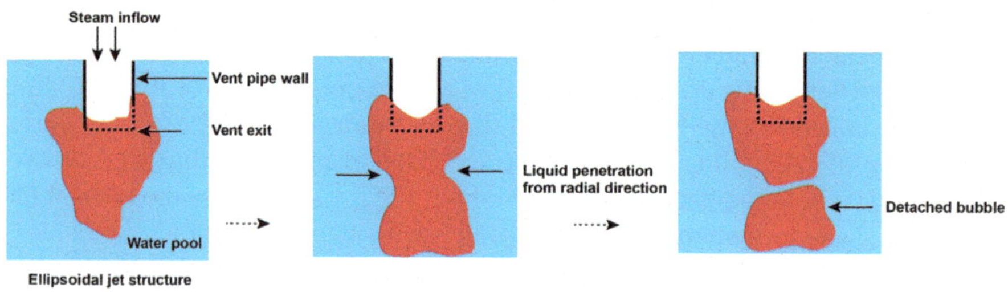

Fig. 13.3 Schematic representation of different phases of oscillatory bubble regime

this situation, bubble detachment is absent due to low condensation rate at the two-phase interface.

(ii) *Ellipsoidal oscillatory bubble regime (zone 2)*

At intermediate steam flow rate (pool temperature remains the same as ellipsoidal jet regime), it can be observed that the vapour momentum is still high enough such that the steam zone is able to exit from the vent pipe. This steam zone translates downwards through the water as an ellipsoid (Fig. 13.3). In this regime, vent pipe encapsulation also occurs as the ellipsoidal zone grows further. At this time instant, due to continuous steam condensation, a local pressure gradient establishes across the two-phase interface which in turn develops a circumferential instability around the steam zone. As time progresses, this circumferential instability grows, and consequently, the surrounding liquid water starts to penetrate into the vapour core from the radial direction. This liquid entrainment within the steam zone finally cuts off the lower part of the ellipsoid. The vapour bubble, thus detached from the main ellipsoid, finally collapses. The whole phenomenon repeats as another ellipsoid will develop.

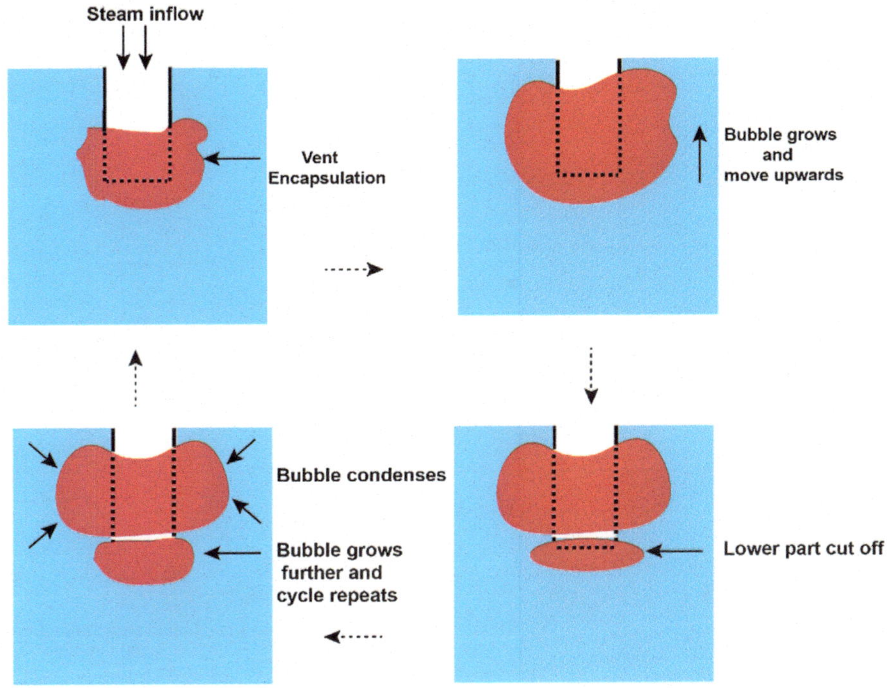

Fig. 13.4 Schematic diagram of oscillatory bubble regime at lowest steam flux and higher temperature

(iii) *Oscillatory bubble regime (zone 3)*

In this regime, the steam injection rate remains at the lowest. However, the pool temperature lies within the range similar as ellipsoidal jet regime and ellipsoidal oscillatory bubble regime. In this scenario, as the pool temperature is high, the steam exits form the vent pipe and a larger encapsulating steam zone is formed (Fig. 13.4). This encapsulated steam zone starts to grow as a cylindrical bubble which has a tendency to move upwards due to buoyancy force. At this time instant, the circumferential instability developed by the continuous condensation process is large enough such that it may cut off the lower portion of the cylindrical bubble. The separated steam zone thus formed floats in the upward direction and condenses. This cycle repeats as the continuous encapsulation leads to form another cylindrical bubble.

(iv) *Oscillatory cone jet (zone 4)*

At higher steam flow rate, if the pool temperature remains lowest, this condensation mode may occur. In this mode, the lowest pool temperature leads to a higher interfacial condensation rate. Thus, a conical-shaped jet is formed as the jet exits from the vent. As this vapour cone propagates through the water in the downward direction, a bubble detachment occurs due to the circumferential instability which appears and penetrates the vapour core from the radial direction (Fig. 13.5).

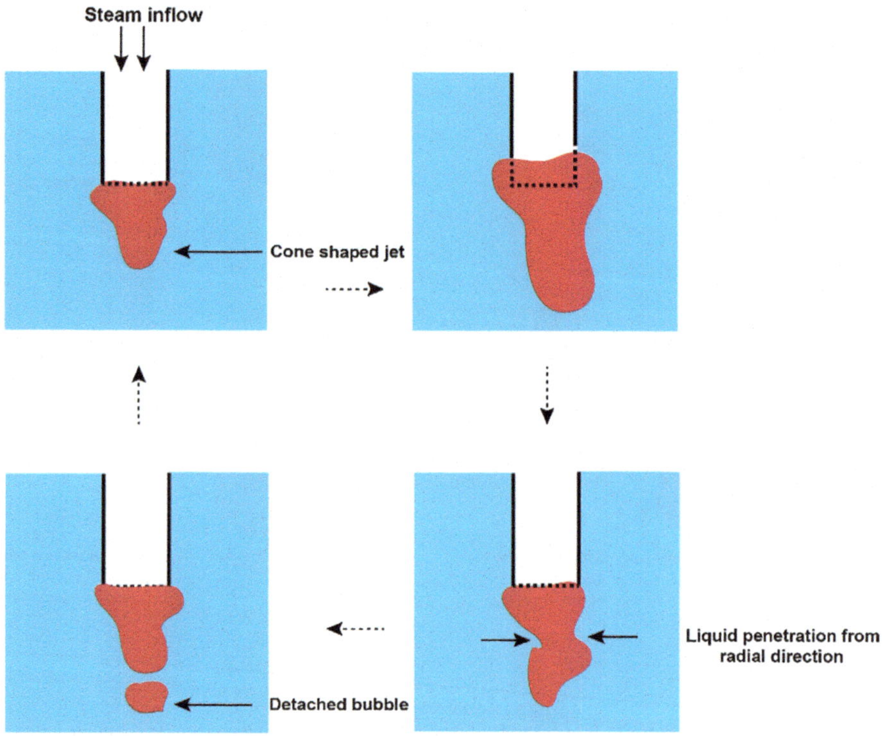

Fig. 13.5 Schematic diagram of different phases of conical jet regime

(v) *Oscillatory bubble regime (zone 5)*

The oscillatory bubble regime may also occur when the steam flow rate lies within the intermediate range and the pool temperature is at the lowest. In this mode, stronger condensation rate dominates over the vapour phase momentum, and hence, initial interfacial instability appears near the pipe exit. As a consequence, bubble detachment occurs near the vent end which causes to cease larger encapsulation. This bubble detachment at the proximity of the vent exit leads to a change in the vapour zone from 'vapour cone' to an 'oscillatory vapour bubble' (Fig. 13.6).

(vi) *Chugging regime*

The chugging condensation mode occurs during low steam mass flux and low pool temperature condition. A typical chugging event can be classified into three different categories depending on the pool water temperature as follows

(I) *External chugging with encapsulating bubble*

This phenomenon is observed when the pool temperature lies within 60–80 °C (Fig. 13.2). In this situation, steam exits from the vent pipe and spreads out. The separated steam zone, thus formed, is able to grow further and can encapsulate the

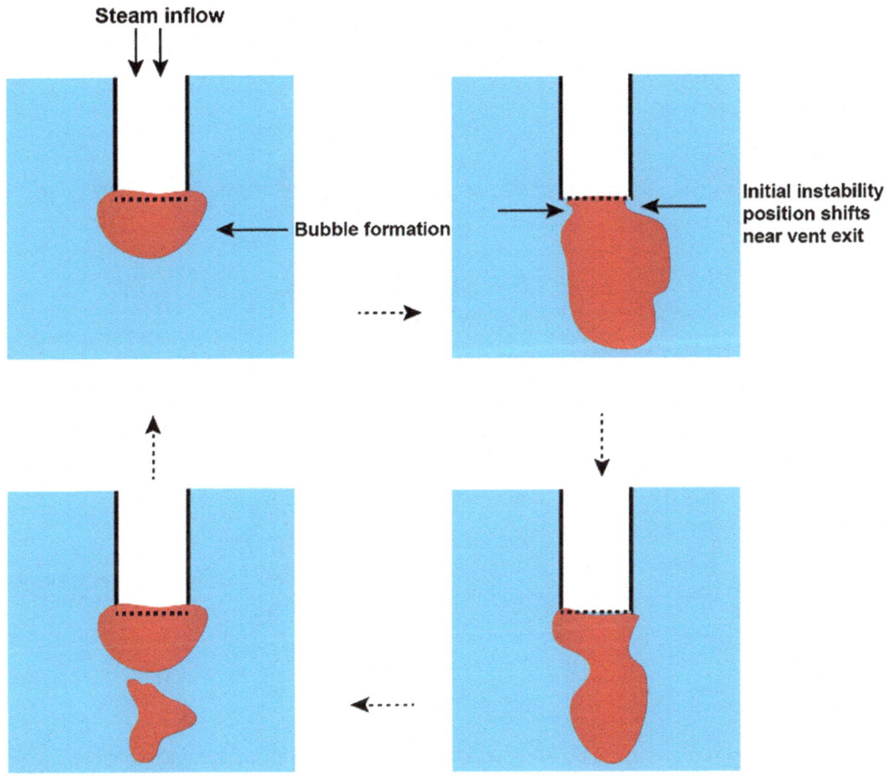

Fig. 13.6 Schematic diagram of oscillatory bubble regime at intermediate steam flux and low pool temperature

vent end as the pool temperature is high enough. At this time instant, the simultaneous condensation process leads to a continuous loss of the steam mass via the two-phase interface. The continuous loss of the steam establishes a local pressure gradient which is attributed to the formation of a disturbance wave over the phase interface. The appearance of this disturbance wave causes an increase in the interfacial area and consequently an enhancement in the interfacial heat transfer rate. The encapsulating bubble attains its maximum size and finally collapses very rapidly as the disturbance wave grows to its maximum. Chan and Lee (1982) observed that this encapsulating bubble collapse event is so rapid and irregular in nature that a mist of smaller bubbles are formed after the collapse is completed. The rapid collapse of the bubble causes a reduction in the pressure near the vent exit which provides the suction force to move the surrounding water into the vent pipe (Fig. 13.7).

As the water sucked up into the vent pipe, water will be heated up which in turn ceases the condensation rate. This results in the further build-up of the steam pressure, and thereon, the next chugging cycle starts as the water is pushed out from the pipe. In this chugging process, the bubble growth process requires more time. Hence, the

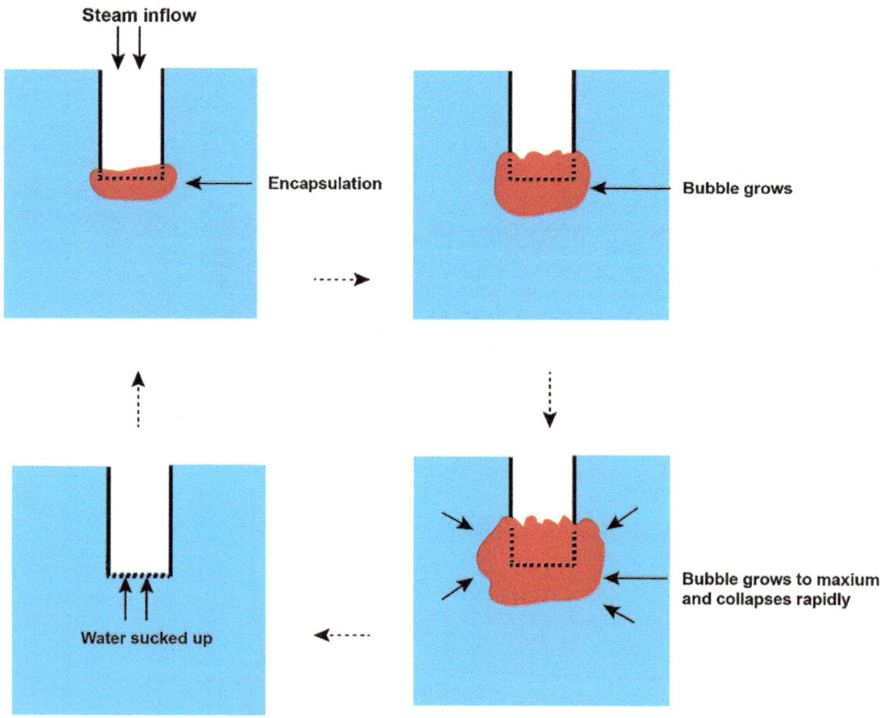

Fig. 13.7 Schematic representation of external chugging mode with encapsulated bubble

frequency of occurrence of the encapsulating bubble chugging is the lowest among all three chugging events.

(II) *External chugging with detached bubble*

At a lower pool temperature (steam flow rate remains the same as in the encapsulating bubble regime), this mode may occur (Fig. 13.8). In this regime, the two-phase interface is still able to exit from the vent pipe and starts to grow as a cylindrical bubble. However, due to higher condensation rate, the bubble growth tends to cease and it starts collapsing. As a consequence, the surrounding water rushes into the bubble (due to the establishment of local pressure difference) and penetrates into the core zone from the radial direction. This leads to the detachment of the cylindrical bubble zone from the mainstream which finally collapses very rapidly into the mist of small bubbles. The bubble collapse creates a low-pressure zone at the proximity of the vent exit which drives the water to be sucked up into the pipe. The chugging cycle repeats as the steam pressure further grows and forces the interface to be pushed out from the vent. This type of chugging events is always associated with the higher chugging heights within the pipe.

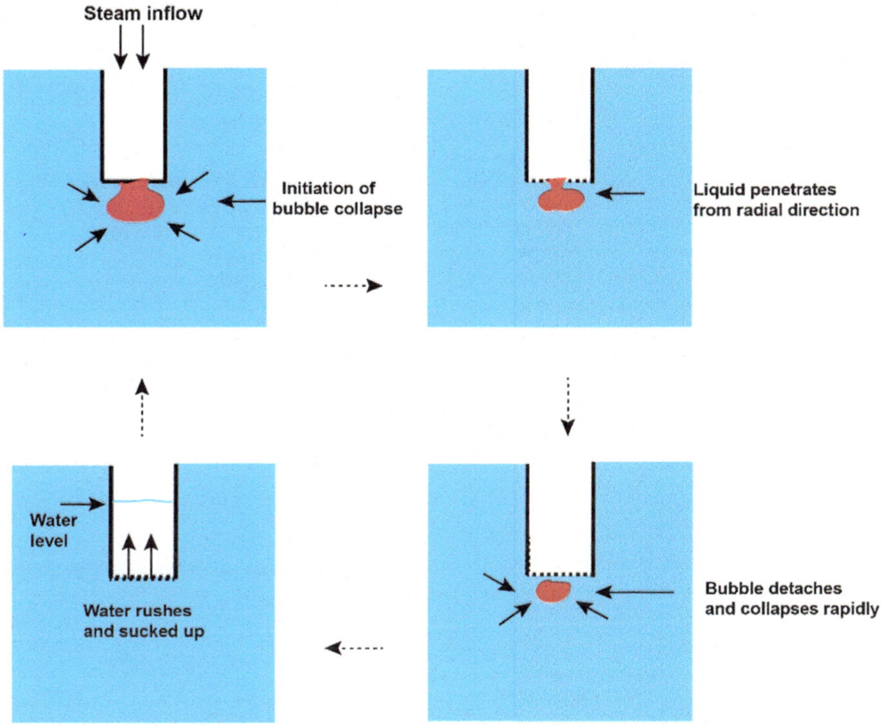

Fig. 13.8 Schematic diagram of external chugging mode with detached bubble

(III) *Internal chugging mode*

This type of chugging regime is encountered when the pool temperature remains lower as compared to the external chugging modes. In this situation, due to the higher condensation rate (as water subcooling is high) water starts entering into the pipe (Fig. 13.9). Therefore, a steam–water counter-current flow is established where steam moves downwards and water enters into the upward direction. This counter-current flow situation between the two phases leads to the establishment of an annular flow regime where a core steam zone is formed which is surrounded by a liquid layer adhered to the pipe wall. The continuous condensation of the steam may introduce disturbances over the phase boundary, and consequently, the interface becomes wavy. As time progresses, the interface wave crest may grow further and can touch the pipe wall. Chan and Lee (1982) called this event as 'bridging'. The bridging phenomenon causes an isolation of a steam bubble from the mainstream. The isolated steam bubble thus formed starts collapsing and creates a local pressure gradient. As a consequence, the surrounding water slug starts accelerating towards the bubble. This water slug acceleration causes compression of the remaining steam bubble which finally collapses very rapidly.

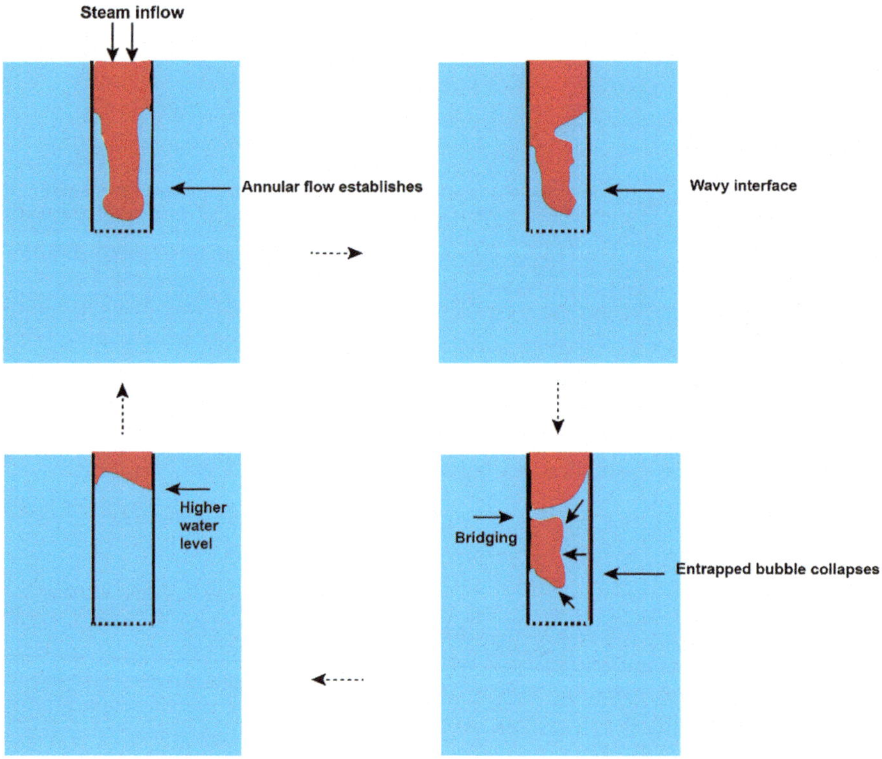

Fig. 13.9 Schematic diagram of internal chugging mode

As the bubble collapses rapidly, the surrounding water slug moves towards the bubble centre for filling up the empty space and, thus, collides. This slug collision is attributed to the generation of the large amplitude pressure peaks. In this type of chugging, the height of the sucked water within the pipe is found to be the highest compared to the other two modes.

Among all the condensation regimes during steam blowdown into water pool, the chugging phenomenon is found to be more chaotic and always associated with the large amplitude, low-frequency pressure transients as the two-phase interface moves in and out from the vent pipe periodically. Therefore, a major thrust of the DCC research is given on the understanding of the chugging phenomena. In the next section, we will focus on the previous researches as well as the state of the art of the chugging condensation mode.

13.2.1.3 Earlier Research on Chugging

It can be inferred from the available literature that a typical chugging event can be divided into a sequence of events (Fig. 13.10), and hence, most of the experimental investigations have been carried out to understand a complete chugging cycle.

The experiments performed in the POOLEX as well as in the upgraded PPOOLEX test facility (Puustinen et al. 2014; Tanskanen 2012; Puustinen and Laine 2008; Laine and Puustinen 2005) at the Lappeenranta University of Technology are considered as one of the most useful databases in the context of chugging research. These two test facilities are the prototype of a typical BWR containment which are built to study the condensation dynamics during steam injection into the subcooled water pool. The initial experimental conditions for both the POOLEX and PPOOLEX test facility are chosen based on the condensation regime map proposed by Chan and Lee (1982) as well as Lahey and Moody (1993). The experimental observations reveal that as the pool temperature decreases, both the bubble size (as observed from Fig. 13.11) and its collapse time are reduced. POOLEX test results also showed that each chugging event is associated with strong pressure peaks inside the blowdown pipe. The pressure peak amplitude rises as the pool temperature decreases. It is reported in the literature (Laine and Puustinen 2005) that among all the POOLEX experimental cases, a pressure peak having an amplitude of 40 bar is found to be the highest. However, it is observed that due to the presence of the non-condensable gases in the dry well, the pressure peak magnitude is found to be lower in the PPOOLEX experiments as compared to the POOLEX tests.

Pellegrini et al. (2016) performed steam blowdown experiments to understand the chugging cycle with and without the presence of non-condensable gases. The authors observed that during pure steam injection into the subcooled water pool, bubble grows to its maximum size within 100 ms. The vapour bubble, thus formed, collapses very rapidly (within 20 ms) due to higher interfacial condensation rate. However, the bubble condensation behaviour is found to be different when the air is introduced within the steam. The presence of the non-condensable gas leads to exit the steam with a reduced partial pressure inside the bubble. The decrease in the steam partial pressure causes a drop in the saturation temperature. As a consequence, the

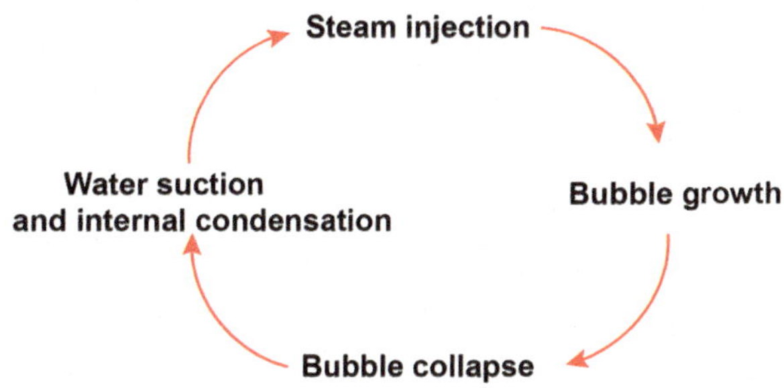

Fig. 13.10 Schematic diagram of a typical chugging cycle

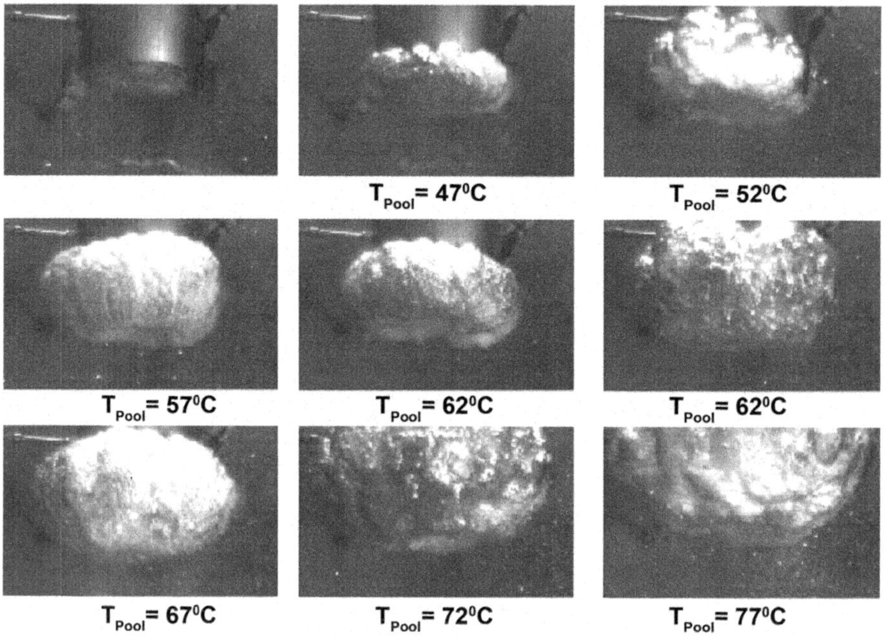

Fig. 13.11 Bubble shape variation in the STB-28 POOLEX test at the vent pipe outlet as a function of pool temperature [used with permission from Tanskanen et al. (2014)]

interfacial heat transfer rate ceases and finally stops as the saturation temperature is equal to the temperature of the surrounding water.

The work of Gregu et al. (2017) also focused on the experimental observation of a complete chugging cycle. Their experiments showed that during the water suction phase in a chugging cycle, high-pressure spikes are observed (as shown in Fig. 13.12). The authors claimed that the occurrence of these high-pressure peaks is driven by the condensation-induced water hammer events within the vent pipe. Li et al. (2015) carried out an experimental work to study the chugging phenomenon in a 'tee-shaped' geometry. Their experiments showed that depending on the water subcooling and steam mass flow rate, water elevation within the vent pipe may vary. The authors concluded that for the same steam mass flux condition, as the water temperature reduces, large chugging events (characterised by the higher water elevation) are observed. Xu et al. (2018) recently performed experiments to investigate different condensation modes for supersonic as well as the subsonic steam injection. From the visualisation of their experimental results, the authors also concluded that a typical chugging event generally occurred for the low steam flux and low pool temperature condition.

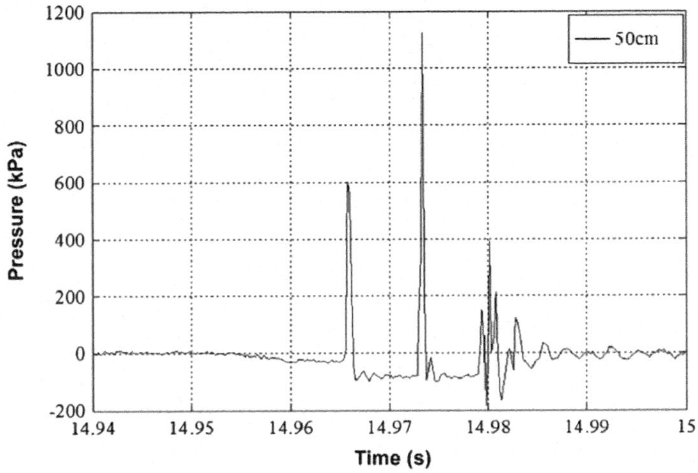

Fig. 13.12 Pressure history due to CIWH at 50 cm away from pipe outlet [used with permission from Gregu et al. (2017)]

13.2.2 Injection of Subcooled Water into a Steam-Filled Region

This is another situation for the occurrence of direct contact condensation phenomenon. In a nuclear reactor, subcooled water injection into a steam-filled zone is found to be a very efficient way of heat removal during transient as well as in the accidental situations. During a loss-of-coolant accident (LOCA) in a PWR, subcooled water is injected from the emergency core cooling system (ECCS) either into the cold leg or hot leg of the reactor. This situation may also be encountered very often in the steam generator feed water line of a PWR during reactor transient operating condition (Jones et al. 1979).

This type of steam–water DCC scenario, however, under certain conditions may cause violent condensation which can lead to the generation of the fast (of the order of acoustic timescale) pressure transients. The amplitude of these pressure peaks may attain tens of bars or even higher than hundred bars (Milivojevic et al. 2014) and, hence, gained attention in the nuclear safety research. The following section will focus on the underlying mechanism of the occurrence of violent condensation events as well as the state of the art of the DCC phenomenon during subcooled water injection into a steam-filled zone.

13.2.2.1 Theoretical Background

In this type of DCC event, as the subcooled water starts entering into the steam-filled line, a stratified flow regime may establish between the two phases (as shown in

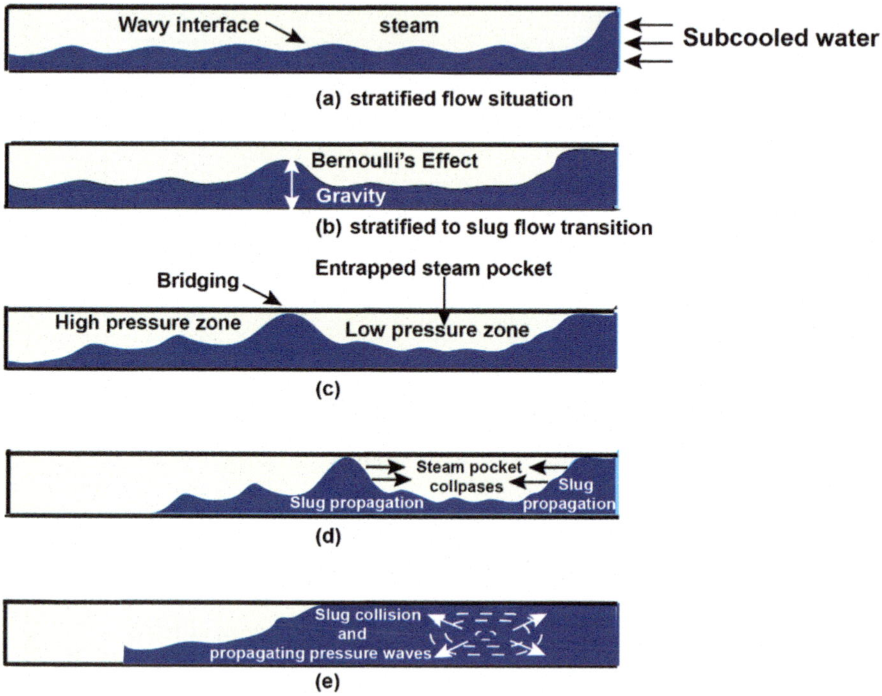

Fig. 13.13 Mechanism of the occurrence of condensation-induced water hammer in a horizontal pipe geometry

Fig. 13.13a). The steam–water direct contact leads to the continuous steam condensation over the two-phase interface. The continuous loss of steam due to condensation allows the fresh steam to rush into the condensation zone and filling up the empty space. This establishes a relative velocity difference between the two phases, and consequently, the phase interface becomes wavy due to Kelvin–Helmholtz instability (Mahapatra et al. 2017; Drazin 2002; Chandrasekhar 1961).

The fresh steam velocity (as well as the relative velocity between two phases) over the interface becomes substantial as the interfacial condensation rate is sufficiently high. The higher steam velocity will induce a 'Bernoulli effect' which results in the interface wave to grow further enough such that the wave crest can touch the pipe wall. This is often termed as 'bridging' (Fig. 13.13). As a consequence of this bridging, a steam section gets isolated and entrapped between the surrounding subcooled water slugs (Fig. 13.13c). The steam pocket, thus entrapped, starts condensing over the phase boundary which leads to the formation of local low-pressure zone. This differential pressure causes slug acceleration in a direction of low-pressure zone, and thus, the remaining steam gets compressed and finally collapses very rapidly. Due to the rapid collapse of the entrapped steam pocket, surrounding water slug flows in a direction (Fig. 13.13d) to fill up the empty space and collides with each other.

The slug collision is attributed to the generation of fast and violent pressure peak which is propagated through the surrounding medium at the sonic speed. This pressure change dynamics under the influence of direct contact condensation is known as condensation-driven or condensation-induced water hammer (CIWH) phenomenon.

13.2.2.2 Brief Review of Earlier Work and the State of the Art of CIWH Research

CIWH is a very complex thermo-hydraulic phenomenon as the dynamic interplay between two-phase instability and phase change process governs the onset of this event. A typical pressure pulse width during CIWH is found to be of the order of millisecond [about 2 ms (Barna et al. 2010a)]. Furthermore, the location of the steam pocket formation and pressure peak amplitude in a CIWH event is stochastic in nature (Urban and Schlüter 2014) and, thus, very difficult to predict. In the following section, we will address the state of the art of the CIWH research.

(i) *Experimental investigations*

The available literature shows that the research on CIWH is quite a few. The group of Griffith and others (Griffith 1997; Bjorge and Griffith 1984) carried out both the experimental and theoretical studies for investigating the CIWH events. The authors performed experiments in both the horizontal and in the inclined pipe geometry. Based on the experimental observations, the authors proposed five necessary conditions for the occurrence of a CIWH event

- The pipe must be horizontal or near horizontal (slope should be less than 2.4°).
- Water subcooling must be greater than 20 °C.
- Pipe length-to-diameter ratio (L/D ratio) should be greater than 24.
- The filling velocity for the cold water, based on full pipe area, must be such that Froude number is less than 1.
- There must be void nearby.
- The system pressure must be great enough so that the water slugs will impact the affected parts of the system at high enough velocity to do damage. This threshold system pressure appears to be somewhere between 100 and 300 psi ($1\text{–}3 \times 10^6$ Pa) depending on the criteria chosen for deciding what constitutes an intolerable steam bubble collapse-induced water hammer (SBCIWH).

Urban and Schlüter (2014) carried out experiments to investigate the underlying cause for the onset of the CIWH events. They performed several experiments in a slightly inclined pipe geometry (1.4° inclination) with identical initial parametric conditions and concluded that CIWH phenomenon is purely stochastic in nature. Table 13.1 shows different parametric conditions as well as the associated CIWH incidents observed in their experiments.

Their observations reveal that even for a low initial system pressure (3.4 bar), maximum pressure peak amplitude can reach up to 134 bar for a particular combination of Froude number (=0.6) and water subcooling (=60 °C). In addition, it is found that

Table 13.1 Parametric conditions and corresponding pressure peak amplitude observed in the experiments of Urban and Schlüter (2014) [used with permission from Urban and Schlüter (2014)]

Exp. no. (–)	Fr (–)	ΔT (K)	$p_{\max}$ (bara)
130115/02	1.6	60	34
130115/06	1.6	60	12
130418/05	1.6	60	61
130418/06	1.6	60	18
$\overline{p} = 31.3 \pm 34.6$ bara; (21.8; 95%; 4)			
130117/06	0.6	60	81
130117/07	0.6	60	50
130117/08	0.6	60	100
130117/09	0.6	60	62
130117/10	0.6	60	63
130129/06	0.6	60	84
130130/03	0.6	60	29
130130/04	0.6	60	61
130130/05	0.6	60	58
130130/06	0.6	60	71
130130/07	0.6	60	88
130201/02	0.6	60	79
130201/03	0.6	60	77
130201/04	0.6	60	134
130201/06	0.6	60	93
130201/09	0.6	60	53
130201/10	0.6	60	31
130207/09	0.6	60	76
$\overline{p} = 71.7 \pm 12.7$ bara; (24.9; 95%; 18)			
130122/06	0.9	60	57
130122/07	0.9	60	19
130122/09	0.9	60	58
130204/06	0.9	60	60
130204/07	0.9	60	62
130206/06	0.9	60	22
$\overline{p} = 46.3 \pm 21.1$ bara; (20.1; 95%; 6)			
130124/06	1.1	60	10
130510/01	1.1	60	48
130510/02	1.1	60	74
$\overline{p} = 44.2 \pm 80.5$ bara; (32.4; 95%; 3)			

Table 13.2 Experimental conditions performed at PMK-2 facility

Experiment	Steam pressure (MPa)	Water temperature (°C)	Water flow rate (kg/s)
AEKI_Exp03	0.98	30	1.20
AEKI_Exp04	1.15	30	0.66
AEKI_Exp05	1.45	25	1.01
AEKI_Exp06	1.50	30	1.66

Reused from Giot (2000)

the CIWH events also appeared for Froude number greater than 1 (Fr = 1.1). These two observations are found to be in contrast to the Griffith's theory for the onset of CIWH phenomenon, and hence, a further refinement is required in the context of CIWH avoiding guidelines.

The most useful database for the analysis of the CIWH phenomenon is the experiments performed in the PMK-2 test facility at KFKI Atomic Energy Research Institute (Giot 2000). PMK-2 test facility is a scaled-down thermo-hydraulic model of the primary circuit of the VVER type of reactor. The water hammer test pipe in this facility is horizontal which have length (2.87 m) to diameter (73 mm) ratio is about 39. Table 13.2 summarises the parametric conditions in different CIWH experiments which are performed in the PMK-2 facility.

The transient pressure history for two different test conditions at the PMK-2 test facility is shown in Fig. 13.14. It is reported that among all the experimental cases, the highest pressure peak amplitude is found to be about 18.35 MPa (Fig. 13.14b). The experimental outcomes clearly reveal that both the inlet water flow rate and water temperature have a strong influence on the pressure peak amplitude during a CIWH event. Furthermore, it can be observed that the onset time of the pressure peak in two experiments is found to be different. This is due to the fact that the location of the vapour pocket formation (due to stratified to slug flow regime transition) within the test pipe has a great uncertainty. Thus, the earlier prediction of the pressure peak magnitude, its occurrence time and location of occurrence is quite difficult.

Wang et al. (2018) recently carried out CIWH experiments in a horizontal pipe geometry for different combinations of steam mass flux and water temperature. Depending on the nature of the pressure signals, the authors classified the CIWH phenomenon mainly into two categories, namely periodic CIWH and non-periodic CIWH. The two-phase flow field behaviour as well as the pressure signal characteristics during periodic and non-periodic CIWH events are shown in Fig. 13.15.

The experiments in the UniBw Facility (Dirndorfer 2017) at the Universität der Bundeswehr München investigated the CIWH events as a function of different non-dimensional parameters such as Jakob number, Reynolds number and Froude number. The experimental results show that with the increase in Jakob number, the probability for the occurrence of intense CIWH events is found to be higher. This is due to the fact that at high Jakob number condensation potential increases which results in the formation of higher amplitude pressure peaks. Furthermore, their experiments

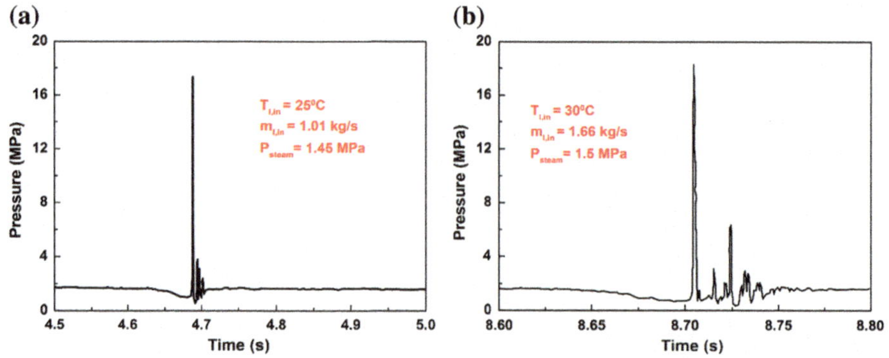

Fig. 13.14 Pressure transient observed in **a** AEKI_Exp05 and **b** AEKI_Exp06 experiments [redrawn with permission from Barna et al. (2010a), Giot (2000)]

reveal that a specific combination of the flow rate (defined by the Froude number) and turbulence (defined by Reynolds number) must coexist for the onset of the water hammer phenomenon.

Hou et al. (2016) carried out experiments and reported that flashing instability-induced water hammer phenomenon can also occur in a natural circulation loop which acts as a passive heat removal system in a nuclear reactor under accidental as well as transient operating conditions.

(ii) *Numerical investigations*

The literature survey reveals that only a few numerical attempts have been made for predicting the CIWH events. The 'WAHA3 code' developed at the Jožef Stefan Institute (Tiselj et al. 2004) is found to be the most dedicated in-house code which is capable of capturing the fast transient phenomena including CIWH. WAHA3 is a one-dimensional code which is based on the six-equation-based two-fluid model approach. To assess the capability of WAHA3, the group of Barna and others (Barna et al. 2015; Barna and Ezsöl 2011; Barna et al. 2010a, b) simulated the CIWH experiments performed at the PMK-2 and ROSA test facility. Their results show that WAHA3 can capture the CIWH pressure transients.

Milivojevic et al. (2014) investigated the DCC-driven water hammer phenomenon in a vertical pipe. The authors developed a one-dimensional in-house code which is based on the homogeneous model of liquid–vapour flow. They used the method of characteristics (MOC) approach. The authors compared (Fig. 13.16) their calculated pressure peak with the experimental data and claimed that the homogeneous model is also able to capture the CIWH events quite satisfactorily. However, the evidence of the flow regime transition as well as the steam pocket formation (which is the primary cause for the onset of a CIWH event) is absent in this work.

In addition to WAHA3, different system codes are also used for the analysis of CIWH phenomenon. The system code 'Analysis of THermal-hydraulics of LEaks and Transients (ATHLET)' is used by Ceuca and Laurinavicius (2015) for investigating

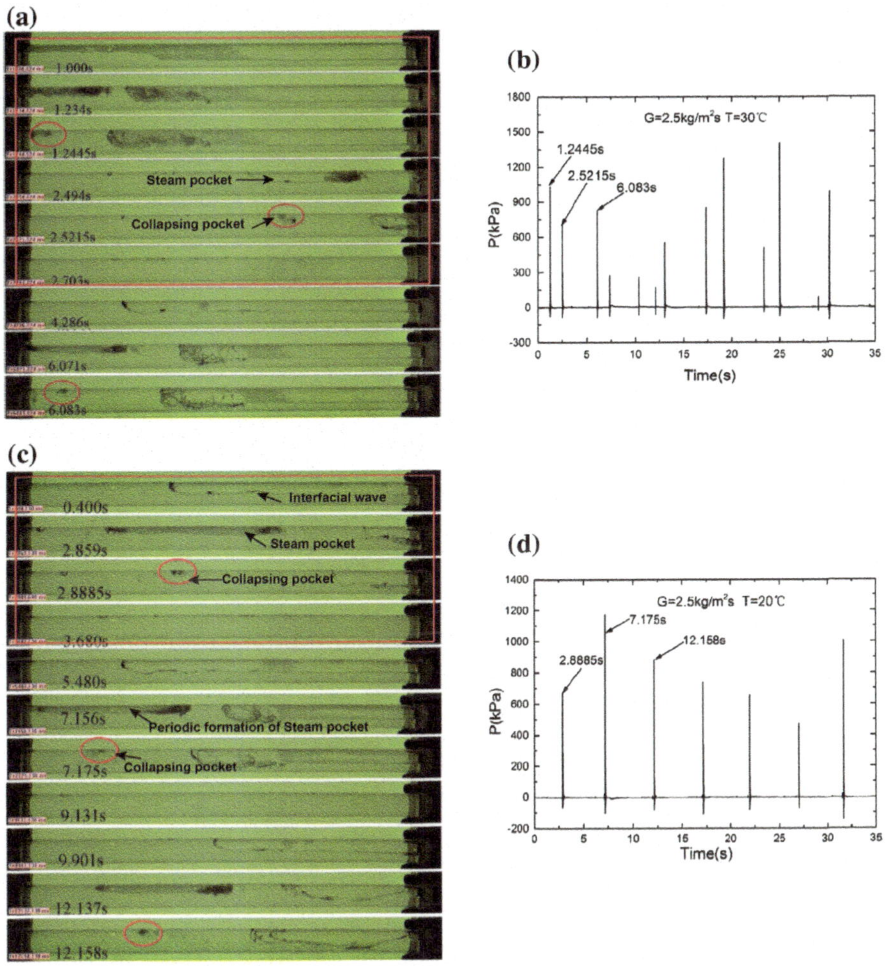

Fig. 13.15 Two-phase flow field behaviour for **a** non-periodic and **c** periodic CWH; pressure characteristics for **b** non-periodic and **d** periodic CIWH [used with permission from Wang et al. (2018)]

the PMK-2 experimental series in the context of CIWH. The authors concluded that the ATHLET code can capture the CIWH events satisfactorily.

The capability of the system code 'Reactor Excursion and Leak Analysis Program (RELAP)' for the prediction of the fast transient phenomena is a debatable issue. The group of Barna et al. (2010a) claimed that RELAP5 is unable to capture the CIWH events. However, Datta et al. (2016) showed that if the spatio-temporal discretisation is fine enough, RELAP5/Mod 3.4 is fairly able to capture the CIWH phenomenon. Figure 13.17 shows the pressure transient comparison obtained by RELAP5 with WAHA3 result and PMK-2 experimental outcomes.

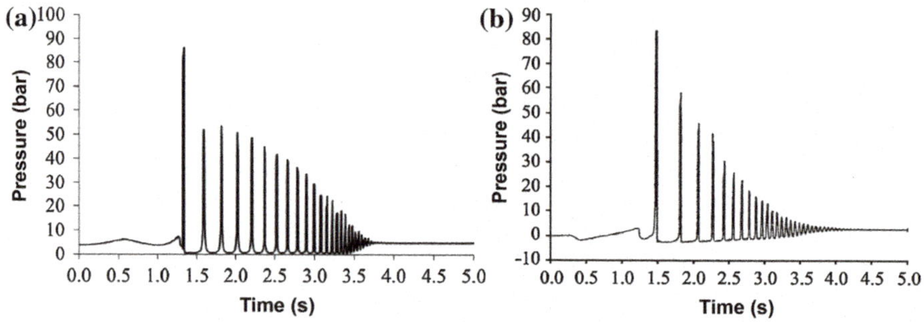

Fig. 13.16 Transient pressure history **a** calculated with homogeneous model and **b** measured from the experiment in a vertical pipe geometry [used with permission from Milivojevic et al. (2014)]

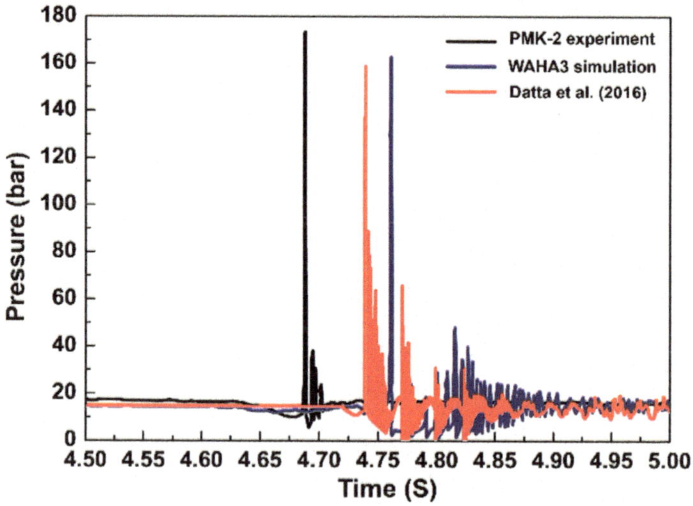

Fig. 13.17 Comparison of the transient pressure history obtained by RELAP5 with WAHA3 simulation and PMK-2 experimental outcomes [redrawn with permission from Datta et al. (2016)]

Their result reveals that the stratified to slug flow regime transition as well as steam pocket formation is well captured by RELAP. The authors also concluded that the location of the steam pocket formation as well as the magnitude of the pressure peak is strongly dependent on the water subcooling. They claimed that the amplitude of the pressure peak increases at higher degree of subcooling. In addition, for the same water injection rate, the peak pressure location within the test pipe moves away from the water inlet section with the increase in subcooling (Fig. 13.18).

This is due to the fact that at higher degree of subcooling, waterfront moves almost as a vertical front through the test section due to higher interfacial condensation rate. As a consequence, the 'bridging' occurs after a sufficient distance within the test section (Fig. 13.19).

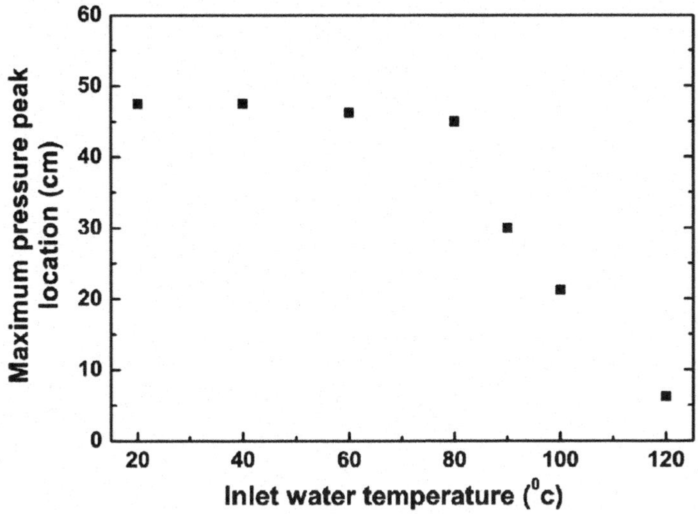

Fig. 13.18 Variation of peak pressure location with inlet water temperature [used with permission from Datta et al. (2016)]

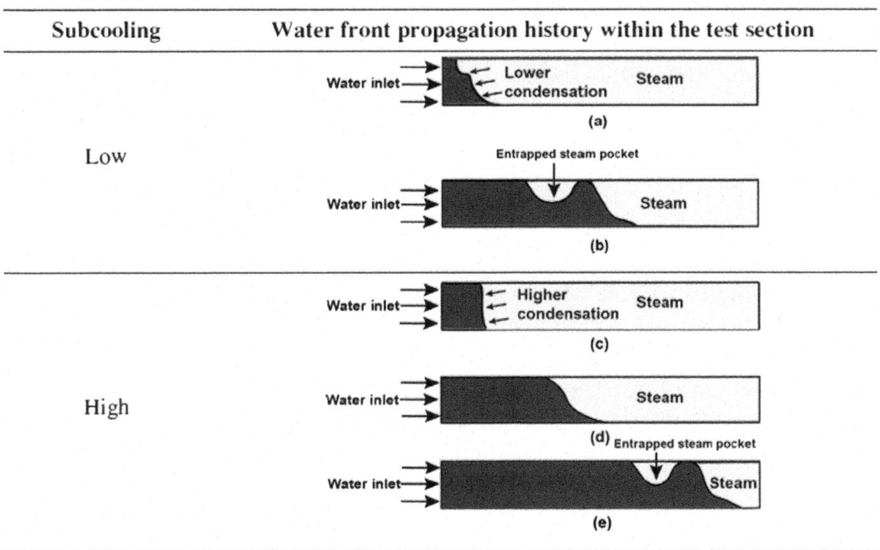

Fig. 13.19 Steam pocket formation location with the variation of water subcooling [used with permission from Datta et al. (2016)]

The effect of water subcooling on the peak pressure occurrence location is also investigated by Urban and Schlüter (2014). Their experiments also showed that the peak pressure location moves away from the water inlet with the increase in water subcooling.

It can be inferred from the above survey that most of the investigations (either with the system code or in-house code) in the context of CIWH focused on the two-fluid modelling approach. However, this approach incorporates several closure laws for accounting different interfacial exchange terms and, hence, introduces modelling uncertainties. Datta et al. (2018) recently investigated the DCC phenomenon during subcooled water injection into a steam-filled zone using pure interface tracking approach (VOF method). The authors used computational fluid dynamics software ANSYS Fluent 14.5 for modelling the DCC event in a two-dimensional horizontal pipe geometry. They assessed the interface mass transfer from the energy jump condition (instead of using any empirical correlations) as follows

$$\dot{m}_{kj} = \frac{\|\boldsymbol{q}''\| \cdot \hat{n}}{h_{kj}} a_{\text{int}} \tag{13.1}$$

where $\|\boldsymbol{q}''\|$ represents the energy jump across the two-phase interface, a_{int} is the interfacial area density, $\hat{n}$ is the unit normal vector (which determines the curvature of the interface), and h_{kj} represents the latent heat of vaporisation.

The energy jump term in Eq. (13.1) is evaluated from the basic conservation law as follows

$$\|\boldsymbol{q}''\| = \boldsymbol{q}''_{k,j} = -\lambda_{k,j} \nabla T \tag{13.2}$$

where λ is the phasic thermal conductivity and ∇T represents the thermal gradient which exists across the two-phase interface. The energy jump term, unit normal vector and interfacial area density in Eq. (13.1) are obtained with different user-defined functions of ANSYS Fluent.

Their study reveals that the pure interface approach can also capture the flow regime transition followed by the steam entrapment phenomenon which may act as a possible source for the onset of a CIWH event. Figure 13.20 shows the liquid volume fraction distribution at a particular time within the test section achieved by the volume of fluid method.

The work of Datta et al. (2017) reported that the VOF method along with interface jump model is also able to capture the vapour bubble collapse in subcooled liquid quite satisfactorily.

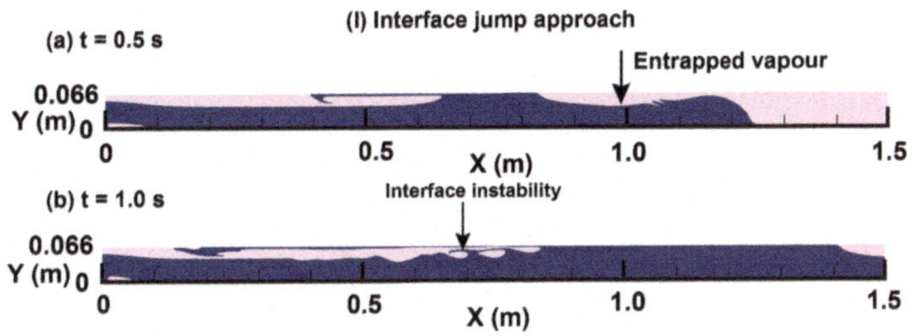

Fig. 13.20 Liquid volume fraction distribution predicted by the interface tracking method for $V_{l,\text{in}} = 3$ m/s and $T_{l,\text{in}} = 20\,°\text{C}$

13.3 Conclusions

In this chapter, steam–water direct contact condensation (DCC) phenomenon is discussed. DCC can occur in two situations—either during steam injection into subcooled water pool or as the subcooled water is injected into a steam-filled region.

During steam injection into the water pool, typically jetting regime is observed if the vapour phase momentum remains in the sonic range. The jet shape (conical, ellipsoidal or divergent) in this situation is characterised by the degree of water subcooling. However, if the momentum of the injected steam remains within the subsonic range, jetting, bubbling and chugging condensation modes are observed. Among these, three condensation modes and chugging events are found to be more chaotic in nature and always associated with the large amplitude pressure spikes which may induce heavy load on the structural components.

On the other hand, during subcooled water injection in a steam-filled pipe may cause immense condensation at the two-phase interface which results in the generation of fast and dangerous pressure transients. The pressure change dynamics under the influence of rapid condensation is typically termed as condensation-induced water hammer (CIWH). The research on CIWH shows that the amplitude of the pressure peaks may attain tens of bars or even higher than hundred bars. It is found that CIWH is stochastic in nature, and the earlier prediction of the pressure peak magnitude and its location of occurrence are very difficult. The available literature reveals that dedicated in-house code such as WAHA3 (based on six-equation-based two-fluid model) is developed for capturing this phenomenon. In addition, different system codes such as RELAP, ATHLET are also used for predicting the CIWH events. It is found that both the in-house codes and the system codes are based on the two-fluid model which introduces modelling uncertainties. Hence, in some of the recent works thrust is given on the prediction of the initial conditions of a CIWH event using pure interfacial tracking approach (volume of fluid method). Observations reveal that the interfacial approach can capture the initial conditions (such as flow regime transition, steam pocket formation) satisfactorily.

Acknowledgements The first author is thankful to the Department of Science and Technology (DST), Government of India, for providing the fellowship under INSPIRE programme. The authors are also grateful to Bhabha Atomic Research Centre (BARC), Mumbai, India, for proving the financial support of the work.

References

Barna IF, Ezsöl G (2011) Multiple condensation induced water hammer events, experiments and theoretical investigations. Kerntechnik 76:231–236

Barna IF, Imre AR, Baranyai G, Ézsöl G (2010a) Experimental and theoretical study of steam condensation induced water hammer phenomena. Nucl Eng Des 240:146–150

Barna IF, Pocsai A, Guba A, Imre AR (2015) Theoretical study of steam condensation induced water hammer phenomena in horizontal pipelines. Kerntechnik 80:420–423

Barna IF, Varga L, Ézsöl G (2010b) Steam condensation induced water hammer simulations for different pipelines in nuclear reactor. In: The 8th international topical meeting on nuclear thermal-hydraulics, operation and safety (NUTHOS-8), Shanghai, China

Bjorge RW, Griffith P (1984) Initiation of water hammer in horizontal and nearly horizontal pipes containing steam and subcooled water. ASME J Heat Transf 106:835–840

Ceuca SC, Laurinavicius D (2015) Experimental and numerical investigations on the direct contact condensation phenomenon in horizontal flow channels and its implications in nuclear safety. Kerntechnik 81:504–511

Chan CK, Lee CKB (1982) A regime map for direct contact condensation. Int J Multiph Flow 8:11–20

Chandrasekhar S (1961) Hydrodynamic and hydromagnetic stability. Dover Publications, INC., New York

Chun M-H, Kim Y-S, Park J-W (1996) An investigation of direct condensation of steam jet in subcooled water. Int Commun Heat Mass Transf 23:947–958

Datta P, Chakravarty A, Ghosh K, Mukhopadhyay A, Sen S (2018) Modeling of steam–water direct contact condensation using volume of fluid approach. Numer Heat Transf, Part A 73:17–33

Datta P, Chakravarty A, Ghosh K, Mukhopadhyay A, Sen S (2017) Modeling aspects of vapour bubble condensation in subcooled liquid using the VOF approach. Numer Heat Transf, Part A 72:236–254

Datta P, Chakravarty A, Ghosh K, Mukhopadhyay A, Sen S, Dutta A, Goyal P (2016) A numerical analysis on the effect of inlet parameters for condensation induced water hammer. Nucl Eng Des 304:50–62

Dirndorfer S (2017) Steam condensation induced water hammer in a vertical up-fill configuration within an integral test facility: experiments and computational simulations. Ph.D. Thesis, Universität der Bundeswehr München, Germany

Drazin PG (2002) Introduction to hydrodynamic stability. Cambridge University Press, Cambridge

Giot M (2000) Two-phase flow water hammer transients and induced loads on materials and structures of nuclear power plants (WAHALoads). https://cordis.europa.eu/docs/publications/6996/69969381-6_en.pdf

Gregu G, Takahashi M, Pellegrini M, Mereu R (2017) Experimental study on steam chugging phenomenon in a vertical sparger. Int J Multiph Flow 88:87–98

Griffith P (1997) Screening reactor steam/water piping systems for water hammer. Massachusetts Institute of Technology, NUREGICR-6519

Hou X, Sun Z, Su J, Fan G (2016) An investigation on flashing instability induced water hammer in an open natural circulation system. Prog Nucl Energy 93:418–430

Jones OC, Saha PJ, Wu BJC, Ginsberg T (1979) condensation induced water hammer in steam generators. Department of Nuclear Energy, Brookhaven National Laboratory, New York, BNL-NUREG-26789

Kim HY, Bae YY, Song C-H, Park JK, Choi SM (2001) Experimental study on stable steam condensation in a quenching tank. Int J Energy Res 25:239–252

Kim Y-S, Park J-W, Song C-H (2004) Investigation of the steam-water direct contact condensation heat transfer coefficients using interfacial transport models. Int Commun Heat Mass Transf 31:397–408

Lahey RT, Moody FJ (1993) The thermal-hydraulics of a boiling water nuclear reactor. American Nuclear Society, La Grange Park

Laine J, Puustinen M (2005) Condensation pool experiments with steam using DN200 blowdown pipe. Lappeenranta University of Technology, Finland, NKS-111, ISBN 187-7893-7171-7891

Li SQ, Wang P, Lu T (2015) CFD based approach for modeling steam-water direct contact condensation in subcooled water flow in a tee junction. Prog Nucl Energy 85:729–746

Liang K-S (1991) Experimental and analytical study of direct contact condensation of steam in water, Ph.D. Thesis, Department of Nuclear Engineering, Massachusetts Institute of Technology, USA

Liang K-S, Griffith P (1994) Experimental and analytical study of direct contact condensation of steam in water. Nucl Eng Des 147:425–435

Mahapatra PS, Datta P, Chakravarty A, Ghosh K, Manna NK, Mukhopadhyay A, Sen S (2017) Molten drop to coolant heat transfer during premixing of fuel coolant interaction. In: Basu S, Agarwal AK, Mukhopadhyay A, Patel C (eds) Application paradigms of droplet and spray transport: paradigms and applications. Springer Nature Singapore Pte Ltd, Singapore

Milivojevic S, Stevanovic V, Maslovaric B (2014) Condensation induced water hammer: numerical prediction. J Fluids Struct 50:416–436

Pellegrini M, Araneo L, Ninokata H, Ricotti M, Naitoh M, Achilli A (2016) Suppression pool testing at the SIET laboratory: experimental investigation of critical phenomena expected in the Fukushima Daiichi suppression chamber. J Nucl Sci Technol 53:614–629

Puustinen M, Laine J (2008) Characterizing experiments of the PPOOLEX test facility. Lappeenranta University of Technology, Finland, NKS-167, ISBN 978-187-7893-7232-7897

Puustinen M, Laine J, Räsänen A, Hujala E (2014) Chugging test with DN100 blowdown pipe in the PPOOLEX facility, NKS-310, ISBN 978-387-7893-7388-7891

Sideman S, Maron D (1982) Direct contact condensation. Adv Heat Transf 15:227–281

Simpson ME, Chan CK (1982) Hydrodynamics of a subsonic vapor jet in subcooled liquid. ASME J Heat Transf 104:271–277

Tanskanen V (2012) CFD modelling of direct contact condensation in suppression pools by applying condensation models of separated flow. Ph.D. Thesis, Lappeenranta University of Technology, Finland

Tanskanen V, Jordan A, Puustinen M, Kyrki-Rajamäki R (2014) CFD simulation and pattern recognition analysis of the chugging condensation regime. Ann Nucl Energy 66:133–143

Tiselj I, Horvat A, Černe G, Gale J, Parzer I, Mavko B, Giot M, Seynhaeve JM, Kucienska B, Lemonnier H (2004) WAHALoads-Two-phase flow water hammer transients and induced loads on materials and structures of nuclear power plants, Deliverable D10: WAHA3 code manual, IJS-DP-8841

Urban C, Schlüter M (2014) Investigations on the stochastic nature of condensation induced water hammer. Int J Multiph Flow 67:1–9

Wang L, Yue X, Chong D, Chen W, Yan J (2018) Experimental investigation on the phenomenon of steam condensation induced water hammer in a horizontal pipe. Exp Thermal Fluid Sci 91:451–458

Wu X, Yan J, Li W, Pan D, Chong D (2009) Experimental study on sonic steam jet condensation in quiescent subcooled water. Chem Eng Sci 64:5002–5012

Wu X, Yan J, Shao S, Cao Y, Liu J (2007) Experimental study on the condensation of supersonic steam jet submerged in quiescent subcooled water: steam plume shape and heat transfer. Int J Multiph Flow 33:1296–1307

Xu Q, Ye S, Chen Y, Chen Q, Guo L (2018) Condensation regime diagram for supersonic and subsonic steam jet condensation in water flow in a vertical pipe. Appl Therm Eng 130:62–73

Chapter 14
A Comprehensive Parametric Modelling for Mixed Convection Film Boiling Analysis on a Vertical Flat Plate

Dipak Chandra Das, Koushik Ghosh and Dipankar Sanyal

Abstract A comparative assessment of existing instability models is carried out to find the appropriate length scale in a computationally inexpensive integral model predicting the heat transfer in film boiling over a vertical flat plate. The use of Kelvin–Helmholtz criterion shows good matching to the limited number of experimental data, whereas for high liquid flow velocity the critical film Reynolds number criterion is found as the best. A generalized model covering the range of both the models is then developed by employing a regression analysis. The generalized model is shown to remain accurate within 10% band over a wide range of parameters.

Keywords Vertical flat plate · Film boiling · Instability length · Regression analysis · Heat transfer coefficient

Nomenclature

g	Acceleration due to gravity
h_{av}	Average heat transfer coefficient
h_{conv}; $\bar{h}_{\text{conv}}$	Convective heat transfer coefficient; averaged
h_{fg}	Latent heat of evaporation
j	Mass flux
Ja_{sub}	Liquid-phase subcooling Jakob number $= c_{pl}(T_{\text{sat}} - T_{\infty})/h_{fg}$
Ja_{sup}	Vapour-phase superheat Jakob number $= c_{pv}(T_w - T_{\text{sat}})/h_{fg}$
k	Thermal conductivity
L_{λ}	Instability length scale
$L_{\lambda-KH}$	Kelvin–Helmholtz instability length scale

D. C. Das
Department of Mechanical Engineering, National Institute of Technology, Agartala 799046, Tripura, India

K. Ghosh · D. Sanyal (✉)
Department of Mechanical Engineering, Jadavpur University, Kolkata 700032, India
e-mail: dipankar.sanyal@jadavpuruniversity.in; dipans26@gmail.com

K. Saha et al. (eds.), *Two-Phase Flow for Automotive and Power Generation Sectors*, Energy, Environment, and Sustainability,
https://doi.org/10.1007/978-981-13-3256-2_14

$L_{\lambda-FRN}$	Length scale based on film Reynolds number
L_c	Characteristic length scale
L_{reg}	Regression length
$\bar{L}_{\text{reg}}$	Non-dimensional regression length
p	Pressure
Re_l	Liquid-phase Reynolds number $= (u_\infty L_c)/\nu_l$
T	Temperature
u, v	Velocity components
x, y	Coordinates

Greek symbols

α	Thermal diffusivity
β	Coefficient of volumetric thermal expansion
δ	Vapour film thickness
δ_l	Liquid momentum boundary layer thickness
δ_t	Liquid thermal boundary layer thickness
ε	Emissivity
μ	Dynamic viscosity
ν	Kinematic viscosity
ρ	Density
$\sigma; \sigma_t$	Stefan–Boltzmann constant; surface tension

Subscripts

eq	Equivalent
l	Liquid
sat	Saturation value
w	Wall of the plate
i	Interface
r	Radiation
v	Vapour
∞	Free stream value

14.1 Introduction

Film boiling has been intensively studied for its various applications, for instance in heat treatment (Zumbrunnen et al. 1989), chemical synthesis (Choi et al. 2011) and severe accident management in nuclear reactors (Liscic 2009; Berthoud and D'Aillon

2009; Dhir and Purohit 1978). Nishio and Ohtake (1993), Kolev (1998), Okkonen (1999), Meduri et al. (2009), Jouhara and Axcell (2009) and Arias (2009) are among the notable contributors, who have improved the understanding of the underlying mechanism of film boiling hydrodynamics and heat transfer over vertical surfaces. The role of the flow, thermal and thermophysical parameters was assessed in great detail. Juric and Tryggvason (1998) could capture the process of phase change in a computational study by using very refined time step and grid spacing. Since the procedure turns out to be very time consuming, there is a major need of developing models that are accurate but computationally inexpensive.

The acceptability of a model is recognized by the comparison of its prediction with available experimental results. Use of soft computing tools like genetic algorithm and regression analysis, for instance, used by Cai et al. (2006) and Ghosh et al. (2013), to build algebraic relations directly from experimental results is an interesting and emerging research trend in heat transfer in general and boiling studies in particular. Yeh et al. (2009), Yun et al. (2005) and Pettersen (2004) carried out regression-based analysis to correlate experimental data of several researchers for phase change and boiling heat transfer.

The advantage of an identification study attains a greater value, when it is applicable for analysing possible situations involving a wide range of parameters. Conventional numerical studies employ internal elements like the scales used for arriving at non-dimensional mathematical model that has wider applicability than a dimensional model. Length scales based on Kelvin–Helmholtz (KH) instability are routinely used in modelling film boiling problems with liquid–vapour phases (Arias 2009). Regression analysis is a powerful tool for identifying correlation in a simple polynomial of non-dimensional numbers providing a reasonable matching with experiments of Meduri et al. (2009). These forms and the scales depend on the problem at hand.

The models pertaining to film boiling over vertical plates and cylinders have been solved, for instance, by Shiotsu and Hama (2000) employing boundary layer approximation of mass and energy conservation principles with appropriate conditions at the liquid–vapour phase interface. In comparison with their experimental results, the solutions of the model obtained by using either integral approach or similarity transformation yielded a gross underestimation of heat transfer coefficient. The main reason behind the modelling limitations was attributed to the choice of the length scale for the analysis.

Nishio and Ohtake (1993) and Bui and Dhir (1985) conceived the liquid–vapour interface under saturated film boiling condition as a wavy pattern with intermediate ripples and bubble release. The measured heat transfer coefficient in a plate much longer than the interfacial wavelength was observed to remain unaltered with axial distance. For subcooled natural convection film boiling, the visual observations made by Vijaykumar and Dhir (1992) supported the findings of Nishio and Ohtake (1993). Based on the observations of interfacial wave pattern, Nishio and Ohtake (1993) performed an inviscid stability analysis for natural convection film boiling under saturated condition. The heat transfer coefficient was estimated for a vertical plate and a cylinder based on a length scale equal to Kelvin–Helmholtz (KH) interface instability length.

Later on, Okkoken (1999) extended the approach for subcooled forced convection film boiling by adopting the same length scale. Kolev (1998) used the KH-type wavelength as the scale for analysing mixed convection film boiling on a vertical plate to predict heat transfer by a simple closure relation. Although the majority of the theoretical work adopted KH-type instability mode, there were variations in the implementation of the same from model to model. While Nishio and Ohtake (1993) and Okkonen (1999) carried out instability analyses along with the integral model, Kolev (1998) used a closure form in his analysis. An alternate model to explain the interfacial instability is to set criterion in terms of a critical film Reynolds number (Hsu and Westwater 1960), beyond which the interface becomes unstable. Kim et al. (2005) carried out a film boiling analysis for a moderate-sized sphere and found this number as 25. These film boiling models are valid for free, forced or mixed convection flow regimes. A stability analysis was carried out by Makishi and Honda (2012) to determine the minimum critical thickness, corresponding to minimum heat flux point of liquid–solid contact, which was compared with the available experimental data.

A mathematical model describing the problem is introduced in Sect. 14.2. In Sect. 14.3, a discussion is given on the length scale that would provide an estimate of heat transfer during film boiling over a vertical plate for a wide range of water and wall temperatures. Two instability models, namely Kelvin–Helmholtz (KH) and film Reynolds number (FRN), are used to implement the length scales in an integral model for estimating the heat transfer over a vertical plate. The predictions of these models have been suitably compared with available experimental results for a wide variation of plate temperature, liquid temperature and flow velocity. The KH model is similar to the one used by Kolev (1998), where the liquid and vapour average velocities are the primary variables. In the ensuing study involving high liquid velocity, the criterion of film instability is set as a critical Reynolds number equal to 25 (Hsu and Westwater 1960).

In Sect. 14.4, a generalized model is proposed, based on the modified length scale of instability. The instability length is modified by a regression analysis, based on the relevant non-dimensional parameters. This length scale is then coupled with the integral analysis, and the model predictions are compared with a large set of experimental data from various researchers. Available experimental results for wide ranges of wall superheat, liquid subcooling and flow velocity have been compared. For the generalized model to be acceptable, it is necessary to be at least as accurate as the existing models in their different ranges of validity. Such a generalized model would provide a powerful design tool not requiring the range-based switching to different models. A conclusion of the achievements is provided in Sect. 14.5.

14.2 Integral Model

Figure 14.1 shows the problem geometry and the coordinate system for mixed convection film boiling analysis for the vertical flat plate. A thin layer of vapour film around the plate surface is formed followed by the liquid thermal and hydrodynamic

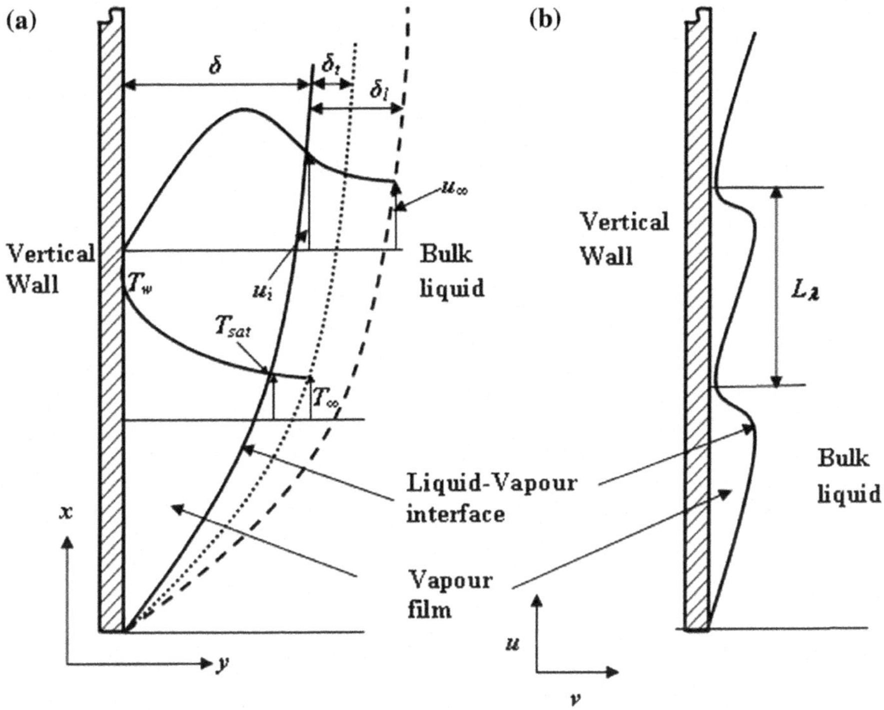

Fig. 14.1 **a** Problem geometry with approximate velocity and temperature profile and **b** pattern of liquid–vapour interfacial wave

boundary layer with usual boundary layer assumption for *Pr* greater than one. A steady-state incompressible laminar flow has been assumed with the viscous dissipation term neglected in the energy equation. In the liquid layer, the effect of natural convection with Boussinesq approximation is assumed. In addition, the effect of induced velocity in the liquid layer is also considered to capture mixed convection. The radiation heat exchange between the surface and the liquid–vapour interface is taken into account by considering the vapour phase to be non-participating and assuming the grey-diffuse enclosure model. All the thermophysical properties of the respective phases are considered at the respective arithmetic mean values of temperature.

Following Das et al. (2014), the conservation equations of mass, momentum and energy in the vapour layer ($0 \leq y \leq \delta$) and liquid layer ($\delta \leq y \leq \delta + \delta_l$) along with the conditions at the boundaries and interfaces under the above assumptions are presented as follows.

$$(\partial u_v/\partial x) + (\partial v_v/\partial y) = 0, \quad (14.1)$$

$$u_v(\partial u_v/\partial x) + v_v(\partial u_v/\partial y) = (\rho_\infty/\rho_v - 1)g + \nu_v(\partial^2 u_v/\partial y^2), \quad (14.2)$$

$$u_v(\partial T_v/\partial x) + v_v(\partial T_v/\partial y) = \alpha_v(\partial^2 T_v/\partial y^2), \tag{14.3}$$

$$(\partial u_l/\partial x) + (\partial v_l/\partial y) = 0, \tag{14.4}$$

$$u_l(\partial u_l/\partial x) + v_l(\partial u_l/\partial y) = \beta_l g(T_l - T_\infty) + \nu_l(\partial^2 u_l/\partial y^2), \tag{14.5}$$

and

$$u_l(\partial T_l/\partial x) + v_l(\partial T_l/\partial y) = \alpha_l(\partial^2 T_l/\partial y^2) \tag{14.6}$$

The boundary conditions are expressed as

$$u_v|_{y=0} = 0; \quad u_v|_{y=\delta} = u_l|_{y=\delta} = u_i; \quad u_l|_{y=\delta+\delta_l} = u_\infty, \tag{14.7a}$$

and

$$T_v|_{y=0} = T_w; \quad T_v|_{y=\delta} = T_l|_{y=\delta} = T_{\text{sat}}\big|_p; \quad T_l|_{y=\delta+\delta_t} = T_\infty \tag{14.7b}$$

Interface mass balance, shear stress and energy balance are denoted as the following three equations

$$j = \rho_v\{u_i(d\delta/dx) - v_v|_{y=\delta}\} = \rho_l\{u_i(d\delta/dx) - v_l|_{y=\delta}\}, \tag{14.8a}$$

$$\mu_v(\partial u_v/\partial y)|_{y=\delta} = \mu_l(\partial u_l/\partial y)|_{y=\delta}, \tag{14.8b}$$

and

$$-k_v \frac{\partial T_v}{\partial y}\bigg|_{y=\delta} + k_l \frac{\partial T_l}{\partial y}\bigg|_{y=\delta} + \varepsilon_{\text{eq}}\sigma(T_w^4 - T_{\text{sat}}^4) = \rho_v h_{fg} \frac{d}{dx}\int_0^\delta u_v dy, \tag{14.8c}$$

where

$$\varepsilon_{\text{eq}} = (1/\varepsilon_w + 1/\varepsilon_i - 1)^{-1}. \tag{14.8d}$$

The governing equations are converted in the integral form with a quadratic profile assumption for velocity and temperature of each phase, and a system of ODEs is formed (Das et al. 2014) with the following non-dimensional parameters

$$\bar{\phi} = \phi/L_\lambda, \quad \bar{u}_i = u_i/u_{\text{ref}}, \tag{14.9a}$$

$$Gr_l = g\beta_l(T_{\text{sat}} - T_\infty)L_\lambda^3/\nu_l^2, \quad Gr_v = g(\rho_l - \rho_v)L_\lambda^3/(\rho_v \nu_v^2), \tag{14.9b}$$

$$Ja_{\text{sub}} = c_{pl}(T_{\text{sat}} - T_\infty)/h_{fg}, \quad Ja_{\text{sup}} = c_{pv}(T_w - T_{\text{sat}})/h_{fg}, \tag{14.9c}$$

$$Re_l = (u_\infty L_\lambda)/\nu_l, \text{ and } Re_v = (u_\infty L_\lambda)/\nu_v, \tag{14.9d}$$

where the generalized length variable ϕ stands for x, δ, δ_l and δ_t as axial location, vapour-phase boundary layer thickness, liquid-phase hydrodynamic boundary layer thickness and liquid-phase thermal boundary layer thickness, respectively, Gr_l is the Grashof number (liquid phase), Gr_v is Grashof number (vapour phase), Ja_{sub} is Jakob number (liquid-phase subcooling), Ja_{sup} is Jakob number (vapour-phase superheat), Re_l is Reynolds number (liquid phase), and Re_v is Reynolds number (vapour phase). The reference velocity is defined as

$$u_{\text{ref}} = u_\infty + \{(\rho_l - \rho_v)/\rho_v\}(g\delta^2/\nu_v). \tag{14.10}$$

The non-dimensional liquid–vapour interfacial velocity ($\bar{u}_i$) is expressed as

$$\begin{aligned}\bar{u}_i = &\{(2/\bar{\delta}_l)(\mu_l/\mu_v) + (\bar{\delta}/2)(Gr_v/Re_v)\}Re_v(Re_v + Gr_v\bar{\delta}^2)^{-1}\} \\ &/\{(2/\bar{\delta}_l)(\mu_l/\mu_v) + (1/\bar{\delta})\}.\end{aligned} \tag{14.11}$$

In terms of these variables ($\bar{\delta}$, $\bar{\delta}_l$ and $\bar{\delta}_t$), with a_2 as an auxiliary parameter for vapour quadratic temperature profile, an ODE system is formed (Das et al. 2014) as the following

$$\begin{bmatrix} m_{11} & m_{12} & 0 & 0 \\ m_{21} & m_{22} & 0 & 0 \\ m_{31} & m_{32} & m_{33} & 0 \\ m_{41} & m_{42} & 0 & m_{44} \end{bmatrix} \begin{bmatrix} d\bar{\delta}/d\bar{x} \\ d\bar{\delta}_l/d\bar{x} \\ d\bar{\delta}_t/d\bar{x} \\ da_2/d\bar{x} \end{bmatrix} = \begin{bmatrix} f_1 \\ f_2 \\ f_3 \\ f_4 \end{bmatrix}. \tag{14.12}$$

This differential equation system has been solved using the ode15s solver in MATLAB.

The energy balance equation near the plate surface can be written as

$$-k_v(\partial T_v/\partial y)_{y=0} = h_{\text{conv}}(T_w - T_{\text{sat}}). \tag{14.13}$$

Following the quadratic temperature profile in the vapour film, the local convective heat transfer coefficient is expressed as

$$h_{\text{conv}} = k_v\{1 + a_2/(L_\lambda\bar{\delta})\}. \tag{14.14}$$

So, the average convective heat transfer coefficient can be expressed as

$$\bar{h}_{\text{conv}} = k_v \int_0^1 \{1 + a_2/(L_\lambda\bar{\delta})\}d\bar{x}. \tag{14.15}$$

Using the linearized form given by Bromley (1950), the average heat transfer coefficient (h_{av}) has been determined for $\bar{h}_{\text{conv}} > h_r$ as

$$h_{av} = \bar{h}_{\text{conv}} + 3h_r/4, \tag{14.16}$$

where

$$h_r = \varepsilon_{eq}\sigma(T_w^4 - T_{sat}^4)/(T_w - T_{sat}). \tag{14.17}$$

14.3 Instability Length Models

Two types of length scales (L_λ) are popularly used for identifying the liquid–vapour interface for determining the heat transfer coefficient from Eq. (14.19), pertaining to the geometry described in Fig. 14.1. These are Kelvin–Helmholtz (KH) and film Reynolds number (FRN) instability lengths. An interface is seen as a vertically repeating wave pattern that arises due to a small perturbation.

From the derivation of Nishio and Ohtake (1993), which was later implemented by Okkonen (1999) and Kolev (1998), the KH scale is defined as

$$L_{\lambda-KH} = 2\pi\left[2\sigma_t\delta\big|_{x=L_{\lambda-KH}}/\{\rho_v(u_{v-avg} - u_{l-avg})^2\}\right]^{1/2}. \tag{14.18}$$

As discussed by Hsu and Westwater (1960), the interface becomes wavy at a critical value of a film Reynolds number Re_{crit} that can be used to define another length scale as

$$L_{\lambda-FRN} = Re_{crit}\mu_v/(\rho_v u_i). \tag{14.19}$$

Based on an analysis of Kim et al. (2005) on film boiling over a sphere, this critical value is fixed at 25 in the present analysis. The KH or FRN model is dependent on the results of the integral model implicitly, as indicated by Eqs. (14.18) and (14.19). The output of the ODE systems from the integral model has been solved first, and the average and interfacial velocities are determined to find out the respective length scale with numerical iteration to find out the instability length following Eqs. (14.18) and (14.19).

14.3.1 Results for Various Instability Models

The comparison of various instability models along with their coupling with the present integral model is made through validation against available experimental results of mixed convection film boiling for the vertical flat plate. In Fig. 14.2, predicted heat transfer from the present models is compared with the experimental data of Bui and Dhir (1985). The experiment was performed for a 6.3-cm-wide and 10.3-cm-high isothermal vertical surface made of copper under saturated natural convection film boiling of water. The KH scale model predicts the heat transfer data quite satisfactorily with the overall mean deviation from experimental data within 3.8%.

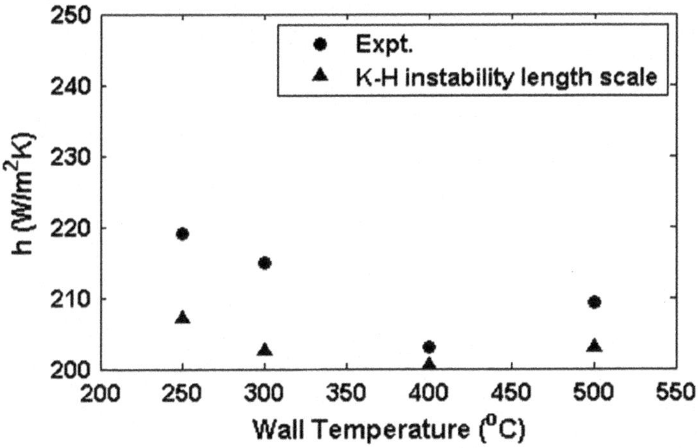

Fig. 14.2 Comparison of the present model prediction against the experiment of Bui and Dhir (1985)

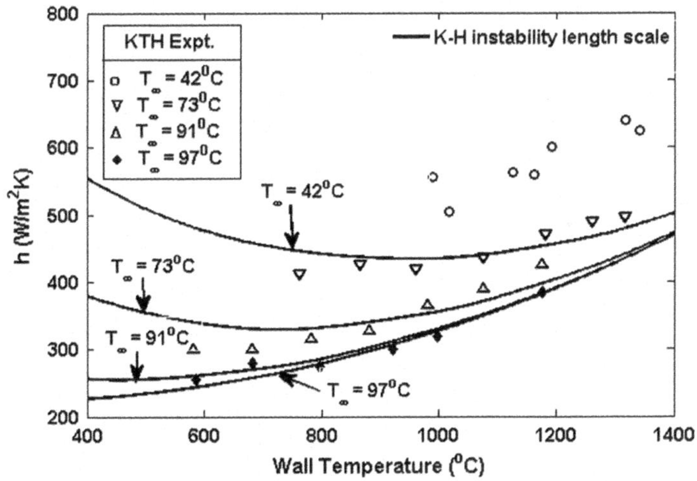

Fig. 14.3 Comparison of the present model prediction with KTH experiment (Okkonen et al. 1996)

Figure 14.3 depicts the comparison of the KH model with the available KTH experimental data due to Okkonen et al. (1996) on film boiling over a vertical flat plate for natural convection with vertical tube of 1.5 m in length and 26 mm in diameter. The KH model gives agreeable results of the predicted heat transfer coefficients with respect to the experimental data for water subcooling up to 9 °C. However, for 58 °C water subcooling, or water temperature of 42 °C, the model underpredicts heat transfer data. The maximum deviation between the model prediction and the experimental result is 25%.

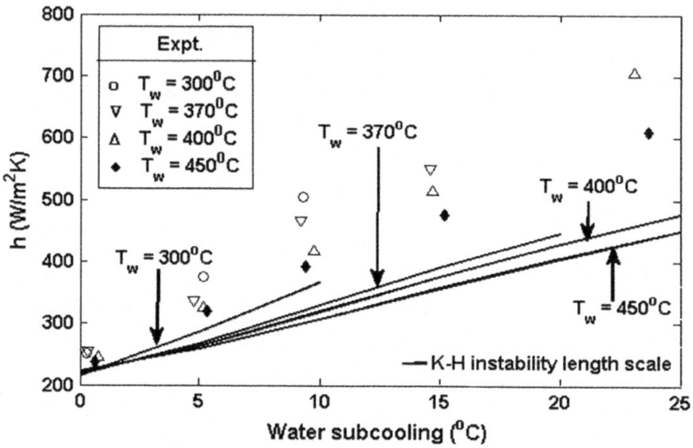

Fig. 14.4 Comparison of the present model prediction with the experiment of Meduri et al. (2009)

Figures 14.4 and 14.5 show the comparison of film boiling heat transfer coefficient between the present model and the experimental data of Meduri et al. (2009) and Jouhara and Axcell (2009). These experiments on a vertical flat plate were performed under the mixed convection film boiling condition with free stream velocities of 0.35 m/s (Meduri et al. 2009) and 0.12 m/s (Jouhara and Axcell 2009), respectively. The overall mean errors of the KH model predictions with respect to these experimental data were, respectively, 22 and 12%. The effect of radiation in both of these cases is far less than that of KTH experiment, which may be the reason for a better agreement with the experimental data. For the KH scale model, the deviation of the predicted heat transfer coefficient with respect to the experimental data increases with the increase in water subcooling.

Figure 14.6 shows the comparison of the present model with the experiment of Shiotsu and Hama (2000) performed for a vertical cylinder at 2.94 bar for saturated mixed convection film boiling with water flow velocities 0–2.65 m/s. Except for stagnant liquid, KH model predicts much lower values. For velocities of 2.32 and 2.65 m/s, critical film Reynolds number model gives a closer match with the experimental data within 8.9 and 5%, respectively.

14.4 Regression Analysis and Validation

The results of the KH instability model show poor agreement for the predicted film boiling heat transfer at large wall superheat and water subcooling and for high liquid velocity. At low superheat and saturated conditions, the KH model is suitable. It is evident from the preceding section that there is a need for the FRN model at high liquid velocity for such prediction. As observed from the results and work of

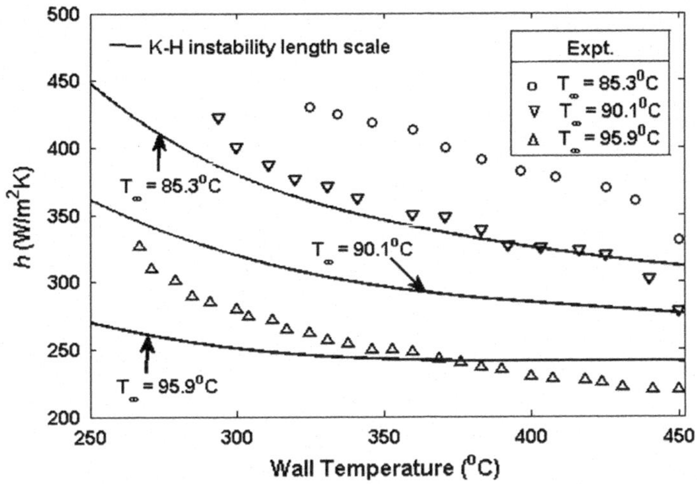

Fig. 14.5 Comparison of model predictions with experimental heat transfer coefficients of Jouhara and Axcell (2009)

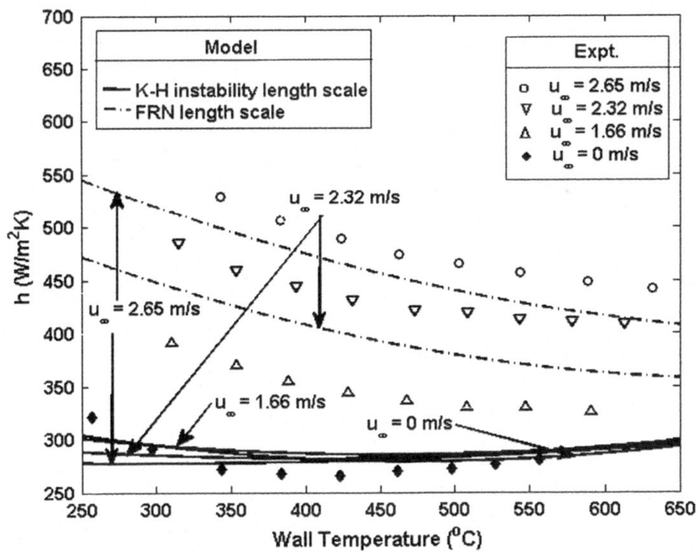

Fig. 14.6 Comparison of the present model prediction against the experiment of Shiotsu and Hama (2000)

Meduri et al. (2009), the instability length is a strong function of the governing non-dimensional parameters for the mixed convection subcooled film boiling. The model developed by Das et al. (2014) shows the dependence of non-dimensional parameters as liquid-phase subcooling Jakob number (Ja_l) (showing the dependence of liquid subcooling), vapour-phase superheat Jakob number (Ja_{sup}) (dependence on wall superheat) and liquid-phase Reynolds number (Re_l) (revealing flow velocity), respectively.

These parameters along with some more dependent non-dimensional length parameter have been selected for developing a multiple-parameter comprehensive regression model. Available experimental data (Meduri et al. 2009; Jouhara and Axcell 2009; Shiotsu and Hama 2000; Bui and Dhir 1985; Okkonen et al. 1996) given in Table 1 involve a wide range of parameters in terms of ranges of wall superheat of 120–1250 K, liquid subcooling of 0–58 K and liquid velocity of 0–2.65 m/s. A total of 69 data have been used for developing the regression model, and the remaining 55 data have been used for the validation purpose.

Liquid-phase Reynolds number in this model is calculated considering characteristic length scale L_c. A length scale is introduced as actual instability length scale L_{act}, which represents the length from the leading edge of the surface at which the predicted heat transfer coefficient from the integral model coincides with the available sets of experimental results. The non-dimensional instability length (as an outcome of the regression analysis) is defined as

$$\bar{L}_{reg} = L_{\lambda-reg}/L_c, \tag{14.20}$$

where L_c is expressed as $[\sigma_t/\{g(\rho_l - \rho_v)\}]$. It is to be noted that this scale is used by Meduri et al. (2009) as the unit wavelength.

A multiple-parameter regression analysis of these four parameters, namely Ja_{sup}, Ja_{sub}, Re_l and $\bar{L}_{reg}$, latter being a non-dimensional length is obtained using MATLAB as follows

$$\bar{L}_{reg} = 7.4699 - 0.00010125 Re_l - 3.8163 Ja_{sup} - 3.2146 Ja_{sub}. \tag{14.21}$$

The average heat transfer coefficient is calculated by using Eq. (14.15), considering the length scale ($L_{\lambda-reg}$) from the regression model.

14.4.1 Results from the Regression Model

The validation of the regression model is presented here in terms of Figs. 14.7, 14.8 and 14.9. Figure 14.7 shows the comparison of the present model based on Kelvin–Helmholtz instability and regression length scales with the available KTH experimental data of Okkonen et al. (1996). As shown in the figure, the heat transfer coefficients based on the regression predict closer agreement with the experimental data. The minimization of the deviation is substantial, particularly at large superheat and high subcooling. However, a marginal deviation from the experiments is observed for a liquid subcooling of 3 °C, where the predicted heat transfer coefficient for the KH length scale is slightly better than that of the length scale based on regression. The maximum deviation of the numerical result from the experimental result is 11.5% for the regression model. Regression model minimizes

Table 14.1 Experimental data used for the regression modelling

Ja_{sup}	Ja_{sub}	Re_l	h_{expt}	$L_{act}L_c^{-1}$	Ja_{sup}	Ja_{sub}	Re_l	h_{expt}	$L_{act}L_c^{-1}$
0.848	0.1071	0	555	3.854	1.001	0.1071	0	563	3.246
1.041	0.1071	0	560	3.246	1.252	0.1071	0	625	2.110
0.609	0.0500	0	412	3.427	0.818	0.0500	0	419	3.064
1.064	0.0500	0	470	2.499	1.222	0.0500	0	495	2.862
0.432	0.0167	0	300	3.887	0.629	0.0167	0	315	3.887
0.840	0.0167	0	365	3.166	1.058	0.0167	0	425	2.805
0.437	0.0056	0	255	4.999	0.645	0.0056	0	274	6.198
0.856	0.0056	0	320	5.798	1.058	0.0056	0	385	4.679
0.174	0.0005	2969	251	4.434	0.174	0.0173	2855	505	5.792
0.236	0.0006	2969	255	3.995	0.236	0.0271	2782	551	6.192
0.308	0.0011	2969	237	5.592	0.308	0.0175	2853	392	4.394
0.308	0.0439	2657	610	7.590	0.263	0.0014	2970	244	4.993
0.263	0.0181	2849	417	5.073	0.263	0.0428	2666	706	6.791
0.196	0.0272	949	430	8.632	0.227	0.0272	949	413	7.348
0.259	0.0272	949	382	7.227	0.286	0.0272	949	365	7.227
0.308	0.0272	949	331	8.030	0.248	0.0272	949	391	7.227
0.174	0.0184	973	400	6.283	0.201	0.0184	973	371	6.093
0.227	0.0184	973	350	6.283	0.256	0.0184	973	327	6.283

(continued)

Table 14.1 (continued)

Ja_{sup}	Ja_{sub}	Re_l	h_{expt}	$L_{act}L_c^{-1}$	Ja_{sup}	Ja_{sub}	Re_l	h_{expt}	$L_{act}L_c^{-1}$
0.286	0.0184	973	320	6.053	0.210	0.0184	973	362	6.093
0.237	0.0184	973	348	5.612	0.248	0.0184	973	338	5.892
0.149	0.0076	1002	310	5.200	0.174	0.0076	1002	280	6.283
0.201	0.0076	1002	257	7.000	0.227	0.0076	1002	248	7.360
0.263	0.0076	1002	230	8.800	0.308	0.0076	1002	220	10.000
0.197	0	28,435	530	2.925	0.273	0	28,435	489	2.966
0.389	0	28,435	458	3.175	0.477	0	28,435	443	3.635
0.171	0	24,894	485	4.052	0.245	0	24,894	445	3.927
0.321	0	24,894	421	3.969	0.389	0	24,894	413	4.011
0.459	0	24,894	409	3.593	0.167	0	17,812	392	5.013
0.278	0	17,812	345	4.930	0.355	0	17,812	330	4.805
0.436	0	17,812	326	3.462	0.118	0	0	321	5.682
0.198	0	0	272	7.938	0.273	0	0	266	7.520
0.345	0	0	272	6.685	0.403	0	0	281	6.058
0.131	0	0	219	6.231	0.355	0	0	209	5.672

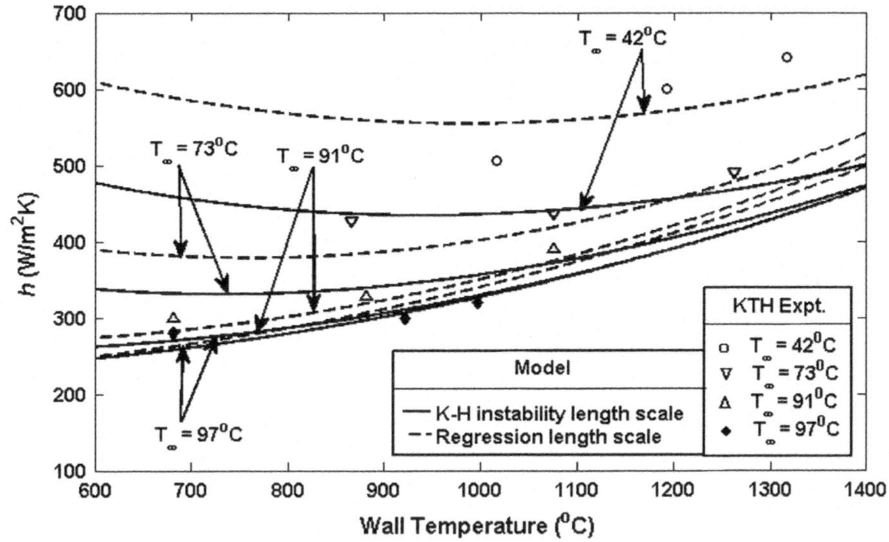

Fig. 14.7 Comparison of the present model prediction with KTH experiment (Okkonen et al. 1996)

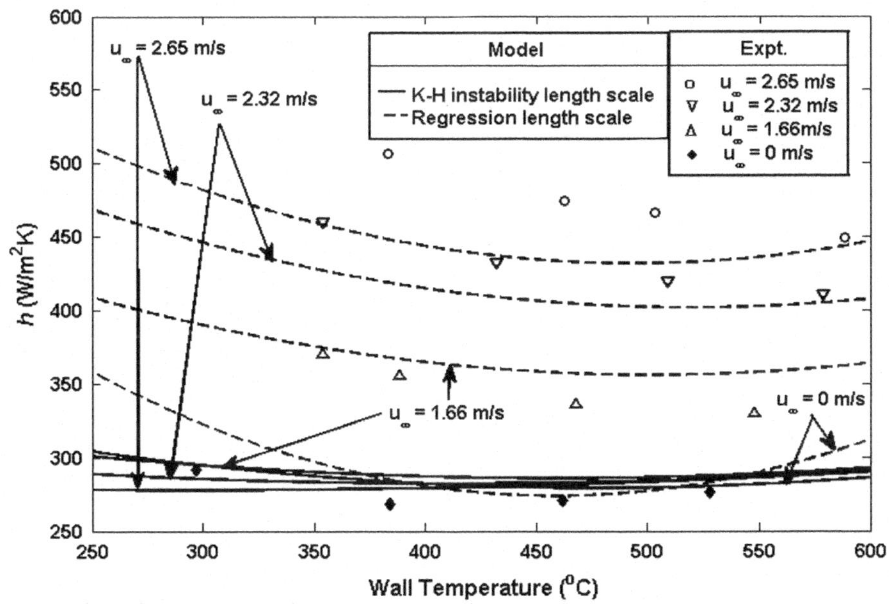

Fig. 14.8 Comparison of the present model prediction against the experiment of Shiotsu and Hama (2000)

the overall mean error to 6% from 11.5%, as compared to the KH model, with respect to overall data of Okkonen et al. (1996).

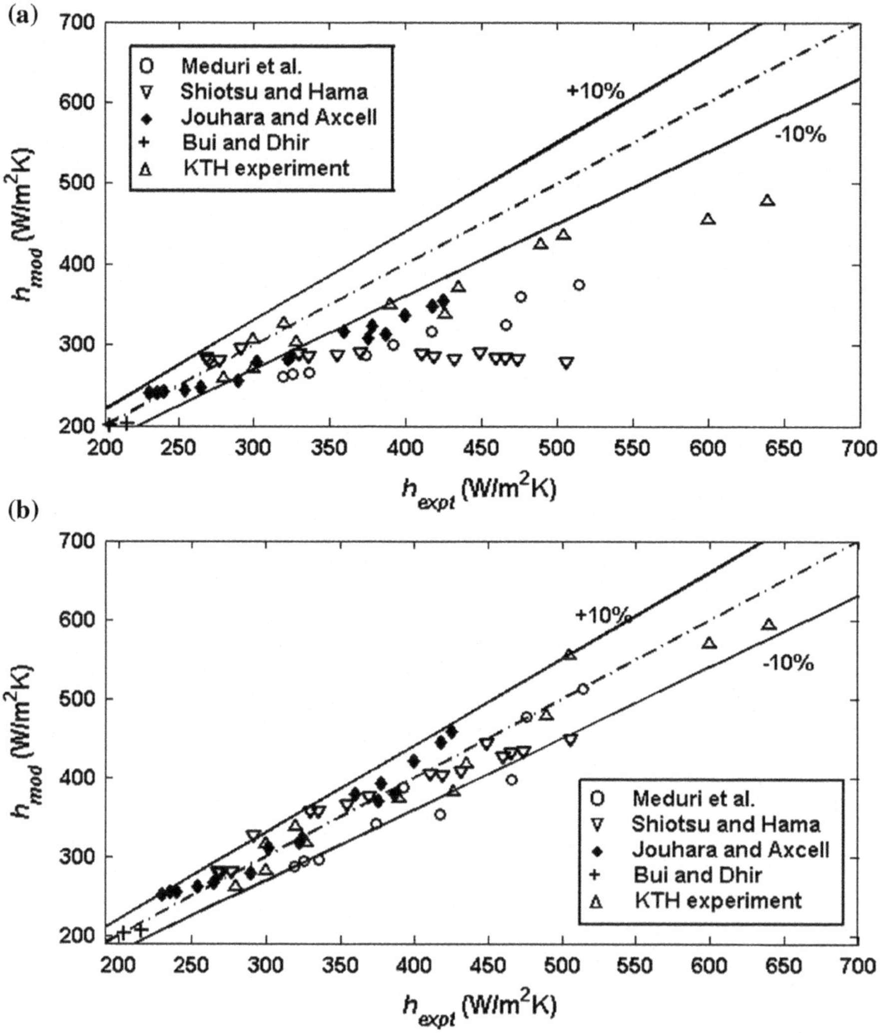

Fig. 14.9 Comparisons of **a** KH model and **b** regression-based model prediction against the experimental data of Meduri et al. (2009), Jouhara and Axcell (2009), Shiotsu and Hama (2000), Bui and Dhir (1985) and Okkonen et al. (1996)

Figure 14.8 represents the comparison of the predicted heat transfer coefficient from the regression and the KH model with the experimental data of Shiotsu and Hama (2000). It is observed that with respect to the experimental data for higher free stream velocity above 1 m/s, the regression model prediction is more satisfactory than KH model prediction. At zero velocity of liquid, the regression model gives slightly better result than the KH model. Regression model reduces the overall mean error from 23.3% (the KH model) to 5% for overall data of Shiotsu and Hama (2000).

To compare the total data of regression-based single comprehensive model with the experimental data, the KH model prediction is compared first with overall experimental data (Meduri et al. 2009; Jouhara and Axcell 2009; Shiotsu and Hama 2000; Bui and Dhir 1985; Okkonen et al. 1996) used in the present work. The prediction of regression-based modified model is compared next.

Figure 14.9a shows that 71% of the total 55 experimental test data lie outside 10% error band, when the KH model is used as an instability length in the integral analysis. The predicted heat transfer coefficient data from regression-based model, on the other hand, show a remarkable improvement as and only 8% data lie outside the 10% error band with the corresponding experimental data as revealed by Fig. 14.9b.

14.5 Conclusion

The common instability models (KH and FRN) are coupled in an integral-based film boiling analysis for the vertical flat plate under the mixed convection film boiling. In addition, a multiple-parameter regression-based refined instability model is proposed to predict film boiling heat transfer with an objective of achieving closer agreement to experiments over a range wider than those provided by the existing models. The salient findings of the present work are as follows:

1. The KH model predicts results closer to the experiments, for both natural convection and mixed convection film boiling except for very high subcooling, high superheat and high flow velocity.
2. At very high flow velocity, the FRN model gives excellent agreement with experiments.
3. The regression-based model reduces the experimental error, particularly at high wall superheat and liquid velocity.
4. The majority of data (more than 90%) lie within the 10% band of experimental ones, when regression-based model is used.

Acknowledgements The financial support provided for this work by CSIR, India, and BARC, Mumbai, India, is gratefully acknowledged. The authors express their gratitude towards Deb Mukhopadhyay, RSD, BARC, Mumbai, India, for his encouragement and support in film boiling research.

References

Arias FJ (2009) Self-induced thermocapillary convection in film boiling heat transfer from a vertical surface in saturated conditions and viscous regimen. Int J Heat Fluid Flow 30:911–915

Berthoud G, D'Aillon LG (2009) Film boiling heat transfer around a very high temperature thin wire immersed into water at pressure from 1 to 210 bar: Experimental results and analysis. Int J Therm Sci 48:1728–1740

Bromley LA (1950) Heat transfer in stable film boiling. Chem Eng Prog 46(5):221–227

Bui TD, Dhir VK (1985) Film boiling heat transfer on an isothermal vertical surface. Trans ASME J Heat Trans 107:764–771

Cai W, Pacheco-Vega A, Sen M, Yang KT (2006) Heat transfer correlations by symbolic regression. Int J Heat Mass Trans 49:4352–4359

Choi SR, Evangelista JW, Avedisian CT, Tsang W (2011) Experimental study of chemical conversion of methanol and ethylene glycol in a film boiling reactor. Int J Heat Mass Trans 54:500–511

Das DC, Ghosh K, Sanyal D, Meignen R (2014) A novel approach for modeling mixed convection film boiling for a vertical flat plate. Numer Heat Trans A-Appl 66:1112–1130

Dhir VK, Purohit GP (1978) Subcooled film-boiling heat transfer from spheres. Nucl Eng Des 47(1):49–66

Ghosh S, Pratihar DK, Maiti B, Das PK (2013) Automatic classification of vertical counter-current two-phase flow by capturing hydrodynamic characteristics through objective descriptions. Int J Multiph Flow 52:102–120

Hsu YY, Westwater JW (1960) Approximate theory for film boiling on vertical surfaces. Chem Eng Prog Symp Ser 56(30):15–24

Jouhara H, Axcell BP (2009) Film boiling heat transfer and vapour film collapse on spheres, cylinders and plane surfaces. Nucl Eng Des 239:1885–1900

Juric D, Tryggvason G (1998) Computations of boiling flows. Int J Multiphas Flow 24:387–410

Kim CS, Suh KY, Rempe JL, Cheung FB, Kim SB (2005) Effect of interfacial wavy motion on film boiling heat transfer from isothermal downward-facing hemispheres. Nucl Eng Des 235:2141–2154

Kolev NI (1998) Film boiling on vertical plates and spheres. Exp Therm Fluid Sci 18:97–115

Liscic B (2009) Heat transfer control during quenching. Mater Manuf Process 24:879–886

Makishi O, Honda H (2012) Examination of minimum-heat-flux-point condition for film boiling on a sphere in terms of the limiting liquid superheat and the critical vapor film thickness. Int J Heat Mass Trans 55:2377–2383

Meduri PK, Warrier GR, Dhir VK (2009) Wall heat flux partitioning during subcooled forced flow film boiling of water on a vertical surface. Int J Heat Mass Trans 52:3534–3546

Nishio S, Ohtake H (1993) Vapor-film-unit model and heat transfer correlation for natural-convection film boiling with wave motion under subcooled conditions. Int J Heat Mass Trans 36(10):2541–2552

Okkonen T (1999) Film boiling on a vertical high-temperature surface: focusing on melt jet-water interactions. Nucl Eng Des 189:273–297

Okkonen T, Wennerstrom H, Hedberg S, Blomstrand J, Sehgal BR, Frid W (1996) Film boiling on a long vertical surface under high heat flux and water subcooling conditions, heat transfer, Houston. AIChE Symp Ser 92(310):294–303

Pettersen J (2004) Flow vaporization of CO_2 in microchannel tubes. Exp Therm Fluid Sci 28:111–121

Shiotsu M, Hama K (2000) Film boiling heat transfer from a vertical cylinder in forced flow of liquids under saturated and subcooled conditions at pressures. Nucl Eng Des 200:23–38

Vijaykumar R, Dhir VK (1992) An experimental study of subcooled film boiling on a vertical surface—hydrodynamic aspects. Trans ASME J Heat Trans 114:161–168

Yeh C, Chen C, Chen Y (2009) Heat transfer analysis of a loop heat pipe with biporous wicks. Int J Heat Mass Trans 52:4426–4434

Yun R, Kim R, Kim MS (2005) Flow boiling heat transfer of carbon dioxide in horizontal mini tubes. Int J Heat Fluid Flow 26:801–809

Zumbrunnen D, Viskanta AR, Incropera FP (1989) The effect of surface motion on forced convection film boiling heat transfer. Trans ASME J Heat Trans 111:760–766

Chapter 15
Numerical Modeling of Boiling

K. Nandi and G. Giustini

Abstract The phenomenon of boiling is visible all around us from cooking to power generation, but despite such all around usages many aspects of boiling are still not very well understood as it is a very complex process and occurs over a wide range of system scales. We often rely on empirical correlations when we want to evaluate different parameters connected with boiling phenomena. Along with the development of empirical correlations for engineering applications, considerable advances are there in understanding the fundamentals of the boiling process. Since the process is very complex and multiple thermal and fluid variables are involved, a complete theoretical model for predicting the boiling heat transfer is yet to be developed. Boiling phenomenon is still being intensively studied and is the focus of research activities in numerous institutions across the world. A better understanding of the physics of boiling can be achieved by either detailed measurements or high-resolution numerical simulation. These two approaches are now complementing each other in understanding the physics of boiling more completely. In recent years, numerical modeling has improved considerably thanks to ever-increasing computational power. With advancing computing capabilities and advent of new numerical techniques for two-phase flow, simulations of boiling heat transfer have become feasible. The main two approaches in numerical simulation of boiling are (i) interpenetrating media approach and (ii) single-fluid approach. In addition to this, some newer techniques like the phase field method and the lattice Boltzmann method have to some extent been used for simulating boiling flows. In this review, we look at the different approaches of numerical simulation of boiling currently being used.

K. Nandi (✉)
Reactor Projects Division, Bhabha Atomic Research Centre, Mumbai, India
e-mail: kausik.nandi@yahoo.com

G. Giustini
Mechanical Engineering Department, Imperial College, London, UK
e-mail: g.giustini12@imperial.ac.uk

K. Saha et al. (eds.), *Two-Phase Flow for Automotive and Power Generation Sectors*, Energy, Environment, and Sustainability,
https://doi.org/10.1007/978-981-13-3256-2_15

15.1 Introduction

According to a definition by Collier and Thome (2001), boiling is 'the process of addition of heat to a liquid in such a way that generation of vapor occurs.' Boiling heat transfer is a very efficient heat transport mechanism and is employed in a wide field of applications. Heat removal by a boiling fluid is encountered in a variety of engineering systems ranging from large nuclear/conventional power plants to cooling of tiny high-performance electronic chips, for the reason it can transfer large heat fluxes across relatively small temperature differences. Due to its intensive use in engineering applications and newer areas of application (e.g, microheatpipes, biochips), research in boiling has intensified over time in the last hundred years (Collier and Thome 2001; Kakaç et al. 1988; Bergles 1988). The energy crunch and its associated environmental consequences have made it a crying need for all appliances to strive for higher thermal efficiency which in turn have led to further efforts to enhance boiling heat transfer.

A comprehensive review of research up to the 1970s has been provided by Bergles (1981a) supplemented by the review by Nishikawa (1987). More recent review articles by Dhir (1998) and Mangalik (2006) summarize the current achievements in the field of boiling heat transfer research.

15.2 Boiling Phenomena

In this section, the phenomenon of boiling and major achievements in boiling research is briefly described. Heat transfer mechanisms during boiling are still not well understood and are an area of research. One of the first comprehensive studies in this area was the pioneering work of Jakob and Fritz (1931); it was followed by Nukiyama (1934) who established the boiling curve for nucleate boiling conditions, i.e., when the generation of vapor is induced via heating of a solid surface. This work has become a kind of benchmark for nucleate boiling investigations. Subsequently, McAdams et al. (1949) extended the nucleate boiling curve to conditions whereby boiling at a surface is induced in a pool of liquid at a temperature below saturation ('subcooled' boiling). With the rapid development in the field of nuclear power and related safety issues, there was explosive growth in the number of publications in the area of nucleate boiling heat transfer. In particular, much attention was devoted to understanding fault conditions leading to uncontrollable boiling modes that can cause damage of the nuclear fuel rods because of critical heat flux (CHF) and predict accurately the heat transfer via circulation of fluids undergoing subcooled boiling (Nishikawa 1987; Bergles 1981b).

In this context, boiling in a stationary pool of liquid at a horizontal surface (pool boiling) and boiling in ducts with a strong imposed liquid flow (flow boiling) are the most studied phenomena. A typical pool boiling curve is shown in Fig. 15.1.

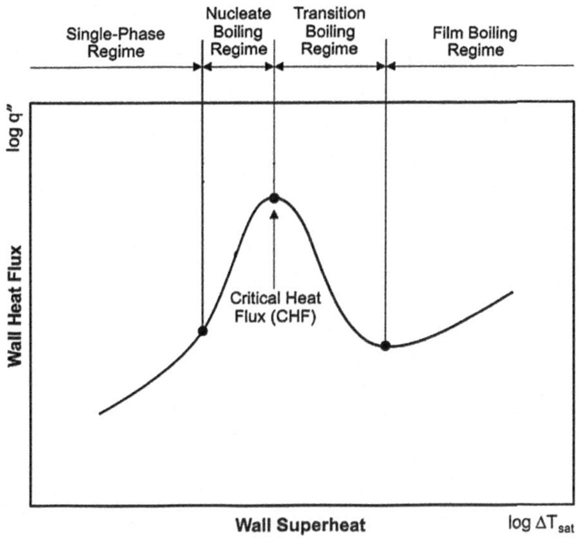

Fig. 15.1 Pool boiling curve

In pool boiling, different regimes of heat transfer are (Collier and Thome 2001; van Stralen and Cole 1979):

(i) Partial nucleate boiling: where individual bubbles form and detach from the heated surface without any interaction. High heat transfer rates are characteristic of this region.
(ii) Nucleate boiling: With the increase of heat flux, numerous bubble nucleation sites are activated and steady columns of vapor bubbles are generated. However, as more and more area of the heating element is blanketed by vapor the required wall superheat increases due to the insulating effect of the vapor and the overall heat removal rate decreases dramatically (a typical instance of CHF).
(iii) Transition to film boiling: When CHF is reached, a large part of the surface of the heater is covered with vapor. With a slight increase in heat flux, the wall temperature increases inordinately, damaging the heating element.
(iv) Stable film boiling: A stable vapor layer is formed in this regime, and the liquid phase is separated from the heated wall.

Numerous experimental studies have been carried out to study the various aspects of boiling: Bubble dynamics, nucleation site density, the formation and evaporation of thin liquid layers beneath growing bubbles are among the most studied topics (Ramaswamy et al. 2002; Cole 1967; Ivey 1967; McHale and Garimella 2010).

In isolated bubbles boiling (i.e., when there is some space between nucleation sites and bubbles do not interact, for example, do not merge or disturb the flow pattern near each other), it is sufficient to focus on any single bubble on the surface and observe local parameters (Duan et al. 2013; Jung and Kim 2014). Figure 15.2 shows the schematic of a single bubble being formed during nucleate boiling. For typical fluids, in low-pressure pool boiling conditions the liquid near the horizontal surface

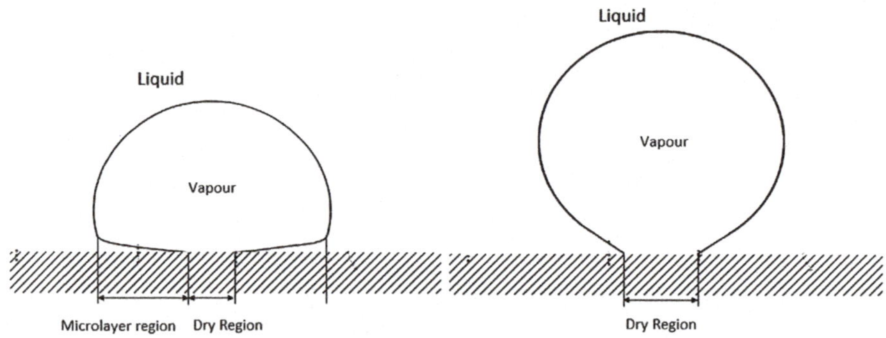

Fig. 15.2 Schematic of single bubble nucleate boiling

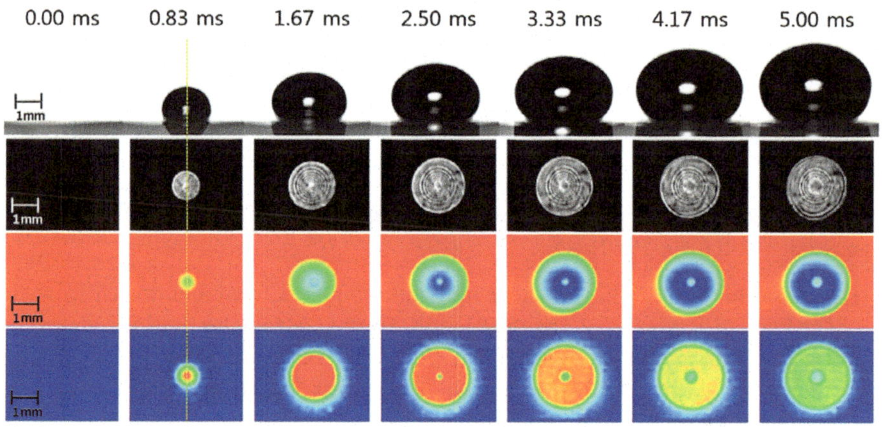

Fig. 15.3 High-speed camera images of a boiling bubble and corresponding liquid–vapor phase boundary, temperature, and heat flux distributions at the boiling surface. Adapted from Jung and Kim (2014)

is highly superheated, which causes rapid bubble expansion (Rayleigh 1917; Plesset and Zwick 1954, 1955; Scriven 1958; Prosperetti and Plesset 1978; Prosperetti 2017). These almost hemispherical bubbles leave behind on the solid surface a liquid layer (microlayer) which then evaporates and contributes itself to bubble expansion (Koffman and Plesset 1983; Jung and Kim 2018). For high-pressure pool boiling, the reduced density ratio (Scriven 1958) reduces the expansion rate and no microlayer is formed (Jung and Kim 2018). In high-pressure boiling, bubble departure diameters are much smaller than in low-pressure conditions. It has been speculated that this could be due to a decrease in wettability of typical metallic surfaces as the temperature increases (Ardron et al. 2017). Figure 15.3 shows images of a boiling bubble taken with the help of a high-speed camera.

In flow boiling conditions, the basic mechanism of heat transfer remains the same. However, due to forced convection the liquid is only slightly superheated near

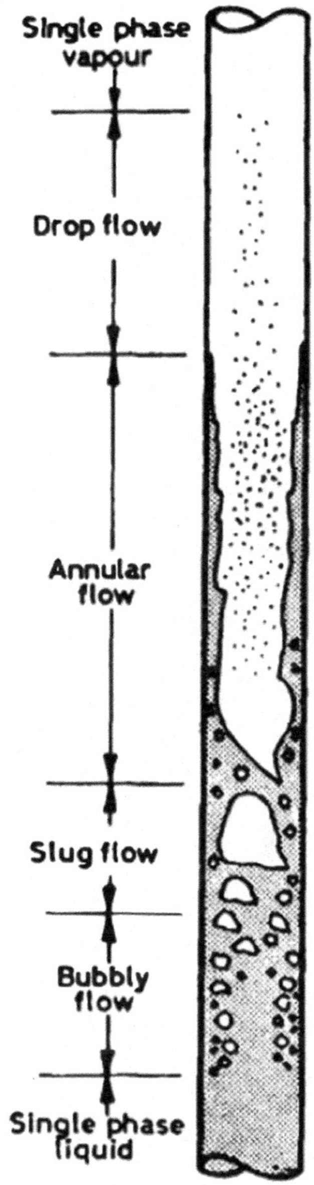

Fig. 15.4 Flow boiling regime in a vertical tube. Adapted from Collier and Thome (2001)

the wall and bubble behavior is dictated mainly by hydrodynamic aspects: lift, drag, buoyancy, surface tension, and wall adhesion forces (Klausner et al. 1993; Thorncroft et al. 1998). Figure 15.4 illustrates a typical flow boiling regime in a vertical tube.

One aspect of boiling phenomena that has always worried designers is CHF. As discussed earlier, when the heat flux is increased, conditions could be reached whereby increasing the power input causes the surface to be entirely covered by a vapor blanket, due to, for example, bubble interactions and 'crowding.' This causes the temperature to increase inordinately following a slight perturbation in the flow parameters (Hewitt 1998a, b). It has become imperative to study the collective behavior of bubbles, which is much more difficult than modeling the behavior of a single bubble. This collective behavior is poorly understood, and the current understanding is empirical or phenomenological in nature. There are different mechanisms for CHF prevalent in published literature, starting with Kutateladzes (1961) pioneering work to the popular work by Zuber (1958). An extensive and very recent two-part review by Liang and Mudawar (2018a, b) is an excellent read on the subject.

Over time measurement techniques have developed and very small length and time scales are now being resolved, which has made the measurement of instantaneous heat transfer beneath a bubble possible with remarkable accuracy. High-speed infrared thermography (Schweizer and Stephan 2009; Wagner et al. 2006) provides a very detailed insight about the transient heat transfer between the heating element and the fluid. Another very important area of boiling research is boiling on microstructured surfaces, as nucleate boiling heat transfer and CHF enhancement are possible via employing carefully engineered surfaces. Several review articles on the enhancement of boiling heat transfer on microstructured surfaces have been reported (Shojaeian and Kosar 2015; Kim 2011; Ahn and Kim 2012; Dong et al. 2015).

As stated earlier, boiling is a complex physical process due to the interaction of a great variety of important parameters. A complete theoretical model which could predict boiling heat fluxes only as function of a given set of input parameters is yet to be developed. Concurrent with experimental studies and empirical correlations, efforts are aimed at understanding the physical mechanisms of boiling in depth. A consensus is still lacking among researchers in this field regarding the dominant mechanism of heat transfer during boiling. Over time different theoretical models have been proposed for boiling heat transfer. One of the first papers that discussed different mechanisms of boiling heat transfer was by Han and Griffith (1965a, b). They described two methods of heat transfer: bulk convection, in which the superheated liquid is removed away from the wall as the bubble detaches and natural convection from the heated wall to the fluid in the space between bubble nucleation locations. Cooper and Lyod (1969) inferred the existence of a thin liquid film beneath boiling bubbles, called a microlayer, which enables the high heat transfer during bubble growth. Kern and Stephan (2003, 2004) developed a theory where they described the mechanisms of the transport of heat from the wall to the fluid. Due to the separation of scales, the model considers the transport phenomena on microscopic and macroscopic scales. A substantial part of the heat from the wall flows through the microlayer where the thermal resistances of the liquid film is negligible which leads to high heat transfer and hence governs the overall heat transfer performance. At a macroscopic scale, the liquid in the vicinity of a rising bubble is set in motion, which results in an enhanced heat transfer and transient heat conduction due to rewetting of the heater surface.

Mechanistic wall heat transfer models proposed by De Valle and Kenning (1985) or Kurul and Podowski (1990) use very crude theory of heat flux partitioning along the lines of the Han and Griffith model, enabling little physical understanding. These models break down when bubbles begin to interact with each other (Basu 2003).

15.3 Numerical Modeling

Boiling flows belong to a subset of a much larger group of flows classified as multiphase flows. Efforts to simulate multiphase flows have been one of the major challenge areas since the inception of computational fluid dynamics (CFD). The main difficulty is solving the Navier–Stokes equations with a deforming phase boundary. In the last two decades, thanks to exponential rise of computing power and development of new numerical tools (Prosperrotti and Trygvassion 2007; Yeoh and Tu 2010), major progress has been achieved in this field such that numerical simulation of boiling is also established as a tool that can complement experimental investigations in order to understand the physics of boiling better. The crux of simulation of multiphase flows is the accurate identification of the interface dynamics through which flow regimes can be defined and momentum and energy transfer mechanisms between the phases can be quantified.

There are mainly two major approaches of numerical simulation of boiling flows: (a) two-fluid models or interpenetrating media approach (Ishii and Hibiki 2011) and (b) single-fluid formulation or interface tracking methods (ITMs) (Tryggvasson et al. 2001). In interpenetrating media approach, each point in the mixture is occupied simultaneously by both phases, and separate conservation equations are required for each field. In single-fluid formalism, the topology and dynamics of the interface are directly simulated by the use of direct interface tracking methods. In the last decade, newer techniques like lattice Boltzmann method and phase field method have been used for the simulation of boiling flows. In the subsequent sections, we will discuss each of these major approaches in detail.

15.3.1 Interpenetrating Media

The main challenge in simulating boiling flows is posed by the requirement of capturing the energy transfer from wall to fluid associate to the formation and release of bubbles at the heat transfer surface. Interpenetrating continua methods predict the evolution of spatially and temporally averaged quantities and provide no means of mechanistic modeling the behavior of bubbles near the wall. Typically, in such circumstances one computational near-wall cell is several times the bubble characteristic dimension.

Various wall boiling models are used in commercial CFD codes (Colombo and Fairweather 2016) with the aim of predicting energy transfer from wall to fluid. All of these approaches rely on heat flux partitioning, and various correlations for wall heat flux partitioning have been proposed in the literature. Mechanistic models based on relevant heat transfer mechanisms occurring during the boiling process are used for the estimation of the wall heat flux as well as the partitioning of the wall heat flux between the liquid and vapor phases. Most numerical simulations of boiling flow are mainly based on the use of these mechanistic models, of which a large majority are extensions of the model proposed by Kurul and Podowski of the Rensselaer Polytechnic Institute (RPI) (Kurul and Podowski 1990). Following Griffith (Han and Griffith 1965a), the wall heat flux is usually partitioned into three heat flux components: convective heat flux, evaporative heat flux, and 'quenching' (i.e., transient conduction to the liquid) heat flux. These models rely on previous knowledge of three unknown parameters, which should be modeled accurately: nucleation density (Na), bubble departure diameter (Db), and bubble departure frequency (f).

The Eulerian–Eulerian two-fluid model represents the most detailed macroscopic formulation of the thermal and hydrodynamic characteristics of any two-phase systems. As noted, the problem with this method is the specification of closure relations for mass, momentum, and energy exchanges across the interface and calculating the corresponding interfacial area. Here, the field equations are expressed by six conservation equations of mass, momentum, and energy. For boiling flow, three equations are used to model the bubbles (i.e., the vapor phase) while the three equations are used to model the liquid phase. The interfacial terms arising out of the averaging of the equations represent the mass, momentum, and energy transfers through the interface between the phases. The existence of these interfacial transfer terms is rather significant as they determine the rate of phase changes, and the degree of thermal non-equilibrium between phases. Most importantly, they provide the necessary closure relations required in two the fluid model.

In the mass conservation equations, mass transfer is accounted between phases due to the evaporation from liquid to bubbles or bubbles being condensing in the bulk liquid (which is at a temperature below saturation). In the momentum conservation equations, the important interfacial effects between the liquid and gas phases due to the drag force as well as other possible so-called non-drag forces in the form of lift, wall lubrication, and turbulent dispersion are incorporated. In the energy conservation equations, the interfacial heat transfer accounts for the phase change due to evaporation/condensation. Also, the prediction of the local bubble sizes in the subcooled liquid flow is strongly influenced by factors like turbulent dispersion, local coolant temperature fluctuations occurring near the heated wall.

The two set of conservation equations governing mass, momentum, and energy can be written as:

Liquid-phase continuity equation

$$\frac{\partial \rho_l \alpha_l}{\partial t} + \nabla \cdot (\rho_l \alpha_l \overline{u_l}) = \Gamma_{lg}$$

Vapor-phase continuity equation

$$\frac{\partial \rho_g \alpha_g}{\partial t} + \nabla \cdot \left(\rho_g \alpha_g \overline{u_g}\right) = \Gamma_{lg}$$

Liquid-phase momentum equation

$$\frac{\partial \rho_l \alpha_l \overline{u}}{\partial t} + \nabla . (\rho_l \alpha_l \overline{u}_l \overline{u}_l) = -\alpha_l \nabla P + \alpha_l \rho_l \overrightarrow{g}$$
$$\nabla . \left[\alpha_l \mu_l^{\mathrm{eff}} \left(\nabla \overline{u}_l + (\nabla \overline{u}_l)^T\right)\right] + \left(\Gamma_{lg} \overline{u}_g - \Gamma_{gl} \overline{u}_l\right) + F_{lg}$$

Vapor-phase momentum equation

$$\frac{\partial \rho_g \alpha_g \overline{u}_g}{\partial t} + \nabla . \left(\rho_g \alpha_g \overline{u}_g \overline{u}_g\right) = -\alpha_g \nabla P + \alpha_g \rho_g \overrightarrow{g}$$
$$+ \nabla . \left[\alpha_g \mu_g^{\mathrm{eff}} \left(\nabla \overline{u}_g + \left(\nabla \overline{u}_g\right)^{\mathrm{T}}\right)\right] + \left(\Gamma_{gl} \overline{u}_l - \Gamma_{gl} \overline{u}_g\right) + F_{gl}$$

Liquid-phase energy equation

$$\frac{\partial \rho_l \alpha_l H_l}{\partial t} + \nabla . (\rho_l \alpha_l \overline{u}_l H_l) = \nabla . \left[\alpha_l \lambda_l \nabla T_l + \frac{\mu_{Tl}}{\mathrm{Pr}_{Tl}} \nabla H_l\right]$$
$$+ \left(\Gamma_{lg} H_g - \Gamma_{gl} H_l\right)$$

Vapor-phase energy equation

$$\frac{\partial \rho_g \alpha_g H_g}{\partial t} + \nabla . \left(\rho_g \alpha_g \overline{u}_g H_g\right) = \nabla . \left[\alpha_g \lambda_g \nabla T_g + \frac{\mu_{T_g}}{\mathrm{Pr}_{T_g}} \nabla H_g\right]$$
$$+ \left(\Gamma_{gl} H_l - \Gamma_{lg} H_g\right)$$

The source term Γ_{lg} represents the mass transfer rate due to evaporation or condensation in the bulk subcooled liquid and $\Gamma_{gl} = -\Gamma_{lg}$. The wall vapor generation rate is modeled in a mechanistic way. Interfacial transfer terms in momentum and energy equations denote transfer from one phase to another. The total interfacial force is the sum of the drag force (Ishii and Zuber 1979), lift force (Drew and Lahey 1979), wall lubrication force (Anglart and Nylund 1996), and turbulence-assisted bubble dispersion force (Antal et al. 1991).

Over time, many researchers (Koncar et al. 2004; Lo 2005) have proposed different techniques for modeling the different interfacial terms; other efforts were directed for the improvement in bubble size modeling and interfacial area concentration modeling. Ishii and Hibiki (2011) were the first to propose a detailed modeling of the interfacial area transport equation. Later, Yeoh and Tu (2005) applied an interfacial area transport equation and bubble number density transport equations in CFD codes for prediction of subcooled boiling flows. Figure 15.5 shows a contour plot of the simulated vapor fraction at the outlet of a pipe during subcooled boiling (Bartolomei and Chanturiya 1967).

Though different models in two-fluid approach have confirmed the ability of numerical simulations in providing detailed predictions for some cases, significant improvements in model accuracy is required for general applicability. Even when codes are built using a mechanistic approach, numerous empirical closure relations

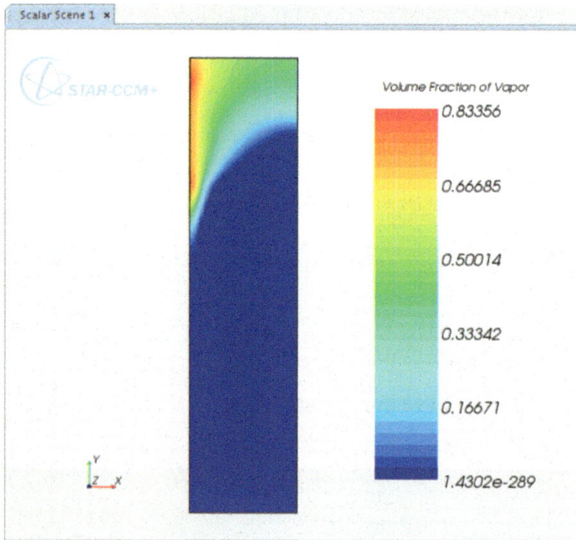

Fig. 15.5 Prediction of subcooled flow boiling (Bartolomei and Chanturiya 1967) using STAR-CCM+(v9.06)

are still required for wall boiling, population balance, and turbulence models. The problem of formulating an all-encompassing closure law arises from the fact that each closure law depends on the specific physical phenomenon.

15.3.2 Single-Fluid Formalism

In single-fluid formalism, the idea is to simulate the whole field as a single fluid, with variable properties changing sharply at the interphase boundary, which is modeled as having zero thickness.

From a hydrodynamic point of view, the main difficulty is caused by the need of accounting for surface terms (e.g., the surface tension force) in the fundamental transport equations, which are derived for stationary fluid control volumes.

From a thermal point of view, capturing the thermodynamic state of the interface in the presence of phase-change processes represents the main difficulty. Interface thermodynamics cannot be modeled from first principles within the framework of continuum mechanics: The continuum description itself is underpinned by underlying thermodynamic hypotheses (e.g., local equilibrium). Hence, it is necessary to make assumptions about the thermodynamic state of the interface. Most single-fluid methods assume that the interface temperature is equal to the equilibrium saturation temperature corresponding to the system pressure.

Interfacial terms (representing surface forces or energies) are modeled as source terms in the momentum equations. They are written as delta functions at the interface. The unsteady Navier–Stokes equations are solved on a fixed where the position of

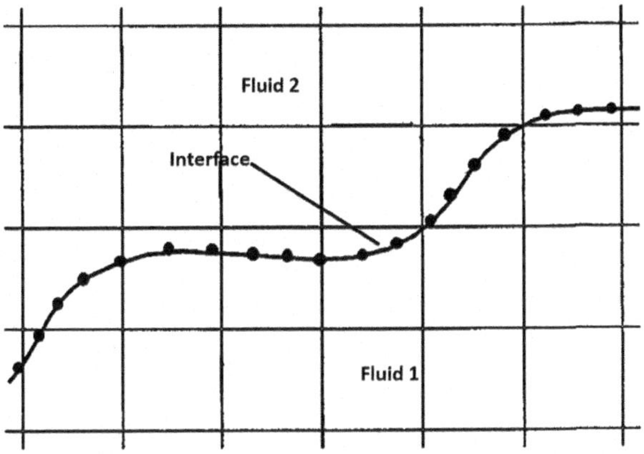

Fig. 15.6 Schematic of two phases of the same fluid separated by an interface marked by particles

the interface, or front, is not known a priori and is a part of the solution. Forces such as surface tension are modeled as volumetric source term. The advection equation is solved in a coupled manner to model the motion of the front. With these methods, there are difficulties in computing the curvature, estimation of the surface tension term, modeling wall adhesion, computing evaporative mass transfer at the interface (Fig. 15.6).

The governing equations in single-fluid formalism are

$$\frac{\partial u_j}{\partial x_j} = \nabla . \vec{V} = 0$$

$$\frac{\partial \rho_m u_i}{\partial t} + \frac{\partial \rho_m u_j u_i}{\partial x_j} = \frac{\partial}{\partial x_j}\left[\mu_m \frac{\partial u_i}{\partial x_j}\right] - \frac{\partial p}{\partial x_i} + \rho_m g_i + F_{st,i}\delta + \frac{\partial}{\partial x_j}\left[\mu_m \frac{\partial u_j}{\partial x_i}\right]$$

The first equation represents volume conservation, whereas the second equation represents momentum equation in conservative form. $F_{st,i}\delta$ is the surface tension force that acts at the interface. The interface location is tracked by solving an equation for the conserved scalar Φ,

$$\frac{\partial \Phi}{\partial t} + \frac{\partial \Phi u_j}{\partial x_j} = 0.$$

Typically, there are mainly three ITMs which are used for the simulation of boiling heat transfer:

- The marker-and-cell (MAC) method (Harlow and Welch 1965) in which the interface is marked by massless particles that are convected by the velocity field and are used to reconstruct the interface position.
- The volume-of-fluid (VOF) method (Hirt and Nichols 1981) in which an advection equation is postulated for predicting the distribution of volume fraction F in space and time. Some geometric properties of the interface are derived from the local

F distribution so as to facilitate evaluation of convective fluxes according to the donor–acceptor.

- The level set (LS) method (Osher and Sethian 1988) which relies in part on the theory of curve and surface evolution (Osher and Fedkiw 2003) and on the link between the front propagation and hyperbolic conservation laws. It is based on the construction of a smooth function, defined everywhere in the computational domain, representing the shortest distance to the front. Negative values correspond to one fluid and positive values to the other. The exact location of the interface corresponds to the zero level of the function.

Besides these three methods, researchers have also developed the arbitrary Lagrangian–Eulerian (ALE) method (Hirt et al. 1974) in which the nodes of the computational mesh may be moved with the continuum in normal Lagrangian fashion or be held fixed in Eulerian manner.

The main difficulty in using the marker-and-cell (MAC) method or the volume-of-fluid (VOF) method is maintaining a sharp boundary between the different fluids and the computation of the surface tension at the interface. Brackbill et al. (1992) proposed the continuum surface force (CSF) model in which the surface force was distributed volumetrically. This to some extent solved the problem of accounting for surface forces in the framework of an Eulerian control volume approach. The problem with the CSF model is its impracticality of computation of the curvature in three spatial dimensions and associated rise of spurious currents; Nandi and Date (2009a, b) formulated a method of calculating curvature from fluid dynamic consideration and were able to reduce the computational effort involved, and it also showed significant reduction of spurious currents.

Tryggvason and co-workers (Tryggvasson et al. 2001; Unverdi and Tryggvason 1992) extended the original MAC method for simulating boiling flows. Esmaeli and Trygvason simulated multimode film boiling on horizontal surfaces and boiling in complex geometries (Esmaeeli and Tryggvason 2004). Their method predicted the interface curvature very accurately which in turn helped improved simulation of very small bubbles. However, microscale heat transfer at the solid–liquid–vapor ('three-phase') contact line at the bubble base, or the transient heat conduction in the solid wall, is not accounted for their model. Figure 15.7 shows the evolution of the interface during film boiling.

Welch and Wilson applied VOF methods to simulate film boiling (Welch and Wilson 2000), implementing a model for phase change suitable for the VOF framework. Subsequently, Welch and Rachidi (2002) extended the model for simulating saturated horizontal film boiling including the conjugate heat transfer with the solid wall. This relaxed the idealization of uniform wall superheat or uniform wall heat flux boundary conditions. Sato and Niceno (2018) used a color function (similar to VOF), to develop a somewhat different numerical approach to simulate nucleate pool boiling, employing an interface-sharpening algorithm. They took into account the conjugate heat transfer between the solid wall and the fluid domains and used their own depletable microlayer model (Sato and Niceno 2015) for computing its vaporization.

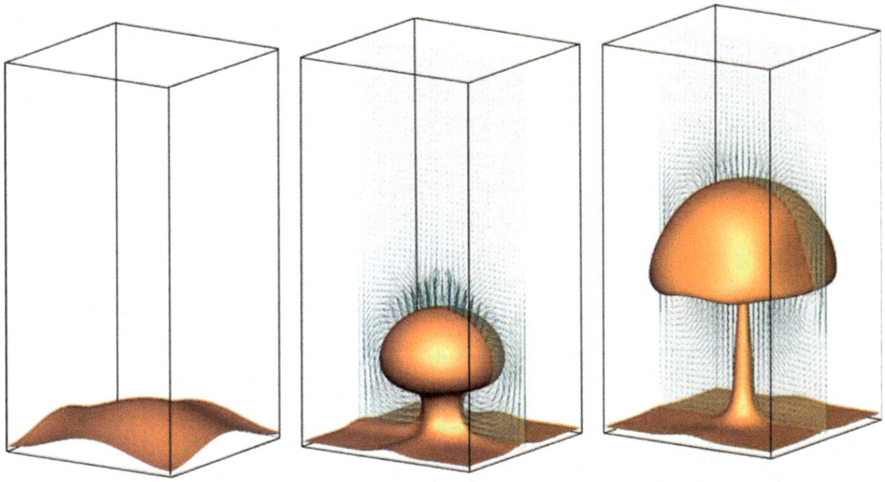

Fig. 15.7 Evolution of a liquid–vapor interface and velocity field during film boiling process (from Esmaeeli and Tryggvason 2004)

Dhir (2001) and co-workers used LS method for various boiling simulations for a wide range of configurations. Son and Dhir (1998) incorporated the effect of phase change into a modified LS method for simulation of film boiling, while Son et al. (1999) developed a microlayer model for estimating single bubble heat transfer associated with nucleate pool boiling. The model accounts for the microscale heat and fluid flow.

VOF methods typically have the problem of generating smeared interfaces, whereas the LS method captures the interface very accurately but leads to violation of mass conservation. A combination of level set–VOF method (CLSVOF) was proposed by Sussman and Puckett (2000), which combined the advantages of both the methods while avoiding the shortcomings. In this method, LS is used only to compute the geometric properties at the interface while the indicator function is advected using the VOF approach. Biswas (Tomar et al. 2005) and Tao (Sun and Tao 2010), among other researchers, used this method for simulation of boiling flows.

Cerne et al. (2001) proposed a coupled Eulerian–Eulerian and VOF model. In the computational domain where the grid resolution is fine enough to allow surface tracking, the VOF method is used and Eulerian two-fluid model is used in regions where the flow is dispersed. Each model uses a separate set of equations suitable for description of the two-phase flow. A ‘switching parameter’ based on the indicator function in the VOF method is used for the transition between the models.

15.3.3 Other Methods

Classically, at a macroscopic scale, an interface between a liquid and its vapor is modeled as a surface of discontinuity with properties like surface tension. However, at a microscopic scale, an interface has nonzero thickness and all forces at the interface are smoothly distributed. Therefore, the general equations of fluid mechanics can be applied to describe a liquid–vapor system with interfaces.

Diffuse interface models provide a way (Anderson et al. 1998) of modeling all forces at the interface as continuum forces and the discontinuities at the interface are smoothed by varying them continuously over thin interfacial layers. Phase field model is one of such models, which is being applied for the calculation of two-phase flows (Jacqmin 1999; Chen and Doolen 1998). This model allows the simulation of interface movement and topological changes on fixed grids. The Navier–Stokes equations are modified by the addition of the continuum forcing term which is a function of composition variable (C) and its chemical potential. The equation for interface advection is replaced by a continuum advective–diffusion equation, with diffusion driven by C's chemical potential gradients, and the liquid–vapor interface is described as a three-dimensional continuous medium across which physical properties have strong but continuous variations.

Phase field methods appear to have several potential advantages over the VOF-LS approach. It can capture interface deformations such as coalescence and interface break-up in an energy-dissipative fashion without losing mass. It is easy to implement in three dimensions and unstructured grids and is free of spurious currents. However, the phase field model also has its drawbacks. One has to accurately model relevant the physical phenomena, and the interface layers have to be very thin, and for this, the numerical phase field interfaces are typically kept four to eight cells wide. But this brings its own problems since large gradients now must be resolved computationally.

In recent decades, the lattice Boltzmann method (LBM) (Chen and Doolen 1998; Mohamad 2011) has emerged as another method tool for solving the Navier–Stokes equations. LBM is based on microscopic models and mesoscopic kinetic equations. It originated from Ludwig Boltzmann's kinetic theory of gases. The fundamental idea is that gases/fluids can be imagined as consisting of a large number of small particles moving with Brownian motion. The exchange of momentum and energy is because of elastic collision between the particles. The LBM simplifies Boltzmann's original idea of gas dynamics by reducing the number of particles and confining them to the nodes of a lattice.

In recent years, the lattice Boltzmann method (LBM) has been applied to simulate multiphase flow, in which the pseudopotential LB model has been quite popular because of automatic phase separation via an inter-particle potential (Frank et al. 2006; Shan and Chen 1993). Gong and Cheng (2015) combined multiphase LBM with an energy equation model to simulate the liquid–vapor phase change.

In the pseudopotential LB approach (Chen et al. 2014; Li et al. 2015), the liquid—vapor interfaces can naturally arise, deform, and migrate without using the interface tracking or interface capturing techniques and hence a big advantage compared to the existing methods.

In the future, ever-increasing computational power and newer computational techniques will enable fundamental understanding of turbulent two-phase flows, which will underpin the development of new boiling heat transfer models. Combining direct numerical simulation (DNS) of turbulence with interface tracking methods for simulating turbulent boiling flows is still not feasible. The computational cost of direct numerical simulation increases linearly with Reynolds number added with this; the complex topological changes of the interface along with its own computational difficulties make it a daunting task. DNS of two-phase flows is understandably still not foreseeable at the moment, but when it is developed it will be the ultimate numerical tool for bridging the gap between scales of simulation and enable decisive advancement of our understanding of the boiling processes.

15.4 Summary

A review of different techniques and models for simulating boiling flows has been presented. Simulation of boiling and two-phase heat transfer poses a number of challenges, and in this brief review, we have tried to show how the challenges were addressed by different researchers using different techniques.

References

Ahn HS, Kim MH (2012) A review on critical heat flux enhancement with nanofluids and surface modification. J Heat Transf 134(2):024001:1–13

Anderson DM, McFadden GB, Wheeler AA (1998) Diffuse interface methods in fluid mechanics. Annu Rev Fluid Mech 30:139–165

Anglart H, Nylund O (1996) CFD application to prediction of void distribution in two-phase bubbly flows in rod bundles. Nucl Sci Eng 163:81–98

Antal SP, Lahey RT Jr, Flaherty JE (1991) Analysis of phase distribution and turbulence in dispersed particle/liquid flows. Chem Eng Commun 174:85–113

Ardron KH, Giustini G, Walker SP (2017) Prediction of dynamic contact angles and bubble departure diameters in pool boiling using equilibrium thermodynamics. Int J Heat Mass Transf 114:1274–1294

Bartolomei GG, Chanturiya VM (1967) Experimental study of true void fraction when boiling subcooled water in vertical tubes. Therm Eng 14(2):123–128

Basu N (2003) Modeling and experiments for wall heat flux partitioning during subcooled flow boiling of water at low pressures. Ph.D. thesis, University of California, Los Angeles

Bergles AE (1981a) Two-phase flow and heat transfer, 1756–1981, Heat Transfer Eng 101–114. 2nd edn. Taylor & Francis, Boca Raton, FL

Bergles AE (1981b) Two-phase flow and heat transfer: 1756–1981. Heat Transf Eng 2(3–4):101–114

Bergles AE (1988) Fundamentals of boiling and evaporation. In: Kakac S et al (eds) Two phase flow heat exchangers: thermal-hydraulic fundamentals and design. Kluwer, The Netherlands, pp 159–200

Brackbill JU, Kothe DB, Zemach C (1992) A continuum method for modeling surface tension. J Comput Phys 100:335

Cerne G, Petelin S, Iztok T (2001) Coupling of the interface tracking and the two-fluid models for simulation of incompressible two phase flows. J Comput Phys 171:776–804

Chen S, Doolen GD (1998) Lattice Boltzmann method for fluid flows. Annu Rev Fluid Mech 30:329–364

Chen L, Kang Q, Mu Y, He YL, Tao WQ (2014) A critical review of the pseudopotential multiphase lattice Boltzmann model: methods and applications. Int J Heat Mass Transf 76:210–236

Cole R (1967) Bubble frequencies and departure volumes at sub atmospheric pressures. AIChE J 13:779–783

Collier JG, Thome JR (2001) Convective boiling and condensation, 3rd edn. Oxford University Press, Oxford

Colombo M, Fairweather M (2016) Accuracy of Eulerian-Eulerian, two-fluid CFD boiling models of subcooled boiling flows. Int J Heat Mass Transf 103:28–44

Cooper MG, Lloyd AJP (1969) The microlayer in nucleate pool boiling. Int J Heat Mass Transf 12:895–913

Del Valle VHM, Kenning DBR (1985) Subcooled flow boiling at high heat flux. Int J Heat Mass Transf 28:1907–1920

Dhir VK (1998) Boiling heat transfer. Annu Rev Fluid Mech 30:365–401

Dhir VK (2001) Numerical simulations of pool-boiling heat transfer. AIChE J 47:813–834

Dong EK, Dong IY, Dong WJ, Moo HK, Ho S (2015) A review of boiling heat transfer enhancement on micro/nanostructured surfaces. Exp Thermal Fluid Sci 66:173–196

Drew DA, Lahey RT Jr (1979) Application of general constitutive principles to the derivation of multi-dimensional two-phase flow equation. Int J Multiph Flow 5:243–264

Duan X, Phillips B, McKrell T, Buongiorno J (2013) Synchronized high-speed video, infrared thermometry, and particle image velocimetry data for validation of interface-tracking simulations of nucleate boiling phenomena. Exp Heat Transf 26:169–197

Esmaeeli A, Tryggvason G (2004) Computations of film boiling. Part I: numerical method. Int J Heat Mass Transf 47:5451–5461

Frank X, Funfschilling D, Midoux N, Li H (2006) Bubbles in a viscous liquid: lattice Boltzmann simulation and experimental validation. J Fluid Mech 546:113–122

Gong S, Cheng P (2015) Numerical simulation of pool boiling heat transfer on smooth surfaces with mixed wettability by lattice Boltzmann method. Int J Heat Mass Transf 80:206–216

Han C-Y, Griffith P (1965a) The mechanism of heat transfer in nucleate pool boiling—part I: bubble initiation, growth and departure. Int J Heat Mass Transf 8:887–904

Han C-Y, Griffith P (1965b) The mechanism of heat transfer in nucleate pool boiling—part II: the heatflux-temperature difference relation. Int J Heat Mass Transf 8:905–914

Harlow FH, Welch JE (1965) Numerical calculation of time-dependent viscous incompressible flow of fluid with free surface. Phys Fluids 8:2182–2189

Hewitt GF (1998a) Boiling, part one. In: Handbook of heat transfer, ed: McGraw-Hill

Hewitt GF (1998b) Boiling, part two. In: Handbook of heat transfer, ed: McGraw-Hill

Hirt CW, Nichols BD (1981) Volume of fluid (VOF) method for the dynamics of free boundaries. J Comput Phys 39:201–225

Hirt CW, Amsden AA, Cook JL (1974) An arbitrary Lagrangian-Eulerian computing method for all speeds. J Comput Phys 14:227–253

Ishii M, Hibiki T (2011) Thermo-fluid dynamics of two phase flow, 2nd edn. Springer, New York

Ishii M, Zuber N (1979) Drag coefficient and relative velocity in bubbly, droplet or particulate flows. AIChE J 25:843–855

Ivey HJ (1967) Relationships between bubble frequency, departure diameter and rise velocity in nucleate boiling. Int J Heat Mass Transf 10:1023–1040

Jacqmin D (1999) Calculation of two-phase Navier-Stokes flows using phase-field modeling. J Comput Phys 155:96–127

Jakob M, Fritz W (1931) Versuche über den Verdampfungsvorgang. Forsch Geb Ingenieurwes 2:435–447

Jung S, Kim H (2014) An experimental method to simultaneously measure the dynamics and heat transfer associated with a single bubble during nucleate boiling on a horizontal surface. Int J Heat Mass Transf 73:365–375

Jung S, Kim H (2018) Hydrodynamic formation of a microlayer underneath a boiling bubble. Int J Heat Mass Transf 120:1229–1240

Kakaç S, Bergles AE, Fernandes EO (1988) Two-phase flow heat exchangers: thermal-hydraulic fundamentals and design. Kluwer, Dordrecht

Kern J, Stephan P (2003) Theoretical model for nucleate boiling heat and mass transfer of binary mixtures. ASME J Heat Transf 125:1106–1115

Kern J, Stephan P (2004) Evaluation of heat and mass transfer phenomena in nucleate boiling. Int J Heat Fluid Flow 25:140–148

Kim H (2011) Enhancement of critical heat flux in nucleate boiling of nanofluids: a state-of-art review. Nanoscale Res Lett 6:415–422

Klausner JF, Mei R, Bernhard DM, Zeng LZ (1993) Vapor bubble departure in forced convection boiling. Int J Heat Mass Transf 36:651–662

Koffman LD, Plesset MS (1983) Experimental observations of the microlayer in vapor bubble growth on a heated solid. J Heat Transf 105:625–632

Koncar B, Kljenak I, Mavko B (2004) Modeling of local two-phase flow parameters in upward subcooled flow at low pressure. Int J Heat Mass Transf 47:1499–1513

Kurul N, Podowski MZ (1990) Multidimensional effects in forced convection subcooled boiling. In: Proceedings of the 9th international heat transfer conference, Jerusalem, Israel, vol 2. Hemisphere Publishing Corporation, New York, pp 21–26

Kutateladze SS (1961) Boiling heat transfer. Int J Heat Mass Transf 4:31–45

Li Q, Kang QJ, Francois MM, He YL, Luo KH (2015) Lattice Boltzmann modeling of boiling heat transfer: the boiling curve and the effects of wettability. Int J Heat Mass Transf 85:787–796

Liang G, Mudawar I (2018a) Pool boiling critical heat flux (CHF)—part 1: review of mechanisms, models, and correlations. Int J Heat Mass Transf 117:1352–1367

Liang G, Mudawar I (2018b) Pool boiling critical heat flux (CHF)—part 2: assessment of models and correlations. Int J Heat Mass Transf 117:1368–1383

Lo S (2005) Modeling multiphase flow with an Eulerian approach. VKI lecture series-industrial two-phase flow CFD. von Karman Institute

Manglik RM (2006) On the advancements in boiling, two-phase flow heat transfer, and interfacial phenomena. J Heat Transf 128:1237–1241

McAdams WH, Kennel WE, Mindon CS, Carl R, Picornel PM, Dew JE (1949) Heat transfer to water with surface boiling. Ind Eng Chem 41:1945–1953

McHale JP, Garimella SV (2010) Bubble nucleation characteristics in pool boiling of a wetting liquid on smooth and rough surfaces. Int J Multiph Flow 36:249–260

Mohamad AA (2011) Lattice Boltzmann method. Springer, London

Nandi K, Date AW (2009a) Formulation of fully implicit method for simulation of flows with interfaces using primitive variables. Int J Heat Mass Transf 52:3217–3224

Nandi K, Date AW (2009b) Validation of fully implicit method for simulation of flows with interfaces using primitive variables. Int J Heat Mass Transf 52:3225–3234

Nishikawa K (1987) Historical developments in the research of boiling heat transfer. JSME Int J Ser III 30(264):897–905

Nukiyama S (1934) The maximum and minimum values of the heat Q transmitted from metal to boiling water under atmospheric pressure. J JSME 37:367–374

Osher S, Fedkiw R (2003) Level set methods and dynamic implicit surfaces. Springer, New York

Osher S, Sethian JA (1988) Fronts propagating with curvature dependent speed: algorithms based on Hamilton-Jacobi formulations. J Comput Phys 79:12–49

Plesset MS, Zwick SA (1954) The growth of vapor bubbles in superheated liquids. J Appl Phys 25:493–500

Plesset MS, Zwick SA (1955) On the dynamics of small vapor bubbles in liquids. J Math Phys 33:308–330
Prosperetti A (2017) Vapor bubbles. Annu Rev Fluid Mech 49:221–248
Prosperetti A, Plesset MS (1978) Vapour-bubble growth in a superheated liquid. J Fluid Mech 85:349–354
Prosperrotti A, Trygvassion G (2007) Computational methods for multiphase flow. Cambridge University Press, New York
Ramaswamy C, Joshi Y, Nakayama W, Johnson WB (2002) High-speed visualization of boiling from an enhanced structure. Int J Heat Mass Transf 45:4761–4771
Rayleigh L (1917) On the pressure developed in a liquid during the collapse of a spherical cavity. Philos Mag Ser 6 34:94–98
Sato Y, Niceno B (2015) A depletable micro-layer model for nucleate pool boiling. J Comput Phys 300:20–52
Sato Y, Niceno B (2018) Pool boiling simulation using an interface tracking method: from nucleate boiling to film boiling regime through critical heat flux. Int J Heat Mass Transf 125:876–890
Schweizer N, Stephan P (2009) Experimental study of bubble behavior and local heat flux in pool boiling under variable gravitational conditions. Multiph Sci Technol 21:329–350
Scriven LE (1958) On the dynamics of phase growth. Chem Eng Sci 10:1–13
Shan X, Chen H (1993) Lattice Boltzmann model for simulating flows with multiple phases and components. Phys Rev E 47:1815
Shojaeian M, Kosar A (2015) Pool boiling and flow boiling on micro- and nanostructured surfaces. Exp Thermal Fluid Sci 63:45–73
Son G, Dhir VK (1998) Numerical simulation of film boiling near critical pressures with a level set method. J Heat Transf 120:183–192
Son G, Dhir VK, Ramanujapu N (1999) Dynamics and heat transfer associated with a single bubble during nucleate boiling on a horizontal surface. J Heat Transf 121:623–631
Sun DL, Tao WQ (2010) A coupled volume-of-fluid and level set (VOSET) method for computing incompressible two-phase flows. Int J Heat Mass Transf 53:645–655
Sussman M, Puckett EG (2000) A coupled level set and volume-of-fluid method for computing 3D and axisymmetric incompressible two-phase flows. J Comput Phys 162:301–320
Thorncroft GE, Klausner JF, Mei R (1998) An experimental investigation of bubble growth and detachment in vertical upflow and downflow boiling. Int J Heat Mass Transf 41:3857–3871
Tomar G, Biswas G, Sharma A, Agrawal A (2005) Numerical simulation of bubble growth in film boiling using a coupled level-set and volume-of-fluid method. Phys Fluids 17:112113
Tryggvasson G, Bunner B, Esmaeeli A, Juric D, Al-Rawahi N (2001) A front tracking method for the computations of multiphase flows. J Comput Phys 169:708–759
Unverdi SO, Tryggvason G (1992) A front-tracking method for viscous, incompressible, multi-fluid flows. J Comput Phys 100:25–37
van Stralen S, Cole R (1979) Boiling phenomena, vol 1. McGraw-Hill, New York
Wagner E, Sodtke C, Schweizer N, Stephan P (2006) Experimental study of nucleate boiling heat transfer under low gravity conditions using TLCs for high resolution temperature measurements. Heat Mass Transf 42:875–883
Welch SWJ, Radichi T (2002) Numerical computation of film boiling including conjugate heat transfer. Numer Heat Transf Part B 42:35–53
Welch SWJ, Wilson J (2000) A volume of fluid based method for fluid flows with phase change. J Comput Phys 160:662–682
Yeoh GH, Tu JY (2005) A unified model considering force balances for departing vapour bubbles and population balance in subcooled boiling flow. Nucl Eng Des 235:1251–1265
Yeoh GH, Tu JY (2010) Computational techniques for multiphase flows. Elsevier, UK
Zuber N (1958) On the stability of boiling heat transfer. Trans ASME 80:711–720

Printed by Printforce, the Netherlands